# *Food and Feed Safety Systems and Analysis*

# *Food and Feed Safety Systems and Analysis*

***Edited by***

Steven C. Ricke
Griffiths G. Atungulu
Chase E. Rainwater
Si Hong Park

ELSEVIER

ACADEMIC PRESS
An imprint of Elsevier

Academic Press is an imprint of Elsevier
125 London Wall, London EC2Y 5AS, United Kingdom
525 B Street, Suite 1800, San Diego, CA 92101-4495, United States
50 Hampshire Street, 5th Floor, Cambridge, MA 02139, United States
The Boulevard, Langford Lane, Kidlington, Oxford OX5 1GB, United Kingdom



**Library of Congress Cataloging-in-Publication Data**
A catalog record for this book is available from the Library of Congress

**British Library Cataloguing-in-Publication Data**
A catalogue record for this book is available from the British Library

ISBN: 978-0-12-811835-1



*Publisher:* Andre G. Wolff
*Acquisition Editor:* Patricia Osborn
*Editorial Project Manager:* Jaclyn A. Truesdell
*Production Project Manager:* Maria Bernard
*Designer:* Matthew Limbert

Typeset by TNQ Books and Journals

# Contents

# List of Contributors


**Soohyoun Ahn** University of Florida, Gainesville, FL, United States

**Clara Assisi** Purdue University, West Lafayette, IN, United States

**Griffiths G. Atungulu** University of Arkansas Division of Agriculture, Fayetteville, AR, United States

**Mushin Aydin** Arkansas State University, Jonesboro, AR, United States

**Rachael E. Blevins** University of Arkansas, Fayetteville, AR, United States

**Todd R. Callaway** University of Georgia, Athens, GA, United States

**Roger Cochrane** Kansas State University, Manhattan, KS, United States

**Nicolae Corcionivoschi** Agri-Food and Biosciences Institute, Northern Ireland, United Kingdom

**Philip G. Crandall** University of Arkansas, Fayetteville, AR, United States

**Morgan L. Davis** University of Arkansas, Fayetteville, AR, United States

**Turki M. Dawoud** University of Arkansas, Fayetteville, AR, United States; King Saud University, Riyadh, Saudi Arabia

**Jia Di** University of Arkansas, Fayetteville, AR, United States

**Andrea J. Etter** Purdue University, West Lafayette, IN, United States

**Peixin Fan** University of Florida, Gainesville, FL, United States

**Ester Grilli** University of Bologna, Ozzano Emilia, Italy

**Jessica C. Hacker** University of Arkansas, Fayetteville, AR, United States

**Shakaree Hale** Alabama Agricultural and Mechanical University, Huntsville, AL, United States

**Anne Huss** BioMatrix International, Princeton, MN, United States

**Armitra Jackson-Davis** Alabama Agricultural and Mechanical University, Huntsville, AL, United States

**Jealae Jackson** Alabama Agricultural and Mechanical University, Huntsville, AL, United States

**Joy Jackson** University of Texas-Arlington, Arlington, TX, United States

**Kwangcheol C. Jeong** University of Florida, Gainesville, FL, United States

**Billie L. Johnson** Simmons Pet Food, Emporia, KS, United States

**Michael G. Johnson** University of Arkansas, Fayetteville, AR, United States

**Cassie Jones** Kansas State University, Manhattan, KS, United States

**Pravin Kaldhone** University of Arkansas, Fayetteville, AR, United States

**Sun Ae Kim** University of Arkansas, Fayetteville, AR, United States

**Alescia King** Alabama Agricultural and Mechanical University, Huntsville, AL, United States

**Ok Kyung Koo** Gyeongsang National University, Jinju, Republic of Korea

**Young M. Kwon** University of Arkansas, Fayetteville, AR, United States

**Shinyoung Lee** University of Florida, Gainesville, FL, United States

**Keshun Liu** United States Department of Agriculture, Agricultural Research Service, Aberdeen, ID, United States

**Aubrey Mendonca** Iowa State University, Ames, IA, United States

**Zeinab Mohammadi-Shad** University of Arkansas Division of Agriculture, Fayetteville, AR, United States; Nestlé Iran P.J.S. Co., Mohammadiyeh City, Iran

**Russell Morgan** Simmons Pet Food, Siloam Springs, AR, United States

**Rkia Moutiq** Iowa State University, Ames, IA, United States

**Mary Muckey** Kansas State University, Manhattan, KS, United States

**Corliss A. O'Bryan** University of Arkansas, Fayetteville, AR, United States

**Gbenga A. Olatunde** University of Arkansas, Fayetteville, AR, United States

**Haley F. Oliver** Purdue University, West Lafayette, IN, United States

**Si Hong Park** University of Arkansas, Fayetteville, AR, United States; Oregon State University, Corvallis, OR, United States

**Sean J. Pendleton** University of Arkansas, Fayetteville, AR, United States

**Andrea Piva** University of Bologna, Ozzano Emilia, Italy

**Chase E. Rainwater** University of Arkansas, Fayetteville, AR, United States

**Steven C. Ricke** University of Arkansas, Fayetteville, AR, United States

**Rafael Rivera** U.S. Poultry & Egg Association, Tucker, GA, United States

**Peter M. Rubinelli** University of Arkansas, Fayetteville, AR, United States

**Joshua C. Saldivar** Stanford University School of Medicine, Stanford, CA, United States

**Richard G. Shields Jr.** Simmons Pet Food, Siloam Springs, AR, United States

**Zhaohao Shi** University of Arkansas, Fayetteville, AR, United States

**Michael L. Sostrin** University of Arkansas, Fayetteville, AR, United States; Walmart Stores, Inc., Bentonville, AR, United States

**Christina L. Swaggerty** United States Department of Agriculture/Agricultural Research Service, College Station, TX, United States

**Lin Teng** University of Florida, Gainesville, FL, United States

**Emalie Thomas-Popo** Iowa State University, Ames, IA, United States

**Dale R. Thompson** University of Arkansas, Fayetteville, AR, United States

**Brittany L. White** Simmons Pet Food, Siloam Springs, AR, United States

**Shantae Wilson** University of Arkansas Division of Agriculture, Fayetteville, AR, United States

**Kristin L. Yearkey** University of Arkansas, Fayetteville, AR, United States

**Chen Zhao** Academy of State Administration of Grain, Beijing, People's Republic of China

**Xuhui Zhuang** Academy of State Administration of Grain, Beijing, People's Republic of China

# Preface

Food safety is comprised of several rapidly changing research fields and public forums (including microbiology, ecology, education, epidemiology, bioinformatics, economics, globalization, among others) with numerous influences at both the scientific and the economic level driving the directions it takes as these fields progress. In addition, the issues associated with food safety that are considered of greatest concern continue to evolve as food production systems become more complicated. Accordingly, food systems have become more interconnected, resulting in many more potential sites for foodborne pathogen contamination that can impact other steps in food production, processing, and retail. Foodborne pathogens possess the ability to intrude into any phase of food production and contaminate the food products that they come in contact with. This includes animal feeds consumed by food animals, resulting in colonization of the gastrointestinal tract and potentially systemic infection of the host animal. Infected animals can become contaminated meat products that in turn can cross-contaminate processing equipment and other meat products that are in contact with this equipment. As technologies such as whole genome sequencing have become more common place, the ability for specifically identifying and subsequently tracking a particular foodborne pathogen has vastly improved the epidemiology and opportunity for a more complete resolution of foodborne disease outbreaks. However, as the data sets grow, issues such as training the next generation of food safety professionals to be proficient in data analytics and cybersecurity management remain challenges for designing academic educational programs to efficiently accomplish this training.

In this book, chapters have been generated to merge the concept of food safety being highly integrated with the more recent developments in scientific research on foodborne pathogens and the new methods developed to study them. Topics are designed to cover food systems and foodborne ecology and how these approaches can be used to not only conduct research on food safety and foodborne pathogens but also implement some of the emerging diagnostic tools. This book will explore and discuss cutting edge research about emerging issues associated with foodborne pathogens and food safety systems at all levels from pet food to live animal production, and finally food processing. Topics are designed to cover food systems and foodborne pathogen ecology and how these approaches can be used to not only conduct research on food safety and foodborne pathogens but also develop educational materials for training incoming professionals in the field on subjects, such as food systems data analytics and cybersecurity.

More specifically, topics in this book will not only cover and update current food safety thinking, but subsections of the book will also explore how to integrate food systems concepts into pet food manufacturing and preharvest food safety practices and potential dietary amendments to restrict foodborne pathogen colonization. This is followed by chapters on postharvest food safety practices, current issues associated with certain foodborne pathogens, and recent developments in interventions to limit their contamination in meat products and food products such as juices. The last section includes chapters on application of emerging technologies such as whole genome sequencing for detection and microbial community profiling. Finally, a chapter on the data analyses aspects of food systems as well as discussion on cross-training education strategies to convey this knowledge to graduate students in food science and computer engineering programs for preparing them to not only understand the biological aspects of foodborne pathogens and food production systems but also develop their ability to integrate these subjects into data analyses and cyberprotection programs. In summary, this book presents food safety as an integrative science among food production systems, the biology of pathogens defined by large data sets, their subsequent analyses, and ultimately the overall interpretation and application in commercial food production.

**Steven C. Ricke**
University of Arkansas
Fayetteville, AR, United States

SECTION 1

# *Animal and Pet Feed Safety*

CHAPTER 1

# Changes in the Food Safety Landscape of Pet Foods in the United States

**Brittany L. White[1], Billie L. Johnson[2], Russell Morgan[1], Richard G. Shields Jr.[1]**
*[1]Simmons Pet Food, Siloam Springs, AR, United States; [2]Simmons Pet Food, Emporia, KS, United States*

## History of Pet Food Types and Formats

In the United States, prepared food for pets actually began as table scraps, and it remains this way in developing countries today. Feeding table scraps potentially presents food safety risks associated with the lack of preservation, especially if the scraps are not fed immediately after preparation. In the 1860s, James Spratt developed Spratt's Dog Cakes which were biscuits for pets, and initial commercial production in the United States began in 1895. This was shortly followed by the introduction of Milk Bone in 1908. Clarence Gaines founded Gaines Food Co. in the mid-1920s with 100-pound bags of dog meal and later introduced a pelleted version called Gaines Krunchons. Gaines is also credited with the first food claiming complete and balanced nutrition, which added nutritional adequacy to the food safety concerns associated with pet food. Meals and pelleted diets tended to mirror processing options available for livestock feeds; however, the development of complete and balanced foods transitioned the formulation to a blend of meat, milk, and cereal grains. The first canned food emerged in the early 1920s with the launch of Ken-L-Ration by the Chappel Brothers as an effort to find a market for horsemeat. This brand was the first to recommend mixed feeding of dry and wet pet food in the 1930s. The canned food ultimately began to dominate the pet food market, using coproducts of numerous meat and fish packers including Wilson Meat Packing, Swift & Company, Armour & Company, and John Morrell & Company. In 1941, canned pet food represented 91% of lbs sold. However, a can shortage during World War II resulted in a transition from cans to glass containers. By 1946 the pet food market had shifted back to dry foods, with wet pet food representing only 15% of the business volume. The initial canned wet pet foods were primarily hermetically sealed in steel cans which were soldered, but the can style ultimately transitioned to drawn and ironed cans of steel and more recently to aluminum and plastic components. Today retortable pouches and semirigid tubs and trays with heat seals compared with conventional metal seams have become quite popular and therefore present new sealing and cooking challenges as overpressure is required to maintain seal integrity.

Food and Feed Safety Systems and Analysis. http://dx.doi.org/10.1016/B978-0-12-811835-1.00001-4

The industry changed dramatically in the 1950s with the launch of the extruded pet food Purina Dog Chow, which almost instantly became the #1 brand nationally. Even in the presence of this shift, wet pet food represented 60% of the volume in 1960, and today represents closer to 25%. The 1980s brought about the addition of chunk-in-gravy style canned food, whereas the loaf format formerly predominated. This required some formulation changes to allow for chunk binding and also gravy viscosity. In the 1990s came the introduction of Purina ONE which was the first product to have actual meat as the #1 ingredient in a dry, extruded formula. As the use of extruders in pet food application expanded, product offerings extended beyond traditional dry kibble to shelf stable, soft moist foods with Gaines-Burgers being the first soft moist offering in the 1960s. Soft moist food formulation actually was an outgrowth of mink diet feeding and required extensive use of hurdle technologies to ensure food safety at higher moistures. In 1981, Kibbles N' Bits was introduced, which represented a new pet food format termed soft dry, which contained dry and soft moist components mixed together in a single offering. This represented new challenges, namely how to prevent moisture migration from the soft moist piece to the dry component. In 1996, product design changes were again required for soft moist cat foods as the Food and Drug Administration (FDA) prohibited use of propylene glycol in cat foods (21 CFR 589.1001) because it had been shown to cause abnormalities in red blood cells (Christopher et al., 1989). Propylene glycol had been the most common humectant used in soft moist formulations as it had an added benefit of being a miticide (EPA, 2004).

There are several other emerging pet food formats that currently represent a much smaller share of the pet food market, including pet treats. Freeze drying is a processing technique that has been growing steadily at 45.5% per year since 2011 (Phillips-Donaldson, 2016). One pet food processing category that has primarily been used for treats products is injection molding, which has origins in the plastics industry. Of interest from a food safety standpoint is Greenies, an injection molded treat that was launched in 1998 by S&M NuTec, LLC. By 2003 Greenies became the #1 treat by value sales, surpassing the long-standing top seller, Milk Bone (http://www.petfoodindustry.com/articles/637-success-and-more). The brand was sold in 2006 amid numerous complaints of choking and digestibility hazards associated with the treat. Another category of pet treats that has emerged in recent years that represents unique food safety challenges is dehydrated jerky treats. In the early 2000s, dehydrated chicken jerky treats for dogs were primarily being produced in China. Between 2007 and 2013, the FDA received 4500 reports of pet illness and 580 reported pet deaths due to consuming jerky treats made in China. After extensive investigation, no specific cause was ever identified. However, in 2013, the state of New York found traces of an antibiotic in the jerky treats made in China, which was not allowed for consumption, so many major brands voluntarily recalled their products being made in China, leaving shelf vacancy that was ultimately filled by jerky treats made in the United States.

## Pet Food Safety Issues Associated With Raw Materials

Alongside changes in product format, the ingredients used in pet food formulations have changed over the years as well. Some of these changes have come either from the need to add "good" ingredients or eliminate "bad" ingredients, when in fact many have both strengths and weaknesses. Some ingredient components that have been traditionally considered as "antinutritional" in animal agriculture because of adverse impacts on growth or food utilization, such as tannins and phytates, may actually have some health-promoting activities (Thompson, 1993).

For the majority of pet food history, ingredients used were familiar, consistent with those used in the animal feed industry, which were derived from domestic grains, rendered meals, or from chemicals synthesized in countries with well-developed feed safety systems. Food safety recalls in pet food were typically the result of lack of research on cumulative toxicity of common feed compounds or mold toxins for pets. Ingredients with a long history of use, such as corn, soy, and wheat, have fallen into disfavor with some pet parents and manufacturers resulting in the need to replace their nutrient or functional contribution with other ingredients that may have long-term effects that are not completely understood. Understanding toxicity of nutrients and ingredient components in both pets and humans is quite challenging as chronic effects of exposure during the lifetime of the pet must be considered. In addition, as with all animals, researchers must also consider ingredient interactions that may either ameliorate or exacerbate the effects of a single compound. The difficulty in understanding the long-term safety of an ingredient is exacerbated by differences in the way in which different species may react that may compound the effects, and utilization of pets in research, especially when there is a risk of causing harm to the pet, has fallen into disfavor.

Use of new and different ingredients in pet foods may be driven by an effort by a brand to gain market share. Brands have often capitalized on "free-from" claims in human and pet foods or have focused highlighted ingredient quality or safety attributes to differentiate themselves from ingredients used in other brands. The individualization of products for pets with special needs has also resulted in a proliferation in the use of unique ingredients. Globalization has caused changes in raw material sourcing as relocation of the ingredient further processing industry and a desire for novel ingredients has required manufacturers to import many ingredients, forcing utilization of unfamiliar materials from unfamiliar suppliers. The humanization of pets has also resulted in use of unique and more expensive ingredients that tend to be more susceptible to economically motivated adulteration (EMA). Because of the species dependence in food safety risks associated with different compounds, the impact of such adulteration can have unintended effects.

### Base Ingredient Selection

With increased recognition of food sensitivities in pets, there has also been increased use of alternative meat and vegetable proteins and carbohydrate sources. In addition, the proliferation of clean label, simple diets with shorter history of use present an increased risk due to increased predominance of a single ingredient in the food compared with formulas that rely on nutrient contributions of a variety of ingredients. This issue has been exacerbated by current trends of "wild" and "paleo" diets for pets, which are based on foods that are eaten by animals in the wild and consist primarily of meats, fish, fruits, and vegetables and excludes grains.

The growing consumer trends of grain free foods and limited ingredient diets tend to use alternative starch sources to replace conventional cereal grains. The risk this presents is two-fold. First, it leads to the sourcing new raw materials outside of the United States and also demands incorporation of ingredients that have a much less comprehensive food safety evaluation. Even some of these ingredients that are familiar, such as potatoes, carry food safety risks such as the presence of glycoalkaloids in coproducts derived from skins (Nicholson et al., 1988; Smith et al., 1996). Alkaloids are also found in tomatoes and eggplant, which have emerged in recent pet food formulations (Friedman et al., 1996).

### *Humanization of Pet Food*

The trend toward humanization of pet foods has also exposed pets to increased quality and/or food safety risks. Following human food trends, there has recently been an interest in utilization of organic, genetically modified (GM)-free, and sustainable ingredients. Meeting these needs requires the use of a more-restricted raw material stream, which also increases susceptibility to EMA. Furthermore, although the intent may not be an impact to food safety, negative consequences can occur if harmful diluents or replacements are used. This was demonstrated by the melamine contamination of wheat gluten and milk in 2007. In a publication by GMA (2010), it was estimated that economic adulteration and counterfeiting of global food costs the food industry 10–15 billion dollars each year. Moore et al. (2012) reported that olive oil, milk, honey, and saffron were the most common adulteration targets in a review of incidents occurring between 1980 and 2010. Fruit juices, especially those of superfruits, have also been known to be subject to dilution with less costly juices. In a Congressional Research Service Report by Renee Johnson (2014), it was also reported that fish and seafood fraud may be widespread in some markets. Even common meats like beef were found in 2013 to be adulterated with horsemeat (Walker et al., 2013).

There has been increased recognition of the presence of several human metabolic defects in pets such as copper storage disease and gluten intolerance in dog breeds and urinary crystals in dogs and cats which in turn results in special dietary needs and restrictions. Many specialized diets with respect to nutrient content are best distributed under the direction of a veterinarian as, although they may be helpful to target animals, they may be inadequate for the general pet population.

Consumption of what would be considered quality human foods by pets can cause a food safety risk. In some cases, it results from species differences in metabolism, whereas in others, it is from the fact that food products intended for humans are not meant to be sole diets, so presence of high levels of some compounds may be well tolerated by humans but not by pets. The toxicity of theobromine in dark chocolates, particularly baking chocolate, is well recognized (Gans et al., 1980). Onions have historically been used in several pet rations; however, it was found that in 1998 that cats fed baby food with onion powder developed Heinz body anemia (Robertson et al., 1998) from the organosulfur compounds present in the onions. In cats this occurred because of the susceptibility of hemoglobin to oxidation (Christopher et al., 1995). Dogs also have reduced tolerance to organosulfur compounds relative to humans because of lower erythrocyte catalase activity. In addition, although the mechanism is not clearly understood, grapes and raisins are also thought to contain harmful compounds, reportedly causing vomiting, diarrhea, and renal disorders. Furthermore, xylitol, a sugar alcohol commonly used in food industry as a nonnutritive sweetener, can cause hepatic failure in dogs (Dunayer and Gwaltney-Brant, 2006). These examples, among others, demonstrate that care must be taken when developing pet foods intended to mimic human food trends, as the consequences can be detrimental to the pet.

### Nutrient Deficiencies and Excesses

An area that is often not considered when designing food safety systems is primary or secondary nutrient deficiency and toxicity; however, this has been the cause of several recalls in recent history. Nutrient deficiencies can result from the lack of primary nutrients in the diet or insufficient bioavailability of the nutrients present in the diet. Insufficient bioavailability can occur both because of the form of the nutrient in the diet or from nutrient interactions, and these are addressed in only very limited scope in AAFCO nutrient profiles (Official Publication, 2016). This is clearly demonstrated by the history of dietary taurine recommendations for cats. In spite of a long history of nutrition research, the fact that taurine was required by the cat to prevent feline central retinal degeneration was not established until 1975 (Hayes et al., 1975). Further research established that taurine status was influenced by dietary potassium level and dietary acidification for urine pH control (Dow et al., 1992). In addition, requirements were higher for canned than for dry diets (Douglass et al., 1991). Inconsistencies remained in taurine status resulting from food consumption which could not be answered by total dietary taurine and it was found that bioavailability in raw materials differed from that of synthetic taurine (Odle et al., 1993). Similarly, copper recommendations in dry and canned pet foods differ (higher in dry diets) and status can be influenced by nutrient balance and raw material bioavailability (Shields, 1998). Copper status is also adversely affected by excess zinc intake (NRC, 2006). Based on some pet research (Morris and Rogers, 1995), copper oxide was ultimately removed as a nutritional source of copper in pet food due to lack of bioavailability. Assurance of nutritional adequacy therefore requires understanding of ingredient nutrient content and bioavailability as well as interactions with other dietary components. When new nutrient requirements are added which require addition of unfamiliar ingredients, there remains a risk of nutrient deficiencies in spite of apparently adequate analyzed dietary content.

In contrast to the aforementioned situation, some sources of nutrients can improve bioavailability relative to standard sources. Shurson et al. (1996) found that stability of vitamins in premixes containing minerals was greater when the latter were provided as organic complexes. Similarly, nutrient retention and reproductive performance have been reported to be higher when dogs were fed mineral complexes compared with inorganic sources (Kuhlman and Rompala, 1998; Lowe and Wiseman, 1998).

Besides nutrient deficiencies, excesses can also be of concern. Thyroid hormones are orally active, and therefore, care must be taken to remove thyroid glands from food and feed ingredients. Hamburger thyrotoxicosis has been found in humans fed ground beef contaminated with bovine thyroid gland (Hedberg et al., 1987). Similarly, Stengel and Neiger (2012) identified hyperthyroidism in dogs fed raw meat or fresh or dried gullets. Hyperthyroidism has also been identified in cats, and Johnson et al. (1992) concluded that thyroid disorders may result from wide swings in dietary iodine levels present in commercial cat diets.

Fat soluble vitamin excesses are also of concern because of the fact that they can accumulate over time. With respect to vitamin D, there is some familiarity with toxicity because of exposure of pets to rodenticides containing high levels of vitamin D. Morris and Rogers (1995) expressed concerns regarding high levels of vitamin D in some cat foods. Subsequent analytical research by this same author, specifically with tuna, suggested that fish viscera and especially liver contain extremely high levels of vitamin D. The concern regarding excess vitamin D has also expanded to dog foods with the reduction in allowed dietary maximums in the 2016 AAFCO Official Publication, likely resulting from the most recent research summary by NRC (2006). Fortunately, some tissues contain high levels of vitamin A, which is known to help ameliorate vitamin D toxicity in some animal species, including dogs (Aburto et al., 1998).

A portion of the vitamin D concern relates to impairment of calcium metabolism. An additional concern for canine diets is excess calcium, for which the maximum allowed levels for large breed puppies was significantly lowered in the most recent AAFCO Official Publication as a result of several articles in peer-reviewed publications. This research demonstrated that large breed puppies may be unable to protect themselves against excessive calcium intake (Hazewinkel et al., 1991).

### *Antinutritional Factors*

Another food safety concern related to nutrient composition of foods is the potential presence of antinutritional factors. Fish and some meat products are known potentially to contain biogenic amines which arise from bacterial fermentation of amino acids. Although histamine (from histidine) has received the most attention, especially with tuna, several others can arise from other amino acids. The term "biogenic amines" refers to the nonvolatile amines, such as cadaverine, putrescine, spermidine, spermine, tyramine, tryptamine, and histamine, produced in fish and shellfish products harvested and not kept at ideal temperature conditions to prevent spoilage. Spoilage microorganisms break down the amino acids and the toxic compounds originate from the decarboxylation of specific free amino acids in fish or shellfish tissue (Rawles et al., 1996). The most well-known health impact of biogenic amines is histamine poisoning, which occurs a few minutes to several hours after the ingestion of foods containing high levels of histamine. Primary symptoms in humans are skin rash, nausea, vomiting, diarrhea, and so forth, which are mimicked in the companion animals symptoms as well. The toxicity of histamine is potentiated by the other biogenic amines present such as putrescine, cadaverine, and spermidine (Radosevich, 2016). A review of biogenic amines in fish species was authored by Shalaby (1996). These biogenic amines appear to be stable to processing (Veciana-Nogues et al., 1997a), and canned cat foods have been found to have measurable levels of histamine (Guraya and Koehler, 1991). There is a multitude of biogenic amines that can be present in these products, so Veciana-Nogues et al. (1997b) recommended using the sum of biogenic amines as a quality attribute.

Some species of fish such as herring, alewives, anchovies, mackerel, and clams also contain thiaminase which degrades thiamine (vitamin B1) and increases the dietary requirement of thiamine, particularly if raw fish sources are used (Smith and Proutt, 1944). This antithiamine effect in the diet can be exacerbated by the preservation of meats using sulfur dioxide as a preservative as it also degrades the vitamin (Studdert and Labuc, 1991). Sodium metabisulfite is another sulfur-containing preservative which has been recognized by FDA in 21 CFR 182.3766 as an antithiamine compound. Many vegetables are treated with sulfites to prevent browning and this must be monitored as well.

Dioxin contamination has affected animal feeds globally on numerous occasions. Dioxins are a group of chemically related compounds that are persistent environmental pollutants. The World Health Organization reports that dioxins are highly toxic and can cause reproductive and developmental problems, damage the immune system, interfere with hormones, and also cause cancer (WHO, 2016). Although not specifically affecting pet foods, the use of ball clay, which contains dioxins, as a conditioning and flow agent in soybean meal did impact ingredients which could have impacted the pet food supply. An instance of dioxin contamination in soybean meal in Mississippi impacted the poultry feed supply in 1997. As a result, ball clay was eliminated by FDA as an allowed conditioning agent in feed ingredients (FDA, 1999). Numerous dioxin contamination cases have also occurred in the European Union (EU) feed supply from contamination of fat sources with industrial oils not designed for food use. An isolated contamination of zinc oxide in some mineral premixes designed for livestock occurred in 2003. With many premix manufacturers preparing products for multiple species, this represented an additional risk to the pet food industry at the time. There was also an incident in 2007 where dioxin contamination was detected in guar gum used in the human food supply in the EU. With the globalization of the raw material supply chain, this presented a true risk to the US supply as this material is a commonly used thickener for both pet and human foods, and, as a result, close monitoring of the supply was initiated.

## Prion Diseases

Prion diseases, also known as transmissible spongiform encephalopathies (TSEs), are neurodegenerative diseases that affect humans and animals and represent another significant food safety concern in pet foods. Even sourcing from countries with effective food safety systems was shaken with the discovery of TSEs including those of bovine origin (bovine spongiform encephalopathy) where the infective agent was not of traditional microbial origin and was not destroyed even with traditional rendering processes. Although no TSE has been identified in dogs, a cat form has been identified (Gruffydd-Jones et al., 1991). This occurrence which started in Western Europe and was later identified in Canada and then in the United States caused trade disruptions and changes in the sources of meat materials used for pet food, including discontinuation of the use of mechanically deboned ruminant materials. FDA published research (Adam and Reeves, 1998; Heller et al., 2001) relating to the presence of pentobarbital in dog foods. The presumption was that this resulted from rendering of euthanized pets as a component of rendered meals and fat. This was not confirmed on species identification and may have in fact resulted from formation of heterocyclic amines or other compounds during processing which were identified as pentobarbital in the assay procedure. With the concerns about TSEs at the time, recycling of animal species components being placed into the food supply of the same species understandably caused concerns.

## Selection of Food Additives

A key aspect of many pet food safety systems is the use of additives as hurdles to slow or inhibit the potential for negative chemical or biochemical reactions. Some additives are used to lower water activity, whereas others may lessen redox potential, lower acidity, control oxidation, or reduce bacterial load (Leistner, 2000). Alternatively, some hurdles such as microbial cultures and prebiotics are added to impart indirect health benefits for the animal. One such additive is propylene glycol, which was once a commonly used humectant in soft moist food formulations. After a long history of use, research results of Bauer et al. (1991) found that consumption of this ingredient by cats resulted in Heinz body anemia. The "no effect" level was found to be well below that required for humectants action; therefore, the FDA banned it from use in cat food formulations. Earlier research with dogs did not result in such effects (Weil et al., 1971), likely because of differences in the susceptibility of cats' and dogs' erythrocytes to oxidative damage.

Besides humectants, the type of antioxidants has also shifted throughout pet food history. Although antioxidants are generally thought to provide protection both in the food and in the animal if absorbed, high levels or specific cellular conditions can result in these same compounds acting as prooxidants (Palozza, 1998). At one time, ethoxyquin was the most commonly used antioxidant in feed and pet food products as it was extremely effective. This compound remains approved today at a maximum usage level of 150 parts per million (ppm) (21 CFR

573.380). In spite of safe use since 1959, in the late 1980s, complaints were received that ethoxyquin consumption could cause allergic reactions, cancer, and reproductive failure. A review of literature by the FDA failed to identify adverse effects from ethoxyquin use at recommended levels in canine diets (Dzanis, 1991). However, Monsanto, the principal manufacturer at the time, agreed to conduct a 3.5-year multigenerational study with Beagle dogs with dietary inclusion levels of ethoxyquin up to 360 ppm. The conclusion of the study, completed in 1996, was that there were no adverse effects other than some slight liver pigmentation increases at the highest doses. As a result of these findings, FDA asked for a voluntary reduction in maximum ethoxyquin levels in dog foods from 150 to 75 ppm to add an additional margin of safety for puppies and lactating females (FDA, 1997). The combined impact of identification of alternatives to ethoxyquin and the growing trend of natural diets has resulted in transition to natural antioxidants. The added trend of GM-free diets and clean labels has resulted in further challenges. This caused a period of extensive research to identify successful alternative antioxidants and resulted in drastic changes in shelf life of products on the market.

### Cross-Contamination Risks

Besides risks associated with specific ingredients, unintentional cross-contamination from batch to batch can occur when foods intended for other species or medicated premixes are made in the same facility. Safran et al. (1993) identified lasalocid toxicity resulting in paralysis of dogs fed diets where cross-contamination occurred of the dog diet with this ionophore that functions as an antibacterial agent. A recent study by Okuma and Hellberg (2015) found evidence of several animal species that were not specifically identified on pet food labels, which was likely at least in part the result of raw material cross-contamination. This also caused a significant expansion of some recalls because of uncertainty of the amount of cross-contamination resulting from either continuous production operations or lack of complete cleanouts between finished product batches. A further risk exists in bulk storage and the transportation of raw materials on their way to the factory, particularly in the case of liquids. The FDA has recognized the latter risk and has included sanitary transportation as a component of the Food Safety Modernization Act (Final Rule on Sanitary Transportation of Human and Animal Food FDA-2013-N-0013 4/2016).

### Mycotoxins

The pet food industry has been notoriously subjected to recalls of pet products due to mycotoxin contamination. One of the first reported outbreaks to occur in manufactured pet food occurred in 1951, but little documented data exist to determine the exact cause of the food poisoning that was seen in 71 dogs (Boermans and Leung, 2007). The earliest documented aflatoxin outbreak dates back to 1974 when hundreds of stray dogs in India died due to the consumption of aflatoxin-contaminated corn (Krishnamachari et al., 1975). In December

2005, 76 dogs were killed from aflatoxin-contaminated pet food in the United States, causing a large recall (Cable News Network, 2005). Understanding the sources and reasons behind mycotoxin production has become a higher priority for pet food manufacturers as mycotoxins, once in the raw grain material cannot be cooked out of the finished pet food product. The US FDA has issued guidance for three types of mycotoxins: aflatoxin, fumonisin, and deoxynivalenol (commonly called vomitoxin, or DON) (Dzanis, 2007). Other mycotoxins (e.g., ochratoxin, zearalenone, T2) also can be troublesome. However, at present, FDA handles contamination issues with mycotoxins, other than the three mentioned previously, on a case-by-case basis. A *Compliance Policy Guide* on "action levels" for aflatoxin was first issued in 1979. The most recent document (FDA Regulatory Guidance for Mycotoxins, 2011) states the advisory levels versus action levels for aflatoxin, fumonisin, and DON in various species (see Table 1.1). This document provides specific maximum tolerable amounts for aflatoxin in various feedstuffs (e.g., corn, cottonseed meal, and peanut products) intended for various species (Dzanis, 2007). Action levels for pet food fall under the "other animal" category, which is set at 20 parts per billion for aflatoxin. Importantly, this level applies to the total aflatoxin content (B1 + B2 + G1 + G2). For fumonisin, the guidance recommends a total fumonisin (FB1 + FB2 + FB3) content in corn and corn by-products intended for pet foods not to exceed 10 ppm. The further stipulation is that corn or corn by-product not exceed more than 50% of the pet food formulation (see Table 1.2). For DON, finished feeds that make up a "complete" diet should not exceed 2 ppm of the toxin, but if the pet food makes up 40% of the diet or less, can have a maximum allowance of 5 ppm (Table 1.3).

**Table 1.1: FDA action limits for aflatoxin in human food, animal feed, and animal feed ingredients.**

| Intended use | Grain or grain by-product | Aflatoxin level (parts per billion, ppb) |
|---|---|---|
| Human consumption | Milk | 0.5 |
| Human consumption | Foods, peanuts and peanut products, brazil, and pistachio nuts | 20 |
| Immature animals | Corn, peanut products, and other animal feeds and ingredients, excluding cottonseed meal | 20 |
| Dairy animals and "other animals" not listed | Corn, peanut products, cottonseed, and other animal feeds and ingredients | 20 |
| Breeding cattle, breeding swine, and mature poultry | Corn and peanut products | 100 |
| Finished swine 100 pounds or greater in weight | Corn and peanut products | 200 |
| Finished beef cattle (i.e., feedlot) | Corn and peanut products | 300 |
| Beef, cattle, swine, or poultry, regardless of age or breeding status | Cottonseed meal | 300 |

*FDA*, Food and Drug Administration.
*Adapted from FDA Regulatory Guidance for Mycotoxins.*

**Table 1.2: FDA advisory limits for fumonisin in human food, animal feed, and animal feed ingredients.**

| Intended use | Grain or grain by-product | Fumonisin level (parts per million, ppm) |
|---|---|---|
| Human consumption | Degermed cornmeal, corn grits (fat content <2.25) | 2 |
| Human consumption | Corn for popcorn | 3 |
| Human consumption | Whole or dry-milled corn products—corn meal, corn grits (fat content >2.25); masa product and corn bran | 4 |
| Horses and rabbits | Corn and corn by-products not to exceed 20% of the diet | 5 (1)[b] |
| Swine and catfish | Corn and corn by-products not to exceed 50% of the diet | 20 (10)[b] |
| Breeding ruminants and poultry[a] | Corn and corn by-products not to exceed 50% of the diet | 30 (15)[b] |
| Ruminants ≥3 months old being raised for slaughter and mink being raised for pelt production | Corn and corn by-products not to exceed 50% of the diet | 60 (30)[b] |
| Poultry being raised for slaughter | Corn and corn by-products not to exceed 50% of the diet | 100 (50)[b] |
| All other species or classes of livestock and pet animals | Corn and corn by-products not to exceed 50% of the diet | 10 (5)[b] |

*FDA*, Food and Drug Administration.
[a]Includes lactating dairy cattle and hens laying eggs for human consumption.
[b]Advisory limits for complete diets.
*Adapted from FDA Regulatory Guidance for Mycotoxins.*

**Table 1.3: FDA action limits for DON (vomitoxin) in human food, animal feed, and animal feed ingredients.**

| Intended use | Grain or grain by-product | DON level (parts per million, ppm) |
|---|---|---|
| Human consumption | Finished wheat products | 1 |
| Swine | Grain and grain by-products not to exceed 20% of diet | 5 (1)[b] |
| Chickens | Grain and grain by-products not to exceed 50% of diet | 10 (5)[b] |
| Ruminating beef and feedlot cattle older than 4 months | Grain and grain by-products[a] | 10 (10)[b] |
| Ruminating dairy cattle older than 4 months | Grain and grain by-products not to exceed 50% of diet[a] | 10 (5)[b] |
| Ruminating beef and feedlot cattle older than 4 months, and ruminating dairy cattle older than 4 months | Distillers grains, brewers grains, gluten feeds, and gluten meals[a] | 30 (10 feed/feedlot; 5 dairy)[b] |
| All other animals | Grain and grain by-products not to exceed 40% of diet | 5 (2)[b] |

*DON*, deoxynivalenol; *FDA*, Food and Drug Administration.
[a]88% dry-matter basis.
[b]Complete diet figures shown within parentheses.
*Adapted from FDA Regulatory Guidance for Mycotoxins.*

Exposure mitigation appears to be the only possibility for risk reduction due to mycotoxins, as environmental conditions contribute to the mold growth on the grains and ultimately the mycotoxin production. Sampling techniques too allow potential for missed risk. Ensuring the raw materials being received are tested for mycotoxins and that suppliers are equally aware of mycotoxin risks is a component that all pet food manufacturers must face to prevent potential food safety hazards.

### Globalization

Besides selection of raw materials, sourcing can have a significant impact on food safety risk. The available options for ingredient sourcing are constantly expanding, and in many cases, significant proportions of an entire raw material supply chain have migrated to countries outside the United States. This has been the case for some vegetable crops as well as many food additives, including vitamins and other fermentation products. Many hydrocolloids and plant protein supplements are also sourced primarily from outside the United States. In addition to economic advantages due to lower labor cost, manufacturers in countries with less well-developed government oversight of food safety systems have resorted to EMA of materials. The most notable example of EMA as previously mentioned occurred in 2007 in which the utilization of melamine scrap in wheat gluten resulted in the deaths of many pets. The use of EMA is a common risk for any ingredient that is in short supply and/or which carries a premium price. This has been noted most recently in human foods in the meat and fish supply sources where materials of lower cost are used to dilute a portion of the premium ingredient (Johnson, 2014).

Another risk associated with globalization is that differences can exist in government-approved additives including antibiotics and pesticides in different countries. Additives used in some countries may not even be included in the US monitoring systems. In addition, the equipment used in raw material processing in some countries may introduce additional food safety risks such as heavy metal contamination. This is why effective raw material food safety monitoring requires an element of verification via thorough facility audits.

## Achieving Commercial Sterility in Pet Food Production and Potential Recontamination

Pet food processing for shelf stable products requires that commercial sterility be achieved. There are a broad array of processing types applied to pet food products including baking, extrusion, retorting, aseptic processing, and high pressure processing as well as air drying and freeze drying. Establishment of appropriate process conditions involves monitoring organisms in the raw material, identification of the most resistant/thermally stable organism, and testing with either this organism or a surrogate (indicator organisms) that have similar lethality curves.

In the case of high-moisture, low-acid foods packed in cans, pouches, or semirigid plastic packaging, destruction of *Clostridium botulinum* spores is the primary objective and most frequently involves application of heat and/or pressure to material packaged in a hermetically sealed container. *C. botulinum* can grow in anaerobic conditions in pH conditions greater than 4.6. The botulinum toxin is secreted on growth of the organism and is a neurotoxin that causes muscular paralysis and intestinal effects (FDA, 2012a). Death associated with botulism can arise from respiratory failure due to inability to move one's diaphragm. The Code of Federal Regulations (21 CFR 113) provides guidelines for processing hermetically sealed foods with pH greater than 4.6. Guidelines include good manufacturing processes, process controls, and processing record requirements. On implementation of the standards that were established in 1979, the incidence of botulism diminished to extremely low levels. As part of these standards, all thermal processes used by a company for high moisture, low-acid foods are filed with the FDA.

Similar to 21 CFR 113 for canned foods established by FDA, the USDA has established standards for processing jerky products. These standards were established following a 2003 outbreak caused by *Salmonella* Kiambu in jerky produced in New Mexico. These are reported as FSIS Compliance Guideline for Meat and Poultry Jerky Produced by Small and Very Small Establishments, revised most recently in August 2014 (USDA FSIS, 2014). This document outlines three options for developing validated jerky processes that are designed to provide at least a 5-log reduction in *Salmonella* organisms. The options include performing challenge studies using *Salmonella* (at a research facility) or a surrogate organism (in a plant), following cook schedules published in a peer-reviewed scientific article, or following specific guidelines outlined in Appendix A of the compliance guideline document that have been demonstrated to achieve sufficient thermal reduction of *Salmonella*.

Many new products in the marketplace are actually pasteurized rather than sterilized, which creates further considerations of transport and storage both in warehouses, retail outlets, and homes. Even with dry sterilized pet foods, storage is often not in the kitchen but rather in the garage where the environment can lead to moisture or odor transmission from other items stored in proximity to the pet food. Packaging format can also be critical in product transport as the material can be subjected to both temperature and pressure changes, which can place stressors on seals.

All processes involve some critical process points with respect to initial conditions (temperature, heat penetration requirements, residual air, packaging component sterility, etc.) and process requirements of time, temperature, pressure, and so forth. Failure to follow any of these increases the risk that commercial sterility is not achieved. The designed scheduled process must be strictly followed. Any deviation from the process must be evaluated by a process authority to determine if the food is safe for consumption.

After successful attainment of commercial sterility, it is important to prevent recontamination with bacteria. In the case of dry pet food, extensive environmental monitoring of drains, air, and various surfaces is typically performed. Proper facility design to maintain positive pressure in postprocessing rooms is also important. For canned foods, maintaining hermetic seals and sterile cooling water is important as microleaks during cooling can allow bacteria to enter the container (Podolak et al., 2010). For this reason, halogens such as chlorine and/or bromine are typically added to cooling water and closely monitored, particularly when recycling of cooling water is practiced or when using crateless retorts where damage to cans may occur. Containers must be allowed to cool at the proper rate to the proper temperature to prevent the growth of thermophilic organisms.

For dry pet food, commercially sterile (postextrusion) product is enrobed with materials that are not subjected to sterilization within the manufacturing plant production system. Examples of such materials include fat, acids, and palatability enhancers. It is therefore important for these raw materials to be produced, shipped, stored, and applied in a manner to attain and maintain commercially sterile status.

## A History of Recalls in Pet Food

Recalls are important as they expose gaps in corporate food safety systems. It is therefore important for all companies to conduct after-action reviews relating to their own recalls and evaluate relative vulnerability to situations resulting in recalls experienced by other manufacturers. The incidence of recalls has been much lower for pet food than human food counterparts, in part resulting from allergen control being part of human food requirements but not pet food. The Reportable Food Registry was established as part of the Food and Drugs Amendments Act of 2007, requiring reporting of food and feed where there is a reasonable probability of serious health consequences on consumption. A 5-year summary was recently published (FDA, 2016a,b,c,d). Of the 1081 primary entrees, 967 involved human food and 114 involved animal food, including pet food. For the latter, incidence of animal food registries ranged from 8.4% to 15% in individual years. Within the human food entrees, bakery products, produce, dairy products, seafood, and spices/seasonings were the highest categories. Undeclared allergens (43.4%), *Salmonella* (36.1%), and *Listeria monocytogenes* (20.4%) accounted for the vast majority of all food and feed entries during the 5-year period. Nutrient imbalance (4%) and undeclared sulfites (2.5%) were the next two highest categories. For animal feeds, during the last 4 years where distribution of animal feeds was reported (95 of the total 114 for the 5-year period), 44% were for *Salmonella*, 23% for nutrient imbalance, 10.5% for drug residue, and 5% for foreign objects.

A review of pet food recalls from 1996 to 2008 was prepared by Rumbeiha and Morrison (2011). The review indicated that, during the defined period, there were only 22 class-I and class-II recalls. Of these recalls, only six were for chemical contaminants including aflatoxin, cholecalciferol, methionine, melamine, and cyanuric acid. The prevalence of pet food recalls has obviously increased in recent years. Most early recalls in pet foods resulted from mold toxin contamination of dry pet foods. Most of these were for aflatoxin contamination; however, vomitoxin was implicated in some recalls. Detection of mold toxins is challenging as they are not uniformly distributed in food, and compositing samples has the opportunity to blunt some contaminated hot spots (Whitaker, 2006).

In 2007, the largest pet food recall in United States history to-date occurred as a result of EMA of imported wheat gluten with melamine scrap. This ultimately led to sweeping changes in regulations, including passage of the Food Safety Modernization Act on January 4, 2011 which includes elements of good manufacturing practices (GMPs), food safety preventive controls, and foreign supplier verification. Perhaps an unintended effect occurred when FDA made it easier for pet owners to directly register complaints without contacting companies first. Although this increased FDA awareness, it also had the impact of dividing complaints between FDA and companies, making it appear that issues were less important.

Also in 2007, there was a wide-reaching recall of both food and pet foods produced in a human food facility for an outbreak of foodborne botulism, which stemmed from a processing issue on one of the canning lines (FDA, 2007). In 2016, there was another incidence of *C. botulinum* contamination of canned soups which did not impact pet foods. There have been no other recent instances of recalls from inadequate thermal processing, likely a result of strict adherence to the FDA regulations for low-acid, canned foods.

Before 2007, there were several recalls resulting from *Salmonella* contamination of pet treats including pig ears, beef hooves, and other chews but not for complete diets. The Centers for Disease Control and Prevention reported that 70 human cases of salmonellosis were linked to *Salmonella* Schwarzengrund contamination of pet food between January 1, 2006 and December 11, 2007 (CDC, 2007). Since this recall, there have been numerous recalls caused by various *Salmonella* organisms both for dry pet food and raw pet food diets, which is an emerging category of pet food in the market. It is uncertain whether the remarkable increase in *Salmonella* recalls represents a true increase in risk or increased monitoring both by pet food companies and the FDA.

In response to the human health risk associated with potentially contaminated pet foods, in March 2012, the FDA issued Guidance for Industry: Testing for *Salmonella* Species in Human Foods and Direct-Human-Contact Animal Foods (FDA, 2012b). This is an important document as it highlights the nonuniformity of distribution of many food safety risks and the importance of using proper sampling methods. It clearly states that it is important to maintain control of product undergoing testing and that any confirmed positive result is valid (i.e., you cannot test your way into food safety). Similarly, the FDA may take enforcement action on lots of food they deem adulterated even in the presence of negative results obtained by the manufacturer.

With the increase in new product formats including chub-style products and raw pet food products, recalls have occurred not only for *Salmonella* but also for *L. monocytogenes* contamination (Nemser et al., 2014). *Listeria* was not associated with pet food recalls before 2007. With further growth of these product categories that have different processing and storage conditions which make them more susceptible to *Listeria* contamination, there may be a continued shift to increased *Listeria* recalls.

Between 2009 and 2011, several recalls occurred in wet cat food for low thiamine levels (FDA, 2016a,b,c,d). This particular nutrient is unstable to high processing temperatures such as those achieved in retorts. In addition, some fish species contain active thiaminase, which breaks down thiamine, and the time period between mixing a batch of pet food and reaching enzyme denaturation temperatures in the retort can allow for extensive thiamine destruction. Additional thiamine-related recalls occurred in 2015 and 2017 in a raw frozen cat food and canned cat food, respectively (FDA, 2017).

There have also been recalls for low or high levels of other vitamins as well as trace minerals, typically resulting from errors made in the production of the vitamin and mineral premixes. In 2006, an error in premixing resulted in excessive vitamin D and caused a product recall of wet pet food. In 2009, there was also a recall for incorrect levels of zinc and potassium from a premix error. Furthermore, in 2010, there was a recall for high levels of vitamin D in dry cat food (Rumbeiha and Morrison, 2011).

As mentioned previously, the list of acceptable additives in each country is not uniform. In 2013, several companies marketing chicken jerky products manufactured in China were recalled because of detection of unapproved antibiotics (FDA, 2015). These might not have been detected as they are not typically reviewed in US monitoring programs. However, the occurrence of numerous complaints of Fanconi syndrome in pets that consumed chicken jerky treats made in China led the FDA to an intense analysis of these products.

Generally, pet food recalls for foreign material contamination have been uncommon. That being said, in 2009 and 2010, there were recalls for plastic contamination of dry and wet pet food, respectively. In addition, in 2014, there was a recall of dry pet food for metal contamination (FDA, 2016a,b,c,d). Although most companies have detection systems in place for metal, plastic detection is quite challenging.

## Changes in the Regulatory Landscape for Pet Food

In response to research findings or product recalls, regulatory agencies often modify regulations accordingly. The changes may relate to formulations and/or required good manufacturing processes, including cleaning and disinfection procedures.

To manufacture organic pet foods, changes are required with respect to cleaning and disinfecting agents which, because of the change, require additional verifications with respect to plant GMP programs.

As mentioned previously, hurdles are critical in formulations containing high moisture levels. With the discovery of propylene glycol safety issues for cats, this ingredient was disallowed in cat food formulations in 1996 (21 CFR 589.1001). This regulatory change required reformulation of soft moist cat foods to use alternate humectants, the most common being glycerin. An unintended consequence of this change was that although water activity could be effectively controlled with alternatives, propylene glycol was an effective miticide, and this effect was not emulated with the alternatives. Mites can have adverse effects on both product integrity and pet health (Kamphues, 1991). More recently, with the transition to glycerin, an alternative supply from the biodiesel industry (crude glycerin) was proposed. This source, however, was found to carry an additional risk of methanol contamination, which carries its own food safety risks. There have also been recalls for mold growth in dry dog food and in soft moist foods because of high water activity, with the latest one being in 2015 (FDA, 2016a,b,c,d).

As a result of recent *Salmonella* recalls starting with the *Salmonella* Schwarzengrund, contamination of dry pet food which affected humans that handled the contaminated food, the FDA issued Compliance Policy Guide Sec. 690.800, *Salmonella* in Animal Feed (FDA, 2013) and defined such food as adulterated. This was the first incidence of pet food safety that considered not only the pet but also the pet parent.

In addition to FDA requirements, most companies in the United States make nutritional adequacy claims on labels based on AAFCO (Official Publication, 2016) recommendations. The most recent changes in the AAFCO Nutrient Profile (recommendations for nutritional adequacy) resulted in addition of new nutrients as well as changes in minimums and maximums. The implementation period was relatively short (less than 2 years), forcing pet food companies to react almost immediately. Some of these changes have resulted in changes in raw material supplies before extensive research to ensure no changes exist with respect to product risks.

In summary, dramatic changes in the types of pet food in the market today, driven by ever-changing consumer demand, have resulted in a shift in food safety and regulatory concerns. Furthermore, the Food Safety and Modernization Act, which was signed into law in 2011, represents the most comprehensive change to food safety regulations in the United States in the past 70 years. This regulation has changed the approach food and pet food companies must take to address food safety concerns. The industry as a whole is now challenged with identifying and mitigating food safety risks, putting focus on the entire ingredient supply chain as opposed to what happens only during production within the walls of the production facility. Given the growing trends of pet food to more closely mimic human food, it is likely that the pet food industry has not seen the end of new formats. Therefore, it is important that the industry continue to adapt and quickly respond to mitigate new food safety risks presented by new ingredients, processes, and packaging types required to meet new demands.

## References


Aburto, A., Edwards Jr., H.M., Britton, W.M., 1998. The influence of vitamin A on the utilization and amelioration of toxicity of cholecalciferol, 25-hydroxycholecalciferol, and 1,25-dihydroxycholecalciferol in young broiler chickens. Poultry Science 77, 585–593.

Adam, L.A., Reeves, V.B., 1998. Procedure for detecting and confirming pentobarbital residues in dog food by gas chromatography/mass spectrometry. Journal of AOAC International 81 (2), 359–368.

Association of American Feed Control Officials (AAFCO), 2016. Official Publication, Champaign, IL.

Bauer, M.C., Weiss, D.J., Perman, V., 1991. Hematological alterations in adult cats fed 6% or 12% propylene glycol. American Journal of Veterinary Research 53 (1), 69–72.

Boermans, H.J., Leung, M.C.K., 2007. Mycotoxins and the pet food industry: toxicological evidence and risk assessment. International Journal of Food Microbiology 119, 95–102.

Cable News Network (CNN), 2005. Company Recalls Tainted Pet Food, 2005. Available from: http://www.cnn.com/2005/US/12/22/dog.deaths/index.html?eref=aol.

Christopher, M.M., Perman, V., Eaton, J.W., 1989. Contribution of propylene glycol-induced Heinz body formation to anemia in cats. Journal of the American Veterinary Medical Association 194 (8), 1045–1056.

Christopher, M.M., Broussard, J.D., Peterson, M.E., 1995. Heinz body formation associated with ketoacidosis in diabetic cats. Journal of Veterinary International Medicine 9 (1), 24–31.

Centers for Disease Control and Prevention (CDC), 2007. *Salmonella* Schwarzengrund Outbreak Investigation. https://www.cdc.gov/salmonella/schwarzengrund.html.

Douglass, G.M., Fern, E.B., Brown, R.C., 1991. Feline plasma and whole blood taurine levels as influenced by commercial dry and canned diets. Journal of Nutrition 121, S179–S180.

Dow, S.W., Fettman, M.J., Smith, K.R., Ching, S.V., Hamar, D.W., Rogers, Q.R., 1992. Taurine depletion and cardiovascular disease in adult cats fed a potassium-depleted acidified diet. American Journal of Veterinary Research 53 (3), 402–405.

Dunayer, E.K., Gwaltney-Brant, S.M., 2006. Acute hepatic failure and coagulapathy associated with xylitol ingestion in eight dogs. Journal of the American Veterinary Medical Association 229 (7), 1113–1117.

Dzanis, D.A., 1991. Safety of ethoxyquin in dog foods. Journal of Nutrition 121, S163–S164.

Dzanis, D.A., 2007. Mycotoxins and Pet Food. Available from: http://www.petfoodindustry.com/articles/591-mycotoxins-petfood.

Environmental Protection Agency (EPA), 2004. Fatty Acid Monoesters with Glycerol or Propanediol Fact Sheet. https://www3.epa.gov/pesticides/chem_search/reg_actions/registration/fs_G-104_20-Oct-04.pdf.

Food and Drug Administration (FDA), 1997. FDA requests that ethoxyquin levels be reduced in dog foods. FDA Veterinary Feed Directive 12 (5).

Food and Drug Administration (FDA), 1999. Guidance for Industry: Dioxin in Anti-caking Agents Used in Animal Feed and Feed Ingredients. Available from: http://www.fda.gov/downloads/AnimalVeterinary/GuidanceComplianceEnforcement/GuidanceforIndustry/ucm052437.pdf.

Food and Drug Administration (FDA), 2007. Clostridium Products (Botulism) Recall. Available from: http://www.fda.gov/NewsEvents/PublicHealthFocus/ucm179201.htm.

Food and Drug Administration (FDA), 2011. FDA Mycotoxin Regulatory Guidance: A Guide for Grain Elevators, Feed Manufacturers, Grain Processors and Exporters. Available from: http://www.ngfa.org/wp-content/uploads/NGFAComplianceGuide-FDARegulatoryGuidanceforMycotoxins8-2011.pdf.

Food and Drug Administration (FDA), 2012a. Bad Bug Book, Foodborne Pathogenic Microorganisms and Natural Toxins, *Clostridium botulinum*, second ed. Available from: http://www.fda.gov/downloads/Food/FoodSafety/FoodborneIllness/FoodborneIllnessFoodbornePathogensNaturalToxins/BadBugBook/UCM297627.pdf.

Food and Drug Administration (FDA), 2012b. Guidance for Industry: Testing for Salmonella Species in Human Foods and Direct-human-contact Animal Foods. Available from: http://www.fda.gov/food/guidanceregulation/guidancedocumentsregulatoryinformation/salmonella/ucm295271.htm.
Food and Drug Administration (FDA), 2013. Compliance Policy Guide Sec 690.800 Salmonella in Food for Animals. Available from: http://www.fda.gov/downloads/ICECI/ComplianceManuals/CompliancePolicyGuidanceManual/UCM361105.pdf.
Food and Drug Administration (FDA), 2015. Jerky Pet Treats – Annual Update. Available from: http://www.fda.gov/AnimalVeterinary/SafetyHealth/ProductSafetyInformation/ucm360951.htm.
Food and Drug Administration (FDA), 2016a. Thermally Processed Low-Acid Food Packaged in Hermetically Sealed Containers. Title 21, Code of Federal Regulations, Part. 113, 2016 ed.
Food and Drug Administration (FDA), 2016b. Sodium Metabisulfite. Title 21 Code of Federal Regulations, Pt. 182.3766. 2016 ed.
Food and Drug Administration (FDA), 2016c. The Reportable Food Registry: A Five Year Overview of Targeting Inspection Resources and Identifying Patterns of Adulteration. Available from: http://www.fda.gov/downloads/Food/ComplianceEnforcement/RFR/UCM502117.pdf.
Food and Drug Administration (FDA), 2016d. Pet Food Recall Products List. Available from: http://www.fda.gov/AnimalVeterinary/SafetyHealth/RecallsWithdrawals/default.htm.
Food and Drug Administration (FDA), 2017. Animal and Veterinary Recalls Archive. Available from: http://www.fda.gov/AnimalVeterinary/SafetyHealth/RecallsWithdrawals/ucm393160.htm.
Friedman, M., Henika, P.R., Mackey, B.E., 1996. Feeding of potato, tomato, and eggplant alkaloids affects food consumption and body and liver weights in mice. Journal of Nutrition 126, 989–999.
Gans, J.H., Korson, R., Cater, M.R., Ackerly, C.C., 1980. Effects of short-term and long-term theobromine administration to male dogs. Toxicology and Applied Pharmacology 53 (3), 481–496.
Grocery Manufacturers Association, 2010. Consumer Product Fraud: Deterrence and Detection. Available from: http://www.gmaonline.org/downloads/research-and-reports/consumerproductfraud.pdf.
Gruffydd-Jones, T.J., Galloway, P.E., Pearson, G.R., 1991. Feline spongiform encephalopathy. Journal of Small Animal Practice 33, 471–476.
Guraya, H.S., Koehler, P.E., 1991. Histamine in cat foods: survey and comparison of methodologies. Veterinary and Human Toxicology 33 (2), 124–128.
Hedberg, C.W., Fishbein, D.B., Janssen, R.S., Meyers, B., McMillen, J.M., MacDonald, K.L., White, K.E., Huss, L.J., Hurwitz, E.S., Farhie, J.R., Simmons, J.L., Braverman, L.E., Ingbar, S.H., Schonberger, L.B., Osterholm, M.T., 1987. An outbreak of thyrotoxicosis caused by the consumption of bovine thyroid gland in ground beef. New England Journal of Medicine 316 (16), 993–998.
Hayes, K.C., Cary, R.E., Schmidt, S.Y., 1975. Retinal degeneration associated with taurine deficiency in the cat. Science 188, 949–951.
Hazewinkel, H.A.W., van den Brom, W.E.R., van Klooster, A.T., Voorhout, G., van Wees, A., 1991. Calcium metabolism in Great Dane dogs fed diets with various calcium and phosphorus levels. Journal of Nutrition 121, S99–S106.
Heller, D.N., Lewis, K.M., Cui, W., 2001. Method for determination of pentobarbital in dry dog foods by gas chromatography/mass spectrometry. Journal of Agricultural and Food Chemistry 49, 4597–4602.
Johnson, L.A., Ford, H.C., Tarttelin, M.F., Feek, C.M., 1992. Iodine content of commercially-prepared cat foods. New Zealand Veterinary Journal 40, 18–20.
Johnson, R., 2014. Food Fraud and "Economically Motivated Adulteration" of Food and Food Ingredients. Congressional Research Service. Available from: https://fas.org/sgp/crs/misc/R43358.pdf.
Kamphues, J., 1991. Mites infestation of feedstuffs for small companion animals. Journal of Nutrition 121, S165.
Krishnamachari, K., Bhat, R., Nagarajan, V., Tilak, T.B., 1975. Hepatitis due to aflatoxicosis. An outbreak in Western India. Lancet 10, 1061–1063.
Kuhlman, G., Rompala, R.E., 1998. The influence of dietary sources of zinc, copper and manganese on canine reproductive performance and hair mineral content. Journal of Nutrition 128, 2603S–2605S.
Leistner, L., 2000. Basic aspects of food preservation by hurdle technology. International Journal of Food Microbiology 55, 181–186.

Lowe, J.A., Wiseman, J., 1998. A comparison of the bioavailability of three dietary zinc sources using four different physiologic parameters in dogs. Journal of Nutrition 128, 2809S–2811S.

Moore, J.C., Ping, J.S., Lipp, M., 2012. Development and application of a database of food ingredient fraud and economically motivated adulteration from 1980 to 2010. Journal of Food Science 77, R118–R126.

Morris, J.G., Rogers, Q.R., 1995. Copper oxide is an ineffective source of copper in queen diets. In: Petfood Forum 95 Proceedings, pp. 107–108.

National Research Council (NRC), 2006. Nutrient Requirements of Dogs and Cats. The National Academies Press, Washington, DC.

Nemser, S.M., Doran, T., Grabenstein, M., McConnell, T., McGrath, T., Pamboukian, R., Smith, A.C., Maya, A., Gregory, D., Sun, K., Yong, L., Sharon, R., Grisel, R., McWilliams Wilson, K., Reimschuessel, R., 2014. Investigation of *Listeria*, *Salmonella*, and Toxigenic *Escherichia coli* in various pet foods. Foodborne Pathogens and Disease 11 (9), 706–709.

Nicholson, J.W.G., Snoddon, P.M., Dean, P.R., 1988. Digestibility and acceptability of potato steam peel by pigs. Canadian Journal of Animal Science 68, 233–239.

Odle, J., Roach, M., Baker, D.H., 1993. Taurine utilization by cats. Journal of Nutrition 123, 1932–1933.

Okuma, T.A., Hellberg, R.S., 2015. Identification of meat species in pet foods using a real-time polymerase chain reaction (PCR) assay. Food Control 50, 9–17.

Palozza, P., 1998. Prooxidant actions of carotenoids in biologic systems. Nutrition Reviews 56 (9), 257–265.

Phillips-Donaldson, D., 2016. Premium Pet Food and Pet Specialty Retail Trends Update. Available from: http://www.petfoodindustry.com/blogs/7-adventures-in-pet-food/post/6103-premium-pet-food-and-pet-specialty-retail-trends-update.

Podolak, R., Stone, W., Black, D.G., 2010. Retort cooling water bacteriological load and possible mitigation strategies for microbial buildup in cooling water. Food Protection Trends 30 (3), 160–167.

Radosevich, J., 2007. Analyzing Amines: Evaluating Potential Toxicity in Petfood Ingredients. Pet Food Industry. Available from: http://www.petfoodindustry.com/articles/521-analyzing-amines.

Rawles, D.D., Flick, G.J., Martin, R.E., 1996. Biogenic amines in fish and shellfish. Advances in Food and Nutrition Research 39, 329–365.

Robertson, J.E., Christopher, M.M., Rogers, Q.R., 1998. Heinz body formation in cats fed baby food containing onion powder. Journal of the American Veterinary Medical Association 212 (8), 1260–1266.

Rumbeiha, W., Morrison, J., 2011. A review of class I and class II pet food recalls involving chemical contaminants from 1996 to 2008. Journal of Medical Toxicology 7, 60–66.

Safran, N., Aizenberg, I., Bark, H., 1993. Paralytic syndrome attributed to lasalocid residues in a commercial ration fed to dogs. Journal of the American Veterinary Medical Association 202 (8), 1273–1275.

Shalaby, A.R., 1996. Significance of biogenic amines to food safety and human health. Food Research International 29 (7), 675–690.

Shields, R., 1998. Vitamin and mineral nutrition update. In: Petfood Forum 98 Proceedings, pp. 15–30.

Shurson, G., Salzer, T., Koehler, D., 1996. Metal-specific amino acid complexes, inorganic trace minerals effect on vitamin stability examined. Feedstuffs 68 (45), 13–23.

Smith, D.B., Roddick, J.G., Jones, J.L., 1996. Potato glycoalkaloids: some unanswered questions. Trends in Food Science & Technology 7, 126–131.

Smith, D.C., Proutt, L.M., 1944. Development of thiamine deficiency in the cat on a diet of raw fish. Proceedings of the Society for Experimental Biology and Medicine 56, 1–3.

Stengel, K.C., Neiger, R., 2012. Dietary hyperthyroidism in dogs. Journal of Small Animal Practice 53 (3), 182–184.

Studdert, V.P., Labuc, R.H., 1991. Thiamine deficiency in cats and dogs associated with feeding meat preserved with sulfur dioxide. Australian Veterinary Journal 68 (2), 54–57.

Thompson, L.U., 1993. Potential health benefits and problems associated with antinutrients in foods. Food Research International 26 (2), 131–149.

US Department of Agriculture Food Safety Inspection Service (USDA), 2014. Compliance Guideline for Meat and Poultry Jerky Produced by Small and Very Small Establishments. Available from: https://meathaccp.wisc.edu/doc_support/asset/Compliance-Guideline-Jerky-2014.pdf.

Veciana-Nogues, M.T., Marine-Font, A., Vidal-Carou, M.C., 1997a. Biogenic amines in fresh and canned tuna. Effects of canning on biogenic amine contents. Journal of Agricultural and Food Chemistry 45, 4324–4328.
Veciana-Nogues, M.T., Marine-Font, A., Vidal-Carou, M.C., 1997b. Biogenic amines as hygienic quality indicators of tuna. Relationships with microbial counts, ATP-related compounds, volatile amines, and organoleptic changes. Journal of Agricultural and Food Chemistry 45, 2036–2041.
Walker, M.J., Burns, M., Burns, D.T., 2013. Horse meat in beef products – species substitution 2013. Journal of the Association of Public Analysts (Online) 41, 67–106.
Weil, C.S., Woodside, M.D., Smyth Jr., H.F., Carpenter, C.P., 1971. Results of feeding propylene glycol in the diet to dogs for two years. Food and Cosmetics Toxicology 9, 479–490.
Whitaker, T.B., 2006. Sampling foods for mycotoxins. Food Additives and Contaminants 23 (1), 50–61.
World Health Organization (WHO), 2016. Dioxins and Their Effects on Human Health. Available from: http://www.who.int/mediacentre/factsheets/fs225/en/.

CHAPTER 2

# Mycotoxin Issues in Pet Food

**Griffiths G. Atungulu[1], Zeinab Mohammadi-Shad[1,2], Shantae Wilson[1]**
[1]*University of Arkansas Division of Agriculture, Fayetteville, AR, United States;* [2]*Nestlé Iran P.J.S. Co., Mohammadiyeh City, Iran*

## Introduction

Mycotoxins are naturally occurring, secondary metabolites of certain fungi that can be detrimental to the health of humans and animals that consume them via contaminated food and feed. The word mycotoxin is derived from the Greek words "myke" meaning fungus and "toxicum" meaning toxin. The United States (US) Food and Drug Administration (FDA) has issued action, guidance, and advisory levels for aflatoxin, fumonisin, and deoxynivalenol (commonly called vomitoxin, or DON), respectively. The trio is not an exhaustive list; other mycotoxins (e.g., ochratoxin and zearalenone) can also be troublesome in pet food. At present, FDA handles contamination issues with mycotoxins other than aflatoxin on a case-by-case basis. The two mycotoxins that frequently affect pets are aflatoxin and vomitoxin.

### Aflatoxin

Aflatoxins are a group of structurally related mycotoxins produced primarily by *Aspergillus flavus* and *Aspergillus parasiticus* fungi (Ashiq, 2014; El-Tras et al., 2011; Hernandez-Mendoza et al., 2009; Klich, 2007; Marroquín-Cardona et al., 2014; Martins et al., 2003; Richard, 2007). The chemical structure of aflatoxin incorporates dihydrofuran and tetrahydrofuran moieties coupled to a substituted coumarin (Leung et al., 2006). There are four principle aflatoxins that are of concern in pet food: aflatoxin B1 (AFB1), aflatoxin B2 (AFB2), aflatoxin G1 (AFG1), and aflatoxin G2 (AFG2). The FDA has issued action levels for aflatoxins which provide specific, maximum tolerable amounts for aflatoxin in a variety of feedstuffs (e.g., corn, cottonseed meal, and peanut products) intended for various animal species. The action level for pet food is set at 20 parts per billion total aflatoxin content (B1 + B2 + G1 + G2). Liver damage and death can result when pets consume food that is contaminated with aflatoxins. Corn and other pet food raw materials are susceptible to aflatoxin-producing fungi that are abundant in soil and air and may contaminate in the field prior to harvest, postharvest in storage, and processing (Hernandez-Mendoza et al., 2009). Aflatoxin contamination has been reported in other commodities including nuts, rice, cottonseeds, spices, figs, and also in many other animal feedstuffs (Ashiq, 2014; Rahimi et al., 2010;

Food and Feed Safety Systems and Analysis. http://dx.doi.org/10.1016/B978-0-12-811835-1.00002-6

Richard, 2007; Rustom, 1997). Generally, pet food raw materials with high amounts of carbohydrates and lipids are more susceptible to aflatoxin contamination because enzymes produced by aflatoxin-producing fungi, *A. flavus*, can hydrolyze these substances (Gallo et al., 2016). Once formed, aflatoxins are relatively stable and resistant to heat and, therefore, are not eliminated by heat processing (Campagnollo et al., 2016).

### *Deoxynivalenol*

Deoxynivalenol (DON, vomitoxin), is generated primarily by *Fusarium culmorum* and *Fusarium graminearum* (Cortinovis et al., 2012; Döll and Dänicke, 2011; Martins et al., 2003; Richard, 2007). The toxin has been classified as type-B nonmacrocyclic trichothecene (Richard, 2007; Streit et al., 2012) and is the most common of the trichothecene mycotoxins detectable in agricultural products (Cortinovis et al., 2012; Martins et al., 2003). Deoxynivalenol contaminates cereal crops worldwide, most commonly affecting wheat, maize, barley, and corn silage (Martins et al., 2003; Streit et al., 2012). Like most mycotoxins, deoxynivalenol is a stable compound that is largely unaffected by most processing methods (Didwania and Joshi, 2013; Magan and Aldred, 2007). The toxin displays stability at temperatures as high as 120–180°C and is soluble in water and polar solvents such as methanol, acetonitrile, and ethyl acetate (Döll and Dänicke, 2011). The FDA advisory level for deoxynivalenol in grains and grain by-products intended for pet food is five parts per million (ppm). As the name suggests, vomitoxin causes severe vomiting and can cause death in pets.

### *Fumonisin*

Fumonisins are a group of at least 15 structurally related toxins (Abbas et al., 1998) produced primarily by *Fusarium verticillioides* and *Fusarium proliferatum* (Martins et al., 2003; Sultana and Hanif, 2009). The fumonisin mycotoxins of which fumonisin group B is the most prevalent and notable one with fumonisin B1 (FB1) and fumonisin B2 (FB2) being esterified with tricarboxylic and polyhydric alcohols (Martins et al., 2003). The toxin can be found in commodities such as maize, and in dried figs, and some plants (Didwania and Joshi, 2013). The FDA has issued guidance levels for fumonisins, and recommends a total fumonisin (FB1 + FB2 + FB3) content in corn and corn by-products intended for pet foods not to exceed 10 ppm. The further stipulation is that corn or corn by-product not exceed more than 50% of the pet food formulation. The International Agency for Research on Cancer classifies FB1 as a group 2B carcinogen, possibly carcinogenic to humans.

### *Ochratoxins*

These are produced by numerous *Aspergillus* and *Penicillium* spp., including *Aspergillus ochraceus* and *Penicillium verrucosum*. The toxins occur primarily in cereal grains but also in animal by-products because of their tight binding with plasma proteins and long half-life in animal tissues. Four homologues of ochratoxin exist, and these include A, B, C, and D with the most prevalent being the A homologue. Ochratoxin A is a potent nephrotoxin in pet species. A 2-week exposure to 0.3 mg of ochratoxin A/kg of body weight is lethal to young male beagle dogs, which exhibit severe kidney damage, anorexia, vomiting, weight loss, tenesmus, intestinal hemorrhage, tonsillitis, dehydration, and prostration. A 1-year exposure to 0.8 mg of ochratoxin A/kg of feed has been shown to cause mild nephropathy.

### *Tremorgens*

These are primarily produced by *Penicillium*, *Aspergillus*, and *Claviceps* (Puschner, 2002). The occurrence of these toxins contrasts that of aflatoxins, ochratoxin A, and the trichothecenes, in that they are rarely found in feed ingredients but are usually produced during food spoilage. Exposure of companion animals, such as dogs, to tremorgenic mycotoxins via consumption of moldy garbage has been reported (Boysen et al., 2002). There are over 20 mycotoxins that are classified as tremorgens. Of the tremorgens, penitrem A, and roquefortine are widely associated with negative effects to companion animals (Beasley et al., 1997; Puschner, 2002). The negative effects of penitrem A in dogs include slight salivation, vomiting, ataxia, and tremors. Oral exposure to 0.175 mg of penitrem A/kg of body weight has been shown to induce muscle tremors in dogs (Hocking et al., 1988).

## *Mycotoxin Prevalence*

The prevalence of mycotoxins in feedstuffs is highly dependent upon climate and weather. Aflatoxins prevail mostly in temperate and tropical regions which can be found in areas of North America and South America, and in South Asia, and Africa. In these regions, the primary pathway for mycotoxins to impact feed is through ingredients such as corn, peanuts, cottonseed, tree nuts, wheat, and rice. Ochratoxins prevail mostly in temperate regions such the Balkans, Northern Europe, and Canada; in these regions, affected crops include barley, corn, wheat, oats, nuts, dried beans, and animal by-products. *Fusarium* mycotoxins prevail in temperate regions of Europe, United States, Canada, Russia, Japan, China, Brazil, and Argentina with the major contaminated feedstuffs being corn, barley, wheat, oats, and soybean.

A survey of recent, documented incidents of mycotoxin contamination in pet food (Table 2.1) indicates that mycotoxins are not uncommon. The reasons for mycotoxin contamination of pet food are multifaceted but originate with ingredients. Grains and grain by-products, which are inherently susceptible to fungal and mycotoxin contamination, are primary ingredients in many pet foods. Grain contamination is unavoidable, to a certain extent, as exposure to the fungi and conditions that promote fungal growth cannot be prevented in the field. Mycotoxin contamination in harvested grains is heterogeneous and, among contaminated kernels, the level of contamination can vary greatly. The heterogeneous nature of mycotoxin contamination in grain introduces sampling issues that directly impact correct lot quantification and subsequent marketing decisions.

Table 2.1: Mycotoxin detection in commercial pet food.

| Year | Location | Samples surveyed | Analytical methods and limits | Mycotoxins detected | References |
|---|---|---|---|---|---|
| 1993 | Iowa (USA) | Two dry dog foods, two dry cat foods, and one rat chow | HPLC with $C_{18}$ SPE column cleanup fumonisins B1 and B2, 200 and 20 μg/kg, respectively | Fumonisins B1 and B2 were detected in all five samples, ranging from 219 to 1410 μg/kg and from 20 to 144 μg/kg, respectively | Hopmans and Murphy (1993) |
| 1994 | North Carolina (USA) | 100 dry dog food samples | ELISA with TLC confirmation aflatoxin B1, 5 μg/kg zearalenone, 100 μg/kg | No sample contained >20 μg of aflatoxin B1/kg; one sample contained 250 μg of zearalenone/kg | Cullen and Newberne (1994) |
| 1997 | Slough, UK | 100 pet food samples:<br>35 dry dog foods<br>35 dry dog cat foods<br>15 domestic bird foods<br>15 wild bird foods | HPLC with immunoaffinity column cleanup aflatoxins B1, B2, G1, and G2, 1.0 μg/kg, respectively<br>Ochratoxin A, 1.0 μg/kg<br>Fumonisins: not specified | Mycotoxins were detected in 16% of the samples, including seven of 30 bird foods tested aflatoxin B1, two samples with 2.1 and 370 μg/kg, respectively Ochratoxin A, 10% of the samples, 1–7 μg/kg fumonisins, 30% of the samples, 90–690 (B1) μg/kg | Scudamore et al. (1997) |
| 2001 | Mexico | 19 dog foods and 16 cat foods | HPLC with immunoaffinity column cleanup aflatoxins B1, B2, G1, G2, M1, M2, and P1, 3–7 μg/kg | Aflatoxins were detected at 89% and 100% of the dog and cat food samples, with mean aflatoxin B1 values of 5.00 and 8.02 μg/kg, respectively | Sharma and Márquez (2001) |
| 2001 | Texas (USA) | 142 wild bird seed samples | ELISA<br>Total aflatoxins, 1 μg/kg | Aflatoxin concentrations ranged from 0 to 2780 μg/kg<br>17% of the samples had aflatoxin concentrations of >100 μg/kg | Henke et al. (2001) |
| 2002 | Alfenas, Brazil | 100 pet food samples:<br>45 dog foods<br>25 cat foods<br>30 bird foods | TLC<br>Aflatoxins B1, B2, G1, and G2, 8, 6, 7, and 5 μg/kg, respectively | Aflatoxins were found in 6.7%, 4.0%, and 26.7% of the dog, cat, and bird food samples, with mean aflatoxin B1 levels of 19, 16, and 110 μg/kg, respectively | Maia and Pereira Bastos de Siqueira (2002) |
| 2002 | Turkey | 18 dog food samples | ELISA with immunoaffinity column cleanup<br>Total aflatoxins, 1.75 μg/kg | Aflatoxins were found in 16.7% of the samples, ranging from 1.75 to 20 μg/kg | Gunsen and Yaroglu (2002) |
| 2003 | Portugal | 60 dry pet food samples:<br>20 dog foods<br>20 cat foods<br>20 domestic bird foods | HPLC with immunoaffinity column cleanup aflatoxins B1, B2, G1, and G2, 1 μg/kg, respectively ochratoxin A, 2 μg/kg fumonisin B1, 10 μg/kg deoxynivalenol, 100 μg/kg | Mycotoxins were detected in only dog food aflatoxins, not detected ochratoxin A, five samples, 2.0–3.6 μg/kg fumonisin B1, three samples, 12.0–24.0 μg/kg deoxynivalenol, three samples, 100.0–130.0 μg/kg | Martins et al. (2003) |

| | | | | | |
|---|---|---|---|---|---|
| 2004 | Poland | 57 brands of standard and therapeutic pet food | HPLC with immunoaffinity column cleanup zearalenone, 5.0 μg/kg | Zearalenone was detected in 84% of the samples mean, 36.2 μg/kg highest levels, 299.5 μg/kg standard), and 158.0 μg/kg (therapeutic) | Zwierzchowski et al. (2004) |
| 2008 | Brazil | 840 raw materials (corn kernels, corn meal, sorghum meal, corn meal, and gluten) and five ready dry pet food (made of the raw materials sampled) | HPLC, aflatoxin B1. 0.1 ng/g | 100% raw materials contaminated (ranging from 0.1 to 24.5 ng/g). The highest aflatoxin contamination in corn and sorghum<br>Nontoxin detected in ready pet food | Campos et al. (2008) |
| 2009 | Argentina | Raw materials and pelletized ready dog food samples | HPLC, aflatoxin B1, and fumonisin B1 | Aflatoxin B1 detected only in wheat and soybean pellet<br>85% ready dog food contained FB1 at levels from 2.0 to 166.9 ng/g | Fernandez Juri et al. (2009) |
| 2015 | Italy | 48 samples of complete extruded dry dog food | UPLC-MS/MS, deoxynivalenol, fumonisin B1 and B2, aflatoxin B1, B2, G1, and G2, ochratoxin A and zearalenone<br>DON and FB1: 1 g/kg; AFB1 and AFB2: 0.5 g/kg; AFG1, AFG2, FB2 and ochratoxin A: 2 g/kg; zearalenone A: 5 g/kg | 15% of all samples contained above the limit of quantification<br>DON in 100% samples AFB1 and AFG1 not detected<br>AFB2 and AFG2 detected in 4% and 8% of samples, respectively<br>Fumonisins and ochratoxin A in 88% and 81% of samples, respectively | Gazzotti et al. (2015) |

*AFB1*, aflatoxin B1; *AFB2*, aflatoxin B2; *AFG1*, aflatoxin G1; *AFG2*, aflatoxin G2; *DON*, deoxynivalenol; *ELISA*, enzyme-linked immunosorbent assay; *FB1*, fumonisin B1; *FB2*, fumonisin B2; *HPLC*, high performance liquid chromatography; *MS*, mass spectrometry; *SPE*, solid phase extraction; *TLC*, thin layer chromatography ; *UPLC-MS/MS*, ultra performance liquid chromatography-tandem mass spectrometer.

*Reprinted with permission from Maxwell, C.K.L., Gabriel, D.-L., Trevor, K.S., 2006. Mycotoxins in pet food: a review on worldwide prevalence and preventative strategies. Journal of Agricultural and Food Chemistry 54, 9623–9635.* 

In addition to misclassification and potential mismarketing of contaminated grains, international trade issues may arise from differing regulatory determinations for acceptable levels of mycotoxins in various food and feedstuffs and the lack of specificity for companion animals. Current recommendations lean more toward farm and production animals than companion animals. Finally, pet food companies must assess the effectiveness of their quality control programs. This is currently being mandated by the Food Safety Modernization Act Preventive Controls for Animal Food rule.

## Pet Food–Related Mycotoxicoses Outbreaks and Recalls

Outbreaks related to pet food date back to as early as 1952 (Devegowda and Castaldo, 2000). Agents of diseases associated with these early outbreaks were not clear at the time they occurred. Subsequent intensive investigations have revealed that fungi are behind occurrence of mycotoxicoses (Bradburn et al., 1994; Asao et al., 1963). There have been at least 10 cases of tremorgenic mycotoxicoses related to dogs in North America, since 1979 (Arp and Richard, 1979; Boysen et al., 2002; Hocking et al., 1988; Lowes et al., 1992; Naude et al., 2002; Puls and Ladyman, 1988; Arp and Richard, 1979; Richard et al., 1981; Wagner et al., 1997; Walter, 2002; Young et al., 2003; Lightfoot and Yeager, 2008). Other countries also continue to struggle to control similar pet food poisoning with mycotoxins (http://pethealthstore.com/articlepdf/foodsafety.pdf) (Pereyra et al., 2008). Table 2.2 lists outbreaks related to mycotoxin contamination of pet food.

Documented reports of acute mycotoxicosis may not represent the scope of the mycotoxin problem. Often overlooked are cases of chronic diseases such as liver and kidney fibrosis and infections resulting from immunosuppression and cancer diseases which stem from mycotoxin poisoning. As more awareness grows on the effect of mycotoxins on pet health, the public response has shifted to quality assurance.

## Mycotoxin Analyses and Related Challenges

### Sampling Issues

Mycotoxins are rarely distributed homogenously in a batch of consignment. The challenge therefore is to obtain a representative sample for analysis. Incorrect sample selection may lead to erroneous results, generally, 90% of the error (Turner et al., 2009). Nonrepresentative sampling may result in a false positive or false negative result which could lead to tremendous socioeconomic effects upon acceptance or rejection of a batch of consignment (Magan, 2004).

An ideal sampling protocol should include (1) objective-oriented analysis to determine why, where, and when to obtain samples and (2) a sampling method to take into account the heterogeneous nature of mycotoxin contamination (Miraglia et al., 2005). The test procedure recommended by the U.S. Department of Agriculture consists of a 5-kg sample, a 250-g

**Table 2.2: Outbreaks related to mycotoxin contamination of pet food.**

| Year | Location | Scale | Diet | Mycotoxin level | References |
|---|---|---|---|---|---|
| 1951–55 | Southeastern USA | 71 food poisoning cases of dogs (with several dead) | A brand of commercial dog food suspected to be made with contaminated corn | Unknown | Bailey and Groth (1959), Newberne et al. (1955) and Seibold and Bailey (1952) |
| 1974 | Rajasthan and Gujarat, India | Reports of dead stray and domestic dogs in about 200 villages (97 people died in the same outbreak) | Contaminated corn | 6.25–15.6 mg of aflatoxins B1 and G1/kg of food sample | Krishnamachari et al. (1975) and Tandon et al. (1977) |
| 1974 | Alabama (USA) | At least three dogs dead and several with anorexia and depression in a kennel after chronic mycotoxin exposure | Cooked cornmeal, meat scraps, and contaminated dry commercial dog food | 60 μg of aflatoxin B1/kg of feed | Greene et al. (1977) |
| 1975 | Queensland, Australia | Three dogs dead after acute and subacute mycotoxin exposure | Mixture of bread loaves (moldy) and canned dog meat | Vomit samples contained 100 and 40 mg of aflatoxins B1 and G1/kg, respectively | Ketterer et al. (1975) |
| 1985 | Georgia (USA) | 13 deaths and several with weakness, icterus, and partial anorexia of 48 dogs in a kennel | Mixture of commercial dog food and restaurant garbage | Unknown | Liggett et al. (1986) |
| 1986 | Georgia (USA) | Nine deaths of 20 dogs in a kennel | Ration of contaminated cornmeal, soybean meal, fat, and a mineral supplement | 465 and 46 μg of aflatoxins B1 and B2/kg of feed, respectively | Liggett et al. (1986) |
| 1987 | Pretoria, South Africa | 10 dogs dead with one acute, seven subacute, and two chronic cases | A brand of contaminated commercial dog food | 100–300 μg of aflatoxins/kg of feed | Bastianello et al. (1987) |
| 1988 | Australia | One dog with severe muscle tremors | Moldy hamburger bun | 35 mg of penitrem A/kg of sample | Hocking et al. (1988) |
| 1991 | Glasgow, UK | One dog with hepatopathy and dermatitis | Damp and moldy biscuit meal | 100 μg of ochratoxin A, 150 μg of citrinin, and 200 μg of sterigmatocystin/kg of meal | Little et al. (1991) |

*Continued*

**Table 2.2: Outbreaks related to mycotoxin contamination of pet food.—cont'd**

| Year | Location | Scale | Diet | Mycotoxin level | References |
|---|---|---|---|---|---|
| 1998 | Texas (USA) | 55 dogs dead with both acute and chronic cases | 17 different formulations of commercial dog food made with two rail cars of nonuniformly contaminated corn in a milling plant in Texas in late summer | 100–300 μg of aflatoxin B1/kg of feed | Garland and Reagor (2001) |
| 2002 | Massachusetts (USA) | Four dogs with ataxia and tremors | Garbage | Unknown level of tremorgenic mycotoxins | Boysen et al. (2002) |
| 2002 | South Africa | Two dogs with vomiting and tremors | Moldy rice | 2.6 mg of penitrem A and 34 mg of roquefortine/kg of rice | Naude et al. (2002) |
| 2005 | Eastern USA. | At least 100 dogs dead | 19 different formulations of commercial dog food made with contaminated corn in a milling plant in South Carolina in summer | Unknown | FDA (2005) and Stenske et al. (2006) |
| 2006 | Korea | Three dogs dead with renal failure | Fungal nephrotoxins in the diet, possibly ochratoxin and citrinin | Unknown | Jeong et al. (2006) |

*Reprinted with permission from Maxwell, C.K.L., Gabriel, D.-L., Trevor, K.S., 2006. Mycotoxins in pet food: a review on worldwide prevalence and preventative strategies. Journal of Agricultural and Food Chemistry 54, 9623–9635. Copyright (2006) American Chemical Society.*

subsample, and a single thin layer chromatography analysis (Whitaker, 2003). This procedure results in a total variance of 630.7, including 521.4 (82.7%) due to the sampling, 59.2 (9.4%) due to subsampling, and 50.1 (7.9%) due to the Thin Layer Chromatography analysis (Whitaker et al., 1974). If the sample size is increased to 20 kg, using a vertical cutter mixer to obtain a 100-g subsample instead of the United States Department of Agriculture/Agricultural Marketing Service 250-g subsample, and replacing Thin Layer Chromatography with high performance liquid chromatography (HPLC), it is possible to reduce the total variance to 176; the variance distribution becomes 130.4 (74.1%) due to sampling, 25.5 (14.5%) due to subsampling, and 20.1 (11.4%) due to analysis (Whitaker et al., 1974).

### Detection Issues

Analytical methods have been developed which are capable of quantifying mycotoxin levels in pet food. However, one analytical protocol cannot be implemented for all mycotoxins because of inherent structural differences of the toxins and matrix interference (Turner et al., 2009; Malachová et al., 2014). A complete analytical protocol includes three steps: sampling, sample preparation, and toxin measurement.

There are different methods for performing quantitative analyses of mycotoxins (Turner et al., 2009). The choice of which method to use depends on structure of mycotoxin to be analyzed and the food matrix it is embedded in. Typically, cleaning steps are necessary to extract and purify the toxins and eliminate interfering materials which may affect accuracy of measurement assay. Three major cleaning methods (Table 2.3) applied in the extraction process

**Table 2.3: Different types of cleaning methods used in analyses of mycotoxins.**

| Method | Basic principle | Advantages | Disadvantages |
|---|---|---|---|
| Liquid–liquid extraction | Use of solvents like hexane and cyclohexane to extract nonpolar contaminants | • Applicable for several toxins<br>• Small scale preparation | • Time consuming<br>• Type of food matrix is important<br>• Type of analyte is important<br>• Losses due to adherence to the glass wall |
| Supercritical fluid extraction | Use of supercritical fluid like $CO_2$ to extract the compound | Work well due to high solvating power | • Not good for routine analysis<br>• High cost<br>• Need for specialized equipment |
| Solid phase extraction | Small disposable cartridge filed with silica gel or bonded phase as stationary phase rinsed with sample loaded in one solvent | • High capacity for binding<br>• Depending on the type of bonding agent can be used for different toxins | • No single cartridge available for all mycotoxins<br>• Each column work at certain conditions (pH, solvent and ion concentration) of the sample |

include liquid–liquid extraction, supercritical fluid extraction, and solid phase extraction (SPE) (Turner et al., 2009). Currently, the most common cleaning method is SPE, and the most popular binding agent used in this method is silica gel (Turner et al., 2009).

Conventional techniques for quantitative measurement of mycotoxins include the following:

1. HPLC coupled with ultraviolet
2. Fluorescence or mass spectrometry detection
3. Gas chromatography coupled with electron capture detection
4. Flame ionization detection or mass detection
5. Thin layer chromatography.

These conventional detection methods use sophisticated equipment and possess very high accuracy. However, the methods require very skilled operators and are not portable, which makes them unsuitable for field-based use (Chauhan et al., 2016; Lu et al., 2016).

Several methods have been developed for rapid detection of mycotoxins in food and feed ingredients. These methods use an enzyme-linked immunosorbent assay (ELISA) or a membrane-based immunoassay for mycotoxin determination (Leslie et al., 2008). These methods are mainly used for screening purposes which means positive samples will still need to be verified using classical methods. The ELISA methods involve few steps and a simple extraction method (Oplatowska-Stachowiak et al., 2016). The methods use specific antibodies for detection of mycotoxins (Vidal et al., 2013).

Research efforts continue to elucidate methods for simplifying mycotoxin detection in pet food. In the literature, several evolutionary detection methods using electrochemical biosensors, immunochips, lateral flow devices, hyperspectral imaging (HSI), and short-wave infrared HSI techniques have been described (Leslie et al., 2008; Guo et al., 2015; Selvaraj et al., 2015; Lu et al., 2016).

## Mycotoxin Management in Pet Food

Most mycotoxins occurring in pet food are chemically and thermally stable compounds. An integrated program of management that considers the raw material supply chain all the way to nutrient extraction by the pet animal is the most effective way to ensure safety along the pet food chain. The ideal preventative strategy should target removing contamination early before formation. Once mycotoxins are formed on the pet food ingredients, efforts should be made to ensure detoxification during processing and/or preventing mycotoxicoses *in vivo* without compromising pet food nutritional quality.

Physical, chemical, and biological methods have been used to overcome negative effects of mycotoxins from contaminated pet food and feedstuffs. The major physical methods include thermal inactivation and irradiation; chemical methods include treatment with acid/base solutions, ozonation, and ammonization; and the biological method involves degradation of the toxins by microorganisms (Diaz and Smith 2005). Methods used by the pet food industry to prevent mycotoxicoses can be categorized into processing techniques, nutrient supplementation, use of mycotoxin-sequestering agents, and microbial deactivation.

Levels of contamination in most cereal-based pet food ingredients can be reduced by sieving, washing, and pearling techniques. Fissured, malformed, and immature kernels as well as debris, dust, or material other grain tend to support proliferation of fungi and by extension support mycotoxin contamination. Using sieves to remove unwanted fractions from kernels may significantly prevent mycotoxin contamination (Trenholm et al., 1991; Abramson et al., 2005). Successful reduction of mycotoxins has also been realized through washing methods (Trenholm et al., 1992). A washing treatment might require that the wetted product be dried which could mean additional expenditure; this may represent significant economic consequences to pet food manufacturers. Dehulling or sequential removal of the outer portions of kernels by abrasive pearling procedures has also been used to reduce mycotoxins (House et al., 2003). Other processing techniques for preventing mycotoxins contaminations in pet food include ozonation and addition of acid-based mold inhibitors (benzoic, acetic, sorbic, and propionic acids) (Hazan et al., 2004; Lambert and Stratford, 1999; Proctor et al., 2004).

The effectiveness of using nutrient supplements, especially neutral amino acids, antioxidants, and polyunsaturated fatty acids, to manage mycotoxin-induced tissue damage and altered behavior has been reported (Table 2.4). In many circumstances, there are still numerous clinical feeding trials that need to be done to determine the efficiency of using these dietary supplements to treat mycotoxicoses in pets.

Mycotoxin-sequestering agents are primarily used to prevent absorption of mycotoxins in the intestines of the pet animals (Diaz and Smith, 2005). The two major categories of mycotoxin-sequestering agents include silicate minerals and hydrated sodium calcium aluminosilicate (HSCAS). Other minor categories include activated charcoal, cholestyramine, chlorophyllin, and yeast cell wall–derived agents. The silicate minerals comprise the largest class of mycotoxin-sequestering agents and include phyllosilicate (mineral clays such as montmorillonite/bentonite) and tectosilicate (zeolites) subclasses. The bentonites are largely used because they possess a high degree of ion exchange capabilities and are primarily effective against aflatoxins. HSCAS from natural zeolite functions to sequester positively charged (cationic) compounds, mainly aflatoxins. Highlights of HSCAS application found in research studies include reports by (Bruinink et al., 1998; Huff et al., 1992). Drawbacks of most mycotoxin-sequestering agents include the fact that they either only adsorb specific mycotoxins, may require a high inclusion rate in animal feed, could cause other health complications, or are just too expensive for industrial applications. In contrast, naturally occurring glucomannan-containing polymers extracted from yeast cell wall may have some merits for practical use in the pet food industry. Glucomannans have a high adsorptive capacity for binding a combination

**Table 2.4: Management of mycotoxins using nutrient supplements.**

| Nutrient supplement | Activity and/or mechanism | References |
|---|---|---|
| Amino acids | Elleviate effects of mycotoxins on the central nervous system. | Smith (1991, 1992) and Cavan et al. (1998) |
| Example: Neutral amino acids, such as leucine, isoleucine, valine, tyrosine, and phenylalanine | Peventing mycotoxin-induced brain uptake of tryptophan and subsequent overt toxicity of T-2 toxin and deoxynivalenol. | |
| Antioxidant | Have therapeutic properties to prevent damage of cell membranes by lipid peroxidation, especially resulting from toxication by aflatoxin B1, fumonisin B1, deoxynivalenol, T-2 toxin, and ochratoxin A | Atroshi et al. (1998), Rizzo et al. (1994), Reddy et al. (2006), Nyandieka and Wakhisi (1993), Chen et al. (2000), and Grosse et al. (1997) |
| Example: Selenium and vitamins A, C, and E | Reported to reduce toxic effects of mycotoxins in vitro and in vivo, prevent aflatoxin-induced carcinogenicity, and reduce DNA adduct formation in the kidney and liver caused by ochratoxin A and zearalenone. | |
| Polyunsaturated fatty acids (PUFA) | Omega-3 PUFA is reported to reduce immunoglobulin A-induced nephropathy. | Shi and Pestka (2006), Jia et al. (2004), Pestka et al. (2002) and Moon and Pestka (2003) |
| Example 1: 3% (w/w) Eicosapentaenoic acid (EPA) | Reported to suppressed development of immunoglobulin A nephropathy caused by deoxynivalenol. | |
| Example 2: Docosahexaenoic acid (DHA) and EPA, but not R-linolenic acid | Reported to have attenuated proinflammatory response and renal damage induced by deoxynivalenol. | |
| Example 3: DHA and EPA | Reduces deoxynivalenol-induced interleukin-6 production and therefore alleviates inflammatory damage to the kidney. | |

of different mycotoxins; this coupled with low inclusion rates may make these compounds attractive to prevent mycotoxicoses in the pet food industry. In their study, Raju and Devegowda (2000) highlighted the use of a polymeric glucomannan mycotoxin absorbent to significantly improve feed intake, body weight gain, and biochemical and hematological parameters of broiler chickens that were victims of aflatoxin B1 (0.3 mg/kg of feed), ochratoxin A (2 mg/kg of feed), and T-2 toxin (3 mg/kg of feed) intake. Similar benefits have been reported by other researchers (Chowdhury et al., 2005a,b; Raymond et al., 2003; Swamy et al., 2002; WHO, 2002), for a variety of mycotoxins and pet animals.

Microorganisms can be used to detoxify mycotoxins into nontoxic metabolites thereby prevent mycotoxicosis in pet food (Piotrowska, 2014; Schatzmayr et al., 2006). Lactic acid bacteria and bifidobacteria are reported to have capability for reducing the bioavailability of aflatoxins through the binding of aflatoxins (Piotrowska, 2014). The aflatoxin-binding property of bacteria is thought to be related to the bacterial cell wall (polysaccharides and peptidoglycan) acting as a binding site for aflatoxin (Bata, 1999; Binder et al., 1996; CAST, 2003).

Mycotoxins can also be decontaminated by binding to the cell wall component of yeast. There is not sufficient information related to mycotoxin's interaction with the yeast cell wall. However, the presence of β-D-glucan in the yeast cell wall has been correlated to removal of mycotoxins such as zearalenone, aflatoxin B1, deoxynivalenol, and ochratoxin A (Armando et al., 2012). The yeast, *Saccharomyces cerevisiae*, can reduce bioavailability of mycotoxins and has also demonstrated antimicrobial activity against some pathogenic bacteria (Armando et al., 2012). *Trichosporon mycotoxinovorans* (*Trichosporon MTV*, 115), a novel yeast strain, was found to be capable of degrading ochratoxin A and zearalenone (Schatzmayr et al., 2006). The detoxifying effectiveness of these strains was demonstrated in yeast and gut tissue cultures (Binder et al., 1996; Schatzmayr et al., 2006). These microorganisms act in the intestinal tract of animals prior to mycotoxin absorption. As a general remark, it should be noted that the property of binding the toxin by microorganisms is strain specific (Hernandez-Mendoza et al., 2009).

Good processing techniques, sequestration agents, nutrient supplementation, and microbial inactivation methods that are used to prevent mycotoxins in pet food should not replace the need to start with high-quality ingredients. How the ingredients are managed from the point of supply, before procurement, and the proceeding storage and handling methods significantly affect their quality. It is important to procure ingredients from reliable suppliers with preprocessing guidelines in place to prevent mycotoxin production on sourced products. For agricultural/plant-based ingredients, various cultural strategies could be used at the farm level to prevent ingredient contamination. These measures range from (1) cultivar selection that targets varieties resistant to mycotoxin-producing fungi, (2) proper cultural practices, such as correct choice of planting and harvest dates, tillage practices, crop rotation, plant population, irrigation, and sanitation schemes to limit proliferation of mycotoxin-producing fungi, and (3) applying crop protection chemicals or biological controls to mitigate mycotoxin contamination.

Innovations that target simultaneous drying and decontamination of raw materials may provide alternative avenues for ensuring safe ingredients for pet food processing. Conventional practices use natural air or heated air for drying pet food raw material and/or ingredients. Unfortunately, these conventional drying methods are not metered to ensure inactivation of all microbial spores that may contaminate the ingredients. As a result, ingredients dried using conventional methods may contain high levels of heat-resistant types of fungi—some of which are toxigenic. Under suitable conditions, spores of these toxigenic types of fungi may produce mycotoxins (Richard, 2007). In past years, Wilson et al. (2016) have undertaken fundamental

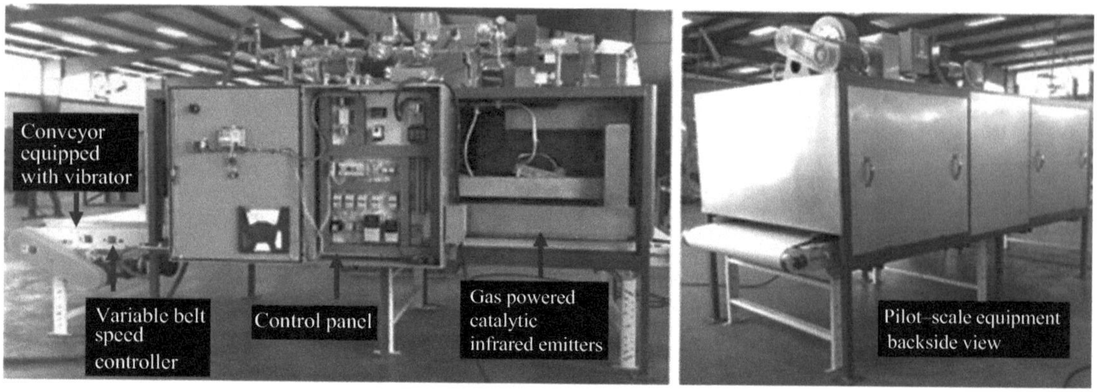


**Figure 2.1**
Pilot scale equipment designed and built with modular characteristics to achieve infrared drying and decontamination processes for grains and other food materials. *Photo courtesy of Food Science Department, University of Arkansas Division of Agriculture, Grain Processing and Engineering group, Fayetteville, Arkansas.*

and applied research to demonstrate the feasibility of infrared heating to improve drying rate of pet ingredients, specifically corn, while simultaneously decontaminating mycotoxin-producing fungi that could inhabit the ingredients. The authors' studies have refined infrared processing parameters including heating intensity, heating duration, and product-to-emitter gap size versus grain initial moisture content to diminish deterioration of the dried product quality. The innovative research conducted at the Food Science Department, University of Arkansas System Division of Agriculture (Fayetteville, Arkansas), has led to design and construction of a scaled up infrared heating unit with modular processing parameters for pet food ingredient processing (Fig. 2.1). The authors' research has highlighted that infrared treatment of high-moisture corn (23% wet basis) significantly reduced drying time to drive the moisture content from 23% to 13% wet basis, that is, the drying time ranged from 9 to 22 minutes depending on infrared intensity and intermittent infrared drying duration. In addition, increasing infrared intensity from 2.39 to 5.55 $kW/m^2$ resulted in average microbial load reductions of 2.6 and 2.9 log colony-forming unit (CFU)/g at 180 and 30 s of intermittent infrared heating, respectively.

## Conclusions

Contamination of pet food ingredients, especially those that are cereal based, can be mitigated by optimal drying and storage management. Rapid drying to reduce moisture content to lower than 14% helps to control fungal growth and insect infestations. Proper ingredients handling is vital to prevention of fungal contamination and mycotoxin formation. Kernels that become damaged due to poor handling tend to support fungal growth which in turn support formation of toxins during storage. Storage factors that contribute to fungal growth and mycotoxin production include temperature, water activity ($a_w$), pH, moisture, and oxygen content and the interaction among these factors with substrates. High temperature, high moisture content, and high water activity will facilitate fungal growth and toxin production. It is important that based on the pet food formulation, the correct storage environment and packaging is selected to maximize the product shelf-life and eventually the safety of pets.

## References


Abbas, H.K., Cartwright, R.D., Shier, W.T., Abouzied, M.M., Bird, C.B., Rice, L.G., Ross, P.F., Sciumbato, G.L., Meredith, F.I., 1998. Natural occurrence of fumonisins in rice with *Fusarium* sheath rot disease. Plant Disease 82 (1), 22–25.

Abramson, D., House, J.D., Nyachoti, C.M., 2005. Reduction of Deoxynivalenol in barley by treatment with aqueous sodium carbonate and heat. Mycopathologia 160 (4), 297–301.

Armando, M.R., Pizzolitto, R.P., Dogi, C.A., Cristofolini, A., Merkis, C., Poloni, V., Dalcero, A.M., Cavaglieri, L.R., 2012. Adsorption of ochratoxin A and zearalenone by potential probiotic *Saccharomyces cerevisiae* strains and its relation with cell wall thickness. Journal of Applied Microbiology 113 (2), 256–264.

Arp, L.H., Richard, J.L., 1979. Intoxication of dogs with the mycotoxin penitrem A. Journal of the American Veterinary Medical Association 175 (6), 565–566.

Asao, T., Buchi, G., Abdel-kader, M., Chang, S.B., Wick, E., Wogan, G.N., 1963. Aflatoxin B and G. Journal of the American Chemical Society 85 (11), 1706–1707.

Ashiq, S., 2014. Natural occurrence of mycotoxins in food and feed: Pakistan perspective. Comprehensive Reviews in Food Science and Food Safety 14 (2), 159–175.

Atroshi, F., Rizzo, A., Westermarck, T., Ali-Vehmas, T., 1998. Effects of tamoxifen, melatonin, coenzyme Q10, and L-carnitine supplementation on bacterial growth in the presence of mycotoxins. Pharmacological Research 38 (4), 289–295.

Bailey, W.S., Groth, A.H., 1959. The relationship of hepatitis X of dogs and moldy corn poisoning of swine. Journal of the American Veterinary Medical Association 134, 483–486.

Bastianello, S.S., Nesbit, J.W., Williams, M.C., Lange, A.L., 1987. Pathological findings in a natural outbreak of aflatoxicosis in dogs. The Onderstepoort Journal of Veterinary Research 54, 635–640.

Bata, A., 1999. Detoxification of mycotoxin-contaminated food and feed by microorganisms. Trends in Food Science & Technology 10 (6–7), 223–228.

Beasley, V.R., Dorman, D.C., Fikes, J.D., Diana, S.G., 1997. Tremorgenic Mycotoxins and Various Staggers Syndromes: A Systems Affected Approach to Veterinary Toxicology. Department of Veterinary Biosciences. College of Veterinary Medicine, University of Illinois, Urbana, IL, pp. 137–142.

Binder, E.M., Heidler, D., Schatzmayr, G., Thimm, N., Fuchs, E., Schuh, M., Krska, R., Binder, J., May 21–25, 2000. Microbial detoxification of mycotoxins in animal feed. In: Proceedings of the 10th International IUPAC Symposium on Mycotoxins and Phycotoxins. Guarujá, Brazil, pp. 271–277.

Boysen, S.R., Rozanski, E.A., Chan, D.L., Grobe, T.L., Fallon, M.J., Rush, J.E., 2002. Tremorgenic mycotoxicosis in four dogs from a single household. Journal of the American Veterinary Medical Association 221 (10), 1441–1444.

Bradburn, N., Coker, R.D., Blunden, G., 1994. The aetiology of Turkey "x" disease. Phytochemistry 35 (3), 817.

Bruinink, A., Rasonyi, T., Sidler, C., 1998. Differences in neurotoxic effects of ochratoxin A, ochracin and ochratoxin-α *in vitro*. Natural Toxins 6 (5), 173–177.

Campagnollo, F.B., Ganev, K.C., Khaneghah, A.M., Portela, J.B., Cruz, A.G., Granato, D., Corassin, C.H., Oliveira, C.A.F., Sant, Ana, A.S., 2016. The occurrence and effect of unit operations for dairy products processing on the fate of aflatoxin M1: a review. Food Control 68, 310–329.

Campos, S.G., Cavaglieri, L.R., Fernández Juri, M.G., Dalcero, A.M., Krüger, C., Keller, L.A.M., Magnoli, C.E., Rosa, C.A.R., 2008. Mycobiota and aflatoxins in raw materials and pet food in Brazil. Journal of Animal Physiology and Animal Nutrition 92 (3), 377–383.

Cavan, K.R., MacDonald, E.J., Smith, T.K., 1988. Potential for dietary amino acid precursors of neurotransmitters to overcome neurochemical changes in acute T-2 toxicosis in rats. Journal of Nutrition 118 (17), 901–907.

Chauhan, R., Singh, J., Sachdev, T., Basu, T., Malhotra, B.D., 2016. Recent advances in mycotoxins detection. Biosensors and Bioelectronics 81, 532–545.

Chen, S.Y., Chen, C.J., Tsai, W.Y., Ahsan, H., Liu, T.Y., Lin, J.T., Santella, R.M., 2000. Associations of plasma aflatoxin B1-albumin adduct level with plasma selenium level and genetic polymorphisms of glutathione S-transferase M1 and T1. Nutrition and Cancer 38 (2), 179–185.

Chowdhury, S.R., Smith, T.K., Boermans, H.J., Woodward, B., 2005a. Effects of feed-borne *Fusarium* mycotoxins on hematology and immunology of laying hens. Poultry Science 84 (12), 1841–1850.

Chowdhury, S.R., Smith, T.K., Boermans, H.J., Woodward, B., 2005b. Effects of feed-borne *Fusarium* mycotoxins on hematology and immunology of turkeys. Poultry Science 84 (11), 1698–1706.

Cortinovis, C., Battini, M., Caloni, F., 2012. Deoxynivalenol and T-2 toxin in raw feeds for horses. Journal of Equine Veterinary Science 32 (2), 72–74.

Council for Agricultural Science and Technology, 2003. Mycotoxins: Risks in Plant, Animal, and Human Systems, Task Force Report 139. U.S. Council for Agricultural Science and Technology (CAST).

Cullen, J.M., Newberne, P.M., 1994. Acute hepatotoxicity of aflatoxins. In: Eaton, D.L., Groopman, J.D. (Eds.), The Toxicology of Aflatoxins. Academic Press, San Diego, CA, pp. 3–26.

Devegowda, G., Castaldo, D., 2000. Mycotoxins: hidden killers in pet foods. Is there a solution? In: Technical Symposium on Mycotoxins. Alltech, Inc., Nicholasville, KY.

Diaz, D.E., Smith, T.K., 2005. Mycotoxin sequestering agents: practical tools for the neutralization of mycotoxins. Journal of Mycotoxin Blue Book 2005, 323–340.

Didwania, N., Joshi, M., 2013. Mycotoxins: a critical review on occurrence and significance. International Journal of Pharmacy and Pharmaceutical Sciences 5 (3), 1014–1019.

Döll, S., Dänicke, S., 2011. The *Fusarium* toxins deoxynivalenol (DON) and zearalenone (ZON) in animal feeding. Preventive Veterinary Medicine 102 (2), 132–145.

El-Tras, W.F., El-Kady, N.N., Tayel, A.A., 2011. Infants exposure to aflatoxin M1 as a novel foodborne zoonosis. Food and Chemical Toxicology 49 (11), 2816–2819.

Fernandez Juri, M.G., Bressan, F., Astoreca, A.L., Magnoli, C.E., 2009. Aflatoxins, fumonisins and toxigenic fungi in raw materials and ready dry dog food in Argentina. Revista Brasileira de Medicina Veterinaria Journal 31 (2), 109–117.

Food and Drug Adminstration, 2005. FDA Issues Consumer Alert on Contaminated Pet Food. U.S. Food and Drug Administration (FDA).

Gallo, A., Solfrizzo, M., Epifani, F., Panzarini, G., Perrone, G., 2016. Effect of temperature and water activity on gene expression and aflatoxin biosynthesis in *Aspergillus flavus* on almond medium. International Journal of Food Microbiology 217, 162–169.

Garland, T., Reagor, J.C., 2001. Chronic canine aflatoxicosis and management of an epidemic. In: de Koe, W.J. (Ed.), Mycotoxins and Phycotoxins in Perspective at the Turn of the Millennium. Ponsen & Looven, Wageningen, The Netherlands, pp. 231–236.

Gazzotti, T., Biagi, G., Pagliuca, G., Pinna, C., Scardilli, M., Grandi, M., Zaghini, G., 2015. Occurrence of mycotoxins in extruded commercial dog food. Animal Feed Science and Technology 202, 81–89.

Greene, C.E., Barsanti, J.A., Jones, B.D., 1977. Disseminated intravascular coagulation complicating aflatoxicosis in dogs. The Cornell Veterinarian 67 (1), 29–49.

Grosse, Y., Chekir-Ghedira, L., Huc, A., Obrecht-Pflumio, S., Dirheimer, G., Bacha, H., Pfohl-Leszkowicz, A., 1997. Retinol, ascorbic acid and alpha tocopherol prevent DNA adduct formation in mice treated with the mycotoxins ochratoxin A and zearalenone. Cancer Letters 114 (1–2), 225–229.

Gunsen, U., Yaroglu, T., 2002. Aflatoxin in dog and horse feeds in Turkey. Veterinary and Human Toxicology 44 (2), 113–114.

Guo, L., Feng, J., Fang, Z., Xu, J., Lu, X., 2015. Application of microfluidic "lab-on-a-chip" for the detection of mycotoxins in foods. Trends in Food Science & Technology 46 (2), 252–263.

Hazan, R., Levine, A., Abeliovich, H., 2004. Benzoic acid, a weak organic acid food preservative, exerts specific effects on intracellular membrane trafficking pathways in *Saccharomyces cerevisiae*. Applied and Environmental Microbiology 70 (8), 4449–4457.

Henke, S.E., Gallardo, V.C., Martinez, B., Bailey, R., 2001. Survey of aflatoxin concentrations in wild bird seed purchased in Texas. Journal of Wildlife Diseases 37 (4), 831–835.

Hernandez-Mendoza, A., Garcia, H.S., Steele, J.L., 2009. Screening of *Lactobacillus casei* strains for their ability to bind aflatoxin B1. Food and Chemical Toxicology 47 (6), 1064–1068.

Hocking, A.D., Holds, K., Tobin, N.F., 1988. Intoxication by tremorgenic mycotoxin (penitrem A) in a dog. Australian Veterinary Journal 65 (3), 82–85.

Hopmans, E.C., Murphy, P.A., 1993. Detection of fumonisins B1, B2, and B3 and hydrolyzed fumonisin B1 in corn-containing foods. Journal of Agricultural and Food Chemistry 41 (10), 1655–1658.

House, J.D., Nyachoti, C.M., Abramson, D., 2003. Deoxynivalenol removal from barley intended as swine feed through the use of an abrasive Pearling procedure. Journal of Agricultural and Food Chemistry 51 (17), 5172–5175.

Huff, W.E., Kubena, L.F., Harvey, R.B., Phillips, T.D., 1992. Efficacy of hydrated sodium calcium aluminosilicate to reduce the individual and combined toxicity of aflatoxin and Ochratoxin A. Poultry Science 71 (1), 64–69.

Jeong, W.-I., Do, S.H., Jeong, D.-H., Chung, J.-Y., Yang, H.-J., Yuan, D.-W., Hong, I.-H., Park, J.-K., Goo, M.-J., Jeong, K.-S., 2006. Canine renal failure syndrome in three dogs. Journal of Veterinary Science 7 (3), 299–301.

Jia, Q., Shi, Y., Bennink, M.B., Pestka, J.J., 2004. Docosahexaenoic acid and eicosapentaenoic acid, but not alpha-linolenic acid, suppress deoxynivalenol-induced experimental IgA nephropathy in mice. Journal of Nutrition 134 (6), 1353–1361.

Ketterer, P.J., Williams, E.S., Blaney, B.J., Connole, M.D., 1975. Canine aflatoxicosis. Australian Veterinary Journal 51 (7), 355–357.

Klich, M.A., 2007. Environmental and developmental factors influencing aflatoxin production by *Aspergillus flavus* and *Aspergillus parasiticus*. Mycoscience 48 (2), 71–80.

Krishnamachari, K.A.V.R., Nagarajan, V., Bhat, R., Tilak, T.B.G., 1975. Hepatitis due to aflatoxicosis. Lancet 305 (7915), 1061–1063.

Lambert, R.J., Stratford, M., 1999. Weak-acid preservatives: modelling microbial inhibition and response. Journal of Applied Microbiology 86 (1), 157–164.

Leslie, J.F., Bandyopadhyay, R., Visconti, A., yopadhyay, V.A., 2008. In: Leslie, J., Bandyopadhyay, R., Visconti, A. (Eds.), Mycotoxins: Detection Methods, Management, Public Health, and Agricultural Trade. CAB International North America, London, UK.

Leung, M.C.K., Díaz-Llano, G., Smith, T.K., 2006. Mycotoxins in pet food: a review on worldwide prevalence and preventative strategies. Journal of Agricultural and Food Chemistry 54 (26), 9623–9635.

Liggett, A.D., Colvin, B.M., Beaver, R.W., Wilson, D.M., 1986. Canine aflatoxicosis: a continuing problem. Veterinary and Human Toxicology 28 (5), 428–430.

Lightfoot, T.L., Yeager, J.M., 2008. Pet bird toxicity and related environmental concerns. Veterinary Clinics of North America: Exotic Animal Practice 11 (2), 229–259.

Little, C.J.L., McNeil, P.E., Robb, J., 1991. Hepatopathy and dermatitis in a dog associated with the ingestion of mycotoxins. Journal of Small Animal Practice 32 (1), 23–26.

Lowes, N.R., Smith, R.A., Beck, B.E., 1992. Roquefortine in the stomach contents of dogs suspected of strychnine poisoning in Alberta. Canadian Veterinary Journal 33 (8), 535–538.

Lu, L., Seenivasan, R., Wang, Y.-C., Yu, J.-H., Gunasekaran, S., 2016. An electrochemical immunosensor for rapid and sensitive detection of mycotoxins fumonisin B1 and deoxynivalenol. Electrochimica Acta 213, 89–97.

Magan, N., 2004. In: Olsen, M. (Ed.), Mycotoxins in Food: Detection and Control. Taylor & Francis, Cambridge, UK.

Magan, N., Aldred, D., 2007. Post-harvest control strategies: minimizing mycotoxins in the food chain. International Journal of Food Microbiology 119 (1–2), 131–139.

Maia, P.P., Pereira Bastos de Siqueira, M.E., 2002. Occurrence of aflatoxins B1, B2, G1 and G2 in some Brazilian pet foods. Food Additives and Contaminants 19 (12), 1180–1183.

Malachová, A., Sulyok, M., Beltrán, E., Berthiller, F., Krska, R., 2014. Optimization and validation of a quantitative liquid chromatography–tandem mass spectrometric method covering 295 bacterial and fungal metabolites including all regulated mycotoxins in four model food matrices. Journal of Chromatography A 1362, 145–156.

Marroquín-Cardona, A.G., Johnson, N.M., Phillips, T.D., Hayes, A.W., 2014. Mycotoxins in a changing global environment – a review. Food and Chemical Toxicology 69, 220–230.

Martins, M.L., Martins, H.M., Bernardo, F., 2003. Fungal flora and mycotoxins detection in commercial pet food. RPCV 98 (548), 179–183.

Moon, Y., Pestka, J.J., 2003. Deoxynivalenol-induced mitogen activated protein kinase phosphorylation and IL-6 expression in mice suppressed by fish oil. Journal of Nutritional Biochemistry 14 (12), 717–726.

Miraglia, M., De Santis, B., Minardi, V., Debegnach, F., Brera, C., 2005. The role of sampling in mycotoxin contamination: an holistic view. Food Additives & Contaminants 22 (1), 31–36.

Naude, T.W., O'Brien, O.M., Rundberget, T., McGregor, A.D.G., Roux, C., Flaoyen, A., 2002. Tremorgenic neuromycotoxicosis in 2 dogs ascribed to the ingestion of penitrem A and possibly roquefortine in rice contaminated with *Penicillium crustosum*: clinical communication. Journal of the South African Veterinary Association 73 (4).

Newberne, J.W., Bailey, W.S., Seibold, H.R., 1955. Notes on a recent outbreak and experimental reproduction of hepatitis X in dogs. Journal of the American Veterinary Medical Association 127, 59–62.

Nyandieka, H.S., Wakhisi, J., 1993. The impact of vitamins A, C, E, and selenium compound on prevention of liver cancer in rats. East African Medical Journal 70 (3), 151–153.

Oplatowska-Stachowiak, M., Sajic, N., Xu, Y., Haughey, S.A., Mooney, M.H., Gong, Y.Y., Verheijen, R., Elliott, C.T., 2016. Fast and sensitive aflatoxin B1 and total aflatoxins ELISAs for analysis of peanuts, maize and feed ingredients. Food Control 63, 239–245.

Pereyra, M.L.G., Carvalho, E.C.Q., Tissera, J.L., Keller, K.M., Magnoli, C.E., Rosa, C.A.R., Dalcero, A.M., Cavaglieri, L.R., 2008. An outbreak of acute aflatoxicosis on a Chinchilla (*Chinchilla Lanigera*) farm in Argentina. Journal of Veterinary Diagnostic Investigation 20 (6), 853–856.

Pestka, J.J., Zhou, H.R., Jia, Q., Timmer, A.M., 2002. Dietary fish oil suppresses experimental immunoglobulin a nephropathy in mice. Journal of Nutrition 132 (2), 261–269.

Piotrowska, M., 2014. The adsorption of ochratoxin A by *Lactobacillus* species. Toxins 6 (9), 2826–2839.

Proctor, A.D., Ahmedna, M., Kumar, J.V., Goktepe, I., 2004. Degradation of aflatoxins in peanut kernels/flour by gaseous ozonation and mild heat treatment. Food Additives and Contaminants 21 (8), 786–793.

Puls, R., Ladyman, E., 1988. Roquefortine toxicity in a dog. Canadian Veterinary Journal 29 (7), 569.

Puschner, B., 2002. Mycotoxins. Veterinary Clinics of North America: Small Animal Practice 32 (2), 409–419.

Rahimi, E., Bonyadian, M., Rafei, M., Kazemeini, H.R., 2010. Occurrence of aflatoxin M1 in raw milk of five dairy species in Ahvaz, Iran. Food and Toxicology 48, 129–131.

Raju, M.V.L.N., Devegowda, G., 2000. Influence of esterified-glucomannan on performance and organ morphology, serum biochemistry and haematology in broilers exposed to individual and combined mycotoxicosis (aflatoxin, ochratoxin and T-2 toxin). British Poultry Science 41 (5), 640–650.

Raymond, S.L., Smith, T.K., Swamy, H.V.L.N., 2003. Effects of feeding a blend of grains naturally contaminated with mycotoxins on feed intake, serum chemistry, and hematology of horses, and the efficacy of a polymeric glucomannan mycotoxin adsorbent. Journal of Animal Science 81 (9), 2123–2130.

Reddy, L., Odhav, B., Bhoola, K., 2006. Aflatoxin B1-induced toxicity in HepG2 cells inhibited by carotenoids: morphology, apoptosis and DNA damage. Journal of Biological Chemistry 387 (1), 87–93.

Richard, J.L., 2007. Some major mycotoxins and their mycotoxicoses—an overview. International Journal of Food Microbiology 119 (1–2), 3–10.

Richard, J.L., Bacchetti, P., Arp, L.H., 1981. Moldy walnut toxicosis in a dog, caused by the mycotoxin, penitrem A. Mycopathologia 76 (1), 55–58.

Rizzo, A.F., Atroshi, F., Ahotupa, M., Sankari, S., Elovaara, E., 1994. Protective effect of antioxidants against free radical-mediated lipid peroxidation induced by DON or T-2 toxin. Zentralbl Veterinarmed A Journal 41 (2), 81–90.

Rustom, I.Y.S., 1997. Aflatoxin in food and feed: occurrence, legislation and inactivation by physical methods. Food Chemistry 59 (1), 57–67.

Schatzmayr, G., Zehner, F., Täubel, M., Schatzmayr, D., Klimitsch, A., Loibner, A.P., Binder, E.M., 2006. Microbiologicals for deactivating mycotoxins. Molecular Nutrition & Food Research 50 (6), 543–551.

Scudamore, K.A., Hetmanski, M.T., Nawaz, S., Naylor, J., Rainbird, S., 1997. Determination of mycotoxins in pet foods sold for domestic pets and wild birds using linked-column immunoassay clean-up and HPLC. Food Additives and Contaminants 14 (2), 175–186.

Seibold, H.R., Bailey, W.S., 1952. An epizootic of hepatitis in the dog. Journal of the American Veterinary Medical Association 121, 201–206.

Selvaraj, J.N., Zhou, L., Wang, Y., Zhao, Y., Xing, F., Dai, X., Liu, Y., 2015. Mycotoxin detection—recent trends at global level. Journal of Integrative Agriculture 14 (11), 2265–2281.

Sharma, M., Márquez, C., 2001. Determination of aflatoxins in domestic pet foods (dog and cat) using immunoaffinity column and HPLC. Animal Feed Science and Technology 93 (1–2), 109–114.

Shi, Y., Pestka, J.J., 2006. Attenuation of mycotoxin-induced IgA nephropathy by eicosapentaenoic acid in the mouse: dose response and relation to IL-6 expression. Journal of Nutritional Biochemistry 17 (10), 697–706.

Smith, T.K., MacDonald, E.J., 1991. Effect of fusaric acid on brain regional neurochemistry and vomiting behavior in swine. Journal of Animal Science 69 (5), 2044–2049.

Smith, T.K., 1992. Recent advances in the understanding of *Fusarium* trichothecene mycotoxicoses. Journal of Animal Science 70 (12), 3989–3993.

Stenske, K.A., Smith, J.R., Newman, S.J., Newman, L.B., Kirk, C.A., 2006. Aflatoxicosis in dogs and dealing with suspected contaminated commercial foods. Journal of the American Veterinary Medical Association 228 (11), 1686–1691.

Streit, E., Schatzmayr, G., Tassis, P., Tzika, E., Marin, D., Taranu, I., Tabuc, C., Nicolau, A., Aprodu, I., Puel, O., Oswald, I.P., 2012. Current situation of mycotoxin contamination and co-occurrence in animal Feed—Focus on Europe. Toxins 4 (12), 788–809.

Sultana, N., Hanif, N.Q., 2009. Mycotoxin contamination in cattle feed and feed ingredients. Pakistan Veterinary Journal 29 (4), 211–213.

Swamy, H., Smith, T., Cotter, P., Boermans, H., Sefton, A., 2002. Effects of feeding blends of grains naturally contaminated with *Fusarium* mycotoxins on production and metabolism in broilers. Poultry Science 81 (7), 966–975.

Tandon, B.N., Krishnamurthy, L., Koshy, A., Tandon, H.D., Ramalingaswami, J.R., Mathur, M.M., Mathur, P.D., 1977. Study of an epidemic of jaundice, presumably due to toxic hepatitis, in Northwest India. Gastroenterology 72 (3), 488–494.

Trenholm, H.L., Charmley, L.L., Prelusky, D.B., Warner, R.M., 1991. Two physical methods for the decontamination of four cereals contaminated with deoxynivalenol and zearalenone. Journal of Agricultural and Food Chemistry 39 (2), 356–360.

Trenholm, H.L., Charmley, L.L., Prelusky, D.B., Warner, R.M., 1992. Washing procedures using water or sodium carbonate solutions for the decontamination of three cereals contaminated with deoxynivalenol and zearalenone. Journal of Agricultural and Food Chemistry 40 (11), 2147–2151.

Turner, N.W., Subrahmanyam, S., Piletsky, S.A., 2009. Analytical methods for determination of mycotoxins: a review. Analytica Chimica Acta 632 (2), 168–180.

Vidal, J.C., Bonel, L., Ezquerra, A., Hernández, S., Bertolín, J.R., Cubel, C., Castillo, J.R., 2013. Electrochemical affinity biosensors for detection of mycotoxins: A review. Biosensors and Bioelectronics 49, 146–158.

Wagner, S.O., Podell, M., Fenner, W.R., 1997. Generalized tremors in dogs: 24 cases (1984–1995). Journal of the American Veterinary Medical Association 211 (6), 731–735.

Walter, S.L., 2002. Acute penitrem A and roquefortine poisoning in a dog. Canadian Veterinary Journal 43 (5), 372–374.

Whitaker, T.B., 2003. Standardisation of mycotoxin sampling procedures: an urgent necessity. Food Control 14 (4), 233–237.

Whitaker, T.B., Dickens, J.W., Monroe, R.J., 1974. Variability of aflatoxin test results. Journal of the American Oil Chemists Society 51 (5), 214–218.

Wilson, S.A., Atungulu, G.G., Olatunde, G., 2016. Drying and decontamination of corn using a pilot-scale continuous-flow radiant heating system. In: 2016 ASABE Annual International Meeting, 162460256, p. 1. http://dx.doi.org/10.13031/aim.20162460256.

World Health Organization, 2002. Evaluation of Certain Mycotoxins in Food. World Health Organization (WHO).
Young, K.L., Villar, D., Carson, T.L., Imerman, P.M., Moore, R.A., Bottoff, M.R., 2003. Tremorgenic mycotoxin intoxication with penitrem A and roquefortine in two dogs. Journal of the American Veterinary Medical Association 222 (1), 52–53.
Zwierzchowski, W., Gajecki, M., Obremski, K., Zielonka, L., Baranowski, M., 2004. The occurrence of zearalenone and its derivatives in standard and therapeutic feeds for companion animals. Polish Journal of Veterinary Sciences 7 (4), 289–293.

# CHAPTER 3

# Emerging Pet Food Drying and Storage Strategies to Maintain Safety

**Gbenga A. Olatunde, Griffiths G. Atungulu**
*University of Arkansas, Fayetteville, AR, United States*

## Introduction

There has been a significant growth in the global pet food market with an estimated projection value of US $91 billion by 2022. Demand for healthier, organic, functional, and natural foods has grown because of increased number of pet owners treating their pets like family. The pet food industry has also experienced increasing numbers of pet food recalls from more than 100 major pet food brands since 2007. Despite the enormous amount of resources being spent by the industry to formulate and manufacture foods, the recall list continues to grow yearly. Hence, basic principles of food storage strategies have to be understood and implemented meticulously to avert this trend. Because pet food industries are relatively new, practical information and scientific guidelines on proper drying, storage, and handling methods of the most common types of pet food are not readily available. Presently, storage conditions and handling procedures are usually left to experience and assumptions. Hence this chapter is intended to examine the storage strategies to minimize nutritional compositions of pet food and then delve into the drying techniques for pet food safety during storage.

## Classification of Pet Food and Main Nutritional Elements

Pet food is comprised of various nutritional and elemental components with differing responses to storage conditions. The very first step in developing an effective strategy to prevent spoilage is to have an in-depth understanding of various classifications and nutritional compositions and their response to storage conditions. The primary components of pet foods include carbohydrate, protein, fats, fiber, vitamins, minerals, and water (Table 3.1). The proportion in a particular pet food can vary with type of pet (dog, cat, rat, etc.), dietary requirements, and brand.

Another important pet food classification based on composition is the amount of moisture content (MC) in the food. MC determines how the food is processed, handled, packaged, stored, and fed to the pet. Dry pet foods are most common due to their convenience of feeding

Food and Feed Safety Systems and Analysis. http://dx.doi.org/10.1016/B978-0-12-811835-1.00003-8

Table 3.1: Composition of some selected pet foods from various brands.

| Brand | Blue ridge | Cesar | Addiction | Blue buffalo | Fromm | Big dog natural | Great life | Wellness |
|---|---|---|---|---|---|---|---|---|
| | Product | | | | | | | |
| Composition (%) | Beef with bones | Cesar classics | New Zealand | Life protection | Gold canned | Chicken supreme | Grain and potato free | Reduced fat grain free |
| Protein | 19.06 | 8 | 8 | 24 | 9 | 30 | 30 | 33 |
| Fiber | 0.99 | 1 | 1.5 | 4 | 1.9 | 2 | 4 | 10 |
| Fat | 9.12 | 3.5 | 6 | 14 | 4 | 17 | 13 | 9 |
| Ash | 1.62 | | | | | | 4 | |
| Calcium | 301 mg/100 g | | 1.5 | 1 | | | 1.27 | 1.9 |
| Moisture | 70.41 | 82 | 78 | 10 | 78 | 12 | 10 | 10 |
| Phosphorus | | 2.4 | 0.9 | | | | 1.17 | 1.3 |
| Sodium | | | 0.3 | | | | 0.5 | |

and storage. For instance, the food can be left for the pet to eat at its own pace without worrying about spoilage within a certain duration of time in comparison with wet foods. Wet foods are mostly stored in can or sealed plastic containers. Unlike dry food, the shelf life of wet food decreases significantly once it is opened and the pet could create a mess while eating.

In addition, MC broadly impacts the proportion of nutritional composition of a particular pet food. Table 3.1 shows the result of the survey of some selected popular dry and wet pet foods with their nutrient composition profile. In general, protein and fat contents in dry foods are significantly higher than in wet foods.

### Main Composition of Pet Food and Primary Sources

Carbohydrates comprise about 30%–70% of the dry food by weight. Common cereal grains such as wheat, corn, rice, barley, and oats are important sources of carbohydrates in pet food. In addition, protein makes up to about 15%–25% on a dry basis of the pet food. The common sources of protein in pet food are chicken, egg, beef, turkey, lamb, soybeans, and fish meal. Fat is another important component of pet food that makes up to about 10%–15% on a dry basis. Fat is a source of energy, taste, and flavor in foods and also is a platform to hold vitamins. Important sources of fats in pet foods include chicken or pork fat, cottonseed oil, vegetable oil, soybean oil, fish oil, safflower oil, and so forth. Finally, vitamins and minerals are also important components in pet food formulations but at a macro level. Minerals that are typically found in pet food include calcium, phosphorus, magnesium, sodium, chloride, sulfur, and potassium. Fruits and vegetables are primary sources of vitamins and minerals.

Each of the nutrients in pet food has specific storage requirements that are necessary to prevent degradation and loss. The goal of the next section is to identify each of these conditions in a broader context of developing an overall drying and storage strategy.

## Storage Strategies to Minimize Nutritional Composition Loss

### Vitamin Potency

Vitamins are complex organic compounds and antioxidants required for normal metabolism, growth, reproduction, and health (Riaz et al., 2009). Vitamins can only be obtained through food and supplements. There are 13 different types of vitamins which are broadly classified into fat soluble and water soluble. Fat-soluble vitamins (vitamin A, D, E, and K) are stored in fat cells and can only be absorbed by fat. They can be obtained from cereal, vegetable, and fruits. Water-soluble vitamins include vitamin B1, B2, B3, B5, B7, B9, B12, and C and are excreted after the body absorbs the required quantity and their sources include whole grain, meat fish meal, soy products, and so forth (Riaz et al., 2009).

Vitamin degradation in feed could be due to many factors such as source, genetics, weather, and storage condition. For instance, Combs and Combs Jr. (1985) investigated the impact of variety and days to maturation of freshly harvested corn that was stored for 5 days at 35°C on vitamin potency and reported that vitamin E values in 42 corn varieties varied from 11.1 to 36.4 IU per kg and days to maturation had minimal impact on vitamin content of corn. Earlier, Underwood (1984) reported that the vitamin content of corn was drastically reduced when the corn crop was harvested too early. When the potency of vitamins source in the raw material is low, the processor has to spend additional resources to supplement the final product, thereby increasing cost. Furthermore, Keskin and Ozkaya (2013) investigated how storage (30-day duration) and *Sitophilus granarius* L. infestation impacts thiamin and riboflavin (vitamin content) of wheat grain stored for 6 months at 30°C, 75% relative humidity (RH). The authors found that vitamins in uninfected samples remain unchanged for the entire period. However, thiamin and riboflavin contents were reduced by 50% and 6% for the infected wheat, respectively. Zia Ur (2006) also investigated the effect of storage on vitamin loss in freshly harvested wheat, maize, and rice grains stored for 6 months at temperatures of 10, 25, and 45°C. The author reported that samples stored at 25 and 45°C had total available lysine and thiamine decreased by 6.5% and 18.5% in wheat, 14.3% and 20.7% in maize, and 23.7% and 34.2% in rice, respectively. In a related investigation, thiamine in high MC corn was found to decrease as the corn was subjected to alternate periods of freezing and thawing (Hoffmann-La, 1991). In summary, to preserve vitamin potency, it is important that the processor pays attention to the raw material quality, storage temperature, and storage RH so as to maintain the vitamins before the subsequent postharvest processing step.

### *Lipid or Fat Rancidity*

Fats consist of several fatty acids that are connected to a glycerol molecule. Fats can be broadly classified based on the structure of the fatty acid as saturated and unsaturated fats. The main difference between the fat types is the number of bonds between the carbon atoms. Saturated fat possesses a single bond, whereas unsaturated fat has at least one double bond between the carbon atoms. The presence of a double bond between carbon atoms implies the loss of the hydrogen atom. Hence, saturated fat has the maximum hydrogen atoms bonded to carbon. Saturated and unsaturated fat share similar functions such as energy storage, absorption, and delivery of fat-soluble vitamins. In addition, saturated fat has the ability to form a rigid cellular membrane, whereas unsaturated fat enhances production of hormones, skin cell growth, and antiinflammatory properties. The main sources of saturated fat in pet food include animal fat, whole milk, and egg, whereas unsaturated fat can be obtained from canola oil, cashew nuts, soybean, sunflower, and corn oil.

The higher the degree of unsaturation, the more prone the fat is to develop an offensive odor as a result of exposure to oxygen to form peroxides (oxidative rancidity). The decomposition of peroxides into a complex mixture of aldehydes, ketones, and other volatile products generates an odor (Jensen et al., 2003; Labuza and Dugan, 1971; Ulu, 2004). The reaction rate depends mostly on environmental and technological factors such as availability of oxygen, moisture, exposed surface area, and temperature (Toci et al., 2013). Fatty acid has been found to increase during storage by Alencar et al. (2010) who investigated changes in fatty acid extracted from soybean stored for 180 days at MCs of 11.2%, 12.8%, and 14.8% and temperatures of 20, 30, or 40°C. Increases in fatty acid during storage could be attributed to lipid hydrolysis by lipase, peroxidase, and phospholipase enzymes present in the grain tissues. Siger and Michalak (2016) investigated the oxidative stability of oil from cold-pressed rapeseed roasted at 140, 160, or 180°C and a roasting duration of 5, 10, or 15 min. The oils obtained were then stored for a 1-year duration. The author found that the roasting process increased the oxidative stability by 50% but after 1 year of storage duration, the quality of oil started to degrade. The main factors that could impact fatty acid were processing temperatures and oil type. Water content appears to be an important factor because water concentration has a great impact on hydrolytic reaction. Fat rancidity can be prevented by using antioxidants such as butylated hydroxyl toluene or lowering the temperature from 5 to −13°F (−15 to −25°C) and preventing fluctuation of temperature during storage. In addition, duration of storage has a significant impact on fat rancidity, and it may not be advisable for a processor to store oil for an extended duration.

Using new drying techniques has also been shown to have potential to slow down rancidity in oil seed. Sharma et al. (2014) explored different pretreatment processes such as drying, microwave (MW) heating, extrusion, cooking, and chemical methods to extend the shelf life of cereal bran. MC, free fatty acids, water activity, and physical quality of bran were evaluated each month for 6 months after treatment. The authors found that the free fatty acid content, moisture, and water activity were stable for the entire period for products that were obtained from the extrusion processing.

### *Protein Denaturation*

Proteins are biomolecules consisting of long chains of amino acids with primary functions that include metabolic reactions, responding to stimuli, DNA replication, muscle mass building, and so forth. The main sources of protein in pet food include meat, grains, and legumes. Proteins from different sources of material differ by the length of the polypeptide chain, the individual amino acids that comprise the respective protein, and the structure of the folded protein.

Protein quality decreases when the shape of protein changes as a result of external pressure; the phenomenon is referred to as denaturation. Kweon et al. (2017) investigated the effect of heat (25, 60, and 90°C) and pressure (atmospheric, 200, 400, and 600 MPa) on soy flour protein denaturation. The authors reported that treatment at 200 and 600 MPa showed a significant effect on denaturation while treatment at 60 and 90°C showed a less-pronounced effect on denaturation. Ledward (1995) earlier reported that high pressure–induced gels-like substance formation during rupturing of hydrogen bond at high temperature could be attributed to hydrophobic and electrostatic interaction at high pressure. Zia Ur (2006) investigated storage effect on nutritional quality of freshly harvested wheat, maize, and rice grains stored at 10, 25, and 45°C for 6 months. The author reported gradual reduction in protein digestibility at 25 and 45°C. Kibar (2015) investigated the crude protein content in wheat and corn stored for 90 days at different temperatures (10, 12, and 14°C). Irrespective of the temperature, the crude protein decreased linearly with storage duration. Using ambient storage conditions and a 12-month study duration, Bhattacharya and Raha (2002) reported that protein content in corn decreased from 13% to 11.2%, while soybean seeds decreased from 40% to 37.6% during storage.

### Carbohydrate Loss

Carbohydrate is an important macronutrient in pet nutrition. The carbohydrates consist primarily of chains of carbon (C), hydrogen (H), and oxygen (O). Generally, carbohydrates can be grouped into three major dietary divisions: sugar (glucose, fructose and galactose), oligosaccharides (maltodextrins, raddinose, stachyose) and polysaccharides (amylose, amylopectin, cellulose, hemicellulose). Most pet food and animal food are typically comprised of sugar (glucose, fructose, and galactose) and starches. However in recent years, producers are gaining more interest in including dietary fiber in pet food formulations. Fibers are insoluble carbohydrates, that is, they are indigestible by pets and human. However they have medical advantages such as weight management, improving colon health, diarrhea, and diabetes mellitus. Common sources of fiber in pet foods include rice hull, soybean hull, peanut hull, and bran.

Carbohydrate loss during storage is significant due to the fact that microorganisms that cause spoilage also rely on carbohydrate in grain for energy. For instance, Bernal-Lugo and Leopold (1992) investigated the effect of storage duration on sucrose and raffinose content of corn stored at 30°C, 75% RH for 100 days. The authors found that the sucrose level was fairly constant within the first 35 days but rapidly decreased to 100 mg/g at 90 days. Similar findings were also reported by Guo et al. (2015) who investigated the effect of postharvest storage duration (3–12 months) and variety of wheat (Shi and Zhong) on digestibility of pig feeds stored under ambient conditions. The authors found that the gross energy and crude protein of the wheat decreased by 2.0% and 12.01%, respectively, while the concentration of neutral detergent fiber, acid detergent fiber, and starch content increased by 30.26%, 19.08%, and 2.46%, respectively. Total nonstarch polysaccharide, total arabinose, total xylose, and total mannose contents decreased by 46.27%, 45.80%, 41.71%, and 75.66%, respectively. The authors concluded that gross energy, digestible energy, and metabolizable energy content were stable during the first 3–6 months but then decreased during the following 6–12 months of storage. Bhattacharya and Raha (2002) reported that total carbohydrate content of corn decreased from 74.7% to 57%, groundnut from 17% to 12.7%, soybean from 21% to 16.8%

when stored for 1 year under natural conditions. The increased deterioration of carbohydrate during storage was attributed to an increasing population of storage fungi. Fungal invasion of stored product has been a major reason for deterioration of grain during storage. Atungulu et al. (2016b) identified MC, storage temperature, and varieties as some of the main factors that should be considered in postharvest storage of cereals.

### Moisture Content

MC of the grain or processed food determines the rate of deterioration and shelf life (Trigo-Stockli and Pedersen, 1994). Moisture is the amount of water content of a particular pet food, frequently expressed on a percentage wet basis (w.b.). There are three classifications of pet food according to MC: dry food (6%–10% w.b.), semimoist food (15%–40% w.b.), and wet food (greater than 60% w.b.) (Zicker, 2008). Until recently, moist and wet food historically constituted a greater segment of the manufactured pet food market. Moist pet foods are always in cans or sealed plastic containers with the food sterilized at approximately 121°C for 3 min so as to kill pathogens such as *Clostridium botulinum* (Zicker, 2008). Moist pet foods are comprised of mostly meat chunk or ground meat. Semimoist foods however use humectants and acidification to control water activity and fungal growth.

Dry food constitutes the major segment of the pet food marketed by the industry. They are produced using the same food materials and processes used in production of human breakfast cereals such as extrusion, baking, flaking, pelleting, and crumbling. Irrespective of the product and processing methods, moisture reduction or drying is a critical step that must be carried out: The extrusion process results in an MC of approximately 25%, and the pelleting process results in 14%–17% (ICMSF, 2005; Rokey et al., 2010).

Similarly, in previous sections, we identified temperature as one of the main factors that influences nutritional loss in stored food. One of the areas where a processor could unknowingly devalue raw material nutritional content is during the drying step. Dry pet foods are mostly from grain sources, and to store them, the harvest MC must be reduced to a safe level. In addition, during food processing, some pet food especially, extruded foods, emerge from the extruder with a high MC which requires further drying. Hence, the next section focuses on emerging drying techniques that could minimize cost and simultaneously prevent deterioration of major nutrients in pet foods.

## Emerging Drying Techniques

### Convective Heated Air Drying

Convective heated air drying is the currently used method for drying most pet food. The process involves heating the air to a specific temperature and then passing the heated air over the material to be dried. The heated air establishes a temperature gradient between the core of the material and the surface in contact. The gradient results in an outward movement of moisture from the core to the surface as the material slowly dries. Typically, air temperature varies between 90 and 180°C, and the speed of the belt used to convey the material dried varies between 40 and 60 m/min (Rokey et al., 2010). The drying rate depends on whether the material is to be dried in single- or multiple-pass, target MC, product size, and density. Drying process parameters may lead to overdrying, tensile stresses with the product which may cause breakage, and in extreme cases burnt product.

Drying high MC material to low moisture range using one-pass convective heated air may not be very safe and could cause case hardening. Heat delivery associated with convective heated air drying does not result in volumetric heating of products. It is not uncommon to have disparity in material phase transitions resulting in product surfaces transiting to a glassy state while the core is still in a rubbery state. Intraproduct state differences (rubbery and glassy) result in stress cracking which may jeopardize quality (Gulati and Datta, 2015; Wang and Brennan, 1995).

Convective heated air dryers are primarily categorized into three types according to the direction of grain and airflow: cross-flow, concurrent flow, and mixed flow dryers. In concurrent flow dryers, both air and grain mass moves in the same direction, whereas in cross-flow dryers, the airflow is perpendicular to the grain flow. In mixed flow dryers, all airflow directions are present, that is, certain sections of the bed experience concurrently, both counter- and cross-airflow directions. The airflow pattern has a significant effect on drying rate, capacity, energy efficiency, and quality of grains. Table 3.2 shows the effect of dryer type on drying-air temperature and stress-cracked kernel in corn (Montross et al., 1999). Hence, the design of convective dryers must minimize or prevent development of a temperature gradient within the grains so as to reduce formation of the stress cracks.

Zhang and Litchfield (1991) investigated the temperature distribution inside a commercial cross-flow dryer for corn and found a large variation in temperature and MC distribution at different depths and layers within the system. Sections where the highest moisture loss occurred exhibited the most stress cracks. Noomhorm et al. (1992) investigated accelerated drying of peanuts in a rotary drier with air temperatures of up to 300°C with the target of achieving surface temperatures of nuts at 70, 90, 110, 130, and 150°C during drying. The authors recommended 70°C with a 3-min exposure time per batch and a 15 min tempering for optimum quality. In respect to the quality as temperature increase, the nut qualities were not

**Table 3.2: Typical operating temperatures of different types of dryers (Montross et al., 1999).**

| Dryer type | Drying-air temperature (°C) | Maximum grain temperature (°C) | Stress crack kernel (%) |
|---|---|---|---|
| Cross-flow | 80–110 | 80–110 | 70–85 |
| Mixed flow | 100–130 | 70–100 | 50–55 |
| Concurrent flow | 200–285 | 60–80 | 30–45 |

as optimal as those that were sun dried. One way to limit over drying of grain in hot-air drying is to have an outlet at the drying zone for the purpose of removing dried grain before they overdry without stopping the entire system (Marks et al., 1993). However, the feasibility of such designs needs careful consideration in order not to inadvertently introduce leakages to the heating zone. Convective heated air-drying methods require special attention, and several combinations of high-temperature and low-temperature drying regimes may hold the key to limit breakage susceptibility and quality loss.

Optimization of the convective dryer for higher performance may not be so far-fetched as a technology, as the level of instrumentation and monitoring control increases, and identification of high-temperature and high-moisture area can now be easily identified and corrected by sensors and probes installed in the bed.

### Microwave Drying

MW is a form of electromagnetic energy within the frequency range of 300 MHz to 300 GHz with a corresponding wavelength of 1 m to 1 mm, respectively (Oghbaei and Mirzaee, 2010; Olatunde et al., 2017). Unlike the convective heated air method, MW heating accords volumetric heating of products. Thus, generated heat has the ability to rapidly penetrate the material thereby providing a uniform temperature rise within the entire material (Atungulu et al., 2016a; Chandrasekaran et al., 2013). The uniform heating prevents formation of temperature and moisture gradients which in turn minimizes stress and case hardening of the dried product.

The most commonly available MW heating appliances are either for domestic use or industrial use and possess frequencies of 2.45 GHz and 915 MHz, respectively. The 2.45 GHz MW is very popular for domestic food heating applications in the United States. Fig. 3.1 shows a

**Figure 3.1**
Industrial-type microwave system: transmitter (1), waveguide (2), heating zone (3), conveyor belt (4), and control panel (5).

typical industrial MW (AMTek Microwaves, Cedar Rapids, IA) located in the Department of Food Science & Division of Agriculture, University of Arkansas, Fayetteville, Arkansas. The industrial MW is comprised of a generator, the control panel, waveguide, and oven. The MW generator converts high-voltage electric energy into MW radiation by a self-excited oscillator (Olatunde et al., 2017). The Waveguide (rectangular pipe) channels the generated electromagnetic field as it is propagated lengthwise while the pipe conveys the MW to the oven where material to be dried is placed.

Optimum processing parameters need to be determined for a particular material to be dried in the MW dryers. Some of the processing parameters include power level, bed thickness, heating duration, initial and final MC. Cui et al. (2003) and Sharma and Prasad (2006) concluded that the combination of MW vacuum and conventional hot-air drying improved the quality of dried garlic with an 80% reduction in drying duration compared with using hot-air drying alone. Using 916 MHz MW, Atungulu et al. (2016a) were able to achieve one-pass drying of rough rice with a specific energy of 600 kJ/kg grain followed by 4 h of tempering at 60°C with minimal reduction in quality. Olatunde et al. (2017) also showed that by changing specific energy and bed thickness, the protein content in medium-grain rough rice changes significantly (Fig. 3.2). Gursoy et al. (2013) used 2.45 GHz MW to dry corn with different initial MCs (18.3%, 26.3%, 34.3% and 42.3% db), power level (70, 175, and 245 W), and

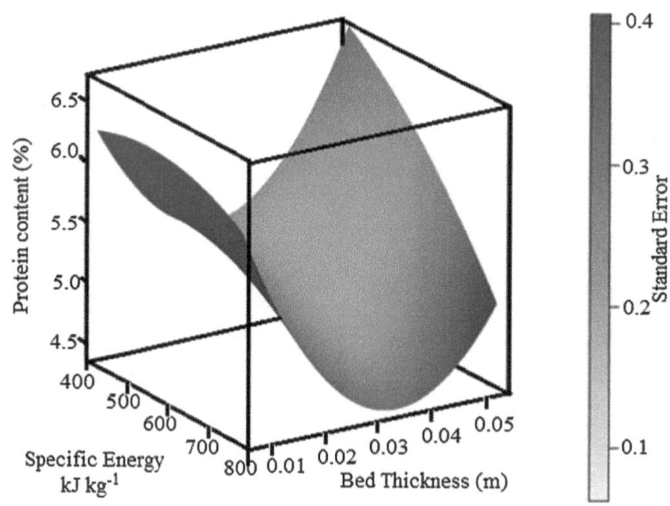


**Figure 3.2**
Surface response of the effect of treatment (thickness and specific energy) on crude protein content of rough rice. *Republished with permission of Elsevier from One-pass drying of rough rice with an industrial 915 MHz microwave dryer: Quality and energy use consideration, Olatunde, G.A., Atungulu, G.G., Smith, D.L., 2017. Biosystems Engineering 155, 33–43. License number: 4038930955949; permission conveyed through Copyright Clearance Center, Inc.*

heating duration (80 and 120 s) and found that MW higher than 175 W significantly increased the stress crack index and reduced germination.

For optimum shape and uniform air-cell distribution, Lee et al. (2000) found that extruded pellets made from corn starch and subsequently dried using MW heating were better than using a convectional drying method. Furthermore, the MW treatment has also been used for other purposes focused on extending shelf life of food material, for example, in food blanching—a process used to enhance and retain food color and inactivate enzymes that could cause spoilage (Chandrasekaran et al., 2013). Singh et al. (1996) blanched marjoram and rosemary in MW and observed maximum retention of color, ascorbic acid, chlorophyll contents, and other quality parameters compared with that of water and steam blanching. In addition, the effect of blanching peanuts in an MW-exposed solution was investigated by Schirack et al. (2007); the authors found that MW saved energy, and inactivated mold spores, but enhanced the stale flavor in peanut at a high temperature as a result of increased concentrations of phenylacetaldehyde, guaiacol, and 2,6-dimethylpyrazine. Other applications of MW in food storage include pasteurization and sterilization (Chandrasekaran et al., 2013). Several authors have demonstrated that moderately low energy is required in the MW drying process (Atungulu et al., 2016a; Olatunde et al., 2017; Tulasidas et al., 1995).

Beyond drying applications, the benefits of MW heating have been reported in multiple other applications including cooking, roasting, sterilization, and pasteurization. MW has potential to extend shelf life, maintain quality, and prevent spoilage of agricultural products. However, there is still concern regarding nonuniformity of drying in batch operations (Cohen and Yang, 1995; Kumar, 2015; Nijhuis et al., 1998; Vadivambal and Jayas, 2010), safety (Osepchuk, 2002; Roberts and Strauss, 2005), cost (Kumar, 2015), and lack of enthusiasm on the part of processors to adopt the technology. The benefits over convective heated air drying such as higher thermal efficiency, reduced drying time, and potential to inactivate harmful, heat-tolerant mold spore could be the key to unlock its potential.

## Infrared Drying

Infrared (IR) radiation can be categorized according to electromagnetic spectra ranges namely 0.75 to 1.4 (near infrared), 1.4 to 3.0 (midinfrared), and 3–1000 μm (far-infrared [FIR]) (Fig. 3.3). Food materials have different IR absorption spectral patterns based on the ability of the molecules to vibrate. For instance, amino acids, polypeptides, and proteins exhibit two strong absorption bands between 3–4 and 6–9 μm, whereas lipids have three strong absorption bands over the entire IR spectrum 3–4, 6, and 9–10 μm, whereas carbohydrates possess two strong absorption bands at 3 and 7–10 μm (Krishnamurthy et al., 2008).

When radiant energy hits the surface of a food material, the atoms and molecules can vibrate, rotate, or yield electrons. In the process, the radiant energy is either absorbed, reflected, or

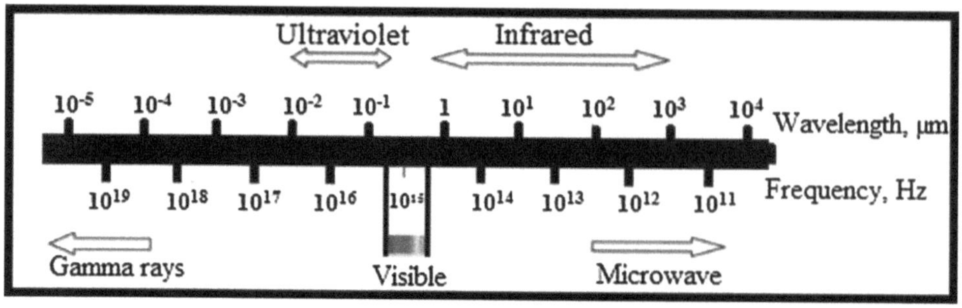


**Figure 3.3**
Electromagnetic wave spectrum.

scattered. The ability to absorb energy varies from material to material and the wavelength of the radiant energy. Incident radiative energy of wavelength ranges from 0.2 to 0.7 µm (ultraviolet and visible rays) results in changes in the electronic state of material molecules. Incident radiative energy of wavelengths in the range of 2.5–1000 µm (FIR) can result in changes in the vibrational state of molecules. Incident radiative energy of wavelength in a range of greater than 1000 µm results in changes to the rotational state of molecules (Decareau, 1985; Krishnamurthy et al., 2008).

IR heating offers many advantages such as high energy flux (heating rate) and considerable energy efficiency. The energy flux has great potential to rapidly dry food material and kill harmful mold spores with minimal reduction in quality (Pan et al., 2008; Wilson et al., 2017). The surface energy flux associated with IR heating is so intense that it penetrates into the material faster than the convective heated air method. The flux has potential to reduce moisture gradients during heating. In addition, because IR does not heat up the air surrounding the product but rather heat the material directly, the process is not constrained by the wet bulb temperature of the surrounding air.

There are several studies that have shown that IR heating is beneficial in food drying and processing. For instance, Wilson et al. (2017) investigated the impact of IR intensities 2.39, 3.78, and 5.55 $kW/m^2$ and duration of exposure (30, 50, and 180 s) on final MC, microbial population, sensory, and pasting qualities of freshly harvested corn at MC of 24% (w. b.). The authors found that when the IR intensity was increased, total molds were reduced with an average value of 2.6 and 2.9 log colony forming units per gram at 30 and 180 s of intermittent heating duration, respectively. In respect to pasting property, the authors concluded that the intensity had no significant impact on final viscosity. However, the intensity at 2.39 $kW/m^2$ and the control had the same effect on the stress crack formation. Low IR treatment temperature and low energy intensity could prevent denaturation of proteins (Ju et al., 2001). In another study, Wilson et al. (2015) reported the potential of IR to rapidly dry freshly harvested corn at initial MCs of 20%, 24%, and 28% wet basis (w. b.) to safe moisture level (Fig. 3.4) while maintaining the corn quality.

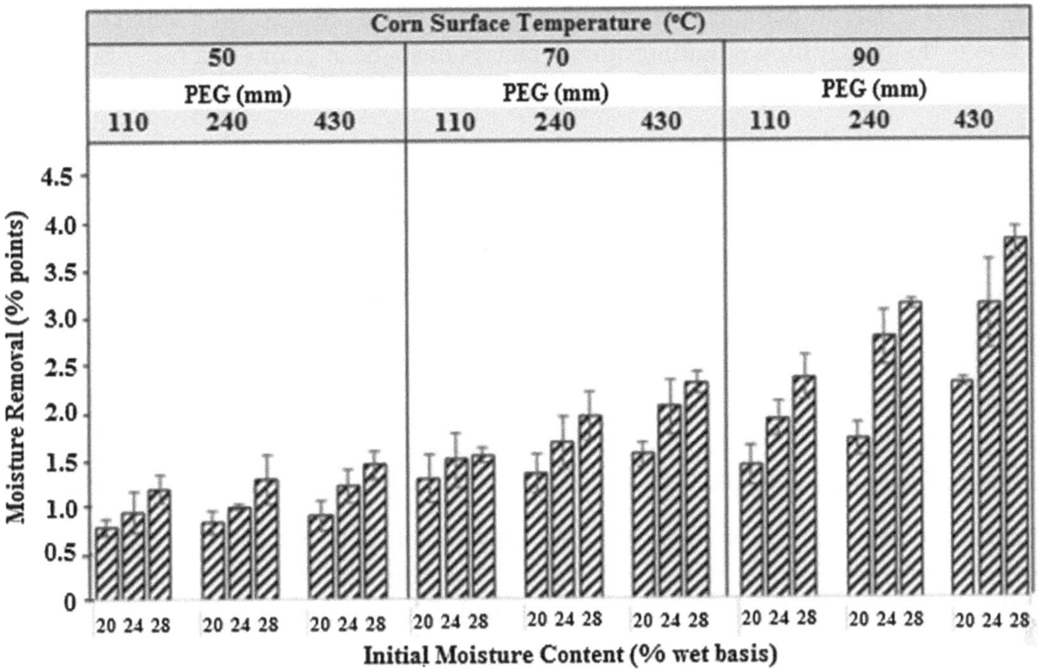


**Figure 3.4**

The effect of initial MC and product-to-emitter-gap size on percentage points of moisture removal when corn kernels at various temperature. MC, moisture content. *From Wilson, S., Atungulu, G., Couch, A., Sadaka, S., 2015. Radiant heating and tempering treatments for improving rate of moisture removal during drying of shelled corn. Applied Engineering in Agriculture 31, 799–808. Copyright 2015 American Society of Agricultural and Biological Engineers. Used with permission.*

Okeyo et al. (2017) also investigated the impact of IR drying on qualities of rough rice of different varieties at an IR intensity of 5.55 kW/m$^2$, heating exposure duration of 30, 50, 90, and 180 s followed by tempering at 60°C for 4 h. The authors found that IR was able to reduce the MC from 23% to 12.5% in one pass. Both final and peak viscosities of treated samples were greater than the control but the treatment had no impact on the sensory and milling quality of rice, except for treatment with a 180-s heating duration. Similar findings were reported by Fasina et al. (1998) that IR heating had no impact on the chemical constituents of barley such as starch, protein, total dietary fiber, ash, and fat. Das et al. (2009) incorporated a vibratory system and investigated the impact of radiation intensity (1509, 2520, 3510, 4520, and 5514 W/m$^2$) versus grain depth (single kernel thickness of 3, 6, 12, and 25 mm) on paddy rice and found that the rate of moisture removal was higher when the bed thickness was smaller. There is still a need to improve the throughput of IR drying process to meet the producers' requirements.

### *Fluidized Bed Drying*

Fluidization is the process of suspending particulate material in a gas stream. When used for drying purposes, fluidized individual particles are exposed to hot gas stream to an improved drying rate. Particle-to-particle contact with a hot gas stream has also been found to be effective in the inactivation of antinutritional factors in soybean grains at a faster rate than the convective heated air method of drying (Dondee et al., 2011; Soponronnarit et al., 2001).

Some of the drawbacks in the industrial scale use of fluidized bed for drying agricultural material are excessive cracking and breakage (Dondee et al., 2011; Wiriyaumpaiwong et al., 2001). The damages are due to high heat transfer rates because the entire surface of the material is exposed to the hot air. Because water migration is controlled by the diffusion of moisture from the core of the grain to the surface, as the temperature increases more than the corresponding air wet bulb temperature, the grain becomes brittle at the surface and is prone to cracking. In addition, high power is required to suspend the entire bed in a gas stream, and as particles collide with each other, the potential of unintended attrition and granulation increases. When MC is high, the system has the potential for defluidization and channel formation (Chou and Chua, 2001; Olatunde et al., 2016).

Some modifications have been made to minimize some of the problems associated with fluidized bed drying. For instance, evaluation of the system designed by Soponronnarit et al. (1995) in which a cross-flow dryer and fluidized bed system was used to dry paddy rice showed that it took 53 min to dry the paddy from 33% to 16%. This drying duration appears to be very long from an industrial perspective. Drying characteristics of shelled corn were studied in a fluidized bed dryer assisted by MW heating by (Momenzadeh et al., 2011). The combination of MW and fluidization resulted in 50% reduction in drying duration compared with MW application alone. Improving the efficiency of a fluidized bed system is still a subject of considerable research. However, application of the fluidized bed drying has found applications in the pharmaceutical industry.

## Conclusions

Drying and storage of pet food requires detailed understanding of the composition of the food to be preserved. More often, one method of treatment may not provide the intended quality of the product. Where necessary, a combination of different unit operations such as for drying and/or storage may hold the key to minimize nutrient losses and optimize processing costs to deliver products with high quality and extended shelf life. Pet food processors need to start with high-quality ingredients, directly at the place of procurement, if they expect to obtain final products with premium quality.

## References

Alencar, E.R.D., Faroni, L.R.D., Peternelli, L.A., Silva, M.T.C.D., Costa, A.R., 2010. Influence of soybean storage conditions on crude oil quality. Revista Brasileira de Engenharia Agrícola e Ambiental 14, 303–308.

Atungulu, G., Smith, D., Wilson, S., Zhong, H., Sadaka, S., Rogers, S., 2016a. Assessment of one-pass drying of rough rice with an industrial microwave system on milling quality. Applied Engineering in Agriculture 32, 417–429.

Atungulu, G.G., Thote, S., Wilson, S., 2016b. Storage of hybrid rough rice—consideration of microbial growth kinetics and prediction models. Journal of Stored Products Research 69, 235–244.

Bernal-Lugo, I., Leopold, A.C., 1992. Changes in soluble carbohydrates during seed storage. Plant Physiology 98, 1207–1210.

Bhattacharya, K., Raha, S., 2002. Deteriorative changes of maize, groundnut and soybean seeds by fungi in storage. Mycopathologia 155, 135–141.

Chandrasekaran, S., Ramanathan, S., Basak, T., 2013. Microwave food processing—a review. Food Research International 52, 243–261.

Chou, S.K., Chua, K.J., 2001. New hybrid drying technologies for heat sensitive foodstuffs. Trends in Food Science & Technology 12, 359–369.

Cohen, J.S., Yang, T.C., 1995. Progress in food dehydration. Trends in Food Science & Technology 6, 20–25.

Combs, S.B., Combs Jr., G.F., 1985. Varietal differences in the vitamin E content of corn. Journal of Agricultural and Food Chemistry 33, 815–817.

Cui, Z.-W., Xu, S.-Y., Sun, D.-W., 2003. Dehydration of garlic slices by combined microwave-vacuum and air drying. Drying Technology 21, 1173–1184.

Das, I., Das, S.K., Bal, S., 2009. Drying kinetics of high moisture paddy undergoing vibration-assisted infrared (IR) drying. Journal of Food Engineering 95, 166–171.

Decareau, R., 1985. Microwaves in the Food Processing Industry, first ed. Academic Press, Atlanta, GA.

Dondee, S., Meeso, N., Soponronnarit, S., Siriamornpun, S., 2011. Reducing cracking and breakage of soybean grains under combined near-infrared radiation and fluidized-bed drying. Journal of Food Engineering 104, 6–13.

Fasina, O., Tyler, R., Pickard, M., 1998. Modelling the infrared radiative heating of agricultural crops. Drying Technology 16, 2065–2082.

Gulati, T., Datta, A.K., 2015. Mechanistic understanding of case-hardening and texture development during drying of food materials. Journal of Food Engineering 166, 119–138.

Guo, P., Li, P., Li, Z., Stein, H., Liu, L., Xia, T., Yang, Y., Ma, Y., 2015. Effects of post-harvest storage duration and variety on nutrient digestibility and energy content wheat in finishing pigs. Asian-Australasian Journal of Animal Sciences 28, 1488–1495.

Gursoy, S., Choudhary, R., Watson, D.G., 2013. Microwave drying kinetics and quality characteristics of corn. International Journal of Agricultural and Biological Engineering 6, 90–99.

Hoffmann-La, R., 1991. Vitamine E for Ruminants. Hoffmann-La Roche Inc., Nurley, New Jersey.

ICMSF, 2005. Micro-organisms in foods 6: microbial ecology of food commodities. In: International Commission on Microbiological Specifications for Foods. Kluwer Academic/Plenum Publishers, Boston, MA.

Jensen, P.N., Sørensen, G., Brockhoff, P., Bertelsen, G., 2003. Investigation of packaging systems for shelled walnuts based on oxygen absorbers. Journal of Agricultural and Food Chemistry 51, 4941–4947.

Ju, Z., Hettiarachchy, N., Rath, N., 2001. Extraction, denaturation and hydrophobic properties of rice flour proteins. Journal of Food Science 66, 229–232.

Keskin, S., Ozkaya, H., 2013. Effect of storage and insect infestation on the mineral and vitamin contents of wheat grain and flour. Journal of Economic Entomology 106, 1058–1063.

Kibar, H., 2015. Determining the functional characteristics of wheat and corn grains depending on storage time and temperature. Journal of Food Processing and Preservation 40, 749–759.

Krishnamurthy, K., Khurana, H.K., Soojin, J., Irudayaraj, J., Demirci, A., 2008. Infrared heating in food processing: an overview. Comprehensive Reviews in Food Science and Food Safety 7, 2–13.

Kumar, Y., 2015. Application of microwave in food drying. International Journal of Engineering Studies and Technical Approach 1, 9–24.

Kweon, M., Slade, L., Levine, H., 2017. Differential scanning calorimetry analysis of the effects of heat and pressure on protein denaturation in soy flour mixed with various types of plasticizers. Journal of Food Science. http://dx.doi.org/10.1111/1750-3841.13616.

Labuza, T.P., Dugan Jr., L., 1971. Kinetics of lipid oxidation in foods. Critical Reviews in Food Science and Nutrition 2, 355–405.

Ledward, D., 1995. High-pressure processing: the potential. In: Ledward, D., Earnshow, R., Hastings, A., Johnston, D. (Eds.), High-Pressure Treatment of Food. Nottingham University Press, Nottingham, UK, pp. 1–5.

Lee, E.Y., Lim, K.I., Lim, J-k., Lim, S.-T., 2000. Effects of gelatinization and moisture content of extruded starch pellets on morphology and physical properties of microwave-expanded products. Journal of Cereal Chemistry 77, 769–773.

Marks, B.P., Maier, D.E., Bakker-Arkema, F.W., 1993. Optimization of a new in-bin counterflow corn drying system. Transactions of the ASAE 36, 529–534.

Momenzadeh, L., Zomorodian, A., Mowla, D., 2011. Experimental and theoretical investigation of shelled corn drying in a microwave-assisted fluidized bed dryer using Artificial Neural Network. Food and Bioproducts Processing 89, 15–21.

Montross, M., Bakker-Arkema, F., Hines, R., 1999. Moisture content variation and grain quality of corn dried in different high-temperature dryers. Transactions of the ASAE 42, 427.

Nijhuis, H., Torringa, H., Muresan, S., Yuksel, D., Leguijt, C., Kloek, W., 1998. Approaches to improving the quality of dried fruit and vegetables. Trends in Food Science & Technology 9, 13–20.

Noomhorm, A., Remakumar, K., Chow Ting, C., 1992. Accelerated drying of groundnuts in batch rotary drier. ASEAN Journal 7, 5–9.

Oghbaei, M., Mirzaee, O., 2010. Microwave versus conventional sintering: a review of fundamentals, advantages and applications. Journal of Alloys and Compounds 494, 175–189.

Okeyo, A., Olatunde, G., Atungulu, G., Sadaka, S., McKay, T., 2017. Infrared drying characteristics of long-grain hybrid, long-grain pureline, and medium-grain rice cultivars. Cereal Chemistry Journal 94, 251–261.

Olatunde, G.A., Atungulu, G.G., Smith, D.L., 2017. One-pass drying of rough rice with an industrial 915 MHz microwave dryer: quality and energy use consideration. Biosystems Engineering 155, 33–43.

Olatunde, G., Fasina, O., McDonald, T., Adhikari, S., Duke, S., 2016. Moisture effect on fluidization behavior of loblolly pine wood grinds. Biomass Conversion and Biorefinery 6, 1–14.

Osepchuk, J.M., 2002. Microwave power applications. IEEE Transactions on Microwave Theory and Techniques 50, 975–985.

Pan, Z., Khir, R., Godfrey, L.D., Lewis, R., Thompson, J.F., Salim, A., 2008. Feasibility of simultaneous rough rice drying and disinfestations by infrared radiation heating and rice milling quality. Journal of Food Engineering 84, 469–479.

Riaz, M.N., Asif, M., Ali, R., 2009. Stability of vitamins during extrusion. Critical Reviews in Food Science and Nutrition 49, 361–368.

Roberts, B.A., Strauss, C.R., 2005. Toward rapid, "green", predictable microwave-assisted synthesis. Accounts of Chemical Research 38, 653–661.

Rokey, G.J., Plattner, B., Souza, E.M.D., 2010. Feed extrusion process description. Revista Brasileira de Zootecnia 39, 510–518.

Schirack, A., Sanders, T., Sandeep, K., 2007. Effect of processing parameters on the temperature and moisture content of microwave-blanched peanuts. Journal of Food Process Engineering 30, 225–240.

Sharma, G., Prasad, S., 2006. Optimization of process parameters for microwave drying of garlic cloves. Journal of Food Engineering 75, 441–446.

Sharma, S., Kaur, S., Dar, B., Singh, B., 2014. Storage stability and quality assessment of processed cereal brans. Journal of Food Science and Technology 51, 583–588.

Siger, A., Michalak, M., 2016. The long-term storage of cold-pressed oil from roasted rapeseed: effects on antioxidant activity and levels of canolol and tocopherols. European Journal of Lipid Science and Technology 118, 1030–1041.

Singh, M., Raghavan, B., Abraham, K., 1996. Processing of marjoram (*Majorana hortensis Moench.*) and rosemary (*Rosmarinus officinalis L.*). Effect of blanching methods on quality. Molecular Nutrition & Food Research 40, 264–266.

Soponronnarit, S., Swasdisevi, T., Wetchacama, S., Wutiwiwatchai, W., 2001. Fluidised bed drying of soybeans. Journal of Stored Products Research 37, 133–151.

Soponronnarit, S., Yapha, M., Pracbayawarakorn, S., 1995. Cross-flow fluidized bed paddy dryer: prototype and commercialization. Drying Technology 13, 2207–2216.

Toci, A.T., Neto, V.J.M.F., Torres, A.G., Farah, A., 2013. Changes in triacylglycerols and free fatty acids composition during storage of roasted coffee. LWT – Food Science and Technology 50, 581–590.

Trigo-Stockli, D., Pedersen, J., 1994. Effect of rice storage conditions on the quality of milled rice. In: Highley, E., Wright, E.J., Banks, H.J., Champ, B.R. (Eds.), Proceedings of the 6th International Working Conference on Stored-Product Protection, Canberra, Australia. CAB International, UK, pp. 17–23.

Tulasidas, T., Raghavan, G., Mujumdar, A., 1995. Microwave drying of grapes in a single mode cavity at 2450 Mhz-11: quality and energy aspects. Drying Technology 13, 1973–1992.

Ulu, H., 2004. Evaluation of three 2-thiobarbituric acid methods for the measurement of lipid oxidation in various meats and meat products. Meat Science 67, 683–687.

Underwood, B.A., 1984. Vitamin A in Animal and Human Nutrition. Accademic Press Inc., Bethesda, MD.

Vadivambal, R., Jayas, D., 2010. Non-uniform temperature distribution during microwave heating of food materials—a review. Food and Bioprocess Technology 3, 161–171.

Wang, N., Brennan, J., 1995. Changes in structure, density and porosity of potato during dehydration. Journal of Food Engineering 24, 61–76.

Wilson, S., Atungulu, G., Couch, A., Sadaka, S., 2015. Radiant heating and tempering treatments for improving rate of moisture removal during drying of shelled corn. Applied Engineering in Agriculture 31, 799–808.

Wilson, S., Okeyo, A., Olatunde, G., Atungulu, G., 2017. Radiant heat treatments for corn drying and decontamination. Journal of Food Processing and Preservation. http://dx.doi.org/10.1111/jfpp.13193.

Wiriyaumpaiwong, S., Soponronnarit, S., Prachayawarakorn, S., 2001. Effect of temperature on soybean quality using spouted bed technique. Kasetsart Journal (Nature of Science) 35, 460–466.

Zhang, Q., Litchfield, J., 1991. Corn quality distribution within a continuous crossflow dryer. Transactions of the ASAE 34, 2516–2520.

Zia Ur, R., 2006. Storage effects on nutritional quality of commonly consumed cereals. Food Chemistry 95, 53–57.

Zicker, S.C., 2008. Evaluating pet foods: how confident are you when you recommend a commercial pet food? Topics in Companion Animal Medicine 23, 121–126.

CHAPTER 4

# Animal Feed Mill Biosecurity: Prevention of Biological Hazards

**Anne Huss[1], Roger Cochrane[2], Mary Muckey[2], Cassie Jones[2]**
[1]*BioMatrix International, Princeton, MN, United States;* [2]*Kansas State University, Manhattan, KS, United States*

## Introduction

Animal food includes feeds for livestock, equine, and poultry, as well as pet foods intended for companion animals. In general, animal food is specifically formulated to ensure that the nutrients required for optimal health and performance are included. To reach these nutrient requirements, animal food is made from plant- and animal-based ingredients. Both ingredient sources have the potential to introduce biological hazards into the finished feeds. These biological hazards can be of bacterial or viral origin, can lead to animal illness or injury, and rarely may be carried into the human food supply. The implementation of the Food Safety Modernization Act (FSMA) requires manufacturers to take an active role to help reduce the risk of foodborne pathogens from being introduced into the animal or human food supply. Although the risk of biological hazard contamination in finished animal food cannot be eliminated completely, multiple interventions can be used to reduce the risk. One important intervention is the implementation of a feed mill biosecurity plan. A feed mill biosecurity plan can help minimize the risk of introduction of a biological hazard resulting in animal and human infections and economic losses. With the world population projected to reach 9.6 billion in 2050, the amount of food needed for this population boom will be 70%–100% more than current production (Meeker and Meisinger, 2015). As the population grows, it is projected that meat consumption worldwide will almost double by 2030 (FAO, 2009). A significant portion of every food animal grown and slaughtered is not consumed by humans and is recycled by the rendering industry into value-added materials. The sustainability of animal agriculture depends on these products, which range from livestock feed to fertilizer, pet food to pharmaceuticals, and lard to lubricants. Most rendered materials are returned to the feed industry as high-energy fats and high-quality protein ingredients that supplement the diet and enable efficient production of beef, veal, pork, poultry, fish, eggs, and milk. A large portion is also used by the pet food industry for inclusion in diets for companion animals. Continued sustainable growth of the livestock industry is reliant on the ability of renderers to reliably transform animal tissue waste to high-quality and safe value-added products. In addition to

Food and Feed Safety Systems and Analysis. http://dx.doi.org/10.1016/B978-0-12-811835-1.00004-X

sustainability, data from animal-production researchers have demonstrated that the quality of finished meat, poultry, and fish products consumed by humans is directly related to animal feeding practices (Capucille et al., 2004; Galtin et al., 2003; Zaghini et al., 2005). Because of the increasing population and demand for animal-based proteins, the ingredients used in animal feeds are fundamentally important in both the quality of the finished goods and safety as related to human health.

## Animal Food

In 2004, the United States produced over 120 million tons of animal feed, consisting of mixes of feed grains, mill by-products, animal proteins, and microingredients (vitamins, minerals, and antibiotics) (Sapkota et al., 2007). Animal food includes feeds produced for consumption by livestock, equine, and poultry animals and pet foods produced for companion animals. Animal food is specifically formulated to provide all nutrients required for the animal to achieve development and growth. In addition, livestock and poultry feeds are typically optimized for peak animal performance to maximize yields for human consumption. Animal feeds are primarily composed of animal- and plant-based ingredients, along with purified minerals and vitamins.

Although the purpose of the animal food industry is to provide consumers with essential meat, poultry, and fish products, a large proportion of the animals are considered inedible by humans. These inedible portions include bones, fat, blood, feathers, and some internal organs (Meeker and Meisinger, 2015). According to information released by the National Renders Association (June 2016), North American renderers collect 60 billion pounds (27 million metric tons) of these raw materials per year. These raw materials are further processed and recycled by renderers into 11 billion pounds (4.9 million metric tons) of fats/oils and 10 billion pounds (4.5 million metric tons) of protein annually. The fats and proteins produced by the rendering industry provide sustainable and high-quality inputs for the livestock and poultry markets along with the growing pet food industry.

Improved sustainability of animal agriculture is the result of rendering, one of the oldest forms of recycling. Without rendering, by-products from food animal agriculture would be deposited in landfills, burned, buried, or inappropriately dumped with large amounts of carbon dioxide, ammonia, and other compounds polluting air and water. Turning waste into value-added products is carried out by rendering plants that are "integrated" with the meat processing companies to process the "captive" by-products generated, or operated by "independent" renderers that collect and process by-products from many different sources, including livestock slaughter and processing facilities, grocery stores, restaurants, and other entities along the meat production chain (Meeker and Meisinger, 2015). During the rendering process, heat is applied, moisture is removed, and fat is separated. However, food safety concerns arise as animals used in the rendering process are natural microbiological reservoirs, including human pathogens *Salmonella* spp., *Clostridium perfringens*, *Listeria monocytogenes*, and *Campylobacter jejuni* (Denton et al., 2005). Continued improvements within the industry have implemented process control monitoring to ensure that proven cook times and temperatures have been reached for inactivation of specific microorganisms deemed to be a food safety hazard (Meeker and Hamilton, 2006).

Adaptation of formal quality and safety control programs, such as the Rendering Code of Practice (National Renderers Association, 2014), has enhanced the safety, quality, and reputation of rendered products by controlling biological, chemical, and physical hazards (Meeker and Meisinger, 2015). Although the rendering industry has an aggressive approach to animal food ingredient quality safety by use of long cook times and high temperatures, contamination with pathogenic microorganisms still occurs. A survey sponsored by the Fats and Proteins Research Foundation evaluated raw materials (animal waste tissues), crax (material from cooking/expelling process), and final rendered products for five human pathogenic bacteria, including *Salmonella* spp. (Troutt et al., 2001). The presence of *Salmonella* spp. was detected in 84.5% of the raw ingredients sampled, 0% of crax samples, and 26.1% of final rendered products (Troutt et al., 2001). These results coincide with other research evaluating the presence of *Salmonella* spp. in final rendered products, including protein meals, meat and bone meals, feather meal, meat meal, and poultry meal (Sapkota et al., 2007; Franco, 2005; Kinley et al., 2010; Moyle, 1966; Denton et al., 2005; Laban et al., 2014).

Notably, *Salmonella* contamination may not necessarily cause illness. In fact, the Food and Drug Administration Compliance Policy Guide 690.800, *Salmonella in Food for Animals* only considers *Salmonella* an adulterant in animal food intended for livestock, poultry, and equine facility if it is known to be pathogenic to that specific species. As an example, *Salmonella* serovar Choleraesuis is pathogenic to swine and therefore considered an adulterant if found in finished swine feed. However, there is a zero-tolerance policy for *Salmonella* of any serotype in pet foods because of the greater potential for pet food to come into direct contact with humans.

Although research has previously linked *Salmonella* to meat and bone meals, evaluation of microbial contamination of rendered fats, specifically chicken fat, beef tallow, or other animal fat products was not included in these surveys. Historically, fats have been thought to be nonconducive of microbial growth due to low-water activity and other required growth factors. Growing evidence is causing this assumption to be questioned.

A survey of common animal feed ingredients by Li et al. (2012) detected *Salmonella* in animal- and plant-derived ingredients. Plant-derived ingredients include cereal grains and oilseeds, including barley, corn, oats, rice, sorghum, and wheat. Oilseeds include sunflower, canola, safflower, and flaxseed along with corn and soybean. Other ingredients include forages (i.e., alfalfa meal and hay, Bermuda coastal grass hay, corn plant, and soybean hay, etc.), plant protein products (i.e., canola meal, cottonseed cakes and meals, peanut meal, safflower meal, soybean meal, etc.), and processed grain by-products (i.e., distillers products, corn gluten, sorghum germ cake and meal, wheat bran, etc.) (Sapkota et al., 2007). During the surveillance period of 2002–06, the prevalence of *Salmonella* in animal-derived ingredients was 66.1%, which was decreased to 41.3% during the period of 2007–09. During those same time periods, plant-derived ingredients were found to be contaminated 11.0% and 10.6% from 2002 to 2006 and 2007–09, respectively (Li et al., 2012; Ge et al., 2013). In addition, cereal grains and oilseeds have been found as sources of pathogen contamination (Ge et al., 2013).

Although the rendering process is effective at pathogen reduction as seen with 0% *Salmonella* spp. in crax (Troutt et al., 2001), it is a point-in-time mitigation technique with no residual activity. Recontamination after processing has been proposed as the primary factor for the presence of *Salmonella* spp. in final rendered products (Troutt et al., 2001; Denton et al., 2005; Kinley et al., 2010). Rendered products recontaminated at the rendering facility have the potential to contaminate feed manufacturing facilities, with the potential to contaminate animal food. Rare instances of human *Salmonella* infection have been traced back to contaminated animal food (Clark et al., 1973; Hirsch and Sapiro-Hirsch, 1958; Knox et al., 1963; Pennington et al., 1968; Semple et al., 1968). Adherence to strict processing temperatures and retention times and use of quality raw ingredients can help reduce the risk of contamination by pathogenic microorganisms. Maintaining hygienic conditions of the processing facility through good manufacturing practices can also help reduce contamination risks. However, all these hurdles offer point-in-time mitigation with no residual activity. Promising research by Cochrane et al. (2015a,b) has shown effective mitigation of *Salmonella* through the addition of chemical additives to rendered animal proteins, including feather meal, avian blood meal, porcine meat and bone meal, and poultry by-product meal. Again, this research did not include rendered animal fats.

Typically, increased fat content and decreased water activity, both exhibiting bacteriostatic properties, are parameters used in foods to prohibit microbial growth. However, two major outbreaks of *Salmonella enterica* infections in the United States were traced back to peanut butter, a high-fat (65%) low-water activity ($a_w < 0.3$) food. Outbreaks involving low-moisture and high-fat foods, such as chocolate and peanut butter, reported low infectious doses of 10–100 *S. enterica* cells, which is considerably lower than the estimated $10^4$ CFU/g required to cause illness in a healthy individual (Kapperud et al., 1990; Blaser and Newman, 1982). Further investigations revealed that *Salmonella* is highly resistant to thermal processing of peanut butter and able to survive over 24 weeks after contamination (Burnett et al., 2000; Ma et al., 2009). Research by Aviles et al. (2013) determined the food matrix itself influences the infectious dose of *Salmonella*, potentially by improving *Salmonella*'s survival in the gastrointestinal tract through cross-protection. Cross-protection is the phenomenon of improved survival in a stressful environment due to previous exposure and adaptation of other stress conditions encountered (Aviles et al., 2013).

With that in mind, there is potential for rendered animal fats to harbor pathogenic bacteria, like *Salmonella* spp., leading to contamination of final products, such as animal and pet foods. An outbreak of human salmonellosis in the United States from 2006 to 2008 was attributed to *Salmonella* serovar Schwarzengrund and traced back to dried dog food. During further investigation of the manufacturing facility responsible for producing the contaminated dog food, a swab collected from the enrobing/flavoring room tested positive for the outbreak strain. In the enrobing/flavoring room, material is sprayed onto the surface of the finished product to enhance palatability (Behravesh et al., 2010). Rendered animal fats are commonly incorporated in animal feed and pet foods for added energy, essential fatty acids, and as a palatant.

The FDA publishes an annual report describing foods that have resulted in Reportable Food Registry (RFR) reports. These are items that any human or animal food, "for which there is a reasonable probability that the use of, or exposure to, such article of food will cause serious adverse health consequences or death to humans or animals" (RFR, FDA, 2015). Of the 114 RFR reports associated with animal food in the five published annual reports, 49 are due to *Salmonella* spp. contamination in pet food (RFR, FDA, 2015). This makes it by far the most substantial hazard quantified by FDA in recent years.

Notably, all 49 RFR reports associated with *Salmonella* spp. were due to contamination in pet food. This is predominantly due to the manner in which FDA considers *Salmonella* spp. as an adulterant. For example, the FDA Compliance Policy Guide 690.800: *Salmonella* in Food for Animals describes FDA's current thinking that any finished pet food contaminated with any serotype of *Salmonella* is considered to be adulterated because of its propensity to be a direct human contact food and potentially impact human health. Alternatively, finished livestock feed is only considered adulterated if it contains a serotype of the bacteria known to cause illness in the animal for which it is intended. Of more than 2600 different serotypes of *Salmonella*, only a few are known to cause illness or injury in specific animal species. Those serotypes and their affected animal species are listed in Table 4.1. The occurrence of these specific serotypes in livestock feed is limited. None of those listed have been linked to an

**Table 4.1: *Salmonella* as a biological hazard of concern in animal food.**[a,b]

| | |
|---|---|
| Poultry | *Salmonella* Pullorum, *Salmonella* Gallinarum, or *Salmonella* Enteritidis |
| Swine | *Salmonella* Choleraesuis |
| Sheep | *Salmonella* Abortusovis |
| Horse | *Salmonella* Abortusequi |
| Dairy and beef | *Salmonella* Newport or *Salmonella* Dublin |
| Pets (including dogs, cats, and aquarium fish) | All serotypes of *Salmonella* |

[a]Additional serotypes will be evaluated on a case-by-case basis.
[b]Animal food will not subsequently undergo a commercial heat step or other commercial process that will kill the *Salmonella*.
*Table adapted from FDA Compliance Policy Guide Sec. 690.800* Salmonella *in Food for Animals, 2013.*

RFR in the six published annual reports available. Furthermore, only *Salmonella* Enteritidis has been demonstrated to be in the 25 most common serotypes found in animal feed or ingredients according to an FDA survey (Li et al., 2012). All other serotypes have a relatively low level of occurrence. Still, presence of any level of the specified serotype, or any serotype at all in pet food, is deemed to be an adulterant in that animal food if it will not undergo further heat processing or a subsequent commercial kill step.

Although *Salmonella* spp. is the primary biological hazard of concern in animal food manufacturing facilities, two other biological hazards have surfaced as concerns in recent years. Before 2015, the pet food industry did not recognize *L. monocytogenes* as a potential hazard in its products. However, seven RFR reports were reported from July 2015 to July 2016, all of which occurred in raw, fresh, and frozen pet foods (FDA, 2015). Survey data collected from 196 raw dog and cat foods revealed that 8% contained *Salmonella* spp. and 16% tested positive for presence of *L. monocytogenes*. Thus facilities producing finished raw, fresh, or frozen pet foods without a thermal kill step are under increased scrutiny to ensure the safety of their products through other control measures including Current Good Manufacturing Practices, other processing controls, supply-chain applied controls, and sanitation controls.

Although *Salmonella* spp. and *L. monocytogenes* have focused on transmission through pet food, another biological hazard has caused harm in swine feed. Recently, feed has been recognized as one of many potential vectors that may transmit porcine epidemic diarrhea virus (PEDV; Dee et al., 2014; Pasick et al., 2014; Schumacher et al., 2016a,b). Biosecurity measures have been implemented at many swine feed mills to prevent the entry of PEDV through ingredients, transportation, and people (Cochrane et al., 2016a,b). Furthermore, dust has been recently shown to cause cross-contamination within feed mills, which may contaminate animal feed manufacturing surfaces (Gebhardt et al., 2016). The transmission of biological hazards has historically been linked to ingredients, but other data support the role of cross-contamination from animal food contact surfaces, such as the interior of conveyors, bucket elevators, bins, and floors (Jones and Richardson, 2004; Jones, 2011; Ge et al., 2013; Dee et al., 2014; Pasick et al., 2014; Jeffery, 2016).

## Manufacture of Animal Feeds

In general, feed mills are responsible for the production of animal feeds, including those consumed by livestock and poultry animals and companion animals. For animal-based agriculture, manufactured feeds provide livestock and poultry animals with optimally formulated feeds to meet all nutrient requirements for optimal growth and performance. These manufactured feeds are specifically formulated to ensure that the animal receives optimal nutrients for each life stage. Optimal nutrition is important for preventing deficiencies, malnutrition, and disease. In addition to providing livestock animals with nutrient dense feeds, feed mills can also be responsible for the manufacture of pet foods for companion animals, including cats and dogs. Although the finished animal feeds can vary greatly between livestock and companion animals, most are generally produced from similar ingredients. These ingredients are derived from plant-based grains and animal-based cuts and by-products. To help ensure the safety of animal food from biological hazards, implementation of a feed mill biosecurity plan is essential.

The process of manufacturing feeds for food-producing animals can vary depending on the intended species, animal age, and animal management practice. Animal food can be fed to animals in the form of mash, crumbles, and pellets of various sizes. Most animal foods intended for livestock and poultry animals in a crumble or pellet form are processed via conditioning (with or without steam) and pelletization. Conditioning temperatures and times can vary. Pelleting can also result in the introduction of high temperatures, pressure, and shear forces.

Pet foods produced for companion animals are also produced in various forms, from kibbles to canned pates to jerky-type treats. Like the process of livestock feed manufacture, pet food kibbles are also conditioned with steam at high temperatures. Unlike livestock feeds, pet food kibbles and some treats are formed through extrusion. Extrusion can also exert a high temperature, high shear, and high pressure.

As for pet foods in canned or hermetically sealed packaging, manufacturers are required to follow the same guidelines as set forth for thermally processed low-acid foods packaged in hermetically sealed containers (21 CFR 113), used for regulation of food canned for human consumption. 21 CFR 113 sets-forth guidelines for production and process controls, including product preparation; establishment of scheduled processes; operations in the thermal processing room; and deviations in processing, venting, or control of critical factors.

### *Biological Hazards*

The FDA defines a hazard as "any biological, chemical (including radiological), or physical agent that has the potential to cause illness or injury in humans or animals" (Federal Register-Roger's review paper). Although not all hazards are biological in nature, we will focus on those of biological importance. Identification of potential biological hazards can come from multiple sources including scientific literature, FDA recalls, and FDA or other regulatory guidance agencies/documents (Cochrane et al., 2016a,b). A database of recalls and withdrawals along with the RFR, both maintained by the FDA, can provide valuable information related to the risk of contamination of animal feeds by biological hazards (Cochrane et al., 2016a,b). As a resource for guidance to industry, a draft list of potentially hazardous contaminants in animal feed and feed ingredients has been released by the FDA in 2006 (https://www.fda.gov/downloads/animalveterinary/safetyhealth/animalfeedsafetysystemafss/ucm053715.pdf). Although not a comprehensive list, potential hazards are categorized as biological, chemical, or physical. Biological contaminants are broken down further into two subcategories: transmissible spongiform encephalopathies and microbiological contaminants. Transmissible spongiform encephalopathies include bovine spongiform encephalopathy and chronic wasting disease. Microbiological containments (bacterial) include *Bacillus* spp., *Clostridium* spp., *Escherichia coli*, *Mycobacterium* spp., *Pseudomonas* spp., *S. enterica* (various serotypes), and *Staphylococcus* spp. (https://www.fda.gov/downloads/animalveterinary/safetyhealth/animalfeedsafetysystemafss/ucm053715.pdf). It is important to emphasize that this list is not comprehensive as viral pathogens of risk to animals are not included, specifically PEDV. In 2013, PEDV was first introduced in the United States and had a devastating impact on the swine industry. Multiple vectors, including transport vehicles, fomites, and aerosols, were identified in PEDV transmission with further research implicating transmission through complete finished swine feeds and feed ingredients (Dee et al., 2014; Pasick et al., 2014; Misener, 2014; Lowe et al., 2014; Alonso et al., 2014). Although PEDV has had a large impact on the swine industry, pathogenesis is limited to swine and does not cause disease in other livestock animals.

Of the biological hazards mentioned, *Salmonella* is the most prevalent across all animal feeds (Cochrane et al., 2016a,b). In addition, *Salmonella* is also the most common foodborne illness reported in humans (Scallan et al., 2011). Furthermore, several outbreaks of human *Salmonella* infections have been traced back to contaminated animal feed (Clark et al., 1973; Hirsch and Sapiro-Hirsch, 1958; Knox et al., 1963; Pennington et al., 1968; Semple et al., 1968). To monitor the prevalence of *Salmonella* in animal feeds, the US FDA initiated the Feed Contaminants Program (2002–09) and *Salmonella* Assignment (2007–09) to sample complete animal feeds, feed ingredients, pet foods, pet treats, and supplements for pets. These results found the prevalence of *Salmonella* overall from 2002 to 2009 to be 12.5% of the samples collected. This included samples collected from complete animal feeds, feed ingredients, pet foods, pet treats, and supplements for pets. Further breakdown of contamination levels in the feed ingredients was separated by category, animal derived and plant derived. Of the contamination found from 2002 to 2006, 66.1% was from animal-derived ingredients while only 11.0% was from plant-derived ingredients. The prevalence of *Salmonella* contamination in animal- and plant-derived ingredients was found to be further reduced from 2007 to 2009, to 41.3% and 10.6%, respectively (Li et al., 2012). Before the surveillance samples collected in the 2000s, the FDA CVM 1994 survey indicated 82% of animal-derived ingredients and 37% of plant-derived ingredients tested were positive for *Salmonella* (McChesney, 1995). Although the reduction in *Salmonella* contamination in various animal feeds and feed ingredients has declined greatly from 1994 to 2009, there is still potential for human illness. Although overall *Salmonella* prevalence is on the decline, it should be noted that the FDA CVM survey in 1994 also indicated that when feed ingredients positive for *Salmonella* were included in complete feeds, those finished feeds were found to be *Salmonella* positive 30% of the time (McChesney, 1995). This would indicate that even reduced prevalence in ingredients can still lead to finished feed contamination.

Food-producing animals have been identified as major reservoirs for many human pathogens, including *Campylobacter* spp., *S. enterica* serotypes, *E. coli* (including O157:H7), *Yersinia* spp., and *Enterococcus* spp. (Mead et al., 1999). However, the surveillance data associated with these pathogens in animal feeds and feed ingredients are far more limited compared with *Salmonella*. A retrospective study by Ge et al. (2013) of samples randomly collected by the FDA in 2002 and 2003 found 39.3% and 86.6% of the 201 samples collected were positive for *E. coli* and *Enterococcus*, respectively. Further qualification of those positive samples was completed and positives were determined to be generic, nonpathogenic *E. coli* (not O157:H7) and generic *Enterococcus*. It was also determined that plant by-products were more frequently contaminated with nonpathogenic strains of *E. coli* and *Enterococcus* compared with *Salmonella*. More specifically, plant-based ingredients had contamination rates of 43% and 89.9% for *E. coli* and *Enterococcus*, respectively. Animal-based ingredients yielded contamination rates of 36.9% and 84.4% for *E. coli* and *Enterococcus*, respectively. *Salmonella* prevalence was 5.1% and 34.4% in plant- and animal-based products, respectively. In addition, all samples collected (201) were negative for *Campylobacter* spp. (Ge et al., 2013).

Again, prevalence studies on animal feed contamination with *E. coli* O157:H7 are limited, with the limited isolation detected in cattle feeds (Dodd et al., 2003; Ge et al., 2013; Davis et al., 2003). Methodology used in prevalence studies may play a role in detection of *E. coli* O157:H7. A comparison of different enrichment medias resulted in detection of *E. coli* O157:H7 in 14.9% of samples, which was significantly higher than that previously reported and would also indicate that more sensitive detection methods may need to be used (Dodd et al., 2003). Further linking animal feed as a vector to animal colonization, research by Davis et al. (2003) found pulsed field gel electrophoresis profiles from *E. coli* O157:H7 isolated from feed closely resembled that from a fecal isolate collected at the same farm. Although feed may contribute to animal colonization of *E. coli* O157:H7, other sources of contamination are also likely. These routes include fecal contamination of feed bunks by other livestock animals or shedding cattle within the pen along with fecal contamination by wild animals (birds, rodents, insects) (Dodd et al., 2003). Independent of how *E. coli* O157:H7 is introduced to the cattle, infected cattle have been identified to significantly increase the risk of transmission of *E. coli* O157:H7 into the human food supply during animal slaughter (LeJeune et al., 2004). To help mitigate this risk, animal slaughter facilities have implemented Hazard Analysis of Critical Control Points procedures during animal harvest (9 CFR 417).

### Food Safety Modernization Act

Because of the risk to human and animal health along with potential economic losses, the FDA has implemented the FSMA. FSMA has shifted the focus of food and feed safety from reactive to proactive. This proactive approach relies on animal feed manufacturers to identify and prevent potential biological hazards from being introduced into finished feeds and pet foods. These biological hazards include bacterial pathogens such as *Salmonella*, pathogenic *E. coli*, and viral pathogens such as PEDV. Furthermore, the FDA Guidance for Industry sets-forth *Salmonella* tolerance policies. This includes a zero-tolerance policy for *Salmonella* of any serotype in pet foods. On the livestock and poultry side, only *Salmonella* serovars considered pathogenic to a specific animal species will have a zero-tolerance level. As an example, *Salmonella* serovar Choleraesuis is pathogenic to swine and therefore considered an adulterant if found in finished swine feed.

## Feed Mill Biosecurity

Biosecurity can be defined as "the implementation of measures that reduce the risk of agents being introduced and spread" (FAO, 2010). This concept of biosecurity as it applies to the production of food animals is not new. As an example, the swine industry in general relies on the implementation of biosecurity protocols to help prevent the introduction of disease to a herd. Implementation of biosecurity protocols for a swine herd is multifaceted and may include facility location and design; control of movement of hogs and people; segregation/isolation practices; feed and waste management; and sanitation (Seaman and Fangman, 2001). Although feed-related biosecurity has traditionally occurred at the farm level, implementation of biosecurity protocols at feed manufacturing facilities can further help reduce the risk of pathogen contamination into the human food supply. Implementation of biosecurity plans at pet food manufacturers can also reduce the risk to human infection from handling contaminated kibbles. As previously mentioned, bacterial pathogens such as *Salmonella* have been the common contaminants in animal feeds and pet foods. Introduction of PEDV into the United States in 2013 further emphasized the devastating effects feed-borne pathogens can have on animal health and subsequent economic losses. Before the introduction of PEDV in the United States, there had been limited demonstrations of viral transmission through feed. However, transmission of bacterial pathogens, specifically *Salmonella*, has been well documented in livestock feeds and pet foods. With increased regulation (FSMA), identification of pathogens, such as *Salmonella*, will result in the contaminated feed to be deemed adulterated and not fit for consumption. These stringent regulatory protocols and increased consumer awareness will require feed mill and pet food manufacturing facilities to implement strict policies for preventing contamination. Extending effective biosecurity protocols to off the farm can improve the safety to consumers. Feed mill biosecurity will play an important role to help control the spread of feed-borne pathogens.

As previously mentioned, the prevalence of *Salmonella* in finished animal feeds and feed ingredients has been extensively evaluated. Unfortunately, prevalence analysis has been focused on evaluating finished feeds and feed ingredients and did not include the collection of environmental samples throughout feed manufacturing facilities. Although identification of high-risk ingredients is important for implementation of preventative plans, contaminated ingredients can potentially spread pathogens throughout the facility. Samples collected from feed mills and on-farm processing equipment in the United Kingdom found that dust, spillage, and aggregated materials at all stages of the operation, including intake pits, ingredient silos and storage areas, transfer augers and elevators, weighing and mixing vessels, finished product bins, and load-out gantries, could harbor pathogens and specifically *Salmonella* in this case (Davies and Wales, 2010). According to research by Jones and Richard (2004), contaminated dust can spread *Salmonella* throughout a feed manufacturing facility. It was confirmed that dust samples were always more heavily contaminated with *Salmonella* and Enterobacteriaceae than feed samples collected from the same area. Contamination of dust collected from pellet coolers were found to be the highest compared with dust collected from ingredient receiving, the mixer, the pellet mill, and load-out. The authors attributed this high level of contamination in the pellet cooler to the increased volume of air pulled through the cooler during operation increasing the likelihood of contamination from surrounding dust and the accumulation of condensation within the cooler allowing pathogen growth (Jones and Richardson, 2004).

To further validate the potential of contaminated dust and aggregate materials to cause facility contamination, a controlled study using *Enterococcus faecium* as a surrogate for *Salmonella* has been reported. A pilot-scale feed manufacturing facility, the Feed Safety Research Center (FSRC) at Kansas State University, specifically designed for the containment and evaluation of pathogen mitigation during the manufacture of feeds, was used for this evaluation. The FSRC allows for containment operations transforming the pilot-scale feed mill into a biosafety level 2 laboratory. During containment operations, a separate air handling system with high-efficiency particulate arrestance filtration and biosafety protocols restrict pathogen exit during operation of equipment. Under containment operations, a standard mash swine diet was inoculated with *E. faecium* in a mixer and pelleted using a pellet mill. Environmental swabs and replicate organism detection and counting plates were collected before manufacture of the contaminated feed (baseline) and immediately after production. Baseline samples indicated the presence of Enterococci in 37.2% of the samples collected throughout the facility. This was not surprising as *Enterococcus* is a ubiquitous organism in nature. After the production of the inoculated feed, 83.7% of the collected samples were found to be positive for Enterococci, a 46.5% increase in contamination levels. A breakdown in the locations of contamination after manufacture indicated that those surfaces with direct contact with the inoculated feed (zone 1) had the highest increase in contamination rates, with 100% of sample locations positive. Surfaces directly adjacent to direct contact surfaces (zone 2) had the next highest level of contamination (83.3%) and all other surfaces (zone 3) the least amount of contamination (76%), although an increase in contamination of 24% was detected (Huss et al., 2015).

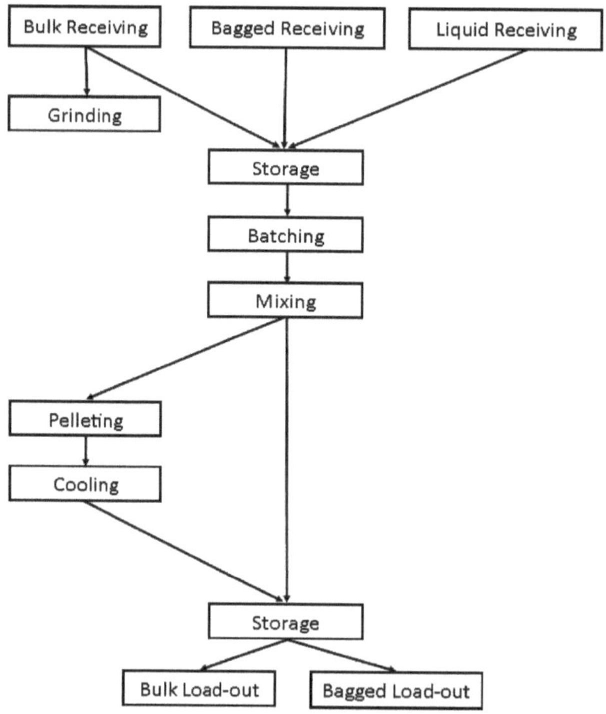


**Figure 4.1**
Flow diagram example of livestock feed manufacturing facility.

## *Hazard Analysis*

### *Identification of Ingredients and Processes*

The first step in developing a feed mill biosecurity plan is to identify all ingredients and processing steps. This is most easily accomplished by creating a flow diagram of all major manufacturing processes within the feed mill (Fig. 4.1). From the flow diagram, potential points of hazard entry and control can easily be identified for consideration into the biosecurity plan. Although facility design of feed mills can vary, common categories in the diagram include receiving of raw ingredients, processing of ingredients, storage of ingredients and finished products, packaging of finished products, loading and delivery of finished goods. Although conveying systems have been identified as an area of high risk, inclusion in the flow diagram may complicate the objectives of the flow diagram (Cochrane et al., 2016a,b; Gebhardt et al., 2016; Schumacher et al., 2016a,b).

### *Hazard Identification*

Potential hazards associated with each ingredient and process should be identified. The list of potential hazards should be as comprehensive as possible with all known or reasonably foreseeable hazards identified for each ingredient and process. FSMA regulations have determined that specific criteria must be considered during hazard identification. This can include product formulation, condition, function, design of facility and equipment, ingredients, transportation, processing procedures, packaging and labeling activities, storage and distribution, intended or reasonably foreseeable use of the feed, sanitation, and other relevant factors (Federal Register 21 CFR 507). As previously mentioned, the FDA has put forth a draft list of potential biological hazards but also recommends considering scientific literature, FDA recalls, and the Reportable Foods Registry to determine the severity and likelihood of identified biological hazards.

### *Hazard Evaluation*

Hazard evaluation requires the hazard severity and probability to be determined. Again, the hazard severity and probability will vary by facility because multiple variables can change from one feed mill to another. Based on the combination of the hazard's severity and probability, a threshold will be set to determine if control of the hazard is required. For example, a feed mill that manufactures feed for both dairy cattle and poultry layers would consider *Salmonella* serovar Enteritidis to be a potentially significant hazard due to the pathogenicity in poultry; while a feed mill manufacturing feed only for dairy cattle may not consider it a significant hazard due to lack of pathogenicity in dairy cattle.

According to FSMA, the determination of severity must take into account the severity of illness or injury if the hazard were to occur. For example, hazards that would cause short-term injury or illness in a single animal would likely have a lower severity assessment than a hazard that would have the potential to cause widespread injury, illness, and/or mortality. Other factors outside the impact of injury or illness severity and the potential number of animals affected can be considered in the overall severity assessment of the hazard. Again, this will vary by facility but may include the value of the animal to the production system or the potential for the hazard to impact human health (Cochrane et al., 2016a,b).

The probability of the hazard to occur must also be taken into consideration during hazard evaluation. Individuals must "assess the probability that the hazard will occur in the absence of preventative controls" according to FSMA regulations (Federal Register 21 CFR 507). Determination of probability can be assessed from FDA recalls or the RFR, as well as facility history.

After hazard severity and probability identification, the resultant combination will determine if specific controls must be set in place. According to FSMA regulations, stringent management components such as monitoring, validation, verification, corrective action, and a recall must be in place for preventative controls.

## *Hazard Mitigation*

After identification and evaluation of hazards, multiple strategies can be implemented to reduce the probability of the biological hazard. Hazard mitigation can include the prevention of hazard entry during ingredient receiving, prevention of hazard entry due to people, prevention of hazard cross-contamination during production, prevention of cross-contamination during load-out and delivery, and proactive reduction of biological hazards.

One of the most effective methods a biosecurity plan can include is the prevention of hazard entry during receiving of all ingredients. Research has shown that entry of contaminated materials into the facility can lead to continued contamination throughout the facility for an extended period (EFSA, 2008).

Entry of hazards via people is one of the most overlooked areas when it comes to implementing a biosecurity plan. Feed mill employees and visitors, including guests, truck drivers, and subcontractors, can inadvertently introduce contaminants into the facility. Most people unknowingly carry contaminated fecal, dirt, or dust on their shoe bottoms and clothing. Foot traffic from individuals who have recently been on a farm or in a feed mill where the hazard is present are at higher risk of tracking the hazard into the facility. Not only is it important to control the food traffic of nonfeed mill employees, monitoring and segregation of feed mill employees should be considered. Hygienic zoning, or preventing employees from going from raw ingredient areas to finished product areas, can help reduce the risk of spreading contamination throughout a facility.

Prevention of hazard cross-contamination during production can also be considered as part of the biosecurity plan. As previously mentioned, contaminated organic material, such as aggregate material and dust, can result in the spread of contamination. This contamination can also persist over time. The accumulation of organic material in processing equipment can lead to cross-contamination of subsequent batches or runs. Because of the design of various types of equipment, including screw conveyers, interiors of coolers and storage bins, and boot pits of bucket elevators, removal of all organic material is not feasible or possible and can lead to increased risk of hazard cross-contamination (Cochrane et al., 2016a,b). To minimize the risk of carryover cross-contamination, flushing and sequencing schedules should be implemented as part of the biosecurity plan. As stated in the FDA Compliance Program Guidance Manual, flushing is defined as "the process of running an ingredient, usually an abrasive-type material such as corn, soybean meal, peanut hulls, etc., through the manufacturing equipment and associated handling equipment after the production of a batch of feed, for the purpose of cleaning out any drug residue." Sequencing is defined as "the preplanned order of production, storage, and distribution of different animal feeds designed to direct drug carryover into subsequent feeds which will not result in unsafe contamination (FDA Compliance Program Guidance Document). To reduce the risk of batch-to-batch drug carryover, feed mills have been using flush and sequence protocols. These concepts can be applied to biological hazards, with the caveat that the efficacy has limited published research. Recent research by K-State has shown limited reduction in PEDV and *Salmonella* contamination in subsequent flush batches after the manufacture of inoculated feed. Both flushing and sequencing should be considered risk-reduction processes and not risk-elimination steps (data not published).

The establishment of a scheduled sanitation and housekeeping program can also help prevent cross-contamination of biological hazards (Cochrane et al., 2016a,b). Collection of aggregate materials and dust from production areas, high-traffic areas, and areas close to feed-contact surfaces can help minimize the spread of potentially contaminated materials. It should also be noted that the use of physical cleaning with air or aerosolizing dust materials can serve to spread contaminated materials to areas not previously contaminated (Cochrane et al., 2016a,b; Huss et al., 2015; Davies and Wales, 2010; Jones and Richardson, 2004). It should also be emphasized that collected aggregate material and dust should be disposed of and not added back into the manufacturing process. This also includes dust from air-collection systems that have been traditionally fed directly back into the feed manufacturing system. All dust particles should be considered high risk and disposed of properly (Cochrane et al., 2016a,b; Huss et al., 2015). Recent research evaluating environmental contamination when manufacturing PEDV-, *Salmonella*-, or *Enterococcus*-infected feed suggests that potentially infective/contaminated dust particles can be dispersed throughout the entire feed manufacturing area (Schumacher et al., 2016; Huss et al., 2015, 2017).

Many times, outbound feed delivery trucks have residual feed or ingredients on exterior, top, and interior compartments. If this residual material is infectious, it could lead to cross-contamination of finished feeds during load-out and delivery. Although inspection of all feed delivery vehicles may be impractical, documentation should be maintained of previous load hauled, shipment lot numbers and locations, and time of loading to improve traceability (Cochrane et al., 2016a,b). If a feed delivery truck is required to make multiple stops with a single load, delivery should be coordinated so feed is delivered to lower risk farms before higher risk farms. The incorporation of truck washes, thermos-assisted decontamination drying, and sanitation methods for feed trucks may also be effective at reducing the risk of cross-contamination of the feed mill, feed, and farms (PIC, 2014).

As mentioned previously, foot traffic can lead to contamination issues and spread contamination. Drivers of feed delivery vehicles should also be taken into consideration when developing a biosecurity plan. Drivers should be instructed to remain in the vehicle when out delivering feed and while loading at the feed mill, whenever possible. Oftentimes, the environment around feed bins on the farm can be unsanitary and contain spilled feed, grain, or other materials. If the driver or truck comes into contact with this material, there is the potential for contaminated material to be transferred back to the feed mill. Ideally, an on-site worker should open all bin lids for feed delivery. If it is not possible for the driver to remain in the vehicle, the driver should wear shoe covers or boots on exiting the vehicle and remove soiled shoe covers/boots on re-entering the vehicle. Drivers should also be instructed to never enter animal barns or have direct contact with animals or their waste materials. Drivers should also avoid areas near exhaust fans, dead animal disposal sites, and livestock contact areas (chutes) to decrease risk of contact with potentially infective materials (Cochrane et al., 2016a,b; PIC, 2014).

A feed mill biosecurity plan will not eliminate all contaminations into a facility, but it can greatly reduce the risk. Aside from preventative procedures, feed mills can proactively mitigate contamination that has entered the facility. This can be accomplished via thermal processing or the introduction of various chemical additives. Typically, pelleting and extrusion of feeds results in high processing temperatures. These high temperatures have been demonstrated to eliminate many biological pathogens, including PEDV and *Salmonella* (Jones and Richardson, 2004; Jones, 2011; Veldman et al., 1995; Stott et al., 1975; Cochrane et al., 2017). Although pelleting or extrusion may be effective at reducing or eliminating pathogens in feed, thermal processing is a point-in-time mitigation step. As pointed out, postprocessing contamination can still occur, thus resulting in recontamination of feed. Many chemical additives have been identified to reduce the risk of pathogen cross-contamination. These additives include formaldehyde, medium chain fatty acids, organic acids, and some essential oils (Cochrane et al., 2016a,b). The efficacy of these additives has been found to vary depending on the feed matrix applied to and inclusion level.

## Conclusions

The very nature of the ingredients that constitute a finished feed can lead to the introduction of various biological hazards. For a feed mill to effectively reduce the risk of finished feeds from being contaminated with a biological hazard, a thorough biosecurity plan must be implemented and followed. The development of an effective and efficient biosecurity plan along with a thorough hazard analysis consists of identification and evaluation of all ingredients and processes followed by hazard mitigation. Hazard mitigation can occur in a multitude of different ways, including thermal processing, inclusion of chemical additives, control over employee flow throughout the facility, preventing the entry of guest, subcontractors, and truck drivers from entering high-risk areas or the facility in general, and the implementation of scheduled sanitation and general good housekeeping. Although a feed mill biosecurity plan cannot eliminate biological hazards from contaminating finished feeds, it can greatly help manage the risk.

## References

Alonso, C., Goede, D.P., Morrison, R.B., Davies, P.R., Rovira, A., Marthaler, D.G., Torremorell, M., 2014. Evidence of infectivity of airborne porcine epidemic diarrhea virus and detection of airborne viral RNA at long distances from infected herds. Veterinary Research 45, 73–78.

Aviles, B., Klotz, C., Smith, T., Williams, R., Ponder, M., 2013. Survival of *Salmonella enterica* serotype Tennessee during simulated gastric passage is improved by low water activity and high fat content. Journal of Food Protection 76 (2), 333–337.

Behravesh, C.B., Ferraro, A., Deasy, M., Dato, V., Moll, M., Sandt, C., Rea, N.K., Rickert, R., Marriott, C., Warren, K., Urdaneta, V., Salehi, E., Villamil, E., Ayers, T., Hoekstra, R.M., Austin, J.L., Ostroff, S., Williams, I.T., Schwarzengrund outbreak investigation Team, 2010. Human *Salmonella* infections linked to contaminated dry dog and cat food, 2006–2008. Pediatrics 126 (3), 477–483.

Blaser, M.J., Newman, L.S., 1982. A review of human salmonellosis. I. Infective dose. Reviews of Infectious Diseases 4, 1096–1106.

Burnett, S.L., Gehm, E.R., Weissinger, W.R., Beuchat, L.R., 2000. Survival of *Salmonella* in peanut butter and peanut butter spread. Journal of Applied Microbiology 89, 472–477.

Capucille, D.J., Moore, M.H., Rogers, G.M., 2004. Growing and finishing performance of steers when fed recycled poultry bedding during the growing period. Journal of Animal Science 82, 3038–3048.

Clark, G.M., Kaufmann, A.F., Gangarosa, E.J., Thompson, M.A., 1973. Epidemiology of an international outbreak of *Salmonella* Agona. Lancet 2, 490–493.

Cochrane, R.A., Dritz, S.S., Woodworth, J.C., Stark, C.R., Huss, A.R., Cano, J.P., Thompson, R.W., Fahrenholz, A.C., Jones, C.K., 2016a. Feed mill biosecurity plans: a systematic approach to prevent biological pathogens in swine feed. The Journal of Swine Health and Production 24 (3), 154–164.

Cochrane, R.A., Huss, A.R., Aldrich, G.C., Stark, C.R., Jones, C.K., 2016b. Evaluating chemical mitigation of *Salmonella* Typhimurium ATCC 14028 in animal feed ingredients. Journal of Food Protection 79 (4), 672–676.

Cochrane, R.A., Schumacher, L.L., Dritz, S.S., Woodworth, J.C., Huss, A.R., Stark, C.R., DeRouchey, J.M., Tokach, M.D., Goodband, R.D., Bai, J., Chen, Q., Zhang, J., Gauger, P.C., Derscheid, R.J., Magstadt, D.R., Main, R.G., Jones, C.K., 2017. Effect of pelleting on survival of porcine epidemic diarrhea virus-contaminated feed. Journal of Animal Science 95, 1170–1178 .

Compliance Policy Guide Sec. 690.800 *Salmonella* in Food for Animals, 78 Fed. Reg. 42526.

21 CFR 507. Current Good Manufacturing Practice, Hazard Analysis, and Risk-Based Preventative Controls for Food Animals, 2015. .

Davies, R.H., Wales, A.D., 2010. Investigation into *Salmonella* contamination in poultry feedmills in the United Kingdom. Journal of Applied Microbiology 109 (4), 1430–1440.

Davis, M.A., Hancock, D.D., Rice, D.H., Call, D.R., DiGiacomo, R., Samadpour, M., Besser, T.E., 2003. Feedstuffs as a vehicle of cattle exposure to *Escherichia coli* O157:H7 and *Salmonella enterica*. Veterinary Microbiology 95 (3), 199–210.

Dee, S., Clement, T., Schelkopf, A., Nerem, J., Knudsen, D., Christopher-Hennings, J., Nelson, E., 2014. An evaluation of contaminated complete feed as a vehicle for porcine epidemic diarrhea virus infection of naïve pigs following consumption via natural feeding behavior: proof of concept. BMC Veterinary Research 10, 176–179.

Denton, J.H., Coon, C.N., Pettigrew, J.E., Parsons, C.M., 2005. Historical and scientific perspectives of same species feeding of animal by-products. Journal of Applied Poultry Research 14, 532–634.

Dodd, C.C., Sanderson, M.W., Sargeant, J.M., Nagaraja, T.G., Oberst, R.D., Smith, R.A., Griffin, D.D., 2003. Prevalence of *Escherichia coli* O157 in cattle feeds in Midwestern feedlots. Applied and Environmental Microbiology 69 (9), 5243–5247.

European Food Safety Authority, 2008. Microbiological risk assessment in feedstuffs for food-producing animals. Scientific opinion of the panel on biological hazard. The EFSA Journal 720, 1–84.

FAO and IFIF, 2010. Good Practices for the Feed Industry—Implementing the Codex Alimentarius Code of Practice on Good Animal Feeding. FAO Animal Production and Health Manual No. 9, Rome.

Federal Register, Hazard analysis and critical control point (HACCP) systems. 9 CFR. Part 417.

Federal Register, 2015. Current good manufacturing practice, hazard analysis and risk-based preventative controls for food for animals. 21 CFR. Part 507.

Federal Register, 2016. Current good manufacturing practice, hazard analysis and risk-based preventative controls for food for animals. 21 CFR. Part 113.

Food and Agriculture Organization of the United Nations (FAO), 2009. How to Feed the World in 2050. http://www.fao.org/fileadmin/templates/wsfs/docs/expert_paper/How_to_Feed_the_World_in_2050.pdf.

Food and Drug Administration, 2015. Reportable Food Registry Fourth Annual Report. fda.gov.

Franco, D.A., 2005. A survey of *Salmonella* serovars and most probable number (MPN) in rendered animal protein meals: inferences for animal and human health. Journal of Environmental Health 67 (6), 18–22.

Galtin, L.A., See, M.T., Hansen, J.A., Odle, J., 2003. Hydrogenated dietary fat improves pork quality of pigs from two lean genotypes. Journal of Animal Science 81, 1989–1997.

Ge, B., LaFon, P.C., Carter, P.J., McDermott, S.D., Abbott, J., Glenn, A., Ayers, S.L., Friedman, S.L., Paige, J.C., Wagner, D.D., Zhao, S., McDermott, P.F., Rasmussen, M.A., 2013. Retrospective analysis of *Salmonella*, *Campylobacter*, *Escherichia coli* and *Enterococcus* in animal feed ingredients. Foodborne Pathogens and Disease 10 (8), 684–691.

Gebhardt, J.T., Woodworth, J.C., Jones, C.K., Gauger, P.C., Tokach, M.D., DeRouchey, J.M., Goodband, R.D., Muckey, M., Cochrane, R.A., Stark, C.R., Bai, J., Chen, Q., Zhang, J., Ramirez, A., Derscheid, R.J., Main, R.G., Dritz, S.S., 2016. Evaluation of the Effects of Flushing Feed Manufacturing Equipment with Chemically Treated Rice Hulls on Likelihood of Porcine Epidemic Diarrhea Virus (PEDV) Transmission by Swine Feed and Feed Manufacturing Equipment. Kansas State University Swine Day. Kansas State University.

Hirsch, W., Sapiro-Hirsch, R., 1958. The role of certain animal feeding stuffs, especially bone meal, in the epidemiology of salmonellosis. Bulletin of Hygiene 33, 647.

Huss, A.R., Cochrane, R.A., Deliephan, A., Stark, C.R., Jones, C.K., 2015. Evaluation of a biological pathogen decontamination protocol for animal feed mills. Journal of Food Protection 78 (9), 1682–1688.

Huss, A.R., Schumacher, L.L., Cochrane, R.A., Poulsen, E., Bai, J., Woodworth, J.C., Dritz, S.S., Stark, C.R., Jones, C.K., 2017. Elimination of porcine epidemic diarrhea virus in an animal feed manufacturing facility. PLoS One 12, e0169612.

Jeffery, A., 2016. The Role of *Salmonella* in Animal Food. Kansas State University, Manhattan, Kan.

Jones, F.T., 2011. A review of practical *Salmonella* control measures in animal feed. Journal of Applied Poultry Research 20, 102–113.

Jones, F.T., Richardson, K.E., 2004. *Salmonella* in commercially manufactured feeds. Poultry Science 83, 384–391.

Kapperud, G., Gustavsen, S., Hellesnes, I., Hansen, A.H., Lassen, J., Hirn, J., Jahkola, M., Montenegro, M.A., Helmuth, R., 1990. Outbreak of *Salmonella* Typhimurium infection traced to contaminated chocolate and caused by a strain lacking the 60-megadalton virulence plasmid. Journal of Clinical Microbiology 28, 2597–2601.

Kinley, B., Rieck, J., Dawson, P., Jiang, X., 2010. Analysis of *Salmonella* and enterococci isolated from rendered animal products. Canadian Journal of Microbiology 56, 65–73.

Knox, W.A., Galbraith, N.C., Lewis, M.J., Hickie, G.C., H Johnston, H., 1963. A milkborne outbreak of food poisoning due to *Salmonella* Heidelberg. Journal of Hygiene 61, 175–185.

Laban, S.E., Moustafa, G.Z., Anwer, W., Badawy, E.M., 2014. Microbial load of poultry by-products following rendering process. Global Veterinaria 12 (6), 756–759.

LeJeune, J.T., Besser, T.E., Rice, D.H., Berg, J.L., Stillborn, R.P., Hancock, D.D., 2004. Longitudinal study of fecal shedding of *Escherichia coli* O157:H7 in feedlot cattle: predominance and persistence of specific clonal types despite massive cattle population turnover. Applied and Environmental Microbiology 70 (1), 377–384.

Li, X., Bethune, L.A., Jia, Y., Lovell, R.A., Proescholdt, T.A., Benz, S.A., Schell, T.C., Kaplan, G., McChesney, D.G., 2012. Surveillance of *Salmonella* prevalence in animal feeds and characterization of the *Salmonella* isolates by serotyping and antimicrobial susceptibility. Foodborne Pathogens and Disease 9 (8), 692–698.

Lowe, J., Gauger, P., Harmon, K., Zhang, J., Connor, J., Yeske, P., Loula, T., Levis, I., Dufresne, L., Main, R., 2014. Role of transportation in spread of porcine epidemic diarrhea virus infection, United States. Emerging Infectious Disease 20 (5), 872–874.

Ma, L., Zhang, G., Gerner-Smidt, P., Mantripragada, V., Ezeoke, I., Doyle, M.P., 2009. Thermal inactivation of *Salmonella* in peanut butter. Journal of Food Protection 72, 1596–1601.

McChesney, D.G., 1995. FDA Survey Results: *Salmonella* Contamination of Finished Feed and Primary Meal Ingredients. Presented at Feed Safety Committee of the United States Animal Health Association, 99th Annual Meeting, Reno, Nevada.

Mead, P.S., Slutsker, L., Dietz, V., 1999. Food-related illness and death in the United States. Emerging Infectious Disease 5, 607–625.

Meeker, D.L., Hamilton, C.R., 2006. An Overview of the Rendering Industry. National Renderers Association Website http://assets.nationalrenderers.org/essential_rendering_overview.pdf.

Meeker, D.L., Meisinger, J.L., 2015. Rendered ingredients significantly influence sustainability, quality, and safety of pet food. Journal of Animal Science 93 (3), 835–847.
Misener, M., 2014. PEDV infected spray dried porcine plasma proteins. In: Allen D. Leman Swine Conference, St. Paul, MN.
Moyle, A.I., 1966. *Salmonellae* in rendering plant by-products. Journal of the American Veterinary Medical Association 149 (9), 1172–1176.
National Renderers Association, 2014. Code of Practice. Alexandria, VA http://www.nationalrenderers.org/biosecurity-appi/code/.
Pasick, J., Berhane, Y., Ojkic, D., Maxie, G., Embury-Hyatt, C., Swekla, K., Handel, K., Fairles, J., Alexandersen, S., 2014. Investigation into the role of potentially contaminated feed as a source of the first-detected outbreaks of porcine epidemic diarrhea in Canada. Transboundary and Emerging Diseases 61 (5), 397–410.
Pennington, J.H., Brooksbank, N.H., Pool, P.M., Seymour, F., 1968. *Salmonella* Virchow in a chicken-packing station and associated rearing units. British Medical Journal 4, 804–806.
PIC North America, 2014. Feed Biosecurity Guidelines: Practical Standards for Mitigating the Risk of Swine Enterica Coronavirus Transmission via Feed. http://www.pic.com/Images/Users/1/BIOSECURITY/FeedBiosecurityGuidelinesPICmultiplicationSep302014(1).pdf.
Sapkota, A.R., Lefferts, L.Y., McKenzie, S., Walker, P., 2007. What do we feed to food-production animals? A review of animal feed ingredients and their potential impacts on human health. Environmental Health Perspectives 115 (5), 663–670.
Scallan, E., Griffin, P.M., Angulo, F.J., Tauxe, R.V., Hoekstra, R.M., 2011. Foodborne illness acquired in the United States—unspecified agents. Emerging Infectious Diseases 17, 16–22.
Schumacher, Cochrane, Evans, Kalivoda, Woodworth, Huss, Stark, Jones, Chen, Main, Zhang, Gauger, Dritz, Tokach, 2016a. Evaluating the effect of manufacturing porcine epidemic diarrhea virus (PEDV)-contaminated feed on subsequent feed mill environmental surface contamination. Journal of Animal Science 94, 77.
Schumacher, L.L., Cochrane, R.A., Woodworth, J.C., Huss, A.R., Stark, C.R., Jones, C.K., et al., 2016b. Utilizing feed sequencing to decrease the risk of porcine epidemic diarrhea virus (PEDV) cross-contamination during feed manufacturing. Journal of Animal Science 94, 76.
Seaman, J.S., Fangman, T.J., November 23, 2001. Biosecurity for Today's Swine Operation. MU Guide: MU Extension. University of Missouri-Columbia.
Semple, A.B., Turner, G.C., Lowry, D.M., 1968. Outbreak of food poisoning caused by *Salmonella* Virchow in spit-roasted chicken. British Medical Journal 4, 801–803.
Stott, J.A., Hodgson, J.E., Chaney, J.C., 1975. Incidence of *Salmonellae* in animal feed and the effect of pelleting on content of Enterobacteriaceae. Journal of Applied 39, 41–46.
Troutt, H.F., Schaeffer, D., Kakoma, I., Pearl, G.G., 2001. Prevalence of Selected Foodborne Pathogens in Final Rendered Products: Pilot Study. Fats and Proteins Research Foundation, Inc. Director's Digest #312.
Veldman, A., Vahl, H.A., Borggreve, G.J., Fuller, D.C., 1995. A survey of the incidence of *Salmonella* species and Enterobacteriaceae in poultry feeds and feed components. The Veterinary Record 136 (7), 169–172.
Zaghini, A., Martelli, G., Roncada, P., Simioli, M., Rizzi, L., 2005. Mannanoligosaccharides and aflatoxin B1 in feed for laying hens: effects on egg quality, aflatoxin B1 and M1 residues in eggs, and aflatoxin B1 levels in liver. Poultry Science 84, 825–832.

## Further Reading

AAFCO, 2015. Official Publication, Association of American Feed Control Officials, Oxford.
United States Department of Health, Human Services/United States Food and Drug Administration, 2006. Draft List of Potentially Hazardous Contaminants in Animal Feed and Feed Ingredients. Rockville, Maryland http://www.fda.gov/downloads/animalveterinary/safetyhealth/animalfeedsafetysystemafss/ucm053715.pdf.

CHAPTER 5

# Physical and Chemical Methods for the Reduction of Biological Hazards in Animal Feeds

**Anne Huss[1], Roger Cochrane[2], Cassie Jones[2], Griffiths G. Atungulu[3]**

*[1]BioMatrix International, Princeton, MN, United States; [2]Kansas State University, Manhattan, KS, United States; [3]University of Arkansas-Fayetteville, Fayetteville, AR, United States*

## Introduction

Animal feeds consist of all food produced for the consumption by livestock and poultry animals and companion animals, including dogs and cats. Animal feed is specifically formulated to ensure that the necessary balance of nutrients is provided to the animal for proper growth, development, and maintenance. The feeding of manufactured animal feeds ensures that the animal receives all the required nutrients and supplements absent from the animal's natural diet. According to the American Feed Industry Association (AAFCO, Association of American Feed Control Officials) over 900 agricultural ingredients and co-products are used to produce animal feeds and pet foods. These ingredients include cereal grains (corn, oats, wheat, etc.), oilseeds and meals (canola meal, soybean meal, etc.), and various processing by-products (corn gluten meal, distillers products, wheat bran, etc.) (http://www.afia.org/howmade). In addition to plant-based ingredients, animal-based products make up a large portion of animal feeds and pet foods. Animal-based proteins can be derived from whole cuts after slaughter or from rendering of portions not used for human consumption (Sapkota et al., 2007). Rendered animal proteins and products include the portions not consumed by humans, including bones, fat, blood, feathers, and some internal organs (Meeker and Hamilton, 2006).

The production of finished feeds ready for consumption by the animal can occur on the farm, in a feed mill, or at a pet food manufacturing facility. In all cases, raw materials are received and included in species-specific formulations. Animal diet formulations take into account the required nutrients of the animal and the nutrients provided by the raw ingredients. The manufacture of raw ingredients into complete finished feeds can include multiple processes. These processes can include grinding, mixing, pelleting, and/or extrusion. Grinding can be done through a roller mill or hammer mill. The goal of grinding is to ensure all components for the finished feed have a similar particle size for uniform distribution during further

Food and Feed Safety Systems and Analysis. http://dx.doi.org/10.1016/B978-0-12-811835-1.00005-1

processing and consumption. Following grinding, all ingredients must be mixed or blended for uniform distribution. Uniformly blended feeds can be further processed by pelleting or extrusion, both thermally intensive processes. During both pelleting and extrusion, the ground ingredients are passed through a mixing chamber and pushed through a die of a desired size. Pelleted and extruded products can increase bulk density and improve handling along with improving protein digestibility. The form of the complete finished diets can vary, from ground mash types to pelleted. Finished pet foods can also be in various forms, from kibbles to pate with gravy to jerky type.

The ingredients in complete animal feeds come from a wide range of raw materials, so it is not surprising that contamination with enteric pathogens, like *Salmonella*, occurs (Ge et al., 2013). The link between the spread of *Salmonella* from contaminated animal feed to both animals and humans causing illness has been well established (Jones, 2011; Crump et al., 2002). Along with *Salmonella*, food-producing animals have been identified as major reservoirs for many human pathogens, including *Campylobacter* spp., *Clostridium* spp., *Escherichia coli* (including O157:H7), *Yersinia* spp., and *Enterococcus* spp. (Mead et al., 1999; Crump et al., 2002). While pathogens of bacterial origin seem to be the most prevalent, viral pathogens should not be overlooked as possible hazards in animal feeds and pet foods. More recently the coronavirus, porcine epidemic diarrhea virus (PEDV) has been found to be transmitted via feed (Dee et al., 2014; Pasick et al., 2014). Although PEDV is not a human pathogen, it generated significant economic losses in North America after its introduction in 2013 (Cochrane et al., 2016a,b). In swine feeds, PEDV was the first substantial viral pathogen to be transmitted by feed and may not be the only. More recently mammalian orthoreovirus has been detected in swine feces and swine blood meal with contaminated blood meal found to be infective (Narayanappa et al., 2015). In addition to swine viruses being reported, a surge of highly pathogenic avian influenza (HPAI) was reported in many poultry flocks throughout the United States in 2015 (APHIS Report). However, no definitive proof-of-concept studies have established feed as a vector for HPAI, as of this publication.

In addition to bacterial and viral pathogens, transmissible spongiform encephalopathies (TSEs) have been identified as potential biological pathogens transmissible in feeds. TSE includes bovine spongiform encephalopathy (BSE) and chronic wasting disease (CWD). TSEs are hypothesized to be caused by an infectious proteinaceous entity called a prion. Prions are composed largely of a protease-resistant misfolding of proteins. BSE, also known as mad cow disease, was first linked to human illness of variant Creutzfeldt-Jakob disease (vCJD), a human neurodegenerative prion disease. vCJD was first described in the United Kingdom in 1995 in two teenagers. Further investigation into vCJD in humans indicated the causative prion of vCJD humans was also the causative prion of BSE in cattle. The introduction of infectious prions into the human population has been attributed to the consumption of contaminated meats. It is proposed that prions are introduced into animals via rendered animal products of diseased animals. Prions may be present in all tissues of diseased animals;

they are known to accumulate in central nervous system tissues, including skull, brains, eyes, parts of the vertebral column, spinal cord, trigeminal ganglia, and dorsal root ganglia. The aforementioned tissues with accumulated prions are referred to as specified risk materials (SRMs) and are banned from use in cattle and other ruminant feeds. However, SRMs are permitted in feeds for nonruminants, including poultry (Crump et al., 2002; Sapkota et al., 2007; Denton et al., 2005).

The prevention of biological hazards into animal feeds and pet foods has put manufacturers under increased scrutiny, especially with the recent implementation of the Food Safety Modernization Act (FSMA). One of the seven rules instituted by FDA is 21 CFR 507: Current Good Manufacturing Practice, Hazard Analysis, and Risk-Based Preventive Controls in Food for Animals. This rule must be followed by nearly all facilities that manufacture, process, pack, or hold animal food. It requires facilities to meet baseline standards for personnel, facility design, sanitation, and manufacturing conditions. In addition, facilities must identify potential hazards and evaluate if they need to be controlled to ensure the resulting animal food is safe.

## Feed Manufacturing

Outside of feed manufacturing processes, overall feed mill and farm practices can influence the safety, from a biological hazard standpoint, of complete finished feeds. Nonprocessing factors include general sanitation of the processing facility, quality attributes of raw ingredients, and other biosecurity measures of the facility. The manufacture of complete finished feeds can occur in a feed mill or on-farm. Independent of location, many of the same processing steps occur. Raw ingredients are weighed or measured, mixed for uniformity, and stored until consumption. Processing of the feed can be minimal, as seen with mash-type diets or more heavily processed through conditioning, pelleting, and extrusion. Although the total elimination of biological hazards may not be possible, feed manufacturers do have options to reduce the risk of contamination in finished feeds. These include physical and chemical methods for the reduction of contamination.

### Physical Methods

Physical methods to reduce the risk of biological hazards in finished feeds can be categorized in two ways: programs or standard operating procedures (SOPs) to prevent the contamination and physical processing methods, like thermal treatment. Although one specific program or SOP will not completely prevent the introduction of biological hazards into finished feed or a feed manufacturing facility, they can contribute to overall risk reduction. Programs and SOPs can include a feed mill biosecurity plan, following current good manufacturing practices (CGMPs), prerequisite supplier programs, sanitation, and pest management.

The implementation of a feed mill biosecurity plan will help reduce the risk of biological hazards from being introduced and spread throughout the feed manufacturing facility. Biosecurity plans are multifaceted and can help control the spread of feedborne diseases. In general, a biosecurity plan requires all aspects of the feed manufacturing process to be evaluated. This includes identifying all ingredients and process steps occurring at the feed manufacturing facility. This allows for the identification of potential points of entry into the facility, opportunities for cross-contamination within the facility, and potential points of control of biological hazards. Following biological hazard identification, all identified hazards must be evaluated for the hazard's severity and probability of occurrence within the facility. Based on the hazard's severity and probability, hazard mitigation steps can be implemented. Hazard mitigation can imply prevention of hazard entry along with potential elimination during further processing or chemical inclusion.

A major component of a feed mill biosecurity plan includes prevention of biological hazard entry into the facility. Preventative measures can range from approved suppliers to sanitation to control people within and outside of the facility. Ingredient receiving is one of the main routes of entry for biological hazards into a feed manufacturing facility. Although real-time testing of all ingredients for all potential biological hazards is not possible, implementation of preventative programs can help reduce the risk of receiving contaminated ingredients. These preventative programs include development of purchase specifications with suppliers. All safety specifications and expectations of ingredients should be clearly communicated with suppliers to ensure the safety of inbound ingredients. Verification of ingredient–supplier protocols and on-site manufacturing facility can also be implemented. Once specifications are in place, it is important to maintain enforcement with routine audits when possible. Ensuring suppliers follow CGMPs can also be another barrier in reducing contamination risks.

The process of ingredient receiving is also at an increased risk of entry of biological hazards. At most facilities, multiple ingredients are received in a day and by multiple inbound trucks. All drivers and visitors should be instructed of appropriate security measures with an emphasis on reducing the amount of foot traffic in and out of the facility and even within the facility. Proper biosecurity plans should address the activities of inbound drivers during receiving. If at all possible, all drivers should not exit the vehicle during delivery. If that is not possible, disposable foot covers for boots should be provided to drivers to reduce the risk of transferring hazards from footwear. While not applicable at all facilities, covers over ingredient pits should remain in place until the truck is ready to unload. In addition, all trucks entering the facility for pit unloading should have all mud and debris removed prior to entering. Documentation is also a key component. When possible, all pertinent information regarding ingredients being delivered should be collected. This includes date, time, and lot number. Additionally, inbound truck information should be collected, including previous loads and the clean-out/disinfection schedule. Other precautions when receiving bulk ingredients through a pit include the use of cone or funneling devices to limit the quantity of material spilled during unloading and prevent spilled material from being swept into the ingredient pit. All spillage should be discarded and not swept into the pit as the area around the pit can easily be contaminated. Special precautions should also be made when receiving high-risk ingredients.

Ingredient receiving in general is a high-risk activity, but movement of people is also a high risk for introduction of contamination and often overlooked. Everyone who enters the feed manufacturing facility, including workers, drivers, subcontractors, and visitors, have the ability to inadvertently carry in biological hazards on clothing and footwear. This risk is even higher if the person has recently been on an animal farm or other feed manufacturing facility. As mentioned previously, drivers should remain in the vehicle whenever possible or wear disposable boot covers. Areas of high risk, like receiving, should be restricted from nonemployee access when possible. Additionally, the foot traffic of mill employees should be monitored. When and where possible, those workers responsible for handling raw ingredients should avoid finished production areas and vice versa. Segregating employees by stage of production can greatly reduce the risk of inadvertent cross-contamination from footwear and clothing.

In the event where biosecurity or other programs fail, biological hazards that do enter the facility can be very difficult to remove from the facility and lead to cross-contamination of finished products (EFSA, 2008). The spread of contamination throughout the facility can occur due to the spread of contaminated dust, foot traffic of employees and visitors, or movement of equipment. Cross-contamination can also occur within equipment that is considered clean. Many times, residual organic material can remain on equipment after production of a product has been completed. Equipment with the highest risk of cross-contamination includes screw conveyers, pellet coolers, storage bins, and bucket elevator boot pits (Cochrane et al., 2016a,b). Residual material found in these high-risk types of equipment is primarily due to their design. The use of flushing and sequencing of diets has been suggested as a method to help reduce the risk of cross-contamination from occurring. Flushing typically consists of running an ingredient, usually an abrasive type of material, through the system between batches to flush out any residual material. Sequencing requires the order of production, storage, and distribution to be preplanned to reduce the carryover of high-risk ingredients to sensitive diets. Flushing and sequencing have been utilized for years by the feed manufacturing industry to reduce the risk of batch-to-batch drug carryover. More recently, these principles have been applied to biological hazard mitigation but have limited reports of efficacy in the literature.

Sanitation practices within the facility can also greatly influence cross-contamination issues during production. Traditionally, floor sweepings from production areas have been added back to feed. Dust and aggregate materials that accumulate in production areas should be properly disposed of and not added back into the feed production system. This also includes all organic material collected from dust collection systems, all floor sweepings, and aggregate materials that accumulate on or around equipment. All dust and collected materials should be considered high risk and disposed of accordingly (Cochrane et al., 2016a,b).

Another avenue of cross-contamination that can occur at a feed manufacturing facility is during loadout and delivery of finished feeds. The principle sequencing used in feed production can also be used during storage, loadout/packaging, and delivery of finished feeds. During the loadout of delivery trucks, cross-contamination could occur due to residual material on truck exteriors and interiors. All organic material on delivery trucks should be removed prior to loadout at the feed mill. In addition, all delivery trucks should maintain a log of shipments to improve traceability. As previously mentioned, truck drivers should remain in the vehicle when possible to avoid tracking potential contaminated materials on footwear. When drivers are required to exit the vehicle, shoe covers should be worn and hands sanitized before reentering the truck. Drivers should also avoid contact with animals and avoid areas around exhaust fans, animal disposal sites, and livestock contact areas, like loadout chutes (Cochrane et al., 2016a,b).

Aside from biosecurity plans and preventative measures, other physical methods of mitigating biological hazards include irradiation and thermal processing. Irradiation using gamma, ultraviolet (UV), or electron beam radiation has been demonstrated to reduce PEDV load by 3-log (Trudeau et al., 2016). Wilson et al. (2015, 2016) demonstrated the feasibility of using infrared radiation to decontaminate raw pet food ingredients; depending on initial moisture content of the ingredients, infrared heating intensity, and tempering duration, up to 3-log of mycotoxigenic fungi was reduced.

Thermal processing of food for humans and animals is commonly used as a method for mitigation of biological hazards. Thermal processing of animal feeds is traditionally done via pelleting, whereas pet foods are commonly produced via extrusion. Thermal processing parameters, including temperature and time, greatly influence the lethality of processing on biological hazards (Jones, 2011). In addition to time and temperature parameters, the feed matrix can also influence biological hazard lethality. It is important to note that thermal processing of animal and pet foods can greatly reduce or eliminate biological hazards but is only a point-in-time strategy and does not reduce or eliminate the possibility of postprocessing contamination.

Pelleting of animal feeds includes three stages: mixing of steam with the mash feed (conditioning), pressing of the conditioned feed through the pellet die (pelleting), and cooling of finished pellets in a pellet cooler. The culmination of these events produces the final pelleted product. During conditioning, temperatures and residence time can be modified based on feed formulation, as some ingredients are heat labile. Other factors influencing the pelleting process include equipment design and maintenance, equipment operating procedures, supporting equipment (steam generator), and environmental conditions. Although most processes can be controlled, not all parameters can be precisely manipulated.

During the conditioning step of pelleting, heat and moisture of the mash are increased and can reduce the microbial load (Jones, 2011). Temperature, residence time at a given temperature, and moisture are the major factors influencing the microbial lethality during the conditioning process. Estimates of reduction of *Salmonella* in animal feed indicate roughly $10^3$ colony-forming units (cfu) per 100 g are destroyed starting at temperatures of 71°C (Stott et al., 1975). Both time and temperature are crucial for reduction or elimination of biological pathogens, like *Salmonella*. Several researchers have determined target temperatures of 80–85°C during conditioning to destroy *Salmonella* (Jones and Richardson, 2004; Veldman et al., 1995). More recently the minimum temperature for destruction of PEDV was noted to be at 54°C during pelleting. Pelleting temperatures of 38 and 43°C were found to not sufficiently destroy PEDV and led to infection in a piglet bioassay (Cochrane et al., 2015a,b). The residence time during conditioning can range from seconds to minutes, depending on the equipment involved. However, with the development of new equipment, conditioning temperatures and times have been enhanced to improve lethality to microbial contaminants (Jones, 2011). The use of expanders in the pelleting process have increased conditioning temperatures to 115–125°C and pressures up to 1200 psi for periods of 10–20 s. The reported reduction in bacterial load with the use of expanders was $10^5$ to $10^6$ cfu/g (Francher et al., 1996). In addition to temperatures reached during conditioning, the physical action of the pellet mill rollers pushing the conditioned feed through the die results in a temperature increase due to friction. This increase in temperature is a minimal amount of 1–2°C. This minimal increase in temperature likely has no impact on destruction of microorganisms as the increased temperature is applied to feed for a fraction of a second (Jones, 2011). The final stage of the pelleting process, cooling, removes heat and moisture via large volumes of air. Air temperature and flow must be closely monitored to ensure proper cooling of pellets. Condensation can form within the pellet cooler when the temperature of the top and walls of the cooler drops below the dew point of the entering air. Generated condensation can create an environment conducive for *Salmonella* or other biological hazard growth (Jones, 2011).

Pet food kibbles are typically manufactured through the process of extrusion, rather than pelleting. Although pelleting and extrusion are similar, extruded diets typically have higher moisture contents and subjected to higher shear and pressure levels. During extrusion, feed is guided by a single or double screw through a barrel with increasing restrictions and forced through a die. This process generates heat and pressure. Sudden expansion of the product also occurs as the product exits the die due to the pressure differential generated. Feed composition, moisture content, processing temperature and pressure, and retention time can all influence product expansion at the die exit. Studies validating extrusion as a kill step for *Salmonella* or other pathogens are limited. For process validation, the specific production equipment should be used to conduct validations. Benchtop or pilot-scale equipment may vary in replication of industrial processes. Generally, industrial or production-scale equipment is not available for controlled validation studies with pathogens. Introduction of pathogens into a pet food manufacturing facility for validation would have a significant impact on food/feed safety. To compensate for the inability to use actual pathogens for testing, surrogate microorganisms that pose no risk to human or animal health can be used. However, not all microorganisms behave in a similar fashion, so translation of processing temperatures and time from the surrogate to actual pathogen may not be correct. To date, *Enterococcus faecium* has been used as a *Salmonella* surrogate during processing of almonds, extrusion of a balanced carbohydrate-protein meal, and a poultry feed (Jeong et al., 2011; Bianchini et al., 2012; Cochrane et al., 2015a,b). Some have also suggested the use of microorganisms with a higher heat tolerance to ensure process robustness. However, the conditions to achieve inactivation of heat-resistant organisms can be misleading and much higher than needed for pathogen destruction or within reason for the feed and feed industries (Okelo et al., 2006, 2008). Of the studies that have evaluated the inactivation of pathogens during extrusion, the temperature and retention times are varied. Evaluation of *Clostridium sporogenes* in a mixture of mechanically deboned turkey and white corn flour by twin-screw extrusion resulted in a 4- to 5-log reduction at 115.6°C and a 2-log reduction at 93.3°C (Li et al., 1993). Inactivation of *Streptococcus thermophiles* in whey protein isolate occurred in twin-screw extrusion with a 4.2-log reduction at 143°C and a 4.9-log reduction at 135°C (Queguiner et al., 1989). Reduction of *Salmonella* in controlled extrusion studies has been variable. Temperatures of 93.3–176.8°C were found to completely eliminate *Salmonella* in feed samples with 25%–35% moisture (Crane et al., 1973). Other studies revealed that the elimination of *Salmonella* serovar Typhimurium occurred at 83°C in feed with 28.5% moisture (Okelo et al., 2006). As with pelleting, extrusion offers only a point-in-time mitigation and does not prevent finished product cross-contamination following processing.

### Chemical Methods

While the goal of implemented biosecurity and preventative programs is to prevent the contamination of biological hazards into feed, however, this is also not possible. Physical methods, like irradiation and thermal processing, have the capacity to mitigate biological hazards once they are present but only offer a point-in-time mitigation. Following finished feed irradiation or thermal processing, cross-contamination at multiple points can lead to contamination. The inclusion of chemical additives, either alone or in combination with other mitigation techniques, has become of more interest and may help decrease the risk of cross-contamination. Common chemical additives include organic acids and formaldehyde, with more recent research evaluating essential oils, medium chain fatty acids, and acidulants like sodium bisulfate.

The most common organic acids incorporated into feed include propionic, formic, lactic, and acetic acids. All these organic acids have been shown to be effective at reducing *Salmonella* in various feed and food matrices (Amado et al., 2013; Anang et al., 2007; Koyuncu et al., 2013; Menconi et al., 2013). Propionic acid has been shown to destroy 90% of the cell population within 1 h and formic acid within 3 h of treatment (Cherrington et al., 1991). The combination of propionic and formic acids in a blend evaluated by Carrique-Mas et al. (2007) performed similarly to a combination of lactic, propionic, formic, and benzoic acids by Cochrane et al. (2016a,b) in the mitigation of *Salmonella* in feed. The proposed mode of action of organic acid treatment to mitigate *Salmonella* contamination suggests that organic acids penetrate the cell membrane and enter the bacterial cell's cytoplasm, where they dissociate causing the pH of the cell to increase, causing cell atrophy (Brul and Coote, 1999). There are further advantages to organic acid treatments compared to formaldehyde because organic acids are proposed to be relatively stable in feed and can occur naturally in living organisms and therefore may have greater consumer appeal when listed on an ingredient label (Wales et al., 2010).

Formaldehyde has been approved as a feed additive to control *Salmonella* and mold in animal feed (Formaldehyde, 2003). Data have demonstrated that formaldehyde is effective at mitigating *Salmonella* in animal feed (Cochrane et al., 2016a,b; EFSA, 2009). However, it requires specialized equipment for successful application, carries potential worker health concerns, and may be perceived negatively by consumers (Jones, 2011).

Many essential oils have been found to have antimicrobial properties, including the essential oils of oregano, rosemary, garlic, turmeric, and capsicum. Oregano and rosemary oils have been found to cause a 1- to 2-log reduction in *Salmonella* in various human foods (Gunduz et al., 2010; Skandamis et al., 2002). Garlic and oregano have both been shown to be effective at mitigating *Salmonella* and have minimum inhibitory concentrations of 729 and 417 ppm and maximal tolerated concentrations of 52 and 104 ppm, respectively (Dussault et al., 2014). Rosemary has also been shown to be effective against *Salmonella* contamination with a minimum inhibitory concentration of 0.3% v/v and minimal bactericidal concentrations of 0.5% v/v against *E. coli* contamination (Jiang et al., 2011). The phenolic compounds in essential oils are proposed as essential to their mode of action as bactericidal compounds (Rasooli et al., 2006). Some essential oils can contain phenol compounds that are thought to interact with and disrupt the cell membrane of bacteria, causing the cell to lose functional properties and leak the inner cell materials (Rasooli et al., 2006). More recently, a combination of garlic oleoresin, turmeric oleoresin, capsicum oleoresin, rosemary extract, and wild oregano essential oils was shown to be effective at mitigation of *Salmonella* Typhimurium in avian blood meal (Cochrane et al., 2016a,b).

Meanwhile, medium chain fatty acids, such as capric and caprylic acid, have been shown to be effective against *E. coli* and *Salmonella* growth (Kim and Rhee, 2013). Caprylic acid added to feed has been shown to decrease the quantity of *Salmonella* colonization in broiler chicks (Johny et al., 2009). A combination of caproic, caprylic, and capric acids has also been shown to be effective in the mitigation of *Salmonella* Typhimurium in feather meal, avian blood meal, porcine meat and bone meal, and poultry by-product meal (Cochrane et al., 2016a,b).

Sodium bisulfate has been used in a variety of food matrices for pH reduction. Sodium bisulfate has also been used in the animal industry for inclusion in feeds to reduce enzyme activity in poultry feed and bacterial loads in bedding of dairy cattle (Kassem et al., 2012; Sun et al., 2008). Additionally, the pet food industry has utilized sodium bisulfate for reduction of struvite formation in cats (Knueven, 2013). Inclusion of sodium bisulfate in various animal feed matrices, including poultry by-product meal and meat and bone meal, was effective at reduction microbial contamination.

Although this is not an exhaustive list of chemical additives capable of mitigating *Salmonella* or other biological hazards in animal feeds, it does give manufacturers an additional option to ensure finished product safety.

## Conclusions

Livestock and companion animal feeds can be manufactured through many processes including grinding, mixing, pelleting, and/or extrusion. These specially formulated blends of plant- and animal-based ingredients provide all the essential nutrients for the intended species. Due to the wide range of raw materials used to make complete feeds, there is potential for contamination of various enteric pathogens, like *Salmonella*. The contamination of livestock animal feeds and pet foods with pathogens can lead to the colonization of livestock and companion animals, ultimately leading to human illness. Contamination is not only limited to bacterial pathogens but can also be contaminated with viral pathogens (PEDV) and TSEs (BSE and CWD). With the recent implementation of the FSMA, animal feeds and pet foods have come under more scrutiny and require manufacturers to specifically identify potential hazards and evaluate control measures to ensure animal food safety. Control measures to ensure the manufacture of safe finished animal foods include physical and chemical intervention methods. Physical intervention can include prevention (implementing a feed mill biosecurity plan), thermal processing (pelleting and/or extrusion), and irradiation (gamma, UV, or electron beam). Chemical interventions can include the use of various chemical additives (organic acids, formaldehyde, medium chain fatty acids, and essential oils). Additionally, a combination of physical and chemical methods can mitigate the risk of contamination of finished feeds even more. Although the risk of contamination by biological hazards in animal feeds cannot be completely eliminated, feed manufacturers have multiple methods to help reduce the risk.

## References

AFIA website. http://www.afia.org/howmade.

Amado, I., Vazquez, J., Fucinos, P., Mendez, J., Pastrana, L., 2013. Optimization of antimicrobial combined effect of organic acids and temperature on foodborne *Salmonella* and *Escherichia coli* in cattle feed by response surface methodology. Foodborne Pathogens and Disease 10, 1030–1036.

Anang, D., Rusul, G., Barker, J., Ling, F., 2007. Effects of lactic acid and lauricidin on the survival of *Listeria monocytogenes*, *Salmonella* Enteritidis, and *Escherichia coli* O157:H7 in chicken breast stored at 4°C. Food Control 18, 961–969.

APHIS, 2015. Epidemiologic and Other Analyses of HPAI-affected Poultry Flocks. July 15, 2015 Report.

Bianchini, A., Stratton, J., Weier, S., Hartter, T., Plattner, B., Rokey, G., Hertzel, G., Gompa, L., Martinez, B., Eskridge, K., 2012. Validation of extrusion as a killing step for *Enterococcus faecium* in a balanced carbohydrate-protein meal by using a response surface design. Journal of Food Protection 75 (9), 1646–1653.

Brul, S., Coote, P., 1999. Preservative agents in foods: mode of action and microbial resistance mechanisms. International Journal of Food Microbiology 50, 1–17.

Carrique-Mas, J., Bedfors, S., Davies, R., 2007. Organic acid and formaldehyde treatment of animal feeds to control *Salmonella*: efficacy and masking during culture. Journal of Applied Microbiology 103, 88–96.

Cherrington, C., Hinton, M., Pearson, G., Chopra, I., 1991. Short-chain organic acids at pH 5.0 kill *Escherichia coli* and *Salmonella* spp. without causing membrace perturbation. Journal of Applied Bacteriology 70, 161–165.

Cochrane, R.A., Stark, C.R., Huss, A.R., Aldrich, G., Knueven, C.J., Jones, C.K., Pitts, J.S., 2015a. *Salmonella* surrogate mitigation in poultry feed using a dry acid powder. In: ADSA-ASAS 2015 Midwest Meeting, Des Moines, IA.

Cochrane, R.A., Schumacher, L.L., Dritz, S.S., Woodworth, J.C., Huss, A.R., Stark, C.R., DeRouchey, J.M., Tokach, M.D., Goodband, R.D., Bai, J., Chen, Q., Zhang, J., Gauger, P.C., Main, R.G., Jones, C.K., 2015b. Effect of thermal mitigation on porcine epidemic diarrhea virus (PEDV)- contaminated feed. Kansas Agricultural Experiment Station Research Reports. 1 (7). http://dx.doi.org/10.4148/2378-5977.1107.

Cochrane, R.A., Huss, A.R., Aldrich, G.C., Stark, C.R., Jones, C.K., 2016a. Evaluating chemical mitigation of *Salmonella* Typhimurium ATCC 14028 in animal feed ingredients. Journal of Food Protection 79 (4), 672–676.

Cochrane, R.A., Dritz, S.S., Woodworth, J.C., Stark, C.R., Huss, A.R., Cano, J.P., Thompson, R.W., Fahrenholz, A.C., Jones, C.K., 2016b. Feed mill biosecurity plans: a systematic approach to prevent biological pathogens in swine feed. Journal of Swine Health and Production 24 (3), 154–164.

Crane, F.M., Hansen, M., Yoder, R., Lepley, K., Cox, P., 1973. Effect of processing feeds on molds, *Salmonella*, and other harmful substances in feeds. In: Proceedings of a Symposium: Effects of Processing on Nutritional Value of Feeds, Gainsville, FL, 11 to 13 January 1972. National Academy of Sciences, Washington, DC.

Crump, J.A., Griffin, P.M., Angulo, F.J., 2002. Bacterial contamination of animal feed and its relationship to human foodborne illness. Clinical Infectious Diseases 35, 859–865.

Dee, S., Clement, T., Schelkopf, A., Nerem, J., Knudsen, D., Christopher-Hennings, J., Nelson, E., 2014. An evaluation of contaminated complete feed as a vehicle for porcine epidemic diarrhea virus infection of naïve pigs following consumption via natural feeding behavior: proof of concept. BMC Veterinary Research 10, 176–179.

Denton, J.H., Coon, C.N., Pettigrew, J.E., Parsons, C.M., 2005. Historical and scientific perspectives of same species feeding of animal by-products. Journal of Applied Poultry Research 14, 532–634.

Dussault, D., Vu, K., Lacroix, M., 2014. In vitro evaluation of antimicrobial activities of various commercial essential oils, oleoresin and pure compounds against food pathogens and application in ham. Meat Science 96, 514–520.

European Food Safety Authority, 2008. Microbiological risk assessment in feedstuffs for food-producing animals. Scientific opinion of the panel on biological hazard. The EFSA Journal 720, 1–84.

European Food Safety Authority, 2009. Available data on notified biocides efficacy under field conditions. EFSA Journal 7 (10), 259.

Formaldehyde, 2003. 21 CFR § 573.460.

Francher, B., Rollins, L.D., Trimbee, B., 1996. Feed processing using the annular gap expander and its impact on poultry performance. Journal of Applied Poultry Research 5, 386–394.

Ge, B., LaFon, P.C., Carter, P.J., McDermott, S.D., Abbott, J., Glenn, A., Ayers, S.L., Friedman, S.L., Paige, J.C., D Wagner, D., Zhao, S., McDermott, P.F., Rasmussen, M.A., 2013. Retrospective analysis of *Salmonella*, *Campylobacter*, *Escherichia coli* and *Enterococcus* in animal feed ingredients. Foodborne Pathogens and Disease 10 (8), 684–691.

Gunduz, G., Gonul, S., Karapinar, M., 2010. Efficacy of sumac and oregano in the inactivation of *Salmonella* Typhimurium on tomatoes. International Journal of Food Microbiology 141, 39–44.

Jeong, S., Marks, B.P., Ryser, E.T., 2011. Quantifying the performance of *Pediococcus* sp. (NRRL B-2354: *Enterococcus faecium*) as a nonpathogenic surrogate for *Salmonella Enteritidis* PT30 during moist-air convection heating of almonds. Journal of Food Protection 74, 603–609.

Jiang, Y., Wu, N., Fu, Y., Wang, W., Luo, M., Zhao, C., Zu, Y., Liu, X., 2011. Chemical composition and antimicrobial activity of the essential oil of Rosemary. Environmental Toxicology and Pharmacology 32 (1), 63–68.

Johny, A., Baskaran, S., Charles, A., Amalaradjou, M., Darre, M., Khan, M., Hoagland, T., Schreiber, D., Donoghue, A., Donoghue, D., Venkitanarayanan, 2009. Prophylactic supplementation of caprylic acid in feed reduces *Salmonella* Enteritidis colonization in commercial broiler chicks. Journal of Food Protection 72 (4), 722–727.

Jones, F.T., Richardson, K.E., 2004. *Salmonella* in commercially manufactured feeds. Poultry Science 83, 384–391.

Jones, F.T., 2011. A review of practical *Salmonella* control measures in animal feed. Journal of Applied Poultry Research 20, 102–113.

Kassem, I.I., Sanad, Y.M., Stonerock, R., Rajashekara, G., 2012. An evaluation of the effect of sodium bisulfate as a feed additive on *Salmonella enterica* serotype Enteritidis in experimentally infected broilers. Poultry Science 91, 1032–1037.

Kim, S., Rhee, M., 2013. Marked synergistic bactericidal effects and mode of action of medium chain fatty acids in combination with organic acids against *Escherichia coli* O157:H7. Applied and Environmental Microbiology 79 (21), 6552–6560.

Knueven, K., 2013. Controlling *Salmonella* in dry pet food. Petfood Industry 55, 72–77.

Koyuncu, S., Anderson, M., Lofstrom, C., Skandamis, P., Gounadaki, A., Zentek, J., Haggblom, P., 2013. Organic acids for control of *Salmonella* in different feed materials. BMC Veterinary Research 9, 81–90.

Li, Y., Hsieh, F., Fields, M.L., Huff, H.E., Badding, S.L., 1993. Thermal inactivation and injury of *Clostridium sporogenes* spores during extrusion of mechanically deboned turkey mixed with white corn flour. Journal of Food Processing and Preservation 17, 391–403.

Mead, P.S., Slutsker, L., Dietz, V., 1999. Food-related illness and death in the United States. Emerging Infectious Diseases 5, 607–625.

Meeker, D.L., Hamilton, C.R., 2006. An Overview of the Rendering Industry. National Renderers Association Website http://assets.nationalrenderers.org/essential_rendering_overview.pdf.

Menconi, A., Reginatto, A., Londero, A., Pumfound, N., Morgan, M., Hargis, B., Tellez, B., 2013. Effect of organic acids on *Salmonella* Typhimurium infection in broiler chickens. International Journal of Poultry Science 12, 72–75.

Narayanappa, A., Sooryanarain, H., Deventhiran, J., Cao, D., Ammayappan Venkatachalan, B., Kambiranda, D., LeRoith, T., Heffron, C., Lindstrom, N., Hall, K., Jobst, P., Sexton, C., Meng, X., Elankumaran, S., 2015. A novel pathogenic mammalian orthoreovirus from diarrheic pigs and swine blood meal in the United States. mBio 6 (3) e00593–e00615.

Okelo, P.O., Joseph, S.W., Wagner, D.D., Wheaton, F.W., Douglass, L.W., Carr, L.E., 2008. Improvements in reduction of feed contamination: an alternative monitor of bacterial killing during feed extrusion. Journal of Applied Poultry Research 17, 219–228.

Okelo, P.O., Wagner, D.D., Carr, L.E., Wheaton, F.W., Douglass, L.W., Joseph, S.W., 2006. Optimization of extrusion conditions for elimination of mesophilic bacteria during thermal processing of animal feed mash. Animal Feed Science and Technology 129, 116–137.

Pasick, J., Berhane, Y., Ojkic, D., Maxie, G., Embury-Hayatt, C., Swekla, K., Alexandersen, S., 2014. Investigation into the role of potentially contaminated feed as a source of the first-detected outbreaks of porcine epidemic diarrhea virus in Canada. Transboundary and Emerging Diseases 4 (61), 397–410.

Queguiner, C., Dumay, E., Cavalier, C., Cheftel, J.C., 1989. Reduction of *Streptococcus thermophiles* in a whey protein isolate by low moisture extrusion cooling without loss of functional properties. International Journal of Food Science and Technology 24, 601–612.

Rasooli, I., Rezaei, M.B., Allameh, A., 2006. Ultrastructural studies on antimicrobial efficacy of thyme essential oils on *Listeria monocytogenes*. International Journal of Infectious Diseases 10 (3), 236–241.

Sapkota, A.R., Lefferts, L.Y., McKenzie, S., Walker, P., 2007. What do we feed to food-production animals? A review of animal feed ingredients and their potential impacts on human health. Environmental Health Perspectives 115 (5), 663–670.

Skandamis, P., Tsigarida, E., Nychas, G., 2002. The effect of oregano essential oil on survival/death of *Salmonella* Typhimurium in meat stored at 5°C under aerobic, VP/MAP conditions. Food Microbiology 19, 97–103.

Stott, J.A., Hodgson, J.E., Chaney, J.C., 1975. Incidence of salmonellae in animal feed and the effect of pelleting on content of Enterobacteriaceae. The Journal of Applied Bacteriology 39, 41–46.

Sun, H., Pan, Y., Zhao, Y., Jackson, W.A., Nuckles, L.M., Malkina, I.L., Arteaga, V.E., Mitloehner, F.M., 2008. Effects of sodium bisulfate on alcohol, amine, and ammonia emissions from dairy slurry. Journal of Environmental Quality 37, 608.

Trudeau, M.P., Verma, H., Sampedro, F., Urriola, P.E., Shurson, G.C., McKelvey, J., Pillai, S.D., Goyal, S.M., 2016. Comparison of thermal and non-thermal processing of swine feed and the use of selected feed additives on inactivation of porcine epidemic diarrhea virus (PEDV). PLoS One 11 (6), e0158128.

Veldman, A., Vahl, H.A., Borggreve, G.J., Fuller, D.C., 1995. A survey of the incidence of *Salmonella* species and Enterobacteriaceae in poultry feeds and feed components. The Veterinary Record 136 (7), 169–172.

Wales, A., Allen, V., Daviess, R., 2010. Chemical treatment of animal feed and water for the control of Salmonella. Foodborne Pathogens and Disease 7 (1).

Wilson, S.A., Atungulu, G.G., Couch, A., Sadaka, S., 2015. Radiant heating and tempering treatments for improving rate of moisture removal during drying of shelled corn. American Society of Biological and agricultural Engineers. Applied Engineering Journal 31 (5), 799–808. http://dx.doi.org/10.13031/aea.31.11243.

Wilson, S.A., Okeyo, A.A., Olatunde, G.A., Atungulu, G.G., 2016. Radiant heat treatments for corn drying and decontamination. Journal of Food Processing and Preservation 2016, e13193. http://dx.doi.org/10.1111/jfpp.13193.

SECTION 2

# *Preharvest Food Safety Systems – Incidence and Control of Foodborne Pathogens*

CHAPTER 6

# Antimicrobial Resistance in Foodborne Pathogens

**Corliss A. O'Bryan, Philip G. Crandall, Steven C. Ricke**
*University of Arkansas, Fayetteville, AR, United States*

## Introduction

Antibiotics, substances that kill or inhibit the growth of microorganisms, are regarded as one of the greatest contributions of the 20th century to medicine and humanity (Davies and Davies, 2010). Antibiotics are used to treat a wide range of infectious diseases caused by bacteria in both animals and humans. The first effective antimicrobials, the sulfonamides, began widespread use in 1937 and before 1940 bacterial resistance to sulfonamides was reported (Davies and Davies, 2010). It is also interesting to note that a bacterial penicillinase was discovered several years before penicillin was placed on the market as a therapeutic (Abraham and Chain, 1940). A compilation of the commonly used antibiotics, their modes of action, and resistance mechanisms are shown in Table 6.1.

Antimicrobial resistance in pathogenic bacteria is not only a growing threat to human health but also a financial burden on the US health care system. In the United States alone, the estimated costs attributed to antibiotic-resistant infections exceed US$20 billion each year (Roberts et al., 2009). Although the quantity of total antimicrobial agents used in the United States is largely unknown, substantial amounts of antimicrobials are used in the production of livestock and poultry animals. Furthermore, the practice of nontargeted blanket application of these drugs in food animals for both therapeutic and nontherapeutic purposes contributed to the prevalence of antimicrobial resistant bacteria among these food animals. Globally, evidence is growing that resistant pathogenic bacteria are entering the human food chain from food animals subjected to regular antibiotic treatments (Anderson et al., 2003; Schroeder et al., 2004; Van Looveren et al., 2001). This is a human health concern because many of the antimicrobials used in animal production and human therapeutics are analogues (Table 6.2). The use of analogues of antibiotics used in human therapeutics may select for resistant phenotypes of foodborne pathogens that could cause more antibiotic-resistant bacterial illnesses in humans. Increasing numbers of foodborne pathogens such as *Salmonella* spp. and *Campylobacter* strains are further endangering human health by exhibiting increased resistance to current antibiotic drug therapies. According to the Centers for Disease Control and Prevention (CDC), 1.4 and 2 million cases of foodborne illnesses are caused by *Salmonella*

Food and Feed Safety Systems and Analysis. http://dx.doi.org/10.1016/B978-0-12-811835-1.00006-3

**Table 6.1: Targets of various classes of antibiotics within the bacterial cell and modes of resistance of bacteria to the antibiotics.**

| Antibiotic class | Examples | Target | Modes of resistance |
|---|---|---|---|
| B-lactams | Penicillins<br>Cephalosporins | Peptidoglycan<br>Biosynthesis | Altered target<br>Enzyme inactivation |
| Aminoglycosides | Gentamicin<br>Streptomycin | Translation | Enzyme inactivation |
| Glycopeptides | Vancomycin | Peptidoglycan<br>Biosynthesis | Altered target |
| Tetracyclines | Tetracycline<br>Minocycline<br>Doxycycline | Translation | Efflux pump<br>Altered target |
| Macrolides | Erythromycin | Translation | Altered target<br>Efflux pump |
| Lincosamides | Clindamycin | Translation | Altered target<br>Efflux pump |
| Fluoroquinolones | Ciprofloxacin<br>Norfloxacin | DNA replication | Altered target |
| Rifamycins | Rifampin | Transcription | Altered target |
| Phenicols | Chloramphenicol | Translation | Altered target<br>Efflux pump |

**Table 6.2: Antibiotics sold or distributed for use in food animals or humans in the United States, 2011.**

| Drug class | Animal total (kg)[a] | % Total | Human total (kg)[b] |
|---|---|---|---|
| Aminoglycosides | 273,536 | 98 | 6,485 |
| Cephalosporins | 27,654 | 5 | 502,561 |
| Ionophores | 4,573,795 | 100 | 0 |
| Lincosamides | 218,140 | 76 | 69,235 |
| Macrolides | 616,274 | 79 | 164,309 |
| Penicillins | 965,196 | 40 | 1,439,930 |
| Sulfas | 493,514 | 51 | 479,484 |
| Tetracyclines | 5,954,361 | 98 | 129,183 |
| NIR[c] | 1,495,959 | 84 | 281,647 |

[a]From FDA, 2012a. Summary Report on Antimicrobials Sold or Distributed for Use in Food-Producing Animals. Available at: http://www.fda.gov/downloads/ForIndustry/UserFees/AnimalDrugUserFeeActADUFA/UCM338170.pdf.
[b]From FDA, 2012b. Estimates of Antibacterial Drug Sales in Human Medicine. Available at: http://www.fda.gov/Drugs/DrugSafety/InformationbyDrugClass/ucm261160.htm.
[c]Not independently reported. Fewer than three distinct providers. Includes aminocoumarins, amphenicols, diaminopyrimidines, flouroquinolones, glycolipids, pleuromutalins, polypeptides, quinoxalines and streptogramin.

and *Campylobacter*, respectively, each year in the United States. Furthermore, certain non-pathogenic commensal bacteria such as *Enterococcus* and *Escherichia coli* in the human and animal gastrointestinal tract may also contaminate food and/or provide a reservoir of antimicrobial resistance to pathogenic strains. Any increase in antibiotic resistance among either these commensal or foodborne pathogenic bacteria will be an increasing health concern for both humans and animals.

## Foodborne Illness and Antimicrobial Resistance

Every year foodborne illnesses affects the lives of millions of Americans and takes a significant economic toll, considering the costs of medical treatment and lost wages in addition to the costs of food recalls that are paid for by the retailer and manufacturer. Overall, it is estimated that the annual total cost of foodborne illness in the United States is about $77.7 billion (Scharff, 2012). Foodborne illnesses caused by *Salmonella* spp. alone result in nearly 400 deaths and a total of $4.4 billion in medical costs and lost productivity in the United States annually (Scallan et al., 2011; Scharff, 2012). An increasing number of foodborne bacteria have been found to be drug resistant, including *Campylobacter*, *Salmonella* Typhi, nontyphoidal salmonellae, and *Shigella* (CDC, 2013). A foodborne outbreak among 156 patients in a hospital in Spain was attributed to a multidrug-resistant (MDR) strain of *Klebsiella pneumonia* (Calbo et al., 2011). Carbapenem-resistant enterobacteriaceae (CRE) are also considered a serious threat in Europe since CRE have been isolated from in pigs, poultry, cattle, and horses as well as humans (EFSA, 2013).

In the United States, 55 foodborne outbreaks which occurred between 1973 and 2011 were attributed to bacteria that were resistant to at least one antibiotic (CSPI, 2013). More than half of these outbreaks were associated with dairy products or ground beef, and the most commonly implicated bacterial genus was *Salmonella* (CSPI, 2013). Nearly 80% of the dairy-related outbreaks were associated with unpasteurized milk or with cheese made from unpasteurized milk (CSPI, 2013). A highly drug-resistant strain of *S.* Kentucky was recently detected in poultry flocks and turkey meat in Europe and in humans in Europe and Canada (Le Hello et al., 2013; Mueller-Doblies et al., 2014). Kentucky is the most common serotype of *Salmonella* found in chickens and ground chicken in the United States (USDA/FSIS, 2016) causing great concern that a drug-resistant strain could become a public health problem in the United States.

### Acquired Antimicrobial Resistance

Many of the currently used antibiotics were developed from compounds that microorganisms produce to inhibit or kill microbial competitors in their environment, for example, penicillin is derived form a chemical made by the fungus *Penicillium*. These substances were made by these organisms long before humans began to make use of them as antibiotics (Abraham and Chain, 1940; Finley et al., 2013). Most antimicrobial agents exert their effects on bacteria by one or more of four basic methods: interference with cell wall synthesis, inhibition of protein synthesis, interference with nucleic acid synthesis, and inhibition of a metabolic pathway (Tenover, 2006). The type of resistance that bacteria developed before antibiotics became commercially available is known as intrinsic resistance. Bacteria can also acquire resistance from other organisms by acquiring the corresponding genes from a plasmid, a transposon, integrons, bacteriophages, by mutation of cellular genes or a combination of these (Hawkey, 1998; Raghunath, 2008; Rice et al., 2003).

Mutations in chromosomes occur only at the rate of one in a population of $10^6$ to $10^8$ microorganisms and come about from errors of replication or as a mistake in the organisms' repair of its damaged DNA (Rice et al., 2003). Mutations can involve several amino acids but may also occur as single-point mutations. Quinolone resistance in *E. coli* is associated with changes in at least seven amino acids in the *gyrA* gene or three amino acids in the *parC* gene (Džidic et al., 2008; Hawkey, 1998; Levy, 1998); however, a single-point mutation in the *rpoB* gene confers complete resistance to rifampin (Rice et al., 2003). When resistance is associated with genes on bacterial chromosomes, resistant microorganisms will tend to spread more slowly, but when resistance determinants are on mobile elements such as plasmids, the resistance will spread quickly within the genus and even unrelated bacterial genera (Džidic et al., 2008).

Resistance genes can be transferred from one bacterium to another by horizontal gene transfer (Bennett, 2008). Genes are transferred mainly by transduction (via bacteriophages and integrons), conjugation (via plasmids and conjugative transposons), or transformation (via incorporation of chromosomal DNA from plasmids into a chromosome). The transferred genes are then incorporated into the recipient's chromosome by recombination or transposition and may cause one or several changes in gene sequence (Alekshun and Levy, 2007; Džidic et al., 2008; Hawkey, 2008). Plasmids are small circular structures of DNA that replicate independently from the host's chromosomal DNA (Alekshun and Levy, 2007). Plasmids contain genes that confer resistance to the main classes of antimicrobial agents (cephalosporins, fluoroquinolones, and aminoglycosides) (Bennett, 2008), toxic heavy metals (mercury, cadmium, silver), as well as virulence determinants that help a cell to survive in the environment of lethal antibiotic doses (Hawkey, 2008). Transposons are small fragments of DNA forming part of a chromosome or plasmid that can spontaneously move to a different location on the plasmid or chromosome (Hawkey, 1998; Kwon et al., 2016). An enzyme, transposase, cuts DNA strands releasing the transposon and opens DNA at another location to allow reinsertion of the transposon DNA (Muñoz-López and García-Pérez, 2010). Transposons are transferred by conjugation, transformation, or transduction and have been used extensively in the laboratory for experimental bacterial mutagenesis (Kwon et al., 2016). An integron is a two-component gene capture and dissemination system, initially discovered in relation to antibiotic resistance but now known to occur in plasmids, chromosomes, and transposons. The first integron component consists of a gene encoding a site-specific recombinase along with a specific site for recombination, whereas the second component comprises fragments of DNA called gene cassettes which can be incorporated intact or undergo shuffling (Domingues et al., 2012a,b). Although integrons themselves are not mobile they have been linked to the spread of antimicrobial resistance because of their ability to excise and integrate resistance gene cassettes from the environment or other bacteria. Integrons have been identified among clinical isolates, in farm animals, and in aquatic environments and thus offer a means for traceability of antimicrobial gene resistance via tracking of specific gene sequences (Li et al., 2012; Lopes et al., 2016).

There are three major mechanisms bacteria employ for transferring genes from one cell to another. The first method is conjugation in which a donor cell forms an attachment (pilus) to a related recipient cell and transfers a copy of its plasmid DNA through the pilus (Stokes and Gillings, 2011). The second method is transduction in which a dormant bacteriophage (prophage) located in the DNA of a bacterium is induced to become active and form bacteriophages that contain pieces of bacterial DNA that were near the viral DNA in the chromosome. These phages can then infect other bacteria and carry along the DNA code for the resistance from the first bacterium (Stokes and Gillings, 2011). Finally, transformation in which pieces of DNA from one bacterium are extruded into the environment and taken up directly by other bacteria (Domingues et al., 2012a,b). All these processes can transfer antibiotic resistance genes from one bacterial cell to another. Conjugation and transduction tend to occur between related bacterial cells, but transformation allows horizontal transfer of DNA among different species. There are numerous examples of the transfer of antibiotic resistance within and between bacterial genera. Just as one example, human fecal wastewater treatment plants are known to provide an opportunity for resistance genes to mix between pathogens, opportunistic pathogens, and environmental bacteria (Szczepanowski et al., 2009). Genome sequencing of some of these bacteria shows that they acquired resistance genes from several sources (Johnning et al., 2013). It is also known that horizontal gene transfer can occur in vivo. Experiments have demonstrated that transfer of antibiotic resistance genes occurs in meal worms and in poultry (Poole and Crippen, 2009).

## Mechanisms of Bacterial Antimicrobial Resistance

There are three main types of biochemical mechanisms that bacteria use for defense against antibiotics. One is to alter target proteins so that antibiotics cannot bind to them. Bacteria can also manufacture enzymes that break down antibiotics or change the structure of the antibiotic to prevent entry into cells. Antimicrobial compounds can also be actively expelled from the bacterial cells. This last mechanism has been found in microorganisms' with multiple drug resistance that is often associated with adenosine triphosphate (ATP)-powered efflux pump systems that actively recognize and pump antibiotics from the cell (Frye and Jackson, 2013).

### Antibiotic Inactivation or Modification

Bacteria can produce enzymes that will break down or change the structure of antibiotic molecules to make the antibiotic less effective or completely ineffective. The three main enzymes that inactivate antibiotics are β-lactamases, aminoglycoside-modifying enzymes, and chloramphenicol acetyltransferases (Castanheira et al., 2014; Hur et al., 2012). The usefulness of the aminoglycoside antibiotics, which includes gentamicin, has been reduced by the surge and dissemination of these host enzymes that provide antibiotic resistance. Aminoglycoside-modifying enzymes can be nucleotidyltransferases, phosphotransferases, or acetyltransferases, and almost every bacteria can manufacture an enzyme to render aminoglycosides inactive (Ramirez and Tolmasky, 2010). The enzymes that inactivate penicillin-type drugs are known as β-lactamases. There is a wide range of β-lactamases, some capable of degrading only a few members of the penicillin family of drugs while others, called extended spectrum β-lactamases (ESBLs), can attack a wide variety of second and third generation β-lactams, including cephalosporins (Thomson and Bonomo, 2005). Examination of approximately 5700 isolates of *E. coli*, *Klebsiella* spp., and *Proteus mirabilis* collected at US hospitals in 2012 revealed that 12.2% carried ESBL genes (Castanheira et al., 2014). Other enzymes, such as esterases, exist that can inactivate antibiotics; erythromycin esterase II is encoded on *E. coli* gene *ereB*, and it hydrolyzes the lactone ring of erythromycin A and oleandomycin rendering these antibiotics ineffective (Kim et al., 2002). Ring-opening epoxidases cause resistance of bacteria to fosfomycin (Džidic et al., 2008).

### *Efflux Pumps*

Efflux pumps are membrane proteins that export antibiotics from inside the cell and maintain a low intracellular concentration, and efflux systems have been discovered in both Gram-positive and Gram-negative bacteria (Džidic et al., 2008). Some efflux systems are specific for certain molecules, but most efflux systems can effectively remove many molecules including dyes and detergents as well as antibiotics used in human and veterinary medicine (Yamasaki et al., 2013). Fernández and Hancock (2012) have written an excellent review on the numerous efflux systems present in many species of bacteria. There are data that indicate that efflux pumps play a role in the increased antibiotic resistance bacteria possess in biofilms (Matsumura et al., 2011; Soto, 2013).

### *Altered Target*

Antibiotics interfere with normal metabolic processes of bacteria by binding to cell wall proteins or other molecules in order to gain entry into cells. This provides yet another way for bacteria to become resistant to the antimicrobial by altering the structure of the target cell wall protein so that the antibiotic cannot bind to it. Methicillin-resistant *Staphylococcus aureus* (MRSA) makes altered cell wall proteins that methicillin cannot bind to which lowers the uptake by the cell (Robinson and Enright, 2004). Fluoroquinolone antibiotics act on the bacterial enzymes DNA gyrase and topoisomerases that function to retain the bacterial chromosome in a supercoiled state. Changes in these enzymes prevent fluoroquinolones from attaching to and inhibiting them and may also confer cross-protection against other antimicrobials (Webber et al., 2013).

## *Reservoirs of Multidrug-Resistant Bacteria and Genes*

Given that antibiotic resistance genes are readily exchanged between bacteria, there is an increasing interest in investigating reservoirs of antibiotic resistance accessible to pathogens. As we will document in this section, research has already verified the presence of MDR bacteria in livestock, companion animals, wild animals, insects, and the natural environment, including soil and surface waters. These MDR bacteria can also be transmitted from person to person (Vanhoof et al., 2012), among animals (Schaer et al., 2010), between animals and humans (Ajiboye et al., 2009), and from insects to animals or humans (Ahmad et al., 2011). Some studies provide evidence for the spread of drug-resistant *E. coli* and *S. aureus* from swine and their environments to humans working in those areas (Alali et al., 2010; Oppliger et al., 2012).

### *Food Animals*

Acquisition of MDR *Salmonella* strains by dairy cattle in the United States has been found to be a common event (Adhikari et al., 2009a,b). The three most common MDR *Salmonella* serovars that the cows acquired were Newport, Typhimurium, and Dublin. New MDR salmonellae were more commonly introduced into large herds of dairy cows or where the practice of off-farm raising of heifers was practiced (Adhikari et al., 2009b). Conversely, a study conducted in Australia determined that more than 70% of the *Salmonella* isolated from calves with diarrhea were susceptible to all drugs tested and less than 3% were resistant to three or more antibiotics (Izzo et al., 2011). The European Food Safety Authority (EFSA) reported that in 2012 the prevalence of MDR *Salmonella* and MRSA in cattle was 34.2% and 33.4%, respectively (EFSA, 2014). Numerous reports of MDR salmonellae in pigs emphasize how widespread this problem is in many countries; *S.* Typhimurium is most often identified as the dominant MDR serovar, but *S.* Infantis and other serovars also express resistance to multiple antibiotics (Bolton et al., 2013; Dahshan et al., 2011; Haley et al., 2012; Keelara et al., 2013; Miller et al., 2011).

Diarrhea in calves can also be attributed to *E. coli*, and more than 80% of virulent and avirulent *E. coli* isolated from calves with diarrhea in one study in the United States were MDR (Barigye et al., 2012). Szmolka et al. (2012) performed a comprehensive phenotyping and genotyping of resistance and virulence genes on *E. coli* from a collection of strains from both healthy and ill poultry, pig, and bovine animals. Regardless of the species of host, genes to confer resistance to three or more antibiotics were present in all the isolates tested. Numerous other MDR organisms have been isolated from food animals worldwide including *S.* Stanley and *E. coli* in Japanese cattle (Dahshan et al., 2011; Ohnishi et al., 2013), *E. coli* in cattle in Nigeria (Amosun et al., 2012), and Shiga toxin producing *E. coli* in cattle in India (Rehman et al., 2013).

### *Companion Animals*

Companion animals and pets share the same environment and remain in close contact with people and can share zoonotic pathogens, such as *Salmonella* spp., *S. aureus*, or *Campylobacter* spp. with humans (Day et al., 2012). More pet owners are now seeking advanced veterinary care for their animals, which often involves the administration of antibiotics. Consequently, it has been determined that dogs, cats, and horses which have spent more than 2 days in a veterinary hospital have more MDR bacteria associated with them than when they entered the hospital (Gibson et al., 2011; Hamilton et al., 2013; Maddox et al., 2011). The European Medicines Agency considered the risk of transferring antibiotic-resistant bacteria from companion animals to humans or the reverse and concluded that the major foodborne zoonotic bacteria *Salmonella* and *Campylobacter* are not a direct hazard for antimicrobial resistance emerging from companion animals, although these may be transmitted by direct contact (EMA, 2015). However, they did determine that there could be a direct hazard of passing MRSA, vancomycin-resistant enterococci, ESBL-producing enterobacteria, and carbapenem-resistant Gram-negative bacteria from companion animals to humans (EMA, 2015).

### Wild Animals and Insects

MDR *E. coli* and *Salmonella* spp. have been detected in a variety of wild mammals, including wolves (Simoes et al., 2012), wild boar (Literak et al., 2010), and mongoose (Pesapane et al., 2013). Wild reptiles and amphibians living in a produce-growing region of central California were also found to harbor MDR bacteria (Gorski et al., 2013). Birds can also harbor MDR bacteria which have been confirmed to be present in ducks (Hasan et al., 2012), seagulls (Poirel et al., 2012), cranes (Kitadai et al., 2012), raptors (Molina-Lopez et al., 2011), and pigeons (Da Silva et al., 2012). Many of these birds are capable of flying over long distances allowing them the opportunity to spread MDR bacteria to new locations; in addition, aquatic birds' droppings may deposit MDR bacteria in surface waters used for drinking and irrigation of food crops.

Insects also may harbor MDR bacteria and are capable of spreading these bacteria around the farm environment and possibly to humans as well. Two poultry farms in the Netherlands were discovered to have house flies and blow flies that harbored ESBL-producing *E. coli* (Blaak et al., 2014). Cockroaches have also been shown to harbor MDR bacteria (Ahmad et al., 2011; Tilahun et al., 2012).

### Environment

Many antibiotics that are used for human and animal treatment are excreted in the urine still fully active and thus can contaminate water and soil and apply low level pressure for the development of antimicrobial resistance (Wagenlehner et al., 2003). When antimicrobial agents are used on the farm, the soil (Walsh and Duffy, 2013) and water (Micallef et al., 2012) may contain numerous antibiotic-resistant strains. These bacteria may also survive in manure and waste lagoons and ultimately spread from these sources to areas outside of the farms (Antunes et al., 2011). Some studies have suggested that the risk of contracting *Salmonella* spp. or *Campylobacter* spp. from recreational waters is as high as or perhaps higher than the risk from eating chicken (Domingues et al., 2012a,b; Evers et al., 2008). Nine percent of the outbreaks caused by *E. coli* O157:H7 in the United States were waterborne (Rangel et al., 2005). Worldwide 120 million cases of gastrointestinal disease annually occur from recreational water activities or by eating raw or lightly cooked shellfish (Shuval, 2003), and multiple outbreaks have been associated with fresh produce (Feng, 2014). Given that the natural environment harbors antibiotic-resistant bacteria, it may provide a pathway for the transmission of antibiotic-resistant bacteria to humans, but the extent to which this actually happens is not yet known.

## *Intervention Methods to Reduce Antibiotic Resistance*

Preventing the emergence of MDR bacteria and protecting food from contamination with these resistant strains are complex problems worldwide. Antibiotic resistance has been declared a crisis by the World Health Organization, CDC, the Institute of Medicine and the Infectious Diseases Society of America. Although there is a lack of data on antibiotic resistance in the United States, the European Union has data on antibiotic use by drug, and resistance data by microorganism for 26 countries (Bartlett et al., 2013). For example, Greece has the highest per capita use of antibiotics, and the Netherlands has among the lowest; the proportion of *S. aureus* isolates that are methicillin resistant is 58% in Greece and 1.6% in the Netherlands (Bartlett et al., 2013).

### *Curtail Antibiotic Use in Agriculture*

More of particular classes of antibiotics (in kilograms) are sold for use in animals than in humans (FDA, 2012a,b). Although some drugs are used to treat sick animals, large amounts have been used in feed to promote growth and increase feed efficiency (Dibner and Richards, 2005; Jones and Ricke, 2003). More judicious use of antibiotics in both veterinary and human medicine has been shown to reduce numbers of resistant bacteria (McEwen and Fedorka-Cray, 2002). In the United States, the Food and Drug Administration (FDA) has issued guidance for drug manufacturers to voluntarily remove the use of drugs for use in production of food animals and to change marketing status from over the counter to prescription by a veterinarian (FDA, 2013). The FDA has also instituted a new rule, the Veterinary Feed Directive, which goes into effect in December 2016 (FDA, 2015). This rule is intended to encourage judicious use of antibiotics in animal agriculture, particularly for drugs that are important in human medicine (FDA, 2015).

Denmark has a large database of microorganisms with information on antibiotic resistance. The use of antibiotics as growth promoters was voluntarily discontinued by Danish growers in 1999. Avoparcin was banned by the Danish government in 1995 resulting in a decrease of glycopeptide-resistant *Enterococcus faecium* in broilers from 72.7% in 1995 to 5.8% in 2000

(Aarestrup et al., 2001). Also in Denmark, from 1995 to 1997 the occurrence of erythromycin resistance among *E. faecium* and *Enterococcus faecalis* isolates from pigs was almost 90%, but after decreases in usage in agriculture, resistance decreased to 46.7% and 28.1% among *E. faecium* and *E. faecalis* isolates from pigs, respectively (Aarestrup et al., 2001). Thus it appears to be possible to reduce the occurrence of antimicrobial resistance in a national population of food animals when the selective antibiotic pressure is removed. However the prevalence of resistant bacteria does not revert to zero when the use of certain antibiotics is ended because reservoirs of resistance genes may persist for long periods. There are several methods that can be utilized for reducing antibiotic usage in livestock. For instance, proper use of substances increase immune function and disease resistance of animals, clinical inspections to identify and treat sick animals before disease spreads to others, maintenance of a hygienic and healthy living environment, or use of laboratory tests to detect animals at risk of developing disease. Strategies using some of these criteria were previously discussed by Trevisi et al. (2014).

### Proper Usage of Antibiotics in Human Medicine

Overuse and misuse of antimicrobials in human medicine occurs when antimicrobials are prescribed for treatment of common viral diseases which are not susceptible to antibiotics or for bacterial infections without testing the sensitivity of the pathogens to the prescribed antibiotic. The use of the procalcitonin level in the blood of patients as a biomarker for infection is a recent development in diagnostic medicine (Schuetz and Mueller, 2016). This test was shown in several well-controlled trials to be successful in allowing physicians to avoid unnecessary use or overuse of antibiotics (Schuetz et al., 2012). Measuring procalcitonin concentrations allows detection of bacteria in the blood (Laukemann et al., 2015), determination of how severe an infection is (Schuetz et al., 2011) and can predict if treatment will or will not be successful (Schuetz et al., 2013). Shorter courses of antibiotics are also effective (Bartlett et al., 2013; Chastre et al., 2009). Colistin is an older drug, having been on the market since 1961, which would be a valuable addition to treatment regimens, but patients are often not administered the correct dose because apparently the recommendations in the manufacturers' package insert are wrong (Velkov et al., 2013). In some developing countries, antibiotics can be purchased over the counter and may therefore be used to treat conditions where they are ineffective. Improved control over the use of antibiotics in human medicine would help control the emergence of MDR.

### Develop New Drugs

Large pharmaceutical companies previously developed new antibiotics in response to the continuing development of pathogens' resistance. Today, pharmaceutical companies no longer invest in the search for new antibiotics because antibiotics are useful for only a short time by people, and they actually cure a disease, prices are generally low, and it is difficult to gain approval from FDA for new drugs (Spellberg et al., 2012). For example, under the way FDA now functions, companies would have to conduct a 2000-patient study on a new drug for treating pneumonia caused by multidrug-resistant bacteria. New methods for dealing with antibiotic development and its related costs are needed. A successful public–private partnership using the combined resources of the Bill & Melinda Gates Foundation, Janssen Pharmaceuticals, and the TB Alliance has produced bedaquiline, the first new FDA-approved drug for tuberculosis in the past 40 years (Gras, 2013). Bedaquiline exerts its effect on the proton pump for ATP synthase, rather than the DNA gyrase which is the target of all other quinolone antibiotics (Deoghare, 2013).

## Conclusions

The importance and value of antibiotics cannot be overestimated; the human population remains totally dependent on them for the treatment of most infectious diseases. Antibiotics are also necessary for prophylactic treatment for persons undergoing organ transplants or other advanced surgeries. New strategies for controlling antibiotic resistance have been recently implemented, and others are being developed but the problem of antibiotic resistance remains a concern. The problem of widespread consumer abuse of antibiotics must also be confronted; often people will demand antibiotics for colds or other viral illness on which antibiotics have no effect other than to increase the risk of antibiotic resistance. We need better diagnostics to allow rapid and specific treatment with a narrow spectrum antibiotic rather than taking the traditional approach of immediately prescribing a broad-spectrum antibiotic. In addition, we must continue the search for new antibiotics with novel mechanisms of action. Once these new antibiotics are identified and the most appropriate use for it is determined, the use should be restricted to that narrow usage only.

## References

Aarestrup, F.M., Seyfarth, A.M., Emborg, H.D., Pedersen, K., Hendriksen, R.S., Bager, F., 2001. Effect of abolishment of the use of antimicrobial agents for growth promotion on occurrence of antimicrobial resistance in fecal enterococci from food animals in Denmark. Antimicrobial Agents and Chemotherapy 45, 2054–2059.

Abraham, E.P., Chain, E., 1940. An enzyme from bacteria able to destroy penicillin. Reviews in Infectious Diseases 10, 677–678.

Adhikari, B., Besser, T.E., Gay, J.M., Fox, L.K., Davis, M.A., Cobbold, R.N., Berge, A.C., McClanahan, R., Hancock, D.D., 2009a. Introduction of new multidrug-resistant *Salmonella enterica* strains into commercial dairy herds. Journal of Dairy Science 92, 4218–4228.

Adhikari, B., Besser, T.E., Gay, J.M., Fox, L.K., Davis, M.A., Cobbold, R.N., Berge, A.C., Hancock, D.D., 2009b. The role of animal movement, including off-farm rearing of heifers, in the interherd transmission of multidrug-resistant *Salmonella*. Journal of Dairy Science 92, 4229–4238.

Ahmad, A., Ghosh, A., Schal, C., Zurek, L., 2011. Insects in confined swine operations carry a large antibiotic resistant and potentially virulent enterococcal community. BMC Microbiology 11, 23. http://dx.doi.org/10.1186/1471-2180-11-23.

Ajiboye, R., Solberg, O., Lee, B., Raphael, E., Debroy, C., Riley, L., 2009. Global spread of mobile antimicrobial drug resistance determinants in human and animal *Escherichia coli* and *Salmonella* strains causing community-acquired infections. Clinical Infectious Diseases 49, 365–371.

Alali, W.Q., Scott, H.M., Norby, B., 2010. Assessing the similarity of antimicrobial resistance phenotypes among fecal *Escherichia coli* isolates from two aggregated occupational cohorts of humans versus swine using cluster analysis and multivariate statistics. Preventive Veterinary Medicine 94, 77–83.

Alekshun, M.N., Levy, S.B., 2007. Molecular mechanisms of antibacterial multidrug resistance. Cell 128, 1037–1050.

Amosun, E., Ojo, O.E., Alao, I.K., Ajuwape, A.T.P., 2012. Antimicrobial resistance among commensal *Escherichia coli* from cattle faeces and beef in Ibadan, Nigeria. African Journal of Biotechnoloogy 11, 12240–12245.

Anderson, A.D., Nelson, J.M., Rossiter, S., Angulo, F.J., 2003. Public health consequences of use of antimicrobial agents in food animals in the United States. Microbial Drug Resistance 9, 373–379.

Antunes, P., Mourao, J., Pestana, N., Peixe, L., 2011. Leakage of emerging clinically relevant multidrug-resistant *Salmonella* clones from pig farms. Journal of Antimicrobials and Chemotherapy 66, 2028–2032.

Barigye, R., Gautam, A., Piche, L.M., Schaan, L.P., Krogh, D.E., Olet, S., 2012. Prevalence and antimicrobial susceptibility of virulent and avirulent multidrug-resistant *Escherichia coli* isolated from diarrheic neonatal calves. American Journal of Veterinary Research 73, 1944–1950.

Bartlett, J.G., Gilbert, D., Spellberg, B., 2013. Seven ways to preserve the miracle of antibiotics. Clinical Infectious Diseases 56, 1445–1450.

Bennett, P.M., 2008. Plasmid encoded antibiotic resistance: acquisition and transfer of antibiotic resistance genes in bacteria. British Journal of Pharmacology 153, 347–357.

Blaak, H., Hamidjaja, R.A., van Hoek, A.H.A.M., de Heer, L., Husman, A.M.D., Schets, F.M., 2014. Detection of extended-spectrum Beta-lactamase (ESBL)-producing *Escherichia coli* on flies at poultry farms. Applied and Environmental Microbiology 80, 239–246.

Bolton, D.J., Ivory, C., McDowell, D., 2013. A study of *Salmonella* in pigs from birth to carcass: serotypes, genotypes, antibiotic resistance and virulence profiles. International Journal of Food Microbiology 160, 298–303.

Calbo, E., Freixas, N., Xercavins, M., Riera, M., Nicolas, C., Monistrol, O., del mar Solé, M., Sala, M.R., Vila, J., Garau, J., 2011. Foodborne nosocomial outbreak of SHV1 and CTX-M-15-producing *Klebsiella pneumoniae*: epidemiology and control. Clinical Infectious Diseases 52, 743–749.

Castanheira, M., Farrell, S.E., Krause, K.M., Jones, R.N., Sader, H.S., 2014. Contemporary diversity of beta-lactamases among enterobacteriaceae in the nine US census regions and ceftazidime-avibactam activity tested against isolates producing the most prevalent beta-lactamase groups. Antimicrobial Agents and Chemotherapy 58, 833–838.

CDC, 2013. Antibiotic Resistance Threats in the United States. Available at: http://www.cdc.gov/drugresistance/threat-report-2013/.

Chastre, J., Wolff, M., Fagon, J.Y., Chevret, S., Thomas, F., Wermert, D., Clementi, E., Gonzalez, J., Jusserand, D., Asfar, P., Perrin, D., Fieux, F., Aubas, S., 2009. Comparison of 8 vs. 15 days of antibiotic therapy for ventilator-associated pneumonia in adults: a randomized trial. JAMA 209, 2588–2598.

CSPI, 2013. Antibiotic Resistance in Foodborne Pathogens. Available at: https://cspinet.org/new/pdf/outbreaks_antibiotic_resistance_in_foodborne_pathogens_2013.pdf.

Dahshan, H., Abd-El-Kader, M.A., Chuma, T., Moriki, H., Okamoto, K., 2011. Re-emergence of multi-drug resistant *Salmonella enterica* serovar Stanley from cattle. Veterinary Research Communications 35, 55–60.

Da Silva, V., Caçador, N.C., da Silva Cdos, S., Fontes, C.O., Garcia, G.D., Nicoli, J.R., Diniz, C.G., 2012. Occurrence of multidrug-resistant and toxic-metal tolerant enterococci in fresh feces from urban pigeons in Brazil. Microbes and Environments 27, 179–185.

Davies, J., Davies, D., 2010. Origins and evolution of antibiotic resistance. Microbiology and Molecular Biology Reviews 74 (3), 417–433.

Day, M.J., Breitschwerdt, E., Cleaveland, S., Karkare, U., Khanna, C., Kirpensteijn, J., Kuiken, T., Lappin, M.R., McQuiston, J., Mumford, E., Myers, T., Palatnik-de-Sousa, C.B., Rubin, C., Takashima, G., Thiermann, A., 2012. Surveillance of zoonotic infectious disease transmitted by small companion animals. Emerging Infectious Diseases 18. http://dx.doi.org/10.3201/eid.1812.120664.

Deoghare, S., 2013. Bedaquiline: a new drug approved for treatment of multidrug-resistant tuberculosis. Indian Journal of Pharmacology 45 (5), 536–537.

Dibner, J.J., Richards, J.D., 2005. Antibiotic growth promoters in agriculture: history and mode of action. Poultry Science 84, 634–643.

Domingues, A.R., Pires, S.M., Halasa, T., Hald, T., 2012a. Source attribution of human campylobacteriosis using a meta-analysis of case-control studies of sporadic infections. Epidemiology and Infection 140, 970–981.

Domingues, A.R., Pires, S.M., Halasa, T., Hald, T., 2012b. Source attribution of human salmonellosis using a meta-analysis of case-control studies of sporadic infections. Epidemiology and Infection 140, 959–969.

Džidic, S., Šuškovic, J., Kos, B., 2008. Antibiotic resistance mechanisms in bacteria: biochemical and genetic aspects. Food Technology and Biotechnology 46, 11–21.

EFSA, 2013. Scientific opinion on carbapenem resistance in food animal ecosystems. EFSA Journal 11, 3501. http://dx.doi.org/10.2903/j.efsa.2013.3501.

EFSA, 2014. The European Union summary report on antimicrobial resistance in zoonotic and indicator bacteria from humans, animals and food in 2012. EFSA Journal 12, 3590. http://dx.doi.org/10.2903/j.efsa.2014.3590.

EMA, 2015. Reflection Paper on the Risk of Antimicrobial Resistance Transfer from Companion Animals. Available at: http://www.ema.europa.eu/docs/en_GB/document_library/Scientific_guideline/2015/01/WC500181642.pdf.

Evers, E.G., van der Vels-Klerx, H.J., Nauta, M.J., Schijven, J.F., Havelaar, A.H., 2008. *Campylobacter* source attribution by exposure assessment. International Journal of Risk Assessment and Management 8, 174–189.

FDA, 2012a. Summary Report on Antimicrobials Sold or Distributed for Use in Food-Producing Animals. Available at: http://www.fda.gov/downloads/ForIndustry/UserFees/AnimalDrugUserFeeActADUFA/UCM338170.pdf.

FDA, 2012b. Estimates of Antibacterial Drug Sales in Human Medicine. Available at: http://www.fda.gov/Drugs/DrugSafety/InformationbyDrugClass/ucm261160.htm.

FDA, 2013. New Animal Drugs and New Animal Drug Combination Products Administered in or on Medicated Feed or Drinking Water of Food-producing Animals: Recommendations for Drug Sponsors for Voluntarily Aligning Product Use Conditions with GFI #209. Available at: http://www.fda.gov/downloads/AnimalVeterinary/GuidanceComplianceEnforcement/GuidanceforIndustry/UCM299624.pdf.

FDA, 2015. Veterinary Feed Directive. Available at: http://www.fda.gov/AnimalVeterinary/DevelopmentApprovalProcess/ucm449019.htm.

Feng, P., 2014. Shiga toxin-producing *Escherichia coli* (STEC) in fresh produce–A food safety dilemma. Microbiology Spectrum 2 (4). http://dx.doi.org/10.1128/microbiolspec.EHEC-0010-2013.

Fernández, L., Hancock, R.E.W., 2012. Adaptive and mutational resistance: role of porins and efflux pumps in drug resistance. Clinical Microbiology Reviews 25, 661–681.

Finley, R.L., Collignon, P., Larsson, D.G.J., McEwen, S.A., Li, X.Z., Gaze, W.H., Reid-Smith, R., Timinouni, M., Graham, D.W., Topp, E., 2013. The scourge of antibiotic resistance: the important role of the environment. Clinical Infectious Diseases 57, 704–710.

Frye, J.G., Jackson, C.R., 2013. Genetic mechanisms of antimicrobial resistance identified in *Salmonella enterica*, *Escherichia coli*, and *Enterococcus* spp. isolated from U.S. food animals. Frontiers in Microbiology 4, 135. http://dx.doi.org/10.3389/fmicb.2013.00135.

Gibson, J., Morton, J.M., Cobbold, R.N., Filippich, L.J., Trott, D.J., 2011. Risk factors for dogs becoming rectal carriers of multidrug-resistant *Escherichia coli* during hospitalization. Epidemiology and Infection 139, 1511–1521.

Gorski, L., Jay-Russell, M.T., Liang, A.S., Walker, S., Bengson, Y., Govoni, J., Mandrell, R.E., 2013. Diversity of pulsed-field gel electrophoresis pulsotypes, serovars, and antibiotic resistance among *Salmonella* isolates from wild amphibians and reptiles in the California central coast. Foodborne Pathogens and Disease 10, 540–548.

Gras, J., 2013. Bedaquiline for the treatment of pulmonary multidrug resistant tuberculosis in adults. Drugs of Today 49, 353–361.

Haley, C.A., Dargatz, D.A., Bush, E.J., Erdman, M.M., Fedorka-Cray, P.J., 2012. *Salmonella* prevalence and antimicrobial susceptibility from the national animal health monitoring system swine 2000 and 2006 studies. Journal of Food Protection 75, 428–436.

Hamilton, E., Kruger, J.M., Schall, W., Beal, M., Manning, S.D., Kaneene, J.B., 2013. Acquisition and persistence of antimicrobial-resistant bacteria isolated from dogs and cats admitted to a veterinary teaching hospital. Journal of the American Veterinary Medical Association 243, 990–1000.

Hasan, B., Sandegren, L., Melhus, A., Drobni, M., Hernandez, J., Waldenström, J., Alam, M., Olsen, B., 2012. Antimicrobial drug-resistant *Escherichia coli* in wild birds and free-range poultry, Bangladesh. Emerging Infectious Diseases 18, 2055–2058.

Hawkey, P.M., 1998. The origins and molecular basis of antibiotic resistance. British Medical Journal 317, 657–660.

Hawkey, P.M., 2008. Molecular epidemiology of clinical significant antibiotic resistance genes. British Journal of Pharmacology 153, 406–413.

Hur, J., Jawale, C., Lee, J.H., 2012. Antimicrobial resistance of *Salmonella* isolated from food animals: a review. Food Research International 45, 819–830.

Izzo, M.M., Mohler, V.L., House, J.K., 2011. Antimicrobial susceptibility of *Salmonella* isolates recovered from calves with diarrhoea in Australia. Australian Veterinary Journal 89, 402–408.

Johnning, A., Moore, E.R.B., Svensson-Stadler, L., Shouche, Y.S., Larsson, D.G.J., Kristiansson, E., 2013. Acquired genetic mechanisms of a multiresistant bacterium isolated from a treatment plant receiving wastewater from antibiotic production. Applied and Environmental Microbiology 79, 7256–7263.

Jones, F.T., Ricke, S.C., 2003. Observations on the history of the development of antimicrobials and their use in poultry feeds. Poultry Science 82, 613–617.

Keelara, S., Scott, H.M., Morrow, W.M., Gebreyes, W.A., Correa, M., Nayak, R., Stefanova, R., Thakur, S., 2013. Longitudinal study of distributions of similar antimicrobial-resistant *Salmonella* serovars in pigs and their environment in two distinct swine production systems. Applied and Environmental Microbiology 79, 5167–5178.

Kim, Y.-K., Cha, C.-J., Cerniglia, C.E., 2002. Purification and characterization of an erythromycin esterase from an erythromycin-resistant *Pseudomonas* sp. FEMS Microbiology Letters 210, 239–244.

Kitadai, N., Obi, T., Yamashita, S., Murase, T., Takase, K., 2012. Antimicrobial susceptibility of *Escherichia coli* isolated from feces of wild cranes migrating to Kagoshima, Japan. Journal of Veterinary Medical Science 74, 395–397.

Kwon, Y.M., Ricke, S.C., Manda, R.K., 2016. Transposon sequencing: methods and expanding applications. Applied Microbiology and Biotechnology 100, 31–43.

Laukemann, S., Kasper, N., Kulkarni, P., Steiner, D., Rast, A.C., Kutz, A., Felder, S., Haubitz, S., Faessler, L., Huber, A., Fux, C.A., Mueller, B., Schuetz, P., 2015. Can we reduce negative blood cultures with clinical scores and blood markers? Results from an observational cohort study. Medicine 94 (49), e2264. http://dx.doi.org/10.1097/MD.0000000000002264.

Le Hello, S., Harrois, D., Bouchrif, B., Sontag, L., Elhani, D., Guibert, V., Zerouali, L., Weill, F.X., 2013. Highly drug-resistant *Salmonella enterica* serotype Kentucky ST198-X1: a microbiological study. Lancet Infectious Diseases 13, 672–679.

Levy, S.B., March 1998. The challenge of antibiotic resistance. Scientific American 278, 46–53.

Li, J., Wang, T., Shao, B., Shen, J., Wang, S., Wu, Y., 2012. Plasmid-mediated quinolone resistance genes and antibiotic residues in wastewater and soil adjacent to swine feedlots: potential transfer to agricultural lands. Environmental Health Perspectives 120, 8. http://dx.doi.org/10.1289/ehp.1104776.

Literak, I., Dolejska, M., Radimersky, T., Klimes, J., Friedman, M., Aarestrup, F.M., Hasman, H., Cizek, A., 2010. Antimicrobial resistant faecal *Escherichia coli* in wild mammals in central Europe: multiresistant *Escherichia coli* producing extended-spectrum beta-lactamases in wild boars. Journal of Applied Microbiology 108, 1702–1711.

Lopes, G.V., Michael, G.B., Cardoso, M., Schwarz, S., 2016. Antimicrobial resistance and class 1 integron-associated gene cassettes in *Salmonella enterica* serovar Typhimurium isolated from pigs at slaughter and abattoir environment. Veterinary Microbiology 194, 84–92.

Maddox, T., Williams, N.J., Clegg, P.D., O'Donnell, A.J., Dawson, S., Pinchbeck, G.L., 2011. Longitudinal study of antimicrobial resistant commensal *Escherichia coli* in the faeces of horses in an equine hospital. Preventative Veterinary Medicine 100, 134–145.

Matsumura, K., Furukawa, S., Ogihara, H., Morinaga, Y., 2011. Roles of multidrug efflux pumps on the biofilm formation of *Escherichia coli* K-12. Biocontrol Sciences 16, 69–72.

McEwen, S.A., Fedorka-Cray, P.J., 2002. Antimicrobial use and resistance in animals. Clinical Infectious Diseases 34 (Suppl. 3), S93–S106.

Micallef, S., Goldstein, R.E.R., George, A., Kleinfelter, L., Boyer, M.S., McLaughlin, C.R., Estrin, A., Ewing, L., Jean-Gilles Beaubrun, J., Hanes, D.E., Kothary, M.H., Tall, B.D., Razeq, J.H., Joseph, S.W., Sapkota, A.R., 2012. Occurrence and antibiotic resistance of multiple *Salmonella* serotypes recovered from water, sediment and soil on mid-Atlantic tomato farms. Environmental Research 114, 31–39.

Miller, A.J., Twomey, D.F., Davies, R.H., Teale, C.J., Williamson, S.M., Reichel, R., Featherstone, C.A., Cook, A.J., Snow, L.C., Armstrong, J.D., 2011. *Salmonella* serovars and antimicrobial resistance patterns on a sample of high seroprevalence pig farms in England and Wales (2003–2008). Zoonoses and Public Health 58, 549–559.

Molina-Lopez, R.A., Valverdú, N., Martin, M., Mateu, E., Obon, E., Cerdà-Cuéllar, M., Darwich, L., 2011. Wild raptors as carriers of antimicrobial-resistant *Salmonella* and *Campylobacter* strains. Veterinary Record 168, 565b. http://dx.doi.org/10.1136/vr.c7123.

Mueller-Doblies, D., Carrique-Mas, J.J., Davies, R.H., 2014. A study of the dynamics of *Salmonella* infection in turkey breeding, rearing and finishing houses with special reference to elimination, persistence and introduction of *Salmonella*. Avian Pathology 43 (2), 146–154.

Muñoz-López, M., García-Pérez, J.L., 2010. DNA transposons: nature and applications in genomics. Current Genomics 11, 115–128.

Ohnishi, M., Okatani, A.T., Esaki, H., Harada, K., Sawada, T., Murakami, M., Marumo, K., Kato, Y., Sato, R., Shimura, K., Hatanaka, N., Takahashi, T., 2013. Herd prevalence of Enterobacteriaceae producing Ctx-M-type and Cmy-2-lactamases among Japanese dairy farms. Journal of Applied Microbiology 115, 282–289.

Oppliger, A., Moreillon, P., Charrière, N., et al., 2012. Antimicrobial resistance of *Staphylococcus aureus* strains acquired by pig farmers from pigs. Applied and Environmental Microbiology 78, 8010–8014.

Pesapane, R., Ponder, M., Alexander, K., 2013. Tracking pathogen transmission at the human-wildlife interface: banded mongoose and *Escherichia coli*. Ecohealth 10, 115–128.

Poirel, L., Potron, A., De La Cuesta, C., Cleary, T., Nordmann, P., Munoz-Price, L.S., 2012. Wild coastline birds as reservoirs of broad-spectrum-beta-lactamase-producing enterobacteriaceae in Miami Beach, Florida. Antimicrobial Agents and Chemotherapy 56, 2756–2758.

Poole, T., Crippen, T., 2009. Conjugative plasmid transfer between *Salmonella enterica* Newport and *Escherichia coli* within the gastrointestinal tract of the lesser mealworm beetle, *Alphitobius diaperinus* (Coleoptera: Tenebrionidae). Poultry Science 88, 1553–1558.

Raghunath, D., 2008. Emerging antibiotic resistance in bacteria with special reference to India. Journal of Bioscience 33 (4), 593–603.

Ramirez, M.S., Tolmasky, M.E., 2010. Aminoglycoside modifying enzymes. Drug Resistance Updates 13 (6), 151–171.

Rangel, J.M., Sparling, P.H., Crowe, C., Griffin, P.M., Swerdlow, D.L., 2005. Epidemiology of *Escherichia coli* O157:H7 outbreaks, United States, 1982–2002. Emerging Infectious Diseases 11, 603–609.

Rehman, M.U., Rashid, M., Sheikh, J.A., Wani, S.A., Farook, S., 2013. Multi-drug resistance among Shiga toxin producing *Escherichia coli* isolated from bovines and their handlers in Jammu Region, India. Veterinary World 6, 655–658.

Rice, L.B., Sahm, D., Binomo, R.A., 2003. Mechanisms of resistance to antibacterial agents. In: Murray, P.R., Baron, E.J., Jorgensen, J.H., Phaller, M.A., Yolken, R.H. (Eds.), Manual of Clinical Microbiology. ASM Press, Washington, D.C.

Roberts, R.R., Hota, B., Ahmad, I., Scott, R.D., Foster, S.D., Abbasi, F., Schabowski, S., Kampe, L.M., Ciavarella, G.G., Supino, M., Naples, J., Cordell, R., Levy, S.B., Weinstein, R.A., 2009. Hospital and societal costs of antimicrobial-resistant infections in a Chicago teaching hospital: implications for antibiotic stewardship. Clinical Infectious Diseases 49, 1175–1184.

Robinson, D.A., Enright, M.C., 2004. Multilocus sequence typing and the evolution of methicillin-resistant *Staphylococcus aureus*. Clinical Microbiology and Infection 10, 92–97.

Scallan, E., Hoekstra, R.M., Angulo, F.J., Tauxe, R.V., Widdowson, M.A., Roy, S.L., Jones, J.L., Griffin, P.M., 2011. Foodborne illness acquired in the United States – major pathogens. Emerging Infectious Disease 17, 7–15.

Schaer, B., Aceto, H., Rankin, S., 2010. Outbreak of salmonellosis caused by *Salmonella enterica* serovar Newport MDR-AmpC in a large animal veterinary teaching hospital. Journal of Veterinary Internal Medicine 24, 1138–1146.

Scharff, R., 2012. Economic burden from health losses due to foodborne illness in the United States. Journal of Food Protection 75, 123–131.

Schroeder, C.M., White, D.G., Meng, J., 2004. Retail meat and poultry as a reservoir of antimicrobial-resistant *Escherichia coli*. Food Microbiology 21, 249–255.

Schuetz, P., Briel, M., Christ-Crain, M., Stolz, D., Bouadma, L., Wolff, M., Luyt, C.E., Chastre, J., Tubach, F., Kristoffersen, K.B., Wei, L., Burkhardt, O., Welte, T., Schroeder, S., Nobre, V., Tamm, M., Bhatnagar, N., Bucher, H.C., Mueller, B., 2012. Procalcitonin to guide initiation and duration of antibiotic treatment in acute respiratory infections: an individual patient analysis. Clinical Infectious Disease 55, 651–662.

Schuetz, P., Maurer, P., Punjabi, V., Desai, A., Amin, D.N., Gluck, E., 2013. Procalcitonin decrease over 72 hours in US critical care units predicts fatal outcome in sepsis patients. Critical Care 17, R115. http://dx.doi.org/10.1186/cc12787.

Schuetz, P., Mueller, B., 2016. Procalcitonin in critically ill patients: time to change guidelines and antibiotic use in practice. Lancet Infectious Diseases 16, 758–760.

Schuetz, P., Suter-Widmer, I., Chaudri, A., Christ-Crain, M., Zimmerli, W., Mueller, B., 2011. Prognostic value of procalcitonin in community-acquired pneumonia. European Respiratory Journal 37, 384–392.

Shuval, H., 2003. Estimating the global burden of thalassogenic diseases: human infectious diseases caused by wastewater pollution of the marine environment. Journal of Water and Health 1, 53–64.

Simoes, R., Ferreira, C., Goncalves, J., Alvares, F., Rio-Maior, H., Roque, S., Brandão, R., Martins da Costa, P., 2012. Occurrence of virulence genes in multidrug-resistant *Escherichia coli* isolates from Iberian wolves (*Canis lupus signatus*) in Portugal. European Journal of Wildlife Research 58, 677–684.

Soto, S.M., 2013. Role of efflux pumps in the antibiotic resistance of bacteria embedded in a biofilm. Virulence 4, 223–229.

Spellberg, B., Brass, E.P., Lewis, R.J., et al., 2012. Recommendations on the conduct of superiority and organism-specific clinical trials of antibacterial agents for the treatment of infections caused by drug-resistant bacterial pathogens. Clinical Infectious Diseases 55, 1031–1046.

Stokes, H.W., Gillings, M.R., 2011. Gene flow, mobile genetic elements and the recruitment of antibiotic resistance genes into Gram-negative pathogens. FEMS Microbiology Reviews 35, 790–819.

Szczepanowski, R., Linke, B., Krahn, I., Gartemann, K.H., Gützkow, T., Eichler, W., Pühler, A., Schlüter, A., 2009. Detection of 140 clinically relevant antibiotic-resistance genes in the plasmid metagenome of wastewater treatment plant bacteria showing reduced susceptibility to selected antibiotics. Microbiology 155, 2306–2319.

Szmolka, A., Anjum, M.F., LaRagione, R.M., Kaszanyitzky, E.J., Nagy, B., 2012. Microarray based comparative genotyping of gentamicin resistant *Escherichia coli* strains from food animals and humans. Veterinary Microbiology 156, 110–118.

Tenover, F.C., 2006. Mechanisms of antimicrobial resistance in bacteria. American Journal of Medicine 119 (Suppl. 6A), S3–S10.

Thomson, J.M., Bonomo, R., 2005. The threat of antibiotic resistance in Gram-negative pathogenic bacteria: β-lactams in peril!. Current Opinion in Microbiology 8, 518–524.

Tilahun, B., Worku, B., Tachbele, E., Terefe, S., Kloos, H., Legesse, W., 2012. High load of multi-drug resistant nosocomial neonatal pathogens carried by cockroaches in a neonatal intensive care unit at Tikur Anbessa Specialized Hospital, Addis Ababa, Ethiopia. Antimicrobial Research and Infection Control 1, 12. http://dx.doi.org/10.1186/2047-2994-1-12.

Trevisi, E., Zecconi, A., Cogrossi, S., Razzuoli, E., Grossi, P., Amadori, M., 2014. Strategies for reduced antibiotic usage in dairy cattle farms. Research in Veterinary Science 96, 229–233.

USDA/FSIS, 2016. Serotypes Profile of *Salmonella* Isolates from Meat and Poultry Products. January 1998 through December 2014 Available at: http://www.fsis.usda.gov/wps/portal/fsis/topics/data-collection-and-reports/microbiology/annual-serotyping-reports.

Vanhoof, R., Gillis, P., Stevart, O., Boland, C., Vandenberg, O., Fux, F., Collard, J.M., Bertrand, S., 2012. Transmission of multiple resistant *Salmonella* Concord from internationally adopted children to their adoptive families and social environment: proposition of guidelines. European Journal of Clinical Microbiology and Infectious Diseases 31, 491–497.

Van Looveren, M., Daube, G., De Zutter, L., Dumont, J.M., Lammens, C., Wijdooghe, M., Vandamme, P., Jouret, M., Cornelis, M., Goossens, H., 2001. Antimicrobial susceptibilities of *Campylobacter* strains isolated from food animals in Belgium. Journal of Antimicrobial and Chemotherapy 48, 235–240.

Velkov, T., Roberts, K.D., Nation, R.L., Thompson, P.E., Li, J., 2013. Pharmacology of polymyxins: new insights into an "old" class of antibiotics. Future Microbiology 8, 711–724.

Wagenlehner, F.M.E., Wydra, S., Onda, H., Kinzig-Schippers, M., Sörgel, F., Naber, K.G., 2003. Concentrations in plasma, urinary excretion, and bactericidal activity of linezolid (600 milligrams) versus those of ciprofloxacin (500 milligrams) in healthy volunteers receiving a single oral dose. Antimicrobial Agents and Chemotherapy 47, 3789–3794.

Walsh, F., Duffy, B., 2013. The culturable soil antibiotic resistome: a community of multidrug resistant bacteria. PLoS One 8, e65567. http://dx.doi.org/10.1371/journal.pone.0065567.

Webber, M.A., Ricci, V., Whitehead, R., Patel, M., Fookes, M., Ivensc, A., Piddock, L.J.V., 2013. Clinically relevant mutant DNA gyrase alters supercoiling, changes the transcriptome, and confers multidrug resistance. mBio 4, 4. http://dx.doi.org/10.1128/mBio.00273-13.

Yamasaki, S., Nagasawa, S., Fukushima, A., Hayashi-Nishino, M., Nishino, K., 2013. Cooperation of the multidrug efflux pump and lipopolysaccharides in the intrinsic antibiotic resistance of *Salmonella enterica* serovar Typhimurium. Journal of Antimicrobial Chemotherapy 68, 1066–1070.

CHAPTER 7

# *The First 30 Years of Shiga Toxin–Producing* Escherichia coli *in Cattle Production: Incidence, Preharvest Ecology, and Management*

**Christina L. Swaggerty[1], Nicolae Corcionivoschi[2], Steven C. Ricke[3], Todd R. Callaway[4]**

*[1]United States Department of Agriculture/Agricultural Research Service, College Station, TX, United States; [2]Agri-Food and Biosciences Institute, Northern Ireland, United Kingdom; [3]University of Arkansas, Fayetteville, AR, United States; [4]University of Georgia, Athens, GA, United States*

## *Background*

*Escherichia coli* is a Gram-negative, facultative anaerobic, rod-shaped bacterium that is commonly found in the gastrointestinal (GI) tract of warm-blooded animals. The core genome of all *E. coli* strains has approximately 2200 genes, and each strain is supplemented with an "accessory genome" (Chaudhuri et al., 2010; Touchon et al., 2009), providing *E. coli* opportunities to alter its genetic composition for new types to emerge (Franz et al., 2014). There are more than 700 known serotypes, most of which are commensal; however, some serovars range from mildly to highly pathogenic in humans and are capable of causing death, especially in children, the elderly, and the immunocompromised (Karmali et al., 2010).

There are four basic classes of enterovirulent *E. coli* that can occur in food and water supplies (Kaper et al., 2004; Nataro and Kaper, 1998). Enterotoxigenic *E. coli* strains are the leading cause of travelers' diarrhea and a major cause of diarrheal disease in lower income countries. Enteropathogenic *E. coli* are common causes of infantile diarrhea and are important pathogens infecting children worldwide because of their high prevalence in both the community and hospital settings. Enteroinvasive *E. coli* produce a syndrome identical to shigellosis and are characterized by profuse diarrhea that may contain blood and mucus. Finally, there are enterohemorrhagic *E. coli* (EHEC) which are also referred to as verocytotoxin *E. coli* (VTEC) or Shiga toxin–producing *E. coli* (STEC). All three terms (EHEC, VTEC, and STEC) are used interchangeably in the literature and describe *E. coli* strains that acquired toxin genes

**Food and Feed Safety Systems and Analysis. http://dx.doi.org/10.1016/B978-0-12-811835-1.00007-5**

from *Shigella* via a gene transfer event (Kaper et al., 2004; Karmali et al., 2010) but hereafter will be referred to as STEC. Three major virulence genes (*$stx_1$*, *$stx_2$*, and *eae*) are often identified from human clinical cases of STEC-induced enteritis (Tostes et al., 2017). The STEC are very virulent with as few as 4 to 10 organisms capable of causing human illness characterized by severe gastroenteritis, enterocolitis, bloody diarrhea, and weight loss (Kothary and Babu, 2001). In severe STEC cases (typically seen in children), the toxins can cause hemolytic-uremic syndrome, which can lead to acute renal failure and death (Kothary and Babu, 2001; Nataro and Kaper, 1998).

There are several STEC, but the most widely recognized in the United States is *E. coli* O157:H7. Presently, STEC are a significant cause of bacterial-derived foodborne illness in the United States (CDC, 2016) and in 2014 resulted in more than 175,000 human illness 2400 hospitalizations, and 20 deaths with direct and indirect costs exceeding $1 billion per year (Scallan et al., 2011; Scharff, 2010). Of those illnesses, more than 63,000 and all fatalities were due to *E. coli* O157:H7, whereas the remaining 113,000 were from non-O157:H7 STEC infections (Scallan et al., 2011). Historically, most STEC-induced foodborne illnesses are associated with handling and/or consumption of contaminated or undercooked beef products (White et al., 2016).

Because cattle along with other ruminant animals are natural reservoirs for STEC, including *E. coli* O157:H7 (Karmali et al., 2010), a complete and integrated approach including preharvest and postharvest intervention strategies must be incorporated to limit human foodborne illnesses. The emergence of STEC as a leading cause of foodborne illness has had a profound impact on the beef industry both economically and how they approach food safety to ensure that a safe and wholesome product reaches the consumer. Since 1982 when *E. coli* O157:H7 was first recognized as a major foodborne pathogen, the cattle industry has spent in excess of $2B USD to combat this STEC (Kay, 2003).

Despite its initial recognition in 1982, it was not until 1993 that *E. coli* O157:H7 became regarded as a very important and threatening pathogen after a multistate outbreak associated with consumption of contaminated, undercooked ground beef patties from a national fast food chain (Bell et al., 1994; Rangel et al., 2005) that resulted in over 700 illnesses that left four individuals dead and nearly 200 of those infected with permanent injury. As a result, in 1994, the Food Safety Inspection Service (FSIS) declared *E. coli* O157:H7 an adulterant in raw ground beef. Since then, additional non-O157 serogroups of STEC have been linked to human illness outbreaks and include O26, O45, O103, O111, O121, and O145 (Bettelheim, 2007; Fremaux et al., 2007; Scallan et al., 2011); and in 2011, this group commonly referred to as "the big six" was also declared adulterants in raw, nonintact beef (FSIS, 2011). As might be expected, some of these non-O157 serotypes are more virulent to humans than others and O26 appears to come from a highly virulent lineage (Gabrielsen et al., 2015). A different STEC (O104) was linked to a European outbreak that sickened over 4000 and resulted in 50 deaths (Rahal et al., 2015). Since this declaration, emphasis has shifted to understanding general STEC ecology, rather than focusing solely on *E. coli* O157:H7 (Gill and Gill, 2010).

Initial control measures implemented by the cattle industry to combat STEC were introduced at the postharvest stage in the processing plants and targeted pathogen reduction on hides and carcasses (Koohmaraie et al., 2005) because O157:H7 and the other "big six" serotypes have been isolated from cattle feces. Postharvest intervention strategies that are focused on reducing or eliminating fecal contamination from cattle hides and carcasses are extremely important, especially in the summer months when the incidence of STEC and associated human illnesses peak (Lal et al., 2012; Naumova et al., 2007; Wells et al., 2009). Each postharvest intervention reduces pathogen load, but no single step can effectively eliminate 100% of the pathogen load; therefore a "multiple hurdle" approach that incorporates hide treatments, careful evisceration, carcass washes, steam pasteurization, and a test-and-hold policy has been incorporated with the approach that each step throughout the process will further reduce STEC numbers and/or other potential bacterial pathogens entering the food supply (Baker et al., 2016a,b; Koohmaraie et al., 2005). Although postharvest pathogen-reduction strategies have been largely successful at reducing direct foodborne illness, these processing interventions are not perfect (Arthur et al., 2007; Barkocy-Gallagher et al., 2003), in large part, because human exposure also includes indirect routes (LeJeune and Kersting, 2010; Nastasijevic, 2011). To further reduce the incidence of human illness and ensure a safe and wholesome food supply, research into preharvest pathogen-reduction strategies, interventions, and development of animal management programs for use in live cattle has grown in recent years (LeJeune and Wetzel, 2007; Oliver et al., 2008; Sargeant et al., 2007).

## *Ecology of Shiga Toxin–Producing* Escherichia coli *and Gastrointestinal Colonization*

Because cattle are a natural reservoir for *E. coli* O157:H7 (and some non-O157 STEC), this coevolution has uniquely equipped these bacteria to survive in the GI tract as commensal organisms (Law, 2000). Although *E. coli* O157:H7 can live in the rumen of cattle (Rasmussen et al., 1999), or anywhere along the GI tract, the terminal rectum is the primary site of colonization (Naylor et al., 2003; Smith et al., 2009). Cattle are not adversely impacted by the toxins produced by STEC because they do not have the appropriate toxin receptors on the cells of their GI tract and hence are "immune" to the deleterious cytotoxic effects that are observed in humans after colonization by *E. coli* O157:H7 or other STEC (Karmali et al., 2010; Pruimboom-Brees et al., 2000). Furthermore, STEC and/or the Shiga toxins alter the innate and adaptive immune responses in the host cattle by suppressing the development of cellular immunity and altering lymphoproliferative responses (Hoffman et al., 2006), affecting the expression of key cytokines and chemokines by intraepithelial lymphocytes (Moussay et al., 2006), or by downregulating costimulatory molecules that are involved in successful initiation of an adaptive immune response (Menge et al., 2015). Collectively, this means that the natural host-commensal relationship between cattle and STEC enables the organism to persist and therefore potentially be transferred to beef products that reach consumers (Ferens and Hovde, 2011).

Because STEC are persistently found colonizing the ruminant GI tract, it should be no surprise that the bacteria are isolated from fecal deposits (Maule, 2000; Yang et al., 2010) and soils (Bolton et al., 2011; Semenov et al., 2009). To further complicate matters, different STEC strains and an array of virulence genes can be isolated from cattle fecal samples beginning at birth until they reach the feedlot (Hallewell et al., 2016). Favorable conditions within the GI tract and the external environment enable *E. coli* O157:H7 or other STEC to cycle within pens and farms via fecal-oral transmission, which ensures the bacteria recirculates and continues to colonize groups or individual animals (Arthur et al., 2010; Russell and Jarvis, 2001). The frequency and numbers of *E. coli* O157:H7 that individual cattle shed varies greatly, and at times, individual cattle may be shedding over $10^4$ colony-forming units per gram feces and these cattle are referred to as supershedders (Munns et al., 2015). Horizontal transmission to pen/herd mates or reinfection of the same individual can be accentuated when a supershedder(s) is present within the population (Arthur et al., 2009, 2010; Chase-Topping et al., 2008). However, the complexity underlying supershedding cattle and the subsequent host-microbial interaction(s) is not fully understood and remains largely speculative. Thus it is apparent that the overall environment including the farm, facility, and specific pen plays an important role in STEC colonization and recirculation, as well as via direct and indirect transmission to human farm workers/visitors and consumers (Ihekweazu et al., 2012; LeJeune and Kersting, 2010; Smith et al., 2012).

Historically, researchers (including the present authors) assumed non-O157 STEC would behave in a similar physiological and ecological manner as O157. However, studies indicate that in addition to the genetic divergence seen in O157:H7 lineages (Zhang et al., 2007), there are also significant physiological differences between and within non-O157 STEC which may contribute to the ecological niche occupied in the ruminant GI tract by non-O157 serotypes (Bergholz and Whittam, 2007; Free et al., 2012; Fremaux et al., 2007), and in some instances, there is competition between O157 and non-O157 serovars (Martorelli et al., 2015). The non-O157 STEC and O157 also have varying responses to stressors (Mei et al., 2015; Verhaegen et al., 2015) and distinct sensitivities to acidification, irradiation, heat, pressures, and various antimicrobial treatments (Mei et al., 2015; Verhaegen et al., 2015). Although further investigation into these and other physiological differences must be considered as well as their function within the GI microbial population determined, it appears that O157:H7 and other STEC serotypes are well adapted to survive in the GI tract cattle (Chase-Topping et al., 2012; Monaghan et al., 2011; Polifroni et al., 2012) and can be transferred into ground beef products that reach consumers (Bosilevac and Koohmaraie, 2011; Fratamico et al., 2011). Although we now have a better understanding of how *E. coli* O157:H7 behaves in the GI tract and farm environment, overall, very little is known about the ecology of non-O157 STEC in those environments (Monaghan et al., 2011; Polifroni et al., 2012). As such, this chapter focused on what is known about *E. coli* O157:H7. We hypothesize that non-O157:H7 STEC will largely behave in a broadly similar manner to *E. coli* O157:H7 in the GI tract and farm environment; however we advise the reader that theories described herein do not always apply equally to all STEC.

Thus far, attention has focused only on the STEC residing in the GI tract of cattle and being directly passed onto processed beef products entering the food supply that reach the consumer; however, that is just one route of STEC transmission to humans. Other routes of STEC transmission to humans can occur through direct contact with a contaminated animal (either a shedding animal or an animal that has come in contact with another shedder). This route of transmission would be more prevalent among individuals who handle the animals on a daily basis, or through contact at agricultural fairs, open-access farms, or petting zoos (Beecher, 2016; Lanier et al., 2011). Another direct route of STEC transmission is from person-to-person contact with an infected patient. Indirect routes of STEC transmission to humans also happen, such as through contact with contaminated water sources. Canada's most severe *E. coli* O157:H7 outbreak occurred in 2000, when the water supply in Walkerton, ON, became contaminated by farm runoff (Hipel et al., 2003). Furthermore, in 2006, there was a deadly O157:H7 STEC outbreak that was traced back to consumption of prepackaged ready-to-eat spinach that had likely been contaminated by runoff water from a nearby cattle operation (Anonymous, 2006).

## Management Practices and Transportation

Clearly, implementing pathogen-reduction strategies directed toward environmental contamination and exposure routes in live cattle have the potential to reduce human illnesses (Rotariu et al., 2012; Smith et al., 2012) by (1) reducing the number of pathogens entering the processing plant on/in an animal will reduce the burden on the plants and render in-house intervention strategies more effective; (2) reducing horizontal pathogen transfer from infected animals during transport and lairage; (3) lowering the pathogenic bacterial burden in the environment and wastewater streams; and (4) reducing the risk to those in direct contact with animals. The current preharvest intervention strategies to reduce the incidence of human STEC illnesses acquired from beef products have been reviewed (Callaway et al., 2013a,b) but will be briefly highlighted and updated herein.

No single management practice has directly affected colonization or shedding of foodborne pathogens (Ellis-Iversen et al., 2008; Ellis-Iversen and Van Winden, 2008; LeJeune and Wetzel, 2007). Despite the lack of evidence regarding specific reductions in the number of foodborne illnesses, good management practices are critical to ensuring animal health and welfare (Morrow-Tesch, 2001) and good management of cattle is also essential for maximizing animal production and efficiency. Realistically, it is not possible to eliminate all STEC exposures in a cattle operation; however, appropriate management practices would include measures to reduce the frequency of exposure of the cattle to environmental contaminants. A preharvest management program will seek to control STEC exposure in the feed/water, maintain a clean pen environment that is appropriately drained and stocked at a reduced density with similar-aged animals, and incorporate a stringent biosecurity program to exclude wildlife to the best extent possible (FSIS, 2014).

### Biosecurity

Farm biosecurity is critical for maintaining animal health and welfare, especially with respect to animal diseases, but to date, there has been little direct impact demonstrated on foodborne pathogenic bacteria such as *E. coli* O157:H7 (Ellis-Iversen and Van Winden, 2008). Other animals on a farm, including sheep, deer, rodents, insects, birds, and boars, can be STEC carriers and may be linked to an increased risk of *E. coli* O157:H7 shedding (Bolton et al., 2012; Branham et al., 2005; Cernicchiaro et al., 2012; French et al., 2010; Rice et al., 2003; Sánchez et al., 2010; Stacey et al., 2007; Synge et al., 2003). Additional studies indicate flies, other insects, and wild migratory birds can carry STEC and other foodborne pathogens from one location to another (Ahmad et al., 2007; Cernicchiaro et al., 2012; Talley et al., 2009). Although these effects are likely minimal in their direct impact on food safety at a given farm, they are vectors for pathogens to move among "clean" groups of cattle or farms.

### Cattle Grouping

To enhance biosecurity, many farms are closed to entry by animals from other farms to prevent disease transmission, and closed herds can prevent spread of *E. coli* O157:H7 (and other pathogens) from one farm to another (Ellis-Iversen et al., 2008; Ellis-Iversen and Van Winden, 2008). *E. coli* O157:H7 can be considered common to groups of feedlot cattle housed together in pens (Smith et al., 2001), thus keeping groups together throughout their time on a farm, or in a feedlot, without introducing new members to groups that may reduce horizontal transmission between animals.

Age and stocking density are also important factors to consider in management. Young cattle shed more *E. coli* O157:H7 than older cattle (Cobbaut et al., 2009; Cray and Moon, 1995; Smith et al., 2001). Although calves cannot be separated from cows in the beef industry, there is potential benefit in keeping calves that are the same age together to prevent horizontal transmission between animals that are not the same age. Animal density may also play a role in the horizontal spread of *E. coli* O157:H7 and other foodborne pathogens (Vidovic and Korber, 2006). Densely packed animals are more likely to get contaminated by fecal material from nearby pen mates, and higher animal density is linked to increased risk of carriage and shedding of some STEC, including O157:H7 (Frank et al., 2008; Stacey et al., 2007).

### *Bedding and Pen Surfaces*

*E. coli* O157:H7 can live for extended periods in manure, soil, and other organic materials (Maule, 2000; Winfield and Groisman, 2003) and can be transmitted through the environment (Semenov et al., 2009, 2010). Foodborne pathogenic bacteria are readily found in bedding material, which facilitates the spread of the bacteria between cattle (Richards et al., 2006; Wetzel and LeJeune, 2006). Modeling research suggests increasing the frequency of bedding cleaning would reduce the survival of *E. coli* O157:H7 (Vosough Ahmadi et al., 2007).

Feedlot surfaces were thought to contain manure-like bacterial populations, but molecular analyses demonstrate that bacterial communities on feedlot surfaces are divergent from fecal bacterial populations (Durso et al., 2011). However, this finding should not be surprising as the conditions that favor bacterial survival in the GI tract (anaerobic, warm, dark) are different than those found on the surface of a feedlot (aerobic, cooler, sunlit). Overall, bedding and/or pen cleaning will not eliminate STEC from a farm or feedlot, but they could reduce the bacterial numbers and therefore slow the spread of STEC within a herd or among pen mates.

### *Manure Management*

*E. coli* O157:H7 and other STEC survive in manure and can persist for a lengthy period (up to 21 months) (Bolton et al., 2011; Fremaux et al., 2007; Varel et al., 2008). Not all STEC serovars survive comparably in manure, and this variability is likely influenced by the oxidative capacity of each strain (Franz et al., 2011). Manure can be added to soil as a fertilizer and can introduce STEC to an environment and result in STEC uptake directly by plants, including food crops (Franz and Van Bruggen, 2008; Semenov et al., 2009, 2010), and rainfall events can also wash STEC from cattle feces (stored or in fields) into drinking or irrigation water supplies (Cook et al., 2011; Oliveira et al., 2012; Pachepsky et al., 2011). Both scenarios are capable of introducing STEC into the food supply chain and therefore must be addressed during management.

### *Transportation and Lairage*

Handling and transport of cattle to processing plants, feedlots, or other farms may spread *E. coli* O157:H7 because of physical contact or fecal contamination (Mather et al., 2007). Some studies show that transport causes an increase in fecal shedding of *E. coli* O157:H7 (Bach et al., 2004), whereas others find no increase (Reicks et al., 2007; Schuehle Pfeiffer et al., 2009). Reports also indicate transporting cattle more than 100 miles doubles the risk of having STEC-positive hides at slaughter (Dewell et al., 2008) and longer transport times correspond to increased levels of *E. coli* O157:H7 shedding (Bach et al., 2004). Cattle trailers can be important fomites of *E. coli* O157:H7 to uninfected cattle and are frequently positive for *E. coli* O157:H7 when sampled (Cuesta Alonso et al., 2007; Reicks et al., 2007). It has been shown that the incidence of *E. coli* O157:H7 in transport trailers increases the risk of transmission to farms and feedlots from cattle on these trailers (Cuesta Alonso et al., 2007). Studies have shown that the transfer of *E. coli* O157:H7 to hides that occurs in lairage at processing plants accounted for more of the hide and carcass contamination than did the population of cattle leaving the feedlot (Arthur et al., 2008). The exact role of lairage and transport in the spread of STEC (and other pathogens) in cattle is not fully understood but is likely influenced by time and animal density.

### Stress

Stress in animals is not always quantifiable and is quite a complex issue (Rostagno, 2009; Verbrugghe et al., 2012). Exposure of cattle to long-term stress can suppress immune function, rendering them more susceptible to colonization (Carroll and Forsberg, 2007; Kelley, 1980; Salak-Johnson and McGlone, 2007). However, short-term stress associated with weaning, handling, or transport on immune status is less understood. Weaning is stressful to calves and can lead to increased STEC colonization (Chase-Topping et al., 2007; Herriott et al., 1998) but may not directly affect STEC shedding (Edrington et al., 2011; Synge et al., 2003). Heat stress (and methods to alleviate it) can have effects on animal health and productivity (Brown-Brandl et al., 2003), and in some cases, but not all, shedding of *E. coli* O157:H7 and *Salmonella* (Brown-Brandl et al., 2009; Callaway et al., 2006). To date, the effect of stress on colonization or shedding of *E. coli* O157:H7 remains unclear.

## Conclusions

Although most *E. coli* serotypes are not pathogenic, STEC are leading causes of foodborne illness in the United States and around the world. The most well-known and studied STEC is *E. coli* O157:H7; however this serotype does not define the STEC serotype or ecological niche. Humans are sickened by STEC typically through handling or consumption of undercooked beef or through contact with ruminant animals and/or their feces. This relationship occurs because the GI tract of ruminants (including cattle) serves as a natural reservoir of STEC due to a natural coevolution where cattle are unaffected by STEC colonization. Because of the biochemical plasticity and opportunistic nature of *E. coli*, STEC can persist in the fiercely competitive microbial ecosystem of the rumen and GI tract of cattle. Clearly, understanding the ecology of STEC and implementing pathogen-reduction strategies directed toward environmental contamination and exposure routes in live cattle have the potential to reduce human illnesses. The ability of this opportunistic pathogen to colonize cattle asymptomatically but still cause serious illness in human consumers marks STEC as a unique threat to the safety and integrity of our food supply that must be addressed in specific ways, both during meat processing as well as on the farm and feedlot in the live animal.

## References

Ahmad, A., Nagaraja, T.G., Zurek, L., 2007. Transmission of *Escherichia coli* O157:H7 to cattle by house flies. Preventive Veterinary Medicine 80, 74–81.

Anonymous, 2006. Morbidity and Mortality Weekly Report. Ongoing Multistate Outbreak of *Escherichia coli* Serotype O157:H7 Infections Associated with Consumption of Fresh Spinach–United States, September 2006, vol. 55, pp. 1045–1046.

Arthur, T.M., Bosilevac, J.M., Brichta-Harhay, D.M., Guerini, M.N., Kalchayanand, N., Shackelford, S.D., Wheeler, T.L., Koohmaraie, M., 2007. Transportation and lairage environment effects on prevalence, numbers, and diversity of *Escherichia coli* O157:H7 on hides and carcasses of beef cattle at processing. Journal of Food Protection 70, 280–286.

Arthur, T.M., Bosilevac, J.M., Brichta-Harhay, D.M., Kalchayanand, N., King, D.A., Shackelford, S.D., Wheeler, T.L., Koohmaraie, M., 2008. Source tracking of *Escherichia coli* O157:H7 and *Salmonella* contamination in the lairage environment at commercial U.S. beef processing plants and identification of an effective intervention. Journal of Food Protection 71, 1752–1760.

Arthur, T.M., Brichta-Harhay, D.M., Bosilevac, J.M., Kalchayanand, N., Shackelford, S.D., Wheeler, T.L., Koohmaraie, M., 2010. Super shedding of *Escherichia coli* O157:H7 by cattle and the impact on beef carcass contamination. Meat Science 86, 32–37.

Arthur, T.M., Keen, J.E., Bosilevac, J.M., Brichta-Harhay, D.M., Kalchayanand, N., Shackelford, S.D., Wheeler, T.L., Nou, X., Koohmaraie, M., 2009. Longitudinal study of *Escherichia coli* O157:H7 in a beef cattle feedlot and role of high-level shedders in hide contamination. Applied and Environmental Microbiology 75, 6515–6523.

Bach, S.J., Mcallister, T.A., Mears, G.J., Schwartzkopf-Genswein, K.S., 2004. Long-haul transport and lack of preconditioning increases fecal shedding of *Escherichia coli* and *Escherichia coli* O157:H7 by calves. Journal of Food Protection 67, 672–678.

Baker, C.A., Rubinelli, P.M., Park, S.H., Carbonero, F., Ricke, S.C., 2016a. Shiga toxin-producing *Escherichia coli* in food: incidence, ecology, and detection strategies. Food Control 59, 407–419.

Baker, C.A., Rubinelli, P.M., Park, S.H., Ricke, S.C., 2016b. Immuno-based detection of Shiga toxin-producing pathogenic *Escherichia coli* in food—a review on current approaches and potential strategies for optimization. Critical Reviews in Microbiology 42, 656–675.

Barkocy-Gallagher, G.A., Arthur, T.M., Rivera-Betancourt, M., Nou, X., Shackelford, S.D., Wheeler, T.L., Koohmaraie, M., 2003. Seasonal prevalence of shiga toxin-producing *Escherichia coli*, including O157:H7 and non-O157 serotypes, and *Salmonella* in commercial beef processing plants. Journal of Food Protection 66, 1978–1986.

Beecher, C., 2016. From Fairs to Nativity Scenes, Animal Attractions Harbor Danger. Food Safety News Available: http://www.foodsafetynews.com/2016/12/from-fairs-to-nativity-scenes-animal-attractions-harbor-dangers/-.WEbr3XeZPfY.

Bell, B.P., Goldoft, M., Griffin, P.M., Davis, M.A., Gordon, D.C., Tarr, P.I., Bartleson, C.A., Lewis, J.H., Barrett, T.J., Wells, J.G., et al., 1994. A multistate outbreak of *Escherichia coli* O157:H7-associated bloody diarrhea and hemolytic uremic syndrome from hamburgers. The Washington experience. Journal of the American Medical Association 272, 1349–1353.

Bergholz, T.M., Whittam, T.S., 2007. Variation in acid resistance among enterohaemorrhagic *Escherichia coli* in a simulated gastric environment. Journal of Applied Microbiology 102, 352–362.

Bettelheim, K.A., 2007. The non-O157 Shiga-toxigenic (verocytotoxigenic) *Escherichia coli*; under-rated pathogens. Critical Reviews in Microbiology 33, 67–87.

Bolton, D.J., Monaghan, A., Byrne, B., Fanning, S., Sweeney, T., Mcdowell, D.A., 2011. Incidence and survival of non-O157 verocytotoxigenic *Escherichia coli* in soil. Journal of Applied Microbiology 111, 484–490.

Bolton, D.J., O'neill, C.J., Fanning, S., 2012. A preliminary study of *Salmonella*, verocytotoxigenic *Escherichia coli/Escherichia coli* O157 and *Campylobacter* on four mixed farms. Zoonoses and Public Health 59, 217–228.

Bosilevac, J.M., Koohmaraie, M., 2011. Prevalence and characterization of non-O157 shiga toxin-producing *Escherichia coli* isolates from commercial ground beef in the United States. Applied and Environmental Microbiology 77, 2103–2112.

Branham, L.A., Carr, M.A., Scott, C.B., Callaway, T.R., 2005. *E. coli* O157 and *Salmonella* spp. in white-tailed deer and livestock. Current Issues in Intestinal Microbiology 6, 25–29.

Brown-Brandl, T.M., Berry, E.D., Wells, J.E., Arthur, T.M., Nienaber, J.A., 2009. Impacts of individual animal response to heat and handling stresses on *Escherichia coli* and *E. coli* O157:H7 fecal shedding by feedlot cattle. Foodborne Pathogens and Disease 6, 855–864.

Brown-Brandl, T.M., Nienaber, J.A., Eigenberg, R.A., Hahn, G.L., Freetly, H., 2003. Thermoregulatory responses of feeder cattle. Journal of Thermal Biology 28, 149–157.

Callaway, T.R., Edrington, T.S., Loneragan, G.H., Carr, M.A., Nisbet, D.J., 2013a. Current and near-market intervention strategies for reducing Shiga Toxin-producing *Escherichia coli* (STEC) shedding in cattle. Agriculture, Food and Analytical Bacteriology 3, 103–120.

Callaway, T.R., Edrington, T.S., Loneragan, G.H., Carr, M.A., Nisbet, D.J., 2013b. Shiga toxin-producing *Escherichia coli* (STEC) ecology in cattle and management based operations for reducing fecal shedding. Agriculture, Food and Analytical Bacteriology 3, 39–69.

Callaway, T.R., Morrow, J.L., Edrington, T.S., Genovese, K.J., Dowd, S., Carroll, J., Dailey, J.W., Harvey, R.B., Poole, T.L., Anderson, R.C., Nisbet, D.J., 2006. Social stress increases fecal shedding of *Salmonella* Typhimurium by early weaned piglets. Current Issues in Intestinal Microbiology 7, 65–72.

Carroll, J.A., Forsberg, N.E., 2007. Influence of stress and nutrition on cattle immunity. In: Olson, K.C., Hollis, L.C. (Eds.), Veterinary Clinics of North America: Food Animal Practice. Elsevier Saunders, Amsterdam, The Netherlands.

CDC, 2016. Surveillance for Foodborne Disease Outbreaks, United States, 2014. Annual Report. US Department of Health and Human Services, CDC, Atlanta, GA.

Cernicchiaro, N., Pearl, D.L., Mcewen, S.A., Harpster, L., Homan, H.J., Linz, G.M., Lejeune, J.T., 2012. Association of wild bird density and farm management factors with the prevalence of *E. coli* O157 in dairy herds in Ohio (2007–2009). Zoonoses and Public Health 59, 320–329.

Chase-Topping, M., Gally, D., Low, C., Matthews, L., Woolhouse, M., 2008. Super-shedding and the link between human infection and livestock carriage of *Escherichia coli* O157. Nature Reviews in Microbiology 6, 904–912.

Chase-Topping, M.E., Mckendrick, I.J., Pearce, M.C., Macdonald, P., Matthews, L., Halliday, J., Allison, L., Fenlon, D., Low, J.C., Gunn, G., Woolhouse, M.E.J., 2007. Risk factors for the presence of high-level shedders of *Escherichia coli* O157 on Scottish farms. Journal of Clinical Microbiology 45, 1594–1603.

Chase-Topping, M.E., Rosser, T., Allison, L.J., Courcier, E., Evans, J., Mckendrick, I.J., Pearce, M.C., Handel, I., Caprioli, A., Karch, H., Hanson, M.F., Pollock, K.G.J., Locking, M.E., Woolhouse, M.E.J., Matthews, L., Low, J.C., Gally, D.L., 2012. Pathogenic potential to humans of bovine *Escherichia coli* O26, Scotland. Emerging Infectious Diseases 18, 439–448.

Chaudhuri, R.R., Sebaihia, M., Hobman, J.L., Webber, M.A., Leyton, D.L., Goldberg, M.D., Cunningham, A.F., Scott-Tucker, A., Ferguson, P.R., Thomas, C.M., Frankel, G., Tang, C.M., Dudley, E.G., Roberts, I.S., Rasko, D.A., Pallen, M.J., Parkhill, J., Nataro, J.P., Thomson, N.R., Henderson, I.R., 2010. Complete genome sequence and comparative metabolic profiling of the prototypical enteroaggregative *Escherichia coli* strain 042. PLoS One 5, e8801.

Cobbaut, K., Berkvens, D., Houf, K., De Deken, R., De Zutter, L., 2009. *Escherichia coli* O157 prevalence in different cattle farm types and identification of potential risk factors. Journal of Food Protection 72, 1848–1853.

Cook, K.L., Bolster, C.H., Ayers, K.A., Reynolds, D.N., 2011. *Escherichia coli* diversity in livestock manures and agriculturally impacted stream waters. Current Microbiology 63, 439–449.

Cray, W.C., Moon, H.W., 1995. Experimental infection of calves and adult cattle with *Escherichia coli* O157:H7. Applied and Environmental Microbiology 61, 1586–1590.

Cuesta Alonso, E.P., Gilliland, S.E., Krehbiel, C.R., 2007. Incidence and toxin production ability of *Escherichia coli* O157:H7 isolated from cattle trucks. Journal of Food Protection 70, 2383–2385.

Dewell, G.A., Simpson, C.A., Dewell, R.D., Hyatt, D.R., Belk, K.E., Scanga, J.A., Morley, P.S., Grandin, T., Smith, G.C., Dargatz, D.A., Wagner, B.A., Salman, M.D., 2008. Impact of transportation and lairage on hide contamination with *Escherichia coli* O157 in finished beef cattle. Journal of Food Protection 71, 1114–1118.

Durso, L.M., Harhay, G.P., Smith, T.P.L., Bono, J.L., Desantis, T.Z., Clawson, M.L., 2011. Bacterial community analysis of beef cattle feedlots reveals pen surface is distinct from feces. Foodborne Pathogens and Disease 8, 647–649.

Edrington, T.S., Carter, B.H., Farrow, R.L., Islas, A., Hagevoort, G.R., Friend, T.H., Callaway, T.R., Anderson, R.C., Nisbet, D.J., 2011. Influence of weaning on fecal shedding of pathogenic bacteria in dairy calves. Foodborne Pathogens and Disease 8, 395–401.

Ellis-Iversen, J., Smith, R.P., Van Winden, S., Paiba, G.A., Watson, E., Snow, L.C., Cook, A.J.C., 2008. Farm practices to control *E. coli* O157 in young cattle—a randomised controlled trial. Veterinary Research 39.

Ellis-Iversen, J., Van Winden, S., 2008. Control of *E. coli* O157 (VTEC) by applied management practices. Cattle Practice 16, 54.

Ferens, W.A., Hovde, C.J., 2011. *Escherichia coli* O157:H7: animal reservoir and sources of human infection. Foodborne Pathogens and Disease 8, 465–487.

Frank, C., Kapfhammer, S., Werber, D., Stark, K., Held, L., 2008. Cattle density and shiga toxin-producing *Escherichia coli* infection in Germany: increased risk for most but not all serogroups. Vector-Borne and Zoonotic Diseases 8, 635–643.

Franz, E., Delaquis, P., Morabito, S., Beutin, L., Gobius, K., Rasko, D.A., Bono, J., French, N., Osek, J., Lindstedt, B.A., Muniesa, M., Manning, S., Lejeune, J., Callaway, T., Beatson, S., Eppinger, M., Dallman, T., Forbes, K.J., Aarts, H., Pearl, D.L., Gannon, V.P.J., Laing, C.R., Strachan, N.J.C., 2014. Exploiting the explosion of information associated with whole genome sequencing to tackle Shiga toxin-producing *Escherichia coli* (STEC) in global food production systems. International Journal of Food Microbiology 187, 57–72.

Franz, E., Van Bruggen, A.H.C., 2008. Ecology of *E. coli* O157:H7 and *Salmonella enterica* in the primary vegetable production chain. Critical Reviews in Microbiology 34, 143–161.

Franz, E., Van Hoek, A.H., Bouw, E., Aarts, H.J.M., 2011. Variability of *Escherichia coli* O157 strain survival in manure-amended soil in relation to strain origin, virulence profile, and carbon nutrition profile. Applied and Environmental Microbiology 77, 8088–8096.

Fratamico, P.M., Bagi, L.K., Cray, W.C., Narang, N., Yan, X., Medina, M., Liu, Y., 2011. Detection by multiplex real-time polymerase chain reaction assays and isolation of shiga toxin-producing *Escherichia coli* serogroups O26, O45, O103, O111, O121, and O145 in ground beef. Foodborne Pathogens and Disease 8, 601–607.

Free, A.L., Duoss, H.A., Bergeron, L.V., Shields-Menard, S.A., Ward, E., Callaway, T.R., Carroll, J.A., Schmidt, T.B., Donaldson, J.R., 2012. Survival of O157:H7 and non-O157 serogroups of *Escherichia coli* in bovine rumen fluid and bile salts. Foodborne Pathogens and Disease 9, 1010–1014.

Fremaux, B., Delignette-Muller, M.L., Prigent-Combaret, C., Gleizal, A., Vernozy-Rozand, C., 2007. Growth and survival of non-O157:H7 Shiga-toxin-producing *Escherichia coli* in cow manure. Journal of Applied Microbiology 102, 89–99.

French, E., Rodriguez-Palacios, A., Lejeune, J.T., 2010. Enteric bacterial pathogens with zoonotic potential isolated from farm-raised deer. Foodborne Pathogens and Disease 7, 1031–1037.

FSIS, 2011. Shiga toxin-producing *Escherichia coli* in certain raw beef products. In: Agriculture DO. Government Publishing Office, Washington DC.

FSIS, 2014. Pre-harvest management controls and intervention option for reducing Shiga toxin-producing *Escherichia coli* shedding in cattle: an overview of current research. In: Agriculture DO. Government Publishing Group, Washington DC.

Gabrielsen, C., Drablos, F., Afset, J.E., 2015. Genome sequences of 11 shiga toxin-producing *Escherichia coli* strains. Genome Announcements 3 e00418–15.

Gill, A., Gill, O.C., 2010. Non-O157 verotoxigenic *Escherichia coli* and beef: a Canadian perspective. Canadian Journal of Veterinary Research 74, 161–169.

Hallewell, J., Reuter, T., Stanford, K., Topp, E., Alexander, T.W., 2016. Monitoring seven potentially pathogenic *Escherichia coli* serogroups in a closed herd of beef cattle from weaning to finishing phases. Foodborne Pathogens and Disease 13, 661–667.

Herriott, D.E., Hancock, D.D., Ebel, E.D., Carpenter, L.V., Rice, D.H., Besser, T.E., 1998. Association of herd management factors with colonization of dairy cattle by shiga toxin-positive *Escherichia coli* O157. Journal of Food Protection 61, 802–807.

Hipel, K.W., Kilgour, D.M., Zhou, N.Z., 2003. Risk analysis of the Walkerton drinking water crisis. Canadian Water Resources Journal 28, 395–419.

Hoffman, M.A., Menge, C., Casey, T.A., Laegreid, W., Bosworth, B.T., Dean-Nystrom, E.A., 2006. Bovine immune response to shiga-toxigenic *Escherichia coli* O157:H7. Clinical and Vaccine Immunology 13, 1322–1327.

Ihekweazu, C., Carroll, K., Adak, B., Smith, G., Pritchard, G.C., Gillespie, I.A., Verlander, N.Q., Harvey-Vince, L., Reacher, M., Edeghere, O., Sultan, B., Cooper, R., Morgan, G., Kinross, P.T.N., Boxall, N.S., Iversen, A., Bickler, G., 2012. Large outbreak of verocytotoxin-producing *Escherichia coli* O157 infection in visitors to a petting farm in South East England, 2009. Epidemiology and Infection 140, 1400–1413.

Kaper, J.B., Nataro, J.P., Mobley, H.L.T., 2004. Pathogenic *Escherichia coli*. Nature Reviews in Microbiology 2, 123–140.

Karmali, M.A., Gannon, V., Sargeant, J.M., 2010. Verocytotoxin-producing *Escherichia coli* (VTEC). Veterinary Microbiology 140, 360–370.

Kay, S., 2003. *E. coli* O157:H7: The costs during the past 10 years. Meat and Poultry News.

Kelley, K.W., 1980. Stress and immune function: a bibliographic review. Annales De Recherches Veterinaires 11, 445–461.

Koohmaraie, M., Arthur, T.M., Bosilevac, J.M., Guerini, M., Shackelford, S.D., Wheeler, T.L., 2005. Post-harvest interventions to reduce/eliminate pathogens in beef. Meat Science 71, 79–91.

Kothary, M.H., Babu, U.S., 2001. Infective dose of foodborne pathogens in volunteers: a review. Journal of Food Safety 21, 49–73.

Lal, A., Hales, S., French, N., Baker, M.G., 2012. Seasonality in human zoonotic enteric diseases: a systematic review. PLoS One 7, e31883–e31893.

Lanier, W.A., Hall, J.M., Herlihy, R.K., Rolfs, R.T., Wagner, J.M., Smith, L.H., Hyytia-Trees, E.K., 2011. Outbreak of Shiga-toxigenic *Escherichia coli* O157:H7 infections associated with rodeo attendance, Utah and Idaho, 2009. Foodborne Pathogens and Disease 8, 1131–1133.

Law, D., 2000. The history and evolution of *Escherichia coli* O157 and other shiga toxin-producing *E. coli*. World Journal of Microbiology & Biotechnology 16, 701–709.

LeJeune, J., Kersting, A., 2010. Zoonoses: an occupational hazard for livestock workers and a public health concern for rural communities. Journal of Agricultural Safety and Health 16, 161–179.

LeJeune, J.T., Wetzel, A.N., 2007. Preharvest control of *Escherichia coli* O157 in cattle. Journal of Animal Science 85, e73–80.

Martorelli, L., Hovde, C.J., Vilte, D.A., Albanese, A., Zotta, E., Ibarra, C., Cantet, R.J., Mercado, E.C., Cataldi, A., 2015. Impact of infection dose and previous serum antibodies against the locus of enterocyte effacement proteins on *Escherichia coli* O157:H7 shedding in calves following experimental infection. Biomed Research International 2015, 290679.

Mather, A.E., Innocent, G.T., Mcewen, S.A., Reilly, W.J., Taylor, D.J., Steele, W.B., Gunn, G.J., Ternent, H.E., Reid, S.W.J., Mellor, D.J., 2007. Risk factors for hide contamination of Scottish cattle at slaughter with *Escherichia coli* O157. Preventive Veterinary Medicine 80, 257–270.

Maule, A., 2000. Survival of vero-cytotoxigenic *Escherichia coli* O157:H7 in soil, water and on surfaces. Journal of Applied Microbiology 88, 71S–78S.

Mei, G.Y., Tang, J., Carey, C., Bach, S., Kostrzynska, M., 2015. The effect of oxidative stress on gene expression of Shiga toxin-producing *Escherichia coli* (STEC) O157:H7 and non-O157 serotypes. International Journal of Food Microbiology 215, 7–15.

Menge, C., Loos, D., Bridger, P.S., Barth, S., Werling, D., Baljer, G., 2015. Bovine macrophages sense *Escherichia coli* Shiga toxin 1. Innate Immunity 21, 655–664.

Monaghan, A.I., Byrne, B., Fanning, S., Sweeney, T., Mcdowell, D., Bolton, D.J., 2011. Serotypes and virulence profiles of non-O157 shiga toxin-producing *Escherichia coli* isolates from bovine farms. Applied and Environmental Microbiology 77, 8662–8668.

Morrow-Tesch, J.L., 2001. Evaluating management practices for their impact on welfare. Journal of the American Veterinary Medical Association 219, 1374–1376.

Moussay, E., Stamm, I., Taubert, A., Baljer, G., Menge, C., 2006. *Escherichia coli* Shiga toxin 1 enhances il-4 transcripts in bovine ileal intraepithelial lymphocytes. Veterinary Immunology and Immunopathology 113, 367–382.

Munns, K.D., Selinger, L.B., Stanford, K., Guan, L., Callaway, T.R., Mcallister, T.A., 2015. Perspectives on super-shedding of *Escherichia coli* O157:H7 by cattle. Foodborne Pathogens and Disease 12, 89–103.

Nastasijevic, I., 2011. STEC O157 in the beef chain - risk assessment and management. CAB Reviews: Perspectives in Agriculture, Veterinary Science, Nutrition and Natural Resources 6, 61–80.

Nataro, J.P., Kaper, J.B., 1998. Diarrheagenic *Escherichia coli*. Clinical Microbiology Reviews 11, 142–201.

Naumova, E.N., Jagai, J.S., Matyas, B., Demaria Jr., A., Macneill, I.B., Griffiths, J.K., 2007. Seasonality in six enterically transmitted diseases and ambient temperature. Epidemiology and Infection 135, 281–292.

Naylor, S.W., Low, J.C., Besser, T.E., Mahajan, A., Gunn, G.J., Pearce, M.C., Mckendrick, I.J., Smith, D.G.E., Gally, D.L., 2003. Lymphoid follicle-dense mucosa at the terminal rectum is the principal site of colonization of enterohaemorrhagic *Escherichia coli* O157:H7 in the bovine host. Infection and Immunity 71, 1505–1512.

Oliveira, M., Viñas, I., Usall, J., Anguera, M., Abadias, M., 2012. Presence and survival of *Escherichia coli* O157:H7 on lettuce leaves and in soil treated with contaminated compost and irrigation water. International Journal of Food Microbiology 156, 133–140.

Oliver, S.P., Patel, D.A., Callaway, T.R., Torrence, M.E., 2008. ASAS centennial paper: developments and future outlook for preharvest food safety. Journal of Animal Science 87, 419–437.

Pachepsky, Y., Shelton, D.R., Mclain, J.E.T., Patel, J., Mandrell, R.E., 2011. Irrigation waters as a source of pathogenic microorganisms in produce. A review. Advances in Agronomy 113, 73–138.

Polifroni, R., Etcheverría, A.I., Sanz, M.E., Cepeda, R.E., Krüger, A., Lucchesi, P.M.A., Fernández, D., Parma, A.E., Padola, N.L., 2012. Molecular characterization of shiga toxin-producing *Escherichia coli* isolated from the environment of a dairy farm. Current Microbiology 1–7.

Pruimboom-Brees, I.M., Morgan, T.W., Ackermann, M.R., Nystrom, E.D., Samuel, J.E., Cornick, N.A., Moon, H.W., 2000. Cattle lack vascular receptors for *Escherichia coli* O157:H7 shiga toxins. Proceedings of the National Academy of Sciences 97, 10325–10329.

Rahal, E.A., Fadlallah, S.M., Nassar, F.J., Kazzi, N., Matar, G.M., 2015. Approaches to treatment of emerging Shiga toxin-producing *Escherichia coli* infections highlighting the O104:H4 serotype. Frontiers in Cellular and Infection Microbiology 5, 24.

Rangel, J.M., Sparling, P.H., Crowe, C., Griffin, P.M., Swerdlow, D.L., 2005. Epidemiology of *Escherichia coli* O157:H7 outbreaks, United States, 1982–2002. Emerging Infectious Diseases 11, 603–609.

Rasmussen, M.A., Wickman, T.L., Cray, W.C., Casey, T.A., 1999. *Escherichia coli* O157:H7 and the Rumen Environment. *E. coli* O157 in Farm Animals. CAB International.

Reicks, A.L., Brashears, M.M., Adams, K.D., Brooks, J.C., Blanton, J.R., Miller, M.F., 2007. Impact of transportation of feedlot cattle to the harvest facility on the prevalence of *Escherichia coli* O157:H7, *Salmonella*, and total aerobic microorganisms on hides. Journal of Food Protection 70, 17–21.

Rice, D.H., Hancock, D.D., Besser, T.E., 2003. Faecal culture of wild animals for *Escherichia coli* O157:H7. Veterinary Record 152, 82–83.

Richards, H.A., Perez-Conesa, D., Doane, C.A., Gillespie, B.E., Mount, J.R., Oliver, S.P., Pangloli, P., Draughon, F.A., 2006. Genetic characterization of a diverse *Escherichia coli* O157:H7 population from a variety of farm environments. Foodborne Pathogens and Disease 3, 259–265.

Rostagno, M.H., 2009. Can stress in farm animals increase food safety risk? Foodborne Pathogens and Disease 6, 767–776.

Rotariu, O., Ogden, I., Macritchie, L., Forbes, K., Williams, A., Cross, P., Hunter, C., Teunis, P., Strachan, N., 2012. Combining risk assessment and epidemiological risk factors to elucidate the sources of human *E. coli* O157 infection. Epidemiology and Infection 140, 1414–1429.

Russell, J.B., Jarvis, G.N., 2001. Practical mechanisms for interrupting the oral-fecal lifecycle of *Escherichia coli*. Journal of Molecular Microbiology and Biotechnology 3, 265–272.

Salak-Johnson, J.L., Mcglone, J.J., 2007. Making sense of apparently conflicting data: stress and immunity in swine and cattle. Journal of Animal Science 85, E81–E88.

Sánchez, S., Martínez, R., García, A., Vidal, D., Blanco, J., Blanco, M., Blanco, J.E., Mora, A., Herrera-León, S., Echeita, A., Alonso, J.M., Rey, J., 2010. Detection and characterisation of O157:H7 and non-O157 Shiga toxin-producing *Escherichia coli* in wild boars. Veterinary Microbiology 143, 420–423.

Sargeant, J.M., Amezcua, M.R., Rajic, A., Waddell, L., 2007. Pre-harvest interventions to reduce the shedding of *E. coli* O157 in the faeces of weaned domestic ruminants: a systematic review. Zoonoses and Public Health 54, 260–277.

Scallan, E., Hoekstra, R.M., Angulo, F.J., Tauxe, R.V., Widdowson, M.-A., Roy, S.L., Jones, J.L., Griffin, P.M., 2011. Foodborne illness acquired in the United States–Major pathogens. Emerging Infectious Diseases 17, 7–15.

Scharff, R.L., 2010. Health-related Costs from Foodborne Illness in the United States. Georgetown University, Washington, DC. Available: http://publichealth.lacounty.gov/eh/docs/ReportPublication/HlthRelatedCostsFromFoodborneIllinessUS.pdf.

Schuehle Pfeiffer, C.E., King, D.A., Lucia, L.M., Cabrera-Diaz, E., Acuff, G.R., Randel, R.D., Welsh Jr., T.H., Oliphint, R.A., Curley Jr., K.O., Vann, R.C., Savell, J.W., 2009. Influence of transportation stress and animal temperament on fecal shedding of *Escherichia coli* O157:H7 in feedlot cattle. Meat Science 81, 300–306.

Semenov, A.M., Kuprianov, A.A., Van Bruggen, A.H.C., 2010. Transfer of enteric pathogens to successive habitats as part of microbial cycles. Microbial Ecology 60, 239–249.

Semenov, A.V., Van Overbeek, L., Van Bruggen, A.H.C., 2009. Percolation and survival of *Escherichia coli* O157:H7 and *Salmonella enterica* serovar Typhimurium in soil amended with contaminated dairy manure or slurry. Applied and Environmental Microbiology 75, 3206–3215.

Smith, B.A., Fazil, A., Lammerding, A.M., 2012. A risk assessment model for *Escherichia coli* O157:H7 in ground beef and beef cuts in Canada: evaluating the effects of interventions. Food Control 29, 364–381.

Smith, D., Blackford, M., Younts, S., Moxley, R., Gray, J., Hungerford, L., Milton, T., Klopfenstein, T., 2001. Ecological relationships between the prevalence of cattle shedding *Escherichia coli* O157:H7 and characteristics of the cattle or conditions of the feedlot pen. Journal of Food Protection 64, 1899–1903.

Smith, D.R., Moxley, R.A., Peterson, R.E., Klopfenstein, T.J., Erickson, G.E., Bretschneider, G., Berberov, E.M., Clowser, S., 2009. A two-dose regimen of a vaccine against type III secreted proteins reduced *Escherichia coli* O157:H7 colonization of the terminal rectum in beef cattle in commercial feedlots. Foodborne Pathogens and Disease 6, 155–161.

Stacey, K.F., Parsons, D.J., Christiansen, K.H., Burton, C.H., 2007. Assessing the effect of interventions on the risk of cattle and sheep carrying *Escherichia coli* O157:H7 to the abattoir using a stochastic model. Preventive Veterinary Medicine 79, 32–45.

Synge, B.A., Chase-Topping, M.E., Hopkins, G.F., Mckendrick, I.J., Thomson-Carter, F., Gray, D., Rusbridge, S.M., Munro, F.I., Foster, G., Gunn, G.J., 2003. Factors influencing the shedding of verocytotoxin-producing *Escherichia coli* O157 by beef suckler cows. Epidemiology and Infection 130, 301–312.

Talley, J.L., Wayadande, A.C., Wasala, L.P., Gerry, A.C., Fletcher, J., Desilva, U., Gilliland, S.E., 2009. Association of *Escherichia coli* O157:H7 with filth flies (Muscidae and Calliphoridae) captured in leafy greens fields and experimental transmission of *E. coli* O157:H7 to spinach leaves by house flies (diptera: Muscidae). Journal of Food Protection 72, 1547–1552.

Tostes, R., Goji, N., Amoako, K., Chui, L., Kastelic, J., Devinney, R., Stanford, K., Reuter, T., 2017. Subtyping *Escherichia coli* virulence genes isolated from feces of beef cattle and clinical cases in Alberta. Foodborne Pathogens and Disease 14, 35–42.

Touchon, M., Hoede, C., Tenaillon, O., Barbe, V., Baeriswyl, S., Bidet, P., Bingen, E., Bonacorsi, S., Bouchier, C., Bouvet, O., Calteau, A., Chiapello, H., Clermont, O., Cruveiller, S., Danchin, A., Diard, M., Dossat, C., Karoui, M.E., Frapy, E., Garry, L., Ghigo, J.M., Gilles, A.M., Johnson, J., Le Bouguenec, C., Lescat, M., Mangenot, S., Martinez-Jehanne, V., Matic, I., Nassif, X., Oztas, S., Petit, M.A., Pichon, C., Rouy, Z., Ruf, C.S., Schneider, D., Tourret, J., Vacherie, B., Vallenet, D., Medigue, C., Rocha, E.P., Denamur, E., 2009. Organised genome dynamics in the *Escherichia coli* species results in highly diverse adaptive paths. PLoS Genetics 5, e1000344.

Varel, V.H., Wells, J.E., Berry, E.D., Spiehs, M.J., Miller, D.N., Ferrell, C.L., Shackelford, S.D., Koohmaraie, M., 2008. Odorant production and persistence of *Escherichia coli* in manure slurries from cattle fed zero, twenty, forty, or sixty percent wet distillers grains with solubles. Journal of Animal Science 86, 3617–3627.

Verbrugghe, E., Boyen, F., Gaastra, W., Bekhuis, L., Leyman, B., Van Parys, A., Haesebrouck, F., Pasmans, F., 2012. The complex interplay between stress and bacterial infections in animals. Veterinary Microbiology 155, 115–127.

Verhaegen, B., De Reu, K., Heyndrickx, M., Van Damme, I., De Zutter, L., 2015. Growth of stressed strains of four non-O157 shiga toxin-producing *Escherichia coli* serogroups in five enrichment broths. Journal of Food Protection 78, 1960–1966.

Vidovic, S., Korber, D.R., 2006. Prevalence of *Escherichia coli* O157 in Saskatchewan cattle: characterization of isolates by using random amplified polymorphic DNA PCR, antibiotic resistance profiles, and pathogenicity determinants. Applied and Environmental Microbiology 72, 4347–4355.

Vosough Ahmadi, B., Frankena, K., Turner, J., Velthuis, A.G.J., Hogeveen, H., Huirne, R.B.M., 2007. Effectiveness of simulated interventions in reducing the estimated prevalence of *E. coli* O157:H7 in lactating cows in dairy herds. Veterinary Research 38, 755–771.

Wells, J.E., Shackelford, S.D., Berry, E.D., Kalchayanand, N., Guerini, M.N., Varel, V.H., Arthur, T.M., Bosilevac, J.M., Freetly, H.C., Wheeler, T.L., Ferrell, C.L., Koohmaraie, M., 2009. Prevalence and level of *Escherichia coli* O157:H7 in feces and on hides of feed lot steers fed diets with or without wet distillers grains with solubles. Journal of Food Protection 72, 1624–1633.

Wetzel, A.N., LeJeune, J.T., 2006. Clonal dissemination of *Escherichia coli* O157:H7 subtypes among dairy farms in Northeast Ohio. Applied and Environmental Microbiology 72, 2621–2626.

White, A., Cronquist, A., Bedrick, E.J., Scallan, E., 2016. Food source prediction of shiga toxin-producing *Escherichia coli* outbreaks using demographic and outbreak characteristics, United States, 1998–2014. Foodborne Pathogens and Disease 13, 527–534.

Winfield, M.D., Groisman, E.A., 2003. Role of nonhost environments in the lifestyles of *Salmonella* and *Escherichia coli*. Applied and Environmental Microbiology 69, 3687–3694.

Yang, H.E., Yang, W.Z., Mckinnon, J.J., Alexander, T.W., Li, Y.L., Mcallister, T.A., 2010. Survival of *Escherichia coli* O157:H7 in ruminal or fecal contents incubated with corn or wheat dried distillers' grains with solubles. Canadian Journal of Microbiology 56, 890–895.

Zhang, Y., Laing, C., Steele, M., Ziebell, K., Johnson, R., Benson, A.K., Taboada, E., Gannon, V.P.J., 2007. Genome evolution in major *Escherichia coli* O157:H7 lineages. BMC Genomics 8, 121–137.

CHAPTER 8

# *The First 30 Years of Shiga Toxin–Producing* Escherichia coli *in Cattle Production: Preharvest Intervention Strategies*

**Christina L. Swaggerty[1], Ester Grilli[2], Andrea Piva[2], Nicolae Corcionivoschi[3], Steven C. Ricke[4], Todd R. Callaway[5]**

*[1]United States Department of Agriculture/Agricultural Research Service, College Station, TX, United States; [2]University of Bologna, Ozzano Emilia, Italy; [3]Agri-Food and Biosciences Institute, Northern Ireland, United Kingdom; [4]University of Arkansas, Fayetteville, AR, United States; [5]University of Georgia, Athens, GA, United States*

## *Background*

Shiga toxin–producing *Escherichia coli* (STEC) are the fourth leading cause of bacterial-derived foodborne illness in the United States (CDC, 2016). Cattle are recognized as a natural reservoir for STEC, including *E. coli* O157:H7 (Karmali et al., 2010), and historically, most STEC-induced foodborne illnesses are associated with handling/consuming contaminated or undercooked beef products (White et al., 2016). In 1994, the Food Safety Inspection Service (FSIS) declared *E. coli* O157:H7 an adulterant in raw ground beef, and in 2011, additional STEC serogroups were declared adulterants (FSIS, 2011). Initial postharvest control measures implemented by the cattle industry to combat STEC were successful; however, to further reduce the incidence of human illness and ensure a safe and wholesome food supply, research into preharvest pathogen reduction and intervention strategies has been pursued (Oliver et al., 2008; Sargeant et al., 2007). Implementation of pathogen-reduction strategies in live cattle has the potential to reduce human illnesses (Rotariu et al., 2012; Smith et al., 2012) by (1) reducing the number of pathogens entering processing plant on/in an animal; (2) reducing horizontal pathogen transfer from infected animals; (3) lowering the pathogenic bacterial burden in the environment and wastewater streams; and (4) reducing the risk to those in direct contact with animals. The current preharvest intervention strategies to reduce the incidence of human STEC illnesses acquired from beef products have been reviewed (Callaway et al., 2013a,b) but will be briefly highlighted and updated herein. A sound management program (described in our previous chapter) in conjunction with preharvest intervention strategies will provide a farm-to-fork approach to reduce the incidence of STEC throughout a cattle operation.

Food and Feed Safety Systems and Analysis. http://dx.doi.org/10.1016/B978-0-12-811835-1.00008-7

## Water and Feed Management

Water and feed management encompass more than merely supplying cattle with clean water and feed as a means to reduce the horizontal transmission of STEC and other foodborne pathogens among the herd. Although these are important aspects of herd management since upwards of 25% of cattle farm water samples contain *E. coli* O157:H7 (Sanderson et al., 2006) and water, troughs are a common source (Polifroni et al., 2012; Wetzel and LeJeune, 2006) indicating that the common-use troughs can be vectors for horizontal transmission of *E. coli* O157:H7. The use of dietary and water amendments can also be used to reduce horizontal transmission of STEC between animals on the same farm or in the same pen. The underlying mechanism(s) behind these treatments are not fully understood, but it has been hypothesized that the impact may be related to increasing fecal pH or altering yeast fermentation end products (Wells et al., 2009). Although the extent of the dietary impact effects is relatively small on the animal, it highlights the potential of dietary composition to impact *E. coli* O157:H7 in the gastrointestinal (GI) tract of cattle.

### Probiotic Approaches to Influence Microbial Ecology in the Gastrointestinal Tract of Cattle

The gut microbiota has the ability to influence numerous physiological, metabolic, and immunological signaling pathways in the host including (1) metabolic/nutritional/energy utilization including vitamin synthesis and short-chain fatty acid use; (2) regulation of the innate and adaptive immune systems by dampening the inflammatory response and induction of immunosuppressive T cells, respectively; (3) epithelial development by stimulating proliferation, angiogenesis, and restitution and then epithelial survival by influencing the cytoprotective effects of pathogen recognition receptor signaling; and (4) competitive exclusion (CE) of pathogens (Neish, 2009). Each of these areas can be impacted by preharvest intervention strategies targeting host microbial ecology as a means to reduce STEC colonization and shedding in cattle. Some of the approaches include the use of direct-fed microbials (DFMs) in the form of probiotics, CE cultures, prebiotics, and colicins. The complex nature of the GI tract, coupled with the long (12–18 months) life span of cattle going into a feedlot, means that any single DFM used in cattle production will likely not eliminate all STEC shedding over the lifetime of the animal; therefore, age-specific strategies should be developed.

### Probiotics

The internationally recognized definition of a probiotic is a group of live microorganisms that, when administered in adequate amounts, confer a health benefit on the host (Sanders, 2008), and furthermore, they must be adequately characterized for content, stability, and health effects. However, because this definition is only "recognized" and not a true "legal" definition, the human and animal markets are inundated with products sold as probiotics, but many may not have undergone the rigorous scientific testing to actually meet the recognized definition of a probiotic. However in recent years, products using either live or dead microorganisms aimed at altering the microbial population of the gut have attracted increased interest as a potential method to reduce the ability of foodborne pathogenic bacteria to inhabit the GI tract of cattle. Oftentimes in agriculture, the term probiotic and DFM may be used interchangeably to describe live or dead microbes and/or their microbial components that are fed to the animal for the purpose of a health benefit (FDA, 1995). In addition to DFM, other probiotic-like approaches are being used more often by the cattle industry and include CE cultures, and prebiotics to reduce *E. coli* O157:H7 populations (McAllister et al., 2011) as a way to enhance food safety.

The precise mechanism(s) behind the health benefits of a probiotic or other similarly marketed product is not fully understood but was originally thought to function by improving the balance of the hosts' microbiota in a way that actually alters the microbial ecology of the GI tract. One potential mechanism is by creating a physical barrier. The DFM/CE bacteria may attach to the intestinal epithelium, which in turn, physically prevents the opportunistic pathogens from attaching to the intestinal wall (Collins and Gibson, 1999; Kim et al., 2008). Another mechanism is the generation of toxic microbial fermentation products by the DFM/CE bacteria. The production of volatile fatty acids can be toxic to some bacterial species (Ricke, 2003; Russell, 1992; Wolin, 1969), and other bacterial products including ethanol, organic acids (OAs), traditional antibiotics, or colicins/bacteriocins that are produced by some intestinal bacteria to eliminate competition within the same environmental niche (Jack et al., 1995; Rodriguez-Palacios et al., 2009). In addition to the role that probiotics may play in altering overall GI ecology, there is increasing evidence that probiotics modulate barrier properties of the intestinal wall in conjunction with host immunity. In conjunction with the host immune response, probiotics promote the transient release of local and systemic bioactive compounds from the intestinal epithelium including mucins, defensins, bacteriocins, and cytokines/chemokines which in turn promote adaptive immune responses such as secretory IgA and regulatory T cells (Corthesy et al., 2007; Neish, 2009; Tellez et al., 2015). Collectively, these mechanisms demonstrate that the complex host-pathogen interaction involved in disrupting the cycle of transmission and colonization of cattle with STEC and emphasize that implementing diverse intervention strategies offers the greatest opportunity to improve preharvest food safety.

### *Direct-Fed Microbials*

DFMs are fed daily to improve ruminal fermentation and production efficiency (Martin and Nisbet, 1992) and are widely used in the cattle industry. Increasingly, companies claim that DFM reduces *E. coli* O157:H7 shedding in cattle, but the results are dependent on DFM composition and animal production stage. A probiotic culture comprised of *Streptococcus bovis* and *Lactobacillus gallinarum* from the rumen of cattle reduced *E. coli* O157 shedding when given to experimentally infected calves (Ohya et al., 2001), and a recent systematic review and meta-analysis of the literature deduced feeding beef cattle a DFM resulted in a significant reduction in *E. coli* O157 fecal shedding (Wisener et al., 2015). However, other studies found no effect on pathogen levels in cattle (Arthur et al., 2010a; Swyers et al., 2011).

A DFM culture of *Lactobacillus acidophilus* derived directly from the rumen of cattle reduced fecal shedding of *E. coli* O157:H7 from 46% to 13% (Ransom et al., 2003), and when this DFM was combined with *Propionibacterium freudenreichii* (a propionate-producing commensal intestinal bacteria), the prevalence of *E. coli* O157:H7 and *Salmonella* was reduced both in the feces and on the hides (Baker et al., 2016; Stephens et al., 2007; Younts-Dahl et al., 2004). Reducing the numbers of *E. coli* O157:H7 in the feces is highly important, as this is a means of horizontal transfer and carcass contamination more so than just the prevalence of *E. coli* O157:H7 in the GI tract (Arthur et al., 2010b). One DFM containing both of the aforementioned strains of bacteria is marketed as Bovamine and is widely used in the cattle industry and is reported to reduce pathogenic bacteria, improve animal performance, and for preharvest use to reduce foodborne pathogens. The cattle industry encompasses diverse production systems (i.e., feedlots, cow-calf, stockers, and dairies), and it is likely that distinct DFM would need to be optimized and implemented for each system to effectively manage/reduce STEC and other foodborne pathogens while improving production efficiency.

### Competitive Exclusion

CE cultures are derived from the animal of interest and comprised of one or more species of nonpathogenic bacteria that can be administered as a means to reduce colonization and/or decrease populations of pathogenic bacteria (Fuller, 1989; Nurmi et al., 1992). An established GI microbial population limits the colonization of transient opportunistic infections (Fuller, 1989), and depending on the production stage of the animal, the goal of CE cultures may be exclusion of pathogens from the naïve gut of a neonatal animal, or displacement of an already established pathogenic bacterial population (Nurmi et al., 1992). Although STEC are not pathogenic to cattle, there is value in using CE cultures as a preharvest intervention strategy to displace/eliminate them from the GI tract because of their ability to cause illness in humans. Numerous studies indicate that CE cultures do reduce *E. coli* O157:H7 and *Salmonella* in the GI tract of cattle (Brashears et al., 2003; Zhao et al., 2003). Administration of a defined population of commensal *E. coli* strains isolated from cattle has been shown to possess the ability to displace an established *E. coli* O157:H7 population from calves (Zhao et al., 1998). Collectively, the efficacy of any product will be multifactorial and further support a "multiple-hurdle" approach to controlling STEC in cattle.

### Prebiotics

Prebiotics are dietary supplements that are nondigestible in the host, that provides a beneficial physiological effect on the host by selectively stimulating favorable growth or metabolic activity of indigenous bacteria in the GI tract (Neish, 2009; Patterson and Burkholder, 2003;

Schrezenmeir and De Vrese, 2001). Fructooligosaccharides are sugars that are not degraded by intestinal enzymes which allow them to pass down to the cecum and colon to become "colonic food" for host bacteria and provide nutrients to the intestinal mucosa (Houdijk et al., 1998; Ricke, 2015; Willard et al., 2000). As seen with CE cultures, prebiotics may provide a competitive advantage to select microorganisms of the host microbiota that may facilitate the exclusion of pathogenic bacteria from the intestine via direct competition for nutrients and/or binding sites (Zopf and Roth, 1996).

### *Colicins*

Select strains of *E. coli* have the ability to produce colicins which are antimicrobial proteins that kill or inhibit growth of other *E. coli* strains, including *E. coli* O157:H7 (Jordi et al., 2001; Schamberger and Diez-Gonzalez, 2002). The concept of using colicins as an intervention strategy to kill foodborne pathogens has been proposed for many years (Joerger, 2003), and large-scale production using recombination protein expression technology now makes it more economically feasible (Stahl et al., 2004). Studies indicate that the sensitivity of various *E. coli* O157:H7 strains to a single colicin is highly variable (Murinda et al., 1998; Schamberger and Diez-Gonzalez, 2002). Because some *E. coli* O157:H7 strains are colicinogenic and produce specific concomitant immunity proteins (Murinda et al., 1998), these strains may be resistant to specific or even broad categories of certain colicins (Alonso et al., 2000) rendering the approach ineffective. Therefore, if colicins are to be used as a preharvest intervention strategy, there must be simultaneous administration of several categories of colicins, and they will need to be protected from gastric and intestinal degradation. One approach to ensure the colicins reach the lower GI tract of cattle proposes feeding colicin-producing *E. coli* in cattle rations (Schamberger and Diez-Gonzalez, 2002; Schamberger et al., 2004; Zhao et al., 1998).

Simultaneous use of CE and prebiotics is referred to as "synbiotics" and may yield a synergistic effect in reducing foodborne pathogens in food animals before slaughter (Bomba et al., 2002). To date, prebiotics have not been widely implemented in cattle because of their expense, and the ability of ruminal microorganisms to degrade a wide variety of typical prebiotic substrates; however as costs change, their inclusion may be economically feasible.

## *Direct Antipathogen Strategies*

The meat industry has made significant progress improving the overall health of herds by using improved breeds and implementing improved husbandry/biosecurity practices. As producers keep up with global consumption, consumer demands and increasing regulations while looking for ways to improve overall robustness, addressing animal welfare issues, and reducing production costs, novel methods to reduce the incidence of foodborne pathogens are becoming increasingly important. In addition to the previously mentioned approaches, other pathogen-reduction strategies have been developed for use in the live animal that directly targets the pathogens. Many of these treatments use the host animal, natural members of the microbial ecosystem, or target an aspect of pathogen physiology to inhibit pathogen survival.

## Antibiotics

All meat-producing industries are being limited in how and when antibiotics can be administered to animals. In 2006, the EU implemented a ban on the use of antibiotics as growth promoters in poultry (Castanon, 2007). It is hypothesized that antibiotics cause a general disruption of the whole microbial ecosystem, and it is this disruption that indirectly affects *E. coli* O157:H7 populations. As such, it is possible that some antibiotics could actually provide STEC strains a competitive advantage in the ruminant GI tract and therefore prove counterproductive in reducing numbers and/or shedding. There are cattle-approved antibiotics, including Neomycin, that can be used to treat enteric infections and have been shown to reduce *E. coli* O157:H7 populations in the gut and on the hides of cattle (Elder et al., 2002; Ransom et al., 2003). However, widespread use of antibiotics to reduce STEC is not recommended for use in any food animals amid concerns of increasing the incidence of antimicrobial resistant bacteria.

## Sodium Chlorate

*E. coli* can respire under anaerobic conditions by reducing nitrate to nitrite via the dissimilatory nitrate reductase enzyme (Stouthamer, 1969), but this enzyme cannot differentiate between nitrate and its analog, chlorate. Chlorate is bactericidal against *E. coli* because when the chlorate is reduced to chlorite, the accumulation of chlorite in the cytoplasm kills bacteria (Stewart, 1988). Studies show that addition of chlorate to feed rations or in the drinking water reduced *E. coli* O157:H7 in cattle, swine, and sheep (Anderson et al., 2002; Callaway et al., 2003; Edrington et al., 2003), and hide treatment reduced fecal and hide populations of *E. coli* (Anderson et al., 2005). Because of the dramatic impact that chlorate has on foodborne pathogenic bacterial populations, it has been suggested that chlorate could be supplemented in the last feeding before cattle are shipped to the slaughterhouse. The use of chlorate to reduce foodborne pathogenic bacteria in food animals has not been approved but is currently under review by the U. S. Food and Drug Administration.

## Bacteriophages

Bacteriophages (phages) are viruses that kill bacteria and have been found in the GI tract of cattle (Callaway et al., 2006; Goodridge, 2010). Phages recognize receptors on the surface of specific bacteria that make them highly selective, which may prove beneficial by allowing us to target a particular to target a particular foodborne pathogen without perturbing the GI microbial ecosystem (Johnson et al., 2008; Lederberg, 1996). Lytic phages take over host biosynthesis, and when intracellular nutrients are depleted, the host bacterium is lyzed and killed. On lysis, the phages are released and continue to seek out their specific host bacteria, which allows the cycle to continue as long as those target bacteria are present. One disadvantage to phages is the rapid development of bacterial resistance to a single phage; consequently much effort is on development of multi-phage cocktails (Stanford and McAllister, personal communication) (Tanji et al., 2005).

Phages can be used to reduce *E. coli* O157:H7 within the gut of cattle before slaughter, and as a hide or environmental decontaminant (Ricke et al., 2012). FSIS issued a letter of no objection to the commercial use of lytic phages for hide washes and surface-cleansing products. Several products are available (OmniLytics' BacWash and Elanco's Finalyse), and company-based research indicates that both products significantly reduce foodborne pathogens including *Salmonella* and *E. coli* O157:H7 from the hide of sprayed cattle and/or trim products, but others show that there is no reduction on hides and carcasses (Arthur et al., 2016). Because of their impact on foodborne pathogens, processors are identifying critical control points to include phage sprays to reduce *E. coli* O157:H7 as the cattle enter the food chain.

With the increasing focus on improving food safety throughout the entire food production cycle, bacteriophages have been evaluated and shown to control experimentally inoculated foodborne pathogenic bacteria, especially *E. coli* O157:H7 in cattle GI tracts (Bach et al., 2009; Callaway et al., 2008; Rozema et al., 2009). Interestingly, several phages have been isolated that are effective against *Salmonella* spp. and *E. coli* O157:H7 (López-Cuevas et al., 2012; Lopez-Velasco et al., 2011; Park et al., 2012), or on diseases that impact animal production or health (Huff et al., 2002; Smith and Huggins, 1987) which offers the hope of phage use as a broad-spectrum food safety improvement. No matter what point in the beef production chain the phages are used in (e.g., hides or in the live animal), they must be carefully selected for (1) action against multiple STEC strains, (2) members of a cocktail must use different receptors to minimize resistance, and (3) must be strictly lytic (i.e., does not transfer genetic material) because phage-mediated transfer is the mechanism by which STEC originally acquired their Shiga toxin genes (Brabban et al., 2005; Law, 2000).

### Cattle Hide Washing

Cattle hides are typically washed to remove visible contamination. Typically, the hide washes contain antimicrobial compounds (e.g., OAs, sodium hydroxide, trisodium phosphate, cetylpyridinium chloride, hypobromous acid, electrolyzed or ozonated water) but are most commonly OAs including lactic or acetic acid (Berry and Cutter, 2000; Loretz et al., 2011). These chemical washes significantly reduce the load of *E. coli* O157:H7 entering the plant on the hide (Arthur et al., 2007, 2010b), thus potentially improving food safety.

### Vaccination

Vaccination is another effective preharvest intervention strategy to reduce the incidence of *E. coli* shedding in cattle. There are two types of *E. coli* vaccines available to the cattle industry and both target different *E. coli* O157:H7 antigens in an attempt to disrupt adherence and replication of the bacteria in the GI tract of cattle (Doyle and Erickson, 2012; Walle et al., 2012). The vaccines are siderophore receptor and porin (SRP) protein vaccines and bacterial extract vaccines (specifically targeting type III–secreted proteins) (FSIS, 2014), and both reduce *E. coli* O157:H7 shedding in feedlot cattle (Cull et al., 2012; Varela et al., 2013; Walle et al., 2012). In addition to the aforementioned proteins serving as vaccine candidates, it may also be possible to immunize with recombinant Shiga toxins (toxoids) to generate neutralizing antibodies to prevent STEC colonization in cattle (Kerner et al., 2015). Despite successful antibody production after vaccination, the mechanism(s) of protection is not fully understood in part because the hosts' immunological response against STEC is not widely understood. A recent study suggested that the hosts' cellular immune response is strain specific and directed toward a TH1 response with elevated interferon gamma and T-bet expression in terminal rectum tissues from colonized cattle in the absence of a general inflammatory response (Corbishley et al., 2014).

Despite being commercially available for several years, vaccination against *E. coli* O157:H7 or any STEC has not been widely implemented by the industry because cattle are not adversely affected by *E. coli*, nor are feed efficiency and production negatively impacted in infected cattle. Estimates suggest that the direct cost of incorporating an *E. coli* vaccine regime over the course of the first 10 years at the feedlot level would be $1–1.8 billion USD depending on the number of doses administered and considering any impact on performance (Tonsor and Schroeder, 2015). Thus far, a market that compensates producers for vaccinating against STEC has not emerged, although foodborne illness resulting from *E. coli*–contaminated ground beef could potentially be reduced by 58% (Hurd and Malladi, 2012). Vaccination is widely accepted in the cattle industry, so it is reasonable to assume that producers will implement this pathogen-reduction strategy if it is economically feasible and can be incorporated into existing production systems. As with any intervention strategy, vaccination alone does not appear to be fully effective at eliminating STEC and should be considered as one facet of a program aimed at reducing the incidence of foodborne pathogens acquired from beef products.

#### *Siderophore Receptor and Porin Protein Vaccines*

Iron is an essential heavy metal required for numerous essential biological processes, including *E. coli* replication (Skaar, 2010); however, the bacteria must acquire the iron from the animal host. One mechanism used by all *E. coli*, including STEC, for iron acquisition is by producing high-affinity iron chelators called siderophores (Ratledge and Dover, 2000), which are small secreted molecules that are capable of scavenging iron from host sources (Skaar, 2010). Recovering iron from the host is a multistep process involving the synthesis and export of siderophores, followed by reception and internalization of the iron-siderophore complex by receptors on the bacterial outer membrane, and lastly iron release into the bacterial cytoplasm (Garenaux et al., 2011). If any step in this process is ablated, the

STEC become iron deprived and die. There are commercially available vaccines directed against these siderophore receptors (SRP vaccines) which are conditionally approved as a means to reduce *E. coli* O157:H7 in cattle. Preliminary studies using such SRP vaccines show promise in generating an antibody response as well as reducing the number of animals shedding *E. coli* O157:H7 (Thornton et al., 2009). Others report highly significant reductions in prevalence and fecal shedding of O157:H7 after administration of multiple doses of an SRP vaccine (Cull et al., 2012; Fox et al., 2009) while not adversely impacting performance (Thomson et al., 2009). Despite the effectiveness of some commercial SRP vaccines at reducing fecal shedding of *E. coli* O157:H7, some studies indicate these vaccines do not confer cross-protection and do not impact fecal shedding of non-O157 adulterant STEC (Cernicchiaro et al., 2014; Paddock et al., 2014).

*Bacterial Extract Vaccines*

The type III secretion system (T3SS) is a protein complex found in numerous gram-negative bacteria, including STEC, that enables the bacteria to successfully penetrate the host cell mucus layer, attach, and subsequently colonize the GI tract by producing effector proteins that aid in pathogen survival and evading host immune mechanisms (Gaytan et al., 2016). Because the T3SS is directly involved in attachment of STEC to the GI tract of cattle, several bacterial proteins from this complex have been identified as vaccine candidates. The hypothesis driving this approach is that cattle vaccinated with this type of subunit vaccine would produce antibodies that would bind to these critical proteins and block the T3SS receptors responsible for attachment and therefore reduce colonization and shedding ultimately lowering the risk of human infection (Doyle and Erickson, 2012).

The first commercially available subunit vaccine–targeting proteins of the T3SS of *E. coli* O157:H7 was Econiche (Bioniche Life Sciences Inc. [TSX:BNC]), and it proved to be very effective at reducing *E. coli* O157:H7 shedding in feedlot cattle (Moxley et al., 2009; Potter et al., 2004; Smith et al., 2009). Despite the effectiveness of Econiche, the product was acquired by another company and has since ceased production (http://www.farmattractions.net/e-coli-vaccine-no-longer-available/). Additional subunit vaccines have been produced that also reduce fecal shedding in experimental-infection models (McNeilly et al., 2010; Sharma et al., 2011). Because STEC share the T3SS, it is possible that a single vaccine that comprised several T3SS proteins may confer some degree of protection against STEC (Asper et al., 2011; McNeilly et al., 2015).

A relatively new technology that creates bacterial ghosts (BGs) is being investigated for its use as vaccine candidates. The BG is generated by E-mediated lysis of gram-negative bacteria, which maintains the cell envelope and leaves all associated structural and morphological features intact while removing all cytoplasmic content rendering the BG immunogenic but not infective (Jaleta et al., 2015). Studies indicate the approach is worth additional consideration as administration of BG reduced *E. coli* O157:H7 populations in calves and also reduced fecal shedding (Khare et al., 2010; Vilte et al., 2012). Others have devised chimeric multiprotein vaccines that can be produced in plants, potentially providing a source of an edible vaccine (Amani et al., 2011) that can be included in cattle rations rather than having to be injected. However, for this approach to be used, the proteins must be protected from the extensive proteolytic nature of the rumen microbial ecosystem, which will obviously add to the complexity and expense of vaccination via the edible vaccine approach.

## Organic Acids and Botanicals

### Organic Acids

OAs are well-known antimicrobial compounds, widely used in animal nutrition both as feed preservatives and as nonantibiotic growth promoters particularly in pigs and poultry. Both applications of OAs are strictly related to their antimicrobial properties as OAs can promote the hygiene of the feed by reducing bacterial and mold load as well as improve the growth of animals by controlling the microbial balance in the GI tract. The mechanisms of action behind the antimicrobial properties of OA have been linked to the "ion model" with the undissociated form (RCOOH) of the acid being able to pass the bacterial cell wall and, once inside the cytosol, to release protons ($H^+$) and anions ($RCOO^-$) disrupting the pH balance of the cell and impairing cell functions such as DNA and protein synthesis (Russell and Diez-Gonzalez, 1998). The inhibitory activity varies depending on the bacterial strain and on the type of acid and its chemical structure and characteristics. Because of their specific antimicrobial properties, OAs attract great attention also from a food safety perspective as a preharvest tool to control specific foodborne pathogens such as *Salmonella*, *Campylobacter*, and *E. coli* in livestock (Grilli and Piva, 2012).

The specific inhibitory activity of OA against *E. coli* O157:H7 has been demonstrated by various experimental studies. Acetic, citric, formic, and lactic acids have been shown to be the most effective in reducing *E. coli* O157:H7 contamination in carcass and meat, and several OA-based products have been implemented for commercial beef processing (Harris et al., 2006; Raftari et al., 2009; Wheeler et al., 2014). Moreover, a study conducted with alfalfa silages contaminated with *E. coli* O157:H7 showed that the addition of propionic acid could inhibit the growth of the pathogen during and after ensiling (Ogunade et al., 2016). In addition, OAs have been frequently shown to reduce the contamination by *E. coli* O157:H7 in fresh produce such as apples and different types of lettuce (Akbas and Olmez, 2007; Park et al., 2011; Sagong et al., 2011). The combination of different OAs can increase the inhibitory effect against *E. coli* O157:H7 as demonstrated in vitro by Kim and Rhee (2013) who reported a marked synergistic effect for citric acid and caprylic acid in reducing the microbial counts of *E. coli* O157:H7 due to irreversible cell membrane damage (Kim and Rhee, 2013).

### Botanicals

Botanicals, also referred to as phytogenics, are a large group of plant-derived compounds that are increasingly gaining interest as feed additives to improve performance of livestock. Particularly, essential oils (EOs) are the most studied phytogenics, and their mechanism of action as antimicrobials is explained by the lipophilic nature of their active compounds, which alters the bacterial cell wall by forming pores and eventually impairs enzymatic activities (Lambert et al., 2001). In ruminant nutrition, EOs are used as feed additives for the modulation of rumen microbial fermentations and consequent reduction in methane and ammonia loss (Cobellis et al., 2016). With respect to *E. coli* O157:H7, oregano and thyme oil and their main phenolic constituents (thymol and carvacrol) were the most effective in reducing the strain in vitro in pure cultures (Burt and Reinders, 2003; Friedman et al., 2002; Mith et al., 2014). Moreover, the addition of carvacrol was able to inhibit *E. coli* O157:H7 not only in a broth assay but also in a rumen fermentation model allowing the reduction of the inoculated *E. coli* O157:H7 to undetectable levels (Rivas et al., 2010).

#### *Organic Acids and Botanicals*

The antimicrobial activity of OA and botanicals can be enhanced by combining them together. In a series of works, Grilli and colleagues showed a marked efficacy of OA and botanicals, when used together, in vitro against *Salmonella* and *Campylobacter*: experiments conducted to test the antimicrobial activity of OA and botanicals through the minimal inhibitory concentration method revealed that OA, when combined with botanicals, such as monoterpenes, were more effective against *Salmonella* Typhimurium or *Campylobacter jejuni* than when tested alone, strongly supporting a synergistic action of OA and botanicals (Grilli and Piva, 2012; Grilli et al., 2013). In addition, OA and botanicals blends were shown to reduce the growth rate and numbers of *E. coli* O157:H7 both in pure cultures and in mixed ruminal fermentations (Grilli et al., 2015a). The mechanism of action proposed at the basis of the synergy between OA and botanicals involves the permeabilizing effect of botanical aromatic compounds that, altering the bacterial membrane, would eventually facilitate the OA entrance and therefore the antimicrobial action (Grilli and Piva, 2012).

However the efficacy of these compounds in vivo is inherently limited by their rapid gastric absorption, or, in the specific case of ruminants, by rumen degradation, which prevents them from reaching the lower gut where *E. coli* colonization and recrudescence mostly occur. To overcome this problem, OA and botanicals need to be protected and lipid microencapsulation is an efficient tool to allow the slow release of active principles in the small and large intestines (Piva et al., 2007). When microencapsulated, in fact, OA and botanicals mixtures can effectively reduce *Campylobacter* and *Salmonella* in the cecum of broilers and pigs by at least 2 Logs (Grilli et al., 2011, 2013, 2015b).

Both OA and botanicals are antimicrobial agents with a well-documented inhibitory activity against *E. coli* O157:H7 in vitro. Results in ruminants are still lacking, but data in pigs and poultry suggest that OA and botanicals, especially when combined together, increase the anti-*E. coli* O157:H7 activity in a synergistic way. Microencapsulation can improve this synergy between OA and botanicals allowing them to be gradually released along the intestine, providing a tool to control *E. coli* O157:H7 colonization.

## Conclusions

Although STEC are regarded in some respects as commensal organisms in the GI tract cattle, they are clearly a significant threat to human consumers and public health. Preharvest controls in cattle offer real potential to reduce STEC dissemination on farms, in the environment, and entering the food chain thereby potentially reducing foodborne pathogen penetration into the food chain. Live-animal management interventions must be incorporated into a coherent, complementary program aimed at reducing pathogens as part of an integrated multiple-hurdle approach that complements postharvest and in-plant pathogen-reduction strategies to minimize pathogen load entering the food supply. Incorporating diverse preharvest intervention strategies throughout a comprehensive program that aims at reducing the incidence of STEC or other foodborne pathogens in live cattle at the farm level offers the best opportunity to reduce the number of illnesses associated with consumption and/or handling of contaminated products.

## References

Akbas, M.Y., Olmez, H., 2007. Inactivation of *Escherichia coli* and *Listeria monocytogenes* on iceberg lettuce by dip wash treatments with organic acids. Letters in Applied Microbiology 44, 619–624.

Alonso, G., Vilchez, G., Vidal, R.L., 2000. How bacteria protect themselves against channel-forming colicins. International Microbiology 3, 81–88.

Amani, J., Mousavi, S.L., Rafati, S., Salmanian, A.H., 2011. Immunogenicity of a plant-derived edible chimeric EspA, Intimin and Tir of *Escherichia coli* O157:H7 in mice. Plant Science 180, 620–627.

Anderson, R.C., Callaway, T.R., Anderson, T.J., Kubena, L.F., Keith, N.K., Nisbet, D.J., 2002. Bactericidal effect of sodium chlorate on *Escherichia coli* concentrations in bovine ruminal and fecal contents in vivo. Microbial Ecology in Health and Disease 14, 24–29.

Anderson, R.C., Carr, M.A., Miller, R.K., King, D.A., Carstens, G.E., Genovese, K.J., Callaway, T.R., Edrington, T.S., Jung, Y.S., Mcreynolds, J.L., Hume, M.E., Beier, R.C., Elder, R.O., Nisbet, D.J., 2005. Effects of experimental chlorate preparations as feed and water supplements on *Escherichia coli* colonization and contamination of beef cattle and carcasses. Food Microbiology 22, 439–447.

Arthur, T.M., Bosilevac, J.M., Brichta-Harhay, D.M., Guerini, M.N., Kalchayanand, N., Shackelford, S.D., Wheeler, T.L., Koohmaraie, M., 2007. Transportation and lairage environment effects on prevalence, numbers, and diversity of *Escherichia coli* O157:H7 on hides and carcasses of beef cattle at processing. Journal of Food Protection 70, 280–286.

Arthur, T.M., Bosilevac, J.M., Kalchayanand, N., Wells, J.E., Shackelford, S.D., Wheeler, T.L., Koohmaraieie, M., 2010a. Evaluation of a direct-fed microbial product effect on the prevalence and load of *Escherichia coli* O157:H7 in feedlot cattle. Journal of Food Protection 73, 366–371.

Arthur, T.M., Brichta-Harhay, D.M., Bosilevac, J.M., Kalchayanand, N., Shackelford, S.D., Wheeler, T.L., Koohmaraie, M., 2010b. Super shedding of *Escherichia coli* O157:H7 by cattle and the impact on beef carcass contamination. Meat Science 86, 32–37.

Arthur, T.M., Kalchayanand, N., Agga, G.E., Wheeler, T.L., Koohmaraie, M., 2016. Evaluation of Bacteriophage Application to Cattle in Lairage at Beef Processing Plants to Reduce *Escherichia coli* O157:H7 Prevalence on Hides and Carcasses. Foodborne Pathogens and Disease.

Asper, D.J., Karmali, M.A., Townsend, H., Rogan, D., Potter, A.A., 2011. Serological response of shiga toxin-producing *Escherichia coli* type III secreted proteins in sera from vaccinated rabbits, naturally infected cattle, and humans. Clinical and Vaccine Immunology 18, 1052–1057.

Bach, S.J., Johnson, R.P., Stanford, K., Mcallister, T.A., 2009. Bacteriophages reduce *Escherichia coli* O157:H7 levels in experimentally inoculated sheep. Canadian Journal of Animal Science 89, 285–293.

Baker, C.A., Rubinelli, P.M., Park, S.H., Ricke, S.C., 2016. Immuno-based detection of Shiga toxin-producing pathogenic *Escherichia coli* in food – a review on current approaches and potential strategies for optimization. Critical Reviews in Microbiology 42, 656–675.

Berry, E.D., Cutter, C.N., 2000. Effects of acid adaptation of *Escherichia coli* O157:H7 on efficacy of acetic acid spray washes to decontaminate beef carcass tissue. Applied and Environmental Microbiology 66, 1493–1498.

Bomba, A., Nemcovǻ, R., Mudronovǻ, D., Guba, P., 2002. The possibilities of potentiating the efficacy of probiotics. Trends in Food Science & Technology 13, 121–126.

Brabban, A.D., Hite, E., Callaway, T.R., 2005. Evolution of foodborne pathogens via temperate bacteriophage-mediated gene transfer. Foodborne Pathogens and Disease 2, 287–303.

Brashears, M.M., Jaroni, D., Trimble, J., 2003. Isolation, selection, and characterization of lactic acid bacteria for a competitive exclusion product to reduce shedding of *Escherichia coli* O157:H7 in cattle. Journal of Food Protection 66, 355–363.

Burt, S.A., Reinders, R.D., 2003. Antibacterial activity of selected plant essential oils against *Escherichia coli* O157:H7. Letters in Applied Microbiology 36, 162–167.

Callaway, T.R., Edrington, T.S., Anderson, R.C., Genovese, K.J., Poole, T.L., Elder, R.O., Byrd, J.A., Bischoff, K.M., Nisbet, D.J., 2003. *Escherichia coli* O157:H7 populations in sheep can be reduced by chlorate supplementation. Journal of Food Protection 66, 194–199.

Callaway, T.R., Edrington, T.S., Brabban, A.D., Anderson, R.C., Rossman, M.L., Engler, M.J., Carr, M.A., Genovese, K.J., Keen, J.E., Looper, M.L., Kutter, E.M., Nisbet, D.J., 2008. Bacteriophage isolated from feedlot cattle can reduce *Escherichia coli* O157:H7 populations in ruminant gastrointestinal tracts. Foodborne Pathogens and Disease 5, 183–192.

Callaway, T.R., Edrington, T.S., Brabban, A.D., Keen, J.E., Anderson, R.C., Rossman, M.L., Engler, M.J., Genovese, K.J., Gwartney, B.L., Reagan, J.O., Poole, T.L., Harvey, R.B., Kutter, E.M., Nisbet, D.J., 2006. Fecal prevalence of *Escherichia coli* O157, *Salmonella, Listeria,* and bacteriophage infecting *E. coli* O157:H7 in feedlot cattle in the southern plains region of the United States. Foodborne Pathogens and Disease 3, 234–244.

Callaway, T.R., Edrington, T.S., Loneragan, G.H., Carr, M.A., Nisbet, D.J., 2013a. Current and near-market intervention strategies for reducing Shiga Toxin-producing *Escherichia coli* (STEC) shedding in cattle. Agriculture, Food and Analytical Bacteriology 3, 103–120.

Callaway, T.R., Edrington, T.S., Loneragan, G.H., Carr, M.A., Nisbet, D.J., 2013b. Shiga toxin-producing *Escherichia coli* (STEC) ecology in cattle and management based operations for reducing fecal shedding. Agriculture, Food and Analytical Bacteriology 3, 39–69.

Castanon, J.I.R., 2007. History of the use of antibiotic as growth promoters in the European poultry feeds. Poultry Science 86, 2466–2471.

CDC, 2016. Surveillance for Foodborne Disease Outbreaks, United States, 2014. Annual Report. US Department of Health and Human Services, CDC, Atlanta, GA.

Cernicchiaro, N., Renter, D.G., Cull, C.A., Paddock, Z.D., Shi, X., Nagaraja, T.G., 2014. Fecal shedding of non-O157 serogroups of Shiga toxin-producing *Escherichia coli* in feedlot cattle vaccinated with an *Escherichia coli* O157:H7 SRP vaccine or fed a *Lactobacillus*-based direct-fed microbial. Journal of Food Protection 77, 732–737.

Cobellis, G., Trabalza-Marinucci, M., Yu, Z., 2016. Critical evaluation of essential oils as rumen modifiers in ruminant nutrition: a review. Science of the Total Environment 545–546, 556–568.

Collins, D.M., Gibson, G.R., 1999. Probiotics, prebiotics, and synbiotics: approaches for modulating the microbial ecology of the gut. American Journal of Clinical Nutrition 69, 1052S–1057S.

Corbishley, A., Ahmad, N.I., Hughes, K., Hutchings, M.R., Mcateer, S.P., Connelley, T.K., Brown, H., Gally, D.L., McNeilly, T.N., 2014. Strain-dependent cellular immune responses in cattle following *Escherichia coli* O157:H7 colonization. Infection and Immunity 82, 5117–5131.

Corthesy, B., Gaskins, H.R., Mercenier, A., 2007. Cross-talk between probiotic bacteria and the host immune system. Journal of Nutrition 137, 781S–790S.

Cull, C.A., Paddock, Z.D., Nagaraja, T.G., Bello, N.M., Babcock, A.H., Renter, D.G., 2012. Efficacy of a vaccine and a direct-fed microbial against fecal shedding of *Escherichia coli* O157:H7 in a randomized pen-level field trial of commercial feedlot cattle. Vaccine 30, 6210–6215.

Doyle, M.P., Erickson, M.C., 2012. Opportunities for mitigating pathogen contamination during on-farm food production. International Journal of Food Microbiology 152, 54–74.

Edrington, T.S., Callaway, T.R., Anderson, R.C., Genovese, K.J., Jung, Y.S., Elder, R.O., Bischoff, K.M., Nisbet, D.J., 2003. Reduction of *E. coli* O157:H7 populations in sheep by supplementation of an experimental sodium chlorate product. Small Ruminant Research 49, 173–181.

Elder, R.O., Keen, J.E., Wittum, T.E., Callaway, T.R., Edrington, T.S., Anderson, R.C., Nisbet, D.J., 2002. Intervention to reduce fecal shedding of enterohemorrhagic *Escherichia coli* O157:H7 in naturally infected cattle using neomycin sulfate. Journal of Animal Science 80 (Suppl. 1), 15 (Abstract).

FDA, 1995. Direct-fed microbial products. In: Services, U. S. D. O. H. A. H.

Fox, J.T., Thomson, D.U., Drouillard, J.S., Thornton, A.B., Burkhardt, D.T., Emery, D.A., Nagaraja, T.G., 2009. Efficacy of *Escherichia coli* O157:H7 siderophore receptor/porin proteins-based vaccine in feedlot cattle naturally shedding *E. coli* O157. Foodborne Pathogens and Disease 6, 893–899.

Friedman, M., Henika, P.R., Mandrell, R.E., 2002. Bactericidal activities of plant essential oils and some of their isolated constituents against *Campylobacter jejuni, Escherichia coli, Listeria monocytogenes*, and *Salmonella enterica*. Journal of Food Protection 65, 1545–1560.

FSIS, 2011. Shiga toxin-producing *Escherichia coli* in certain raw beef products. In: Agriculture, D. O. Government Publishing Office, Washington DC.

FSIS, 2014. Pre-harvest management controls and intervention option for reducing Shiga toxin-producing *Escherichia coli* shedding in cattle: an overview of current reseatch. In: Agriculture, D. O. Government Publishing Group, Washington DC.

Fuller, R., 1989. Probiotics in man and animals. Journal of Applied Bacteriology 66, 365–378.

Garenaux, A., Caza, M., Dozois, C.M., 2011. The Ins and Outs of siderophore mediated iron uptake by extra-intestinal pathogenic *Escherichia coli*. Veterinary Microbiology 153, 89–98.

Gaytan, M.O., Martinez-Santos, V.I., Soto, E., Gonzalez-Pedrajo, B., 2016. Type three secretion system in attaching and effacing pathogens. Frontiers in Cellular and Infection Microbiology 6, 129.

Goodridge, L.D., 2010. Designing phage therapeutics. Current Pharmaceutical Biotechnology 11, 15–27.

Grilli, E., Bari, R., Piva, A., Edrington, T.S., Pitta, D.W., Pinchak, W.E., Nisbet, D.J., Callaway, T.R., 2015a. Organic acid blend with pure botanical product treatment reduces *Escherichia coli* and *Salmonella* populations in pure culture and in in vitro mixed ruminal microorganism fermentations. Foodborne Pathogens and Disease 12, 56–61.

Grilli, E., Foresti, F., Tugnoli, B., Fustini, M., Zanoni, M.G., Pasquali, P., Callaway, T.R., Piva, A., Alborali, G.L., 2015b. Microencapsulated sorbic acid and pure botanicals affect *Salmonella* Typhimurium shedding in pigs: a close-up look from weaning to slaughter in controlled and field conditions. Foodborne Pathogens and Disease 12, 813–819.

Grilli, E., Piva, A., 2012. Organic acids and their role in reduce foodborne pathogens in food animals. In: Callaway, T.R., Edrington, T.S. (Eds.), On-farm Strategies to Control Foodborne Pathogens. Nova Science Pub Inc.

Grilli, E., Tugnoli, B., Formigoni, A., Massi, P., Fantinati, P., Tosi, G., Piva, A., 2011. Microencapsulated sorbic acid and nature-identical compounds reduced *Salmonella* Hadar and *Salmonella* Enteritidis colonization in experimentally infected chickens. Poultry Science 90, 1676–1682.

Grilli, E., Vitari, F., Domeneghini, C., Palmonari, A., Tosi, G., Fantinati, P., Massi, P., Piva, A., 2013. Development of a feed additive to reduce caecal *Campylobacter jejuni* in broilers at slaughter age: from in vitro to in vivo, a proof of concept. Journal of Applied Microbiology 114, 308–317.

Harris, K., Miller, M.F., Loneragan, G.H., Brashears, M.M., 2006. Validation of the use of organic acids and acidified sodium chlorite to reduce *Escherichia coli* O157 and *Salmonella* Typhimurium in beef trim and ground beef in a simulated processing environment. Journal of Food Protection 69, 1802–1807.

Houdijk, J.G.M., Bosch, M.W., Verstegen, M.W.A., Berenpas, H.J., 1998. Effects of dietary oligosaccharides on the growth and faecal characteristics of young growing pigs. Animal Feed Science and Technology 71, 35–48.

Huff, W.E., Huff, G.R., Rath, N.C., Balog, J.M., Xie, H., Moore, P.A., Donoghue, A.M., 2002. Prevention of *Escherichia coli* respiratory infection in broiler chickens with bacteriophage (SPR02). Poultry Science 81, 437–441.

Hurd, H.S., Malladi, S., 2012. An outcomes model to evaluate risks and benefits of *Escherichia coli* vaccination in beef cattle. Foodborne Pathogens and Disease 9, 952–961.

Jack, R.W., Tagg, J.R., Ray, B., 1995. Bacteriocins of gram-positive bacteria. Microbiological Reviews 59, 171–200.

Jaleta, H., Mamo, B., Disassa, H., 2015. Review of bacterial ghost and its application. International Journal of Microbiological Research 6, 200–210.

Joerger, R.D., 2003. Alternatives to antibiotics: bacteriocins, antimicrobial peptides and bacteriophages. Poultry Science 82, 640–647.

Johnson, R.P., Gyles, C.L., Huff, W.E., Ojha, S., Huff, G.R., Rath, N.C., Donoghue, A.M., 2008. Bacteriophages for prophylaxis and therapy in cattle, poultry and pigs. Animal Health Research Reviews 9, 201–215.

Jordi, B.J., Boutaga, K., Van Heeswijk, C.M.E., Van Knapen, F., Lipman, L.J.A., 2001. Sensitivity of shiga toxin-producing *Escherichia coli* (STEC) strains for colicins under different experimental conditions. FEMS Microbiology Letters 204, 329–344.

Karmali, M.A., Gannon, V., Sargeant, J.M., 2010. Verocytotoxin-producing *Escherichia coli* (VTEC). Veterinary Microbiology 140, 360–370.

Kerner, K., Bridger, P.S., Kopf, G., Frohlich, J., Barth, S., Willems, H., Bauerfeind, R., Baljer, G., Menge, C., 2015. Evaluation of biological safety in vitro and immunogenicity in vivo of recombinant *Escherichia coli* Shiga toxoids as candidate vaccines in cattle. Veterinary Research 46, 38.

Khare, S., Alali, W., Zhang, S., Hunter, D., Pugh, R., Fang, F.C., Libby, S.J., Adams, L.G., 2010. Vaccination with attenuated *Salmonella enterica* Dublin expressing *E. coli* O157:H7 outer membrane protein Intimin induces transient reduction of fecal shedding of *E. coli* O157:H7 in cattle. BMC Veterinary Research 6, 35–47.

Kim, J.W., Siletzky, R.M., Kathariou, S., 2008. Host ranges of *Listeria*-specific bacteriophages from the turkey processing plant environment in the United States. Applied and Environmental Microbiology 74, 6623–6630.

Kim, S.A., Rhee, M.S., 2013. Marked synergistic bactericidal effects and mode of action of medium-chain fatty acids in combination with organic acids against *Escherichia coli* O157:H7. Applied and Environmental Microbiology 79, 6552–6560.

Lambert, R.J., Skandamis, P.N., Coote, P.J., Nychas, G.J., 2001. A study of the minimum inhibitory concentration and mode of action of oregano essential oil, thymol and carvacrol. Journal of Applied Microbiology 91, 453–462.

Law, D., 2000. The history and evolution of *Escherichia coli* O157 and other shiga toxin-producing *E. coli*. World Journal of Microbiology & Biotechnology 16, 701–709.

Lederberg, J., 1996. Smaller Fleas…ad infinitum: therapeutic bacteriophage redux. Proceedings of the National Academy of Sciences of the United States of America 93, 3167–3168.

Lóрez-Cuevas, O., Castro-Del Campo, N., León-Félix, J., Valdez-Torres, B., Chaidez, C., 2012. Evaluation of bacteriophage av-08 for simultaneous biocontrol of *Salmonella* Montevideo and *Escherichia coli* O157: H7 in experimentally contaminated chicken skin. Journal of Food Safety 32, 305–310.

Lopez-Velasco, G., Welbaum, G.E., Falkinham Iii, J.O., Ponder, M.A., 2011. Phyllopshere bacterial community structure of spinach (*Spinacia oleracea*) as affected by cultivar and environmental conditions at time of harvest. Diversity 3, 721–738.

Loretz, M., Stephan, R., Zweifel, C., 2011. Antibacterial activity of decontamination treatments for cattle hides and beef carcasses. Food Control 22, 347–359.

Martin, S.A., Nisbet, D.J., 1992. Effect of direct-fed microbials on rumen microbial fermentation. Journal of Dairy Science 75, 1736–1744.

McAllister, T.A., Beauchemin, K.A., Alazzeh, A.Y., Baah, J., Teather, R.M., Stanford, K., 2011. Review: the use of direct fed microbials to mitigate pathogens and enhance production in cattle. Canadian Journal of Animal Science 91, 193–211.

McNeilly, T.N., Mitchell, M.C., Corbishley, A., Nath, M., Simmonds, H., Mcateer, S.P., Mahajan, A., Low, J.C., Smith, D.G., Huntley, J.F., Gally, D.L., 2015. Optimizing the protection of cattle against *Escherichia coli* O157:H7 colonization through immunization with different combinations of H7 Flagellin, Tir, Intimin-531 or EspA. PLoS One 10, e0128391.

McNeilly, T.N., Mitchell, M.C., Rosser, T., Mcateer, S., Low, J.C., Smith, D.G.E., Huntley, J.F., Mahajan, A., Gally, D.L., 2010. Immunization of cattle with a combination of purified intimin-531, EspA and Tir significantly reduces shedding of *Escherichia coli* O157:H7 following oral challenge. Vaccine 28, 1422–1428.

Mith, H., Dure, R., Delcenserie, V., Zhiri, A., Daube, G., Clinquart, A., 2014. Antimicrobial activities of commercial essential oils and their components against food-borne pathogens and food spoilage bacteria. Food Science & Nutrition 2, 403–416.

Moxley, R.A., Smith, D.R., Luebbe, M., Erickson, G.E., Klopfenstein, T.J., Rogan, D., 2009. *Escherichia coli* O157:H7 vaccine dose-effect in feedlot cattle. Foodborne Pathogens and Disease 6, 879–884.

Murinda, S.E., Liu, S.M., Roberts, R.F., Wilson, R.A., 1998. Colicinogeny among *Escherichia coli* serotypes, including O157:H7, representing four closely related diarrheagenic clones. Journal of Food Protection 61, 1431–1438.

Neish, A.S., 2009. Microbes in gastrointestinal health and disease. Gastroenterology 136, 65–80.

Nurmi, E., Nuotio, L., Schncitz, C., 1992. The competitive exclusion concept: development and future. International Journal of Food Microbiology 15, 237–240.

Ogunade, I.M., Kim, D.H., Jiang, Y., Weinberg, Z.G., Jeong, K.C., Adesogan, A.T., 2016. Control of *Escherichia coli* O157:H7 in contaminated alfalfa silage: effects of silage additives. Journal of Dairy Science 99, 4427–4436.

Ohya, T., Akiba, M., Ito, H., 2001. Use of a trial probiotic product in calves experimentally infected with *Escherichia coli* O157. Japan Agricultural Research Quarterly 35, 189–194.

Oliver, S.P., Patel, D.A., Callaway, T.R., Torrence, M.E., 2008. ASAS centennial paper: developments and future outlook for preharvest food safety. Journal of Animal Science 87, 419–437.

Paddock, Z.D., Renter, D.G., Cull, C.A., Shi, X., Bai, J., Nagaraja, T.G., 2014. *Escherichia coli* O26 in feedlot cattle: fecal prevalence, isolation, characterization, and effects of an *E. coli* O157 vaccine and a direct-fed microbial. Foodborne Pathogens and Disease 11, 186–193.

Park, M., Lee, J.H., Shin, H., Kim, M., Choi, J., Kang, D.H., Heu, S., Ryu, S., 2012. Characterization and comparative genomic analysis of a novel bacteriophage, SFP10, simultaneously inhibiting both *Salmonella enterica* and *Escherichia coli* O157:H7. Applied and Environmental Microbiology 78, 58–69.

Park, S.H., Choi, M.R., Park, J.W., Park, K.H., Chung, M.S., Ryu, S., Kang, D.H., 2011. Use of organic acids to inactivate *Escherichia coli* O157:H7, *Salmonella* Typhimurium, and *Listeria monocytogenes* on organic fresh apples and lettuce. Journal of Food Science 76, M293–M298.

Patterson, J.A., Burkholder, K.M., 2003. Application of prebiotics and probiotics in poultry production. Poultry Science 82, 627–631.

Piva, A., Pizzamiglio, V., Morlacchini, M., Tedeschi, M., Piva, G., 2007. Lipid microencapsulation allows slow release of organic acids and natural identical flavors along the swine intestine. Journal of Animal Science 85, 486–493.

Polifroni, R., Etcheverría, A.I., Sanz, M.E., Cepeda, R.E., Krüger, A., Lucchesi, P.M.A., Fernández, D., Parma, A.E., Padola, N.L., 2012. Molecular characterization of shiga toxin-producing *Escherichia coli* isolated from the environment of a dairy farm. Current Microbiology 1–7.

Potter, A.A., Klashinsky, S., Li, Y., Frey, E., Townsend, H., Rogan, D., Erickson, G., Hinkley, S., Klopfenstein, T., Moxley, R.A., Smith, D.R., Finlay, B.B., 2004. Decreased shedding of *Escherichia coli* O157:H7 by cattle following vaccination with type III secreted proteins. Vaccine 22, 362–369.

Raftari, M., Jalilian, F.A., Abdulamir, A.S., Son, R., Sekawi, Z., Fatimah, A.B., 2009. Effect of organic acids on *Escherichia coli* O157:H7 and *Staphylococcus aureus* contaminated meat. Open Microbiology Journal 3, 121–127.

Ransom, J.R., Belk, K.E., Sofos, J.N., Scanga, J.A., Rossman, M.L., Smith, G.C., Tatum, J.D., 2003. Investigation of On-farm Management Practices as Pre-harvest Beef Microbiological Interventions. Centennial, CO National Cattlemen's Beef Association Research Fact Sheet.

Ratledge, C., Dover, L.G., 2000. Iron metabolism in pathogenic bacteria. Annual Review of Microbiology 54, 881–941.

Ricke, S.C., 2003. Perspectives on the use of organic acids and short chain fatty acids as antimicrobials. Poultry Science 82, 632–639.

Ricke, S.C., 2015. Potential of fructooligosaccharide prebiotics in alternative and nonconventional poultry production systems. Poultry Science 94, 1411–1418.

Ricke, S.C., Hererra, P., Biswas, D., 2012. Bacteriophages for potential food safety applications in organic meat production. In: Ricke, S.C., Van Loo, E.J., Johnson, M.G., O'Bryan, C.A. (Eds.), Organic Meat Production and Processing. John Wiley & Sons, Inc., New York.

Rivas, L., Mcdonnell, M.J., Burgess, C.M., O'brien, M., Navarro-Villa, A., Fanning, S., Duffy, G., 2010. Inhibition of verocytotoxigenic *Escherichia coli* in model broth and rumen systems by carvacrol and thymol. International Journal of Food Microbiology 139, 70–78.

Rodriguez-Palacios, A., Staempfli, H.R., Duffield, T., Weese, J.S., 2009. Isolation of bovine intestinal *Lactobacillus plantarum* and *Pediococcus acidilactici* with inhibitory activity against *Escherichia coli* O157 and F5. Journal of Applied Microbiology 106, 393–401.

Rotariu, O., Ogden, I., Macritchie, L., Forbes, K., Williams, A., Cross, P., Hunter, C., Teunis, P., Strachan, N., 2012. Combining risk assessment and epidemiological risk factors to elucidate the sources of human *E. coli* O157 infection. Epidemiology and Infection 140, 1414–1429.

Rozema, E.A., Stephens, T.P., Bach, S.J., Okine, E.K., Johnson, R.P., Stanford, K.I.M., Mcallister, T.A., 2009. Oral and rectal administration of bacteriophages for control of *Escherichia coli* O157:H7 in feedlot cattle. Journal of Food Protection 72, 241–250.

Russell, J.B., 1992. Another explanation for the toxicity of fermentation acids at low pH: anion accumulation versus uncoupling. Journal of Applied Bacteriology 73, 363–370.

Russell, J.B., Diez-Gonzalez, F., 1998. The effects of fermentation acids on bacterial growth. Advances in Microbial Physiology 39, 205–234.

Sagong, H.G., Lee, S.Y., Chang, P.S., Heu, S., Ryu, S., Choi, Y.J., Kang, D.H., 2011. Combined effect of ultrasound and organic acids to reduce *Escherichia coli* O157:H7, *Salmonella* Typhimurium, and *Listeria monocytogenes* on organic fresh lettuce. International Journal of Food Microbiology 145, 287–292.

Sanders, M.E., 2008. Probiotics: definition, sources, selection, and uses. Clinical Infectious Diseases 46 (Suppl. 2), S58–S61 Discussion S144–S151.

Sanderson, M.W., Sargeant, J.M., Shi, X., Nagaraja, T.G., Zurek, L., Alam, M.J., 2006. Longitudinal emergence and distribution of *Escherichia coli* O157 genotypes in a beef feedlot. Applied and Environmental Microbiology 72, 7614–7619.

Sargeant, J.M., Amezcua, M.R., Rajic, A., Waddell, L., 2007. Pre-harvest interventions to reduce the shedding of *E. coli* O157 in the faeces of weaned domestic ruminants: a systematic review. Zoonoses and Public Health 54, 260–277.

Schamberger, G.P., Diez-Gonzalez, F., 2002. Selection of recently isolated colicinogenic *Escherichia coli* strains inhibitory to *Escherichia coli* O157:H7. Journal of Food Protection 65, 1381–1387.

Schamberger, G.P., Phillips, R.L., Jacobs, J.L., Diez-Gonzalez, F., 2004. Reduction of *Escherichia coli* O157:H7 populations in cattle by addition of colicin E7-producing *E. coli* to feed. Applied and Environmental Microbiology 70, 6053–6060.

Schrezenmeir, J., De Vrese, M., 2001. Probiotics, prebiotics, and synbiotics-approaching a definition. American Journal of Clinical Nutrition 73 (Suppl.), 354s–361s.

Sharma, V.K., Dean-Nystrom, E.A., Casey, T.A., 2011. Evaluation of *hha* and *hha sepB* mutant strains of *Escherichia coli* O157:H7 as bacterins for reducing *E. coli* O157:H7 shedding in cattle. Vaccine 29, 5078–5086.

Skaar, E.P., 2010. The battle for iron between bacterial pathogens and their vertebrate hosts. PLoS Pathogens 6, e1000949.

Smith, B.A., Fazil, A., Lammerding, A.M., 2012. A risk assessment model for *Escherichia coli* O157:H7 in ground beef and beef cuts in Canada: evaluating the effects of interventions. Food Control 29, 364–381.

Smith, D.R., Moxley, R.A., Peterson, R.E., Klopfenstein, T.J., Erickson, G.E., Bretschneider, G., Berberov, E.M., Clowser, S., 2009. A two-dose regimen of a vaccine against type III secreted proteins reduced *Escherichia coli* O157:H7 colonization of the terminal rectum in beef cattle in commercial feedlots. Foodborne Pathogens and Disease 6, 155–161.

Smith, H.W., Huggins, R.B., 1987. The control of experimental *E. coli* diarrhea in calves by means of bacteriophage. Journal of General Microbiology 133, 1111–1126.

Stahl, C.H., Callaway, T.R., Lincoln, L.M., Lonergan, S.M., Genovese, K.J., 2004. Inhibitory activities of colicins against *Escherichia coli* strains responsible for postweaning diarrhea and edema disease in swine. Antimicrobial Agents and Chemotherapy 48, 3119–3121.

Stephens, T.P., Loneragan, G.H., Karunasena, E., Brashears, M.M., 2007. Reduction of *Escherichia coli* O157 and *Salmonella* in feces and on hides of feedlot cattle using various doses of a direct-fed microbial. Journal of Food Protection 70, 2386–2391.

Stewart, V.J., 1988. Nitrate respiration in relation to facultative metabolism in enterobacteria. Microbiological Reviews 52, 190–232.

Stouthamer, A.H., 1969. A genetical and biochemical study of chlorate-resistant mutants of *Salmonella typhimurium*. Antoine van Leeuwenhoek 35, 505–521.

Swyers, K.L., Carlson, B.A., Nightingale, K.K., Belk, K.E., Archibeque, S.L., 2011. Naturally colonized beef cattle populations fed combinations of yeast culture and an ionophore in finishing diets containing dried distiller's grains with solubles had similar fecal shedding of *Escherichia coli* O157:H7. Journal of Food Protection 74, 912–918.

Tanji, Y., Shimada, T., Fukudomi, H., Miyanaga, K., Nakai, Y., Unno, H., 2005. Therapeutic use of phage cocktail for controlling *Escherichia coli* O157:H7 in gastrointestinal tract of mice. Journal of Bioscience and Bioengineering 100, 280–287.

Tellez, G., Laukova, A., Latorre, J.D., Hernandez-Velasco, X., Hargis, B.M., Callaway, T.R., 2015. Food-producing animals and their health in relation to human health. Microbial Ecology in Health & Disease 26, 25876.

Thomson, D.U., Loneragan, G.H., Thornton, A.B., Lechtenberg, K.F., Emery, D.A., Burkhardt, D.T., Nagaraja, T.G., 2009. Use of a siderophore receptor and porin proteins-based vaccine to control the burden of *Escherichia coli* O157:H7 in feedlot cattle. Foodborne Pathogens and Disease 6, 871–877.

Thornton, A.B., Thomson, D.U., Loneragan, G.H., Fox, J.T., Burkhardt, D.T., Emery, D.A., Nagaraja, T.G., 2009. Effects of a siderophore receptor and porin proteins-based vaccination on fecal shedding of *Escherichia coli* O157:H7 in experimentally inoculated cattle. Journal of Food Protection 72, 866–869.

Tonsor, G.T., Schroeder, T.C., 2015. Market impacts of *E. coli* vaccination in U.S. feedlot cattle. Agricultural and Food Economics 3, 1–15.

Varela, N.P., Dick, P., Wilson, J., 2013. Assessing the existing information on the efficacy of bovine vaccination against *Escherichia coli* O157: H7-a systematic review and meta-analysis. Zoonoses and Public Health 60, 253–268.

Vilte, D.A., Larzábal, M., Mayr, U.B., Garbaccio, S., Gammella, M., Rabinovitz, B.C., Delgado, F., Meikle, V., Cantet, R.J.C., Lubitz, P., Lubitz, W., Cataldi, A., Mercado, E.C., 2012. A systemic vaccine based on *Escherichia coli* O157:H7 bacterial ghosts (BGs) reduces the excretion of *E. coli* O157:H7 in calves. Veterinary Immunology and Immunopathology 146, 169–176.

Walle, K.V., Vanrompay, D., Cox, E., 2012. Bovine innate and adaptive immune responses against *Escherichia coli* O157:H7 and vaccination strategies to reduce faecal shedding in ruminants. Veterinary Immunology and Immunopathology 152, 109–120.

Wells, J.E., Shackelford, S.D., Berry, E.D., Kalchayanand, N., Guerini, M.N., Varel, V.H., Arthur, T.M., Bosilevac, J.M., Freetly, H.C., Wheeler, T.L., Ferrell, C.L., Koohmaraie, M., 2009. Prevalence and level of *Escherichia coli* O157:H7 in feces and on hides of feed lot steers fed diets with or without wet distillers grains with solubles. Journal of Food Protection 72, 1624–1633.

Wetzel, A.N., LeJeune, J.T., 2006. Clonal dissemination of *Escherichia coli* O157:H7 subtypes among dairy farms in Northeast Ohio. Applied and Environmental Microbiology 72, 2621–2626.

Wheeler, T.L., Kalchayanand, N., Bosilevac, J.M., 2014. Pre- and post-harvest interventions to reduce pathogen contamination in the U.S. beef industry. Meat Science 98, 372–382.

White, A., Cronquist, A., Bedrick, E.J., Scallan, E., 2016. Food source prediction of Shiga toxin-producing *Escherichia coli* outbreaks using demographic and outbreak characteristics, United States, 1998–2014. Foodborne Pathogens and Disease 13, 527–534.

Willard, M.D., Simpson, R.B., Cohen, N.D., Clancy, J.S., 2000. Effects of dietary fructooligosaccharide on selected bacterial populations in feces of dogs. American Journal of Veterinary Research 61, 820–825.

Wisener, L.V., Sargeant, J.M., O'connor, A.M., Faires, M.C., Glass-Kaastra, S.K., 2015. The use of direct-fed microbials to reduce shedding of *Escherichia coli* O157 in beef cattle: a systematic review and meta-analysis. Zoonoses and Public Health 62, 75–89.

Wolin, M.J., 1969. Volatile fatty acids and the inhibition of *Escherichia coli* growth by rumen fluid. Advances in Applied Microbiology 17, 83–87.

Younts-Dahl, S.M., Galyean, M.L., Loneragan, G.H., Elam, N.A., Brashears, M.M., 2004. Dietary supplementation with *Lactobacillus*- and *Propionibacterium*-based direct-fed microbials and prevalence of *Escherichia coli* O157 in beef feedlot cattle and on hides at harvest. Journal of Food Protection 67, 889–893.

Zhao, T., Doyle, M.P., Harmon, B.G., Brown, C.A., Mueller, P.O.E., Parks, A.H., 1998. Reduction of carriage of enterohemorrhagic *Escherichia coli* O157:H7 in cattle by inoculation with probiotic bacteria. Journal of Clinical Microbiology 36, 641–647.

Zhao, T., Tkalcic, S., Doyle, M.P., Harmon, B.G., Brown, C.A., Zhao, P., 2003. Pathogenicity of enterohemorrhagic *Escherichia coli* in neonatal calves and evaluation of fecal shedding by treatment with probiotic *Escherichia coli*. Journal of Food Protection 66, 924–930.

Zopf, D., Roth, S., 1996. Oligosaccharide anti-infective agents. Lancet 347, 1017–1021.

CHAPTER 9

# *Foodborne* Salmonella *in Laying Hens and Egg Production*

**Steven C. Ricke[1], Turki M. Dawoud[1,2], Zhaohao Shi[1], Pravin Kaldhone[1], Young M. Kwon[1]**
[1]*University of Arkansas, Fayetteville, AR, United States;* [2]*King Saud University, Riyadh, Saudi Arabia*

## *Introduction*

Foodborne salmonellosis continues to be an economic and public health issue in the United States (US) as well as internationally (Scallan et al., 2011; Dawoud et al., 2016). Likewise, as global trade expands, this also adds to the impact of foodborne disease for food products transported to their corresponding markets. This occurs by not only increasing the number of potential contamination sites but also adding to the complexity for tracking the origin of foodborne pathogens such as *Salmonella* (Ricke, 2010; Ricke et al., 2015). Foodborne *Salmonella* continues to plague food production with contamination and recontamination potentially occurring at any phase during production all the way from the agricultural fields where produce is grown to the final stages of retail and home food preparation (Carrasco et al., 2012; Erickson, 2010; Hanning et al., 2009; Jarvis et al., 2016). This multitude of potential contamination sources presents a complex and difficult challenge for managing and controlling foodborne *Salmonella* contamination and dissemination. Consequently, control strategies have to be devised that meet the sometimes unique requirements for individual steps in food production, processing, and retail presentation. This becomes particularly true for foods that are marketed in the raw form such as produce commodities that include vegetables and fruits versus foods that are sold in already cooked form such as ready-to-eat meats (Hanning et al., 2009; Koo et al., 2012).

Livestock and meat production have also been problematic sources for *Salmonella* contamination. Both beef and swine have been associated with foodborne *Salmonella* over the years (Arthur et al., 2008; Bacon et al., 2002; Foley et al., 2008). In poultry production, foodborne *Salmonella* serovars remain a consistent contamination threat both for birds in broiler production and laying hens in egg production (Vandeplas et al., 2010; Cox et al., 2011; Finstad et al., 2012; Foley et al., 2011, 2013; Howard et al., 2012). Although broiler production cycles are relatively short term compared with the egg-laying cycles, both types of birds can be susceptible to *Salmonella* colonization (Dunkley et al., 2009; Finstad et al., 2012; Foley et al., 2011,

Food and Feed Safety Systems and Analysis. http://dx.doi.org/10.1016/B978-0-12-811835-1.00009-9

2013; Howard et al., 2012). Certainly, young newly hatched chicks are considered fairly susceptible to *Salmonella* infection, but older birds under certain circumstance are also vulnerable to colonization of the gastrointestinal tract (GIT) and potentially systemic infection (Ricke, 2003a; Ricke et al., 2013).

Consequently, a wide range of intervention strategies have been proposed, developed, and commercially implemented to control *Salmonella* at all phases of poultry production and processing. In live bird production, interventions involve measures employed to limit colonization including prebiotics, probiotics, and vaccines among others as well as administration of biological and chemical antimicrobial agents to reduce *Salmonella* already colonized in the bird GIT (Galiş et al., 2013; Revolledo et al., 2006; Revolledo and Ferreira, 2012; Ricke, 2003b; Ricke et al., 2015; Kollanoor Johny and Venkitanarayanan, 2016). For postharvest interventions, antimicrobials including organic acids, chlorine, and other compounds are administered to poultry carcasses (Ricke et al., 2005; Sofos et al., 2013; Handley et al., 2015). In egg processing, both physical interventions such as in-shell pasteurization and UV (ultraviolet) as well a wide range of chemicals, botanicals, and other sanitizing agents have been examined for reducing *Salmonella*-contaminating egg shell surfaces (Galiş et al., 2013; Latimer et al., 2008; O'Bryan et al., 2016; Upadhyaya et al., 2016). In this chapter, sources of *Salmonella* contamination in egg production and strategies to limit prevalence and reduce contamination will be discussed as well as future directions to achieve further reduction.

## Salmonella *Infections and Poultry*

Historically *Salmonella* Gallinarum and Pullorum, which are host specific to poultry, were considered the primary serovars of economic concern due to the mortalities associated with these serovars when they infected birds (Bäumler et al., 2000; Hitchner, 2004). This prompted rigorous flock eradication programs leading to their demise as primary infecting serovars of poultry (Bäumler et al., 2000). However, as *S.* Gallinarum and Pullorum declined in prevalence with the implementation of the National Poultry Improvement Plan in the 1930s (Hitchner, 2004), *Salmonella enterica* subspecies serovar Enteritidis (*S.* Enteritidis) emerged as the serovar most closely identified with chicken eggs (Bäumler et al., 2000; Braden, 2006; Rabsch et al., 2000; Ricke and Gast, 2014; Saeed et al., 1999; St. Louis et al., 1988). By the late 1970s, *S.* Enteritidis began to receive more attention as an important food safety concern and by the middle of the 1990s it was the most frequent *Salmonella* serovar contributing to cases reported in the US. (Braden, 2006; CDC, 2000; Olsen et al., 2000, 2001; Rabsch et al., 2001).

The infection of poultry and eggs with *Salmonella* is a serious concern for the sake of food safety. While several foodborne *Salmonella* serovars have been associated with poultry infections over the years, most notably *S.* Typhimurium and *S.* Heidelberg, *S.* Enteritidis remained the most closely identified with egg contamination (Foley et al., 2011, 2013;

Howard et al., 2012; Kaldhone et al., 2016; Martelli and Davies, 2012; Saeed et al., 1999; St. Louis et al., 1988). Of all the serovars, *S.* Enteritidis is most closely associated with illness caused by the consumption of eggs and has been found to be capable of both vertical and horizontal transfer among infected chickens. Historically, some methods used by industry may have facilitated the infection of eggs with *S.* Enteritidis, with some research suggesting that one of the primary methods for *Salmonella* contamination was colonization of the reproductive tract and organs before being passed onto the developing egg follicles. Further studies were conducted to better understand how *S.* Enteritidis can infect poultry and eggs and to best prevent illness and ensure consumer safety.

## Salmonella *and the Laying Hen Reproductive Tract*

The main egg contamination mechanism appears to be due to *S.* Enteritidis invasion of the reproductive tract organs after a systemic infection (Gantois et al., 2008; Gast et al., 2009; Upadhyaya et al., 2013). Before eggshell formation and during the formation of eggs, it is possible that formed eggs can be exposed to either descending infections from *S.* Enteritidis colonized ovarian tissue, and infection from upper oviduct tissues, or ascending infections of vaginal and cloacal tissue colonization (Bichler et al., 1996; Gast et al., 2004; Keller et al., 1995; Miyamoto et al., 1997). It is this predilection toward colonization of the laying hen reproductive tract that has been the focus of considerable research efforts to delineate specific mechanisms in *S.* Enteritidis that distinguish it from other *Salmonella* serovars that appear to be less likely to exhibit such host-specific tissue tropism.

The colonization of the reproductive tract tissues by *S.* Enteritidis is certainly a primary factor for contamination of egg contents and may be at least partially responsible for the persistence of *S.* Enteritidis and other *Salmonella* serovars in mature laying flocks (Barnhart et al., 1991; Berchieri Jr. et al., 2001; Davies and Breslin, 2003a,b; Gast et al., 2005; Guard et al., 2010; Van Immerseel, 2010). *S.* Enteritidis can be isolated from reproductive tissues and organs of naturally infected hens without prior intestinal colonization, and some strains can be isolated from oviduct tissue and other tissues within an hour of experimentally infecting laying hens and crop inoculation (Humphrey et al., 1993). In addition, the isolation rate can be considerably higher when extending the incubation time (Humphrey and Whitehead, 1993).

The virulence capabilities of *S.* Enteritidis and its connection to potential tissue tropism for reproductive tissues in laying hens have been a major concern that needs to be understood because evidence from several phenotypic characterization studies have suggested a genetic association (Clavijo et al., 2006; Gantois et al., 2008, 2009; Kang et al., 2006; Lu et al., 2003; Parker et al., 2002; Raspoet et al., 2011; Shah, 2014; Shah et al., 2012a,b; Van Immerseel, 2010). An *S.* Enteritidis PT4 isolate that displayed wrinkled appearance colonies on plates and more tolerance to heat, acid, and hydrogen peroxide stresses was also more invasive to reproductive tissue (Humphrey et al., 1996). Some studies reported that infection with *S.* Enteritidis can induce regression of the reproductive tract (Guard-Bouldin et al., 2004; Guard-Bouldin and Buhr, 2006; Parker et al., 2001, 2002).

The methods by which *S.* Enteritidis colonizes the reproductive tissues and causes transovarian infection raises questions as to what distinguishes this serovar from others to allow it to routinely contaminate eggs. This in turn leads to the identification of which genes are required for adherence and invasion of reproductive tissues. Although both *S.* Enteritidis and *S.* Typhimurium are capable of colonizing chicken reproductive tissues, only *S.* Enteritidis can be isolated from the contents of intact eggs after being laid and is apparently able to survive within intact chicken eggs (Clavijo et al., 2006; De Vylder et al., 2013; Gantois et al., 2008; Keller et al., 1995; Lu et al., 2003; Shah et al., 2012a,b). This suggests that *S.* Enteritidis has distinctive genetic factors compared with other *Salmonella* serovars (Porwollik et al., 2005; Shah et al., 2016; Silva et al., 2012).

Another characteristic of *S.* Enteritidis pathogenesis that makes it unique for laying hens is that the frequent passage and isolation of *S.* Enteritidis from the ovary and the oviducts dramatically increases its ability to contaminate eggs internally without any changes in the invasiveness or colonization of the reproductive organs (Gast et al., 2003, 2009). *S.* Enteritidis has specificity for the hen's reproductive tissue in general and can cause transovarian infection (Braden, 2006; Foley et al., 2011; Galiş et al., 2013; Gantois et al., 2009; Howard et al., 2012; Rabsch et al., 2001; Threlfall et al., 2000). This serovar is known to colonize and infect the ovary and has been commonly isolated from the ovary interstitial tissue surrounding the yolk rather than the follicle yolk (Barrow and Lovell, 1991; Keller et al., 1995).

Chicken ovarian follicles are present at various stages of maturation, primarily differing in the yellow yolk quantity. A mature hen will lay eggs when ovarian follicles become fully developed. Follicle maturity goes into three major stages (Howard et al., 2012; Klein-Hessling, 2007). In stage 1, follicles grow very slowly and may last for several months with no significant amounts of yolk. These follicles are known as small white follicles. In the second stage, a few follicles will develop and subsequently achieve yolk deposition. These follicles will appear yellow and are known as small yellow follicles. In the last stage, follicles from the pool of small yellow follicles are selected and go into a rapid growth stage, leading to mature follicles within 7 days. These follicles go through the oviduct during ovulation, initiating egg formation (Recheis et al., 2000). Generally, the immature follicles exist with multiple granulosa cell layers; however, the granulosa cells can be seen as one layer in the more mature large yellow follicles or F follicles (Howard et al., 2012; Lovell et al., 2003). *S.* Enteritidis has displayed attachment to both developing and mature follicular granulosa cells (Thiagarajan et al., 1994).

Howard et al. (2005) performed an *in vitro* study to compare the ability of *Salmonella* in invading ovarian follicles of laying hens at various maturation phases. By comparing the two most common nontyphodial *Salmonella* serovars, *S.* Typhimurium and S. Enteritidis, they demonstrated that invading intact follicles by *S.* Typhimurium was higher for the immature small white follicles but were less invasive as follicles developed into the more mature small yellow and large yellow follicles (Howard et al., 2005). Howard et al. (2005) concluded that *S.* Enteritidis may possibly use different mechanisms to engage with different follicle patterns during follicle development, although Gantois et al. (2009) has pointed out that not all white follicles will mature and extensive growth in these follicles by *Salmonella* would potentially cause their degeneration prior to maturity. The involvement of attachment and invasion mechanisms in *S.* Enteritidis infection of ovarian follicles may require the ability of two separate genetic properties that need to be recognized and understood.

Dawoud et al. (2011) followed up these initial ovarian follicle *in vitro* invasion assays by comparing five different *S.* Enteritidis strains. After inoculating the freshly collected ovarian follicles, all five *S.* Enteritidis strains were able to invade into the follicles after 2 h of incubation at 37°C with invasion averages ranging from 0.016% to 0.034% but no significant differences were apparent when the strains were compared. These invasion percentages were generally much higher than the negative control *Escherichia coli* K-12 strain, with a mean percentage of invasion of only 0.0003%. While there appears to be minimal strain variation, it remains to be tested whether there are differences among other *Salmonella* serovars.

## *Incidence of* Salmonella *and Management of Laying Hens*

Vertical transmission from parent to offspring via infected hatching eggs was considered the primary transmission modes in birds (Lutful Kabir, 2010), and horizontal transmission can possibly occur via bird-to-bird contact as a result of shedding *Salmonella* in feces by infected birds and inhalation by birds in close proximity. Another route is ingesting *Salmonella*-contaminated materials such as feed and/or contact with other environmental surfaces (Byrd et al., 1998; Liljebjelke et al., 2005; Thakur et al., 2013). Feed can act as a reservoir for pathogens, which can persist under storage conditions for several months and cross-contaminate other sources as the feed is mixed (Maciorowski et al., 2004, 2006, 2007).

The incidence of *S.* Enteritidis infected hens are influenced by two critical aspects, the microbial traits and the management systems in the US commercial egg industries. Cyclic timely collection of contaminated eggs produced by laying hens has been related to acquired-stress from specific procedures such as molting (Holt, 2003). Historically, induced molting using feed withdrawal was the primary method employed by the layer industry to stimulate numerous egg-laying cycles in hens (Dunkley et al., 2009; Golden et al., 2008; Mazzuco et al., 2011; Park et al., 2004; Patwardhan et al., 2011; Ricke, 2003a, 2014; Ricke et al., 2013). The stress associated with feed withdrawal unfortunately was identified as being associated with an increased susceptibility to *S.* Enteritidis infection, with typical increases in intestinal shedding and spreading of *S.* Enteritidis to internal organs such as the liver, spleen, and ovaries (Dunkley et al., 2007; Durant et al., 1999; Gantois et al., 2009; Guard-Petter, 2001; Holt, 2003; Kubena et al., 2005; Ricke, 2003a). Feed withdrawal also increased the susceptibility of tissues such as the crop, ilea, and ceca to the colonization by *S.* Enteritidis (Moore and Holt, 2006). A study by Holt (1995) demonstrated that during the 48 h after application of feed withdrawal molting, systematic invasion of *S.* Enteritidis occurred rapidly and more *S.* Enteritidis were recovered from the livers and spleens of molted hens compared with the unmolted ones.

This susceptibility has also been shown to lead to increased horizontal distribution of *S.* Enteritidis infection to molted hens in nearby cages and more eggs contaminated with the bacterium (Holt, 1995; Holt et al., 1995, 1998). More recent studies demonstrated that horizontal transmission can take place between a group of experimentally infected hens and a susceptible group of laying hens in connection to the housing system (De Vylder et al., 2011). Higher risk was found in systems with continued fecal contact, such as alternative housing systems, because of the contact of chicken manure contaminated with *Salmonella* species. Although the alternative housing systems of having birds in furnished cages or birds in the aviaries exhibited decreased shedding levels and decreased splenic and hepatic colonization, there were no differences in the reproductive tissues, namely, the ovaries and oviducts (De Vylder et al., 2009).

## *Control Programs for* Salmonella *in Laying Hen Flocks—General Concepts*

*S.* Enteritidis represented a major threat to the poultry industry because not only did it contaminate table eggs, but it could also be transmitted vertically from infected breeder flocks. With broiler breeders, this can lead to infection of broilers and the contamination of carcasses (Rasschaert et al., 2007; Van Immerseel et al., 2009). The antigenic attributes and pathogenesis characteristics of *S.* Enteritidis may in part play a role in its increased frequencies and persistence because they share similarities with the restricted avian host pathogens, *S.* Gallinarum and *S.* Pullorum (Foley et al., 2011; Thomson et al., 2008; Uzzau et al., 2000). In some experimental studies, laying hens showed significant reduction of egg production after *S.* Enteritidis infection, although *S.* Enteritidis does not cause severe pathological infections in laying hen flocks or any visible illness (Gast et al., 2002). Because *S.* Enteritidis has multiple reservoirs and vectors including mice and other rodents, minimizing its infections in hen flocks requires a multifaceted effort (Carrique-Mas and Davies, 2008; Carrique-Mas et al., 2009; Garber et al., 2003; Mollenhorst et al., 2005).

Consequently, research efforts in the past half century have been focused not only on development a variety of egg sanitation and disinfectant approaches that could be employed during egg processing, but also on increasing efforts to limit *Salmonella* colonization of the egg-laying hen flocks during egg production. These can be broadly categorized as preharvest and postharvest interventions and involve a wide range of technologies and application approaches. Much of the early work focused on administration of various chemical disinfectants to remove external microbial contamination from the egg shell. However, once it was realized that internal contamination could occur as a function of transovarian infection and subsequent deposition of *Salmonella* during egg formation by the hen, research strategies were altered to develop technologies that would eliminate internally deposited *Salmonella* while minimizing the impact on the integrity of the table shell egg.

Likewise, once it was realized that the laying hen could become vulnerable to *Salmonella* infection during certain stages of her life cycle, research efforts were launched to first identify those stages most likely to be susceptible to infection and subsequently direct intervention development toward those stages of production. Certainly the newly hatched layer chick represented an opportunity for *Salmonella* colonization before establishment and maturing of the gastrointestinal microbiota much in the same way that young broiler chicks are susceptible. In both young broiler and laying hen chicks, administration of preventative measures such as probiotics that could accelerate establishment of a mature GIT microbiota and thus serve as a barrier to *Salmonella* colonization was developed. Likewise, vaccination regimens that helped to trigger the host's immune system to respond to invading *Salmonella* were extensively examined as a means of directly involving the host bird. However, as was demonstrated in the early 2000s, mature laying hens during certain management practices such as molting induced by feed withdrawal could make them vulnerable to *Salmonella* colonization and this in turn required development of specific interventions that would be particularly effective in these adult birds (Holt, 2003; Ricke, 2003a).

In addition, although much of the traditional focus has been directed toward *S.* Enteritidis as the primary serovar most likely to be associated with egg-related salmonellosis, other serovars are beginning to be identified with egg production as well (Foley et al., 2011, 2013; Howard et al., 2012). For example, egg or egg-related products have been identified as vehicles in more than half of the outbreaks caused by *S.* Heidelberg (Chittick et al., 2006; Kaldhone et al., 2016). This signifies the importance of preventive measures for serovars besides *S.* Enteritidis such as *S.* Heidelberg and could mean that colonization properties may differ and intervention strategies geared toward eggs and layer hens that were based on *S.* Enteritidis may have to be altered to account for these differences. The following sections describe in more detail specific preharvest and postharvest control and intervention strategies for laying hen egg production and egg processing.

## *Interventions for Preharvest Egg Production*

To keep harmful bacterial populations below the threat level, poultry growers looked at different options beyond traditional antimicrobial food additives. In general, numerous approaches have been or are being tested and implemented in this area to control *Salmonella* in flocks, but most of this work has focused on *S.* Typhimurium and *S.* Enteritidis and less on other serovars such as *S.* Heidelberg. Among these, probiotics and maternal vaccination seem to be effective to some extent in reducing cecal colonization of *Salmonella*.

Probiotics and prebiotics represent active commercial and biological additive approaches that are currently being extensively examined. A probiotic is a live microbial culture supplement, which alters intestinal microbial balance of the host to benefit the host (Callaway and Ricke, 2012; Callaway et al., 2016; Fuller, 1989; Ricke and Pillai, 1999). Different types of bacteria and yeast have been employed as probiotics, which include *Lactobacillus*, *Bifidobacterium*, *Bacillus*, *Enterococcus*, and *Streptococcus* species (Callaway and Ricke, 2012; Callaway et al., 2016; Nisbet, 2002; Siragusa and Ricke, 2012; Ricke and Saengkerdsub, 2015). Menconi et al. (2011) observed reduction in *S.* Heidelberg populations in chicks and turkey poults after probiotic exposure to *Lactobacillus*. Rahimi et al. (2009) noticed improved gut health and enhanced bird performance among Primalac (direct-feed microbial) fed turkey poults after challenging with *S.* Heidelberg. Different probiotics may act via different mechanisms to reduce the *Salmonella* populations; these include an increase in oxidative burst capacity, competition for nutrients, and synthesis of antibacterial compounds such as fermentation acids, bacteriocins, and hydrogen peroxide (Nisbet, 2002; Farnell et al., 2006; Vandeplas et al., 2010; Ricke, 2003b, 2016; Callaway and Ricke, 2012; Siragusa and Ricke, 2012).

Prebiotics are selectively fermented products that alter the gastrointestinal microbiome to benefit the host (Gibson and Roberfroid, 1995; Ricke, 2015; Roto et al., 2015). They stimulate the immune system, inactivate toxins, and prevent pathogen colonization (Salminen et al., 2010). Fructooligosaccharides (FOS) have been demonstrated to be effective in reducing *Salmonella* prevalence in the chicken gut microbiome (Fukata et al., 1999; Ricke, 2015; Xu et al., 2003). The FOS act by eliciting a positive growth effect of resident bacteria, such as *Bifidobacterium* spp. and *Lactobacillus* spp. In a series of *in vitro* and *in vivo* studies with layer hens and layer hen cecal contents, Donalson et al. (2007, 2008a,b) demonstrated that FOS addition could limit *Salmonella* and alter cecal fermentation. Synbiotics are a combination of probiotic organisms and prebiotic compounds that have also been explored as an intervention strategy (Callaway and Ricke, 2012; Ricke, 2016).

Poultry feed containing plant-based protein meal has shown to be somewhat less significance in preventing colonization and shedding of *S.* Heidelberg (Alali et al., 2011). The same group concluded that reduction in shedding and colonization of *S.* Heidelberg in chicken occurred after treatment with nonpharmaceutical compounds (Alali et al., 2013a). In the same study, Alali et al. (2013a) added lactic acid and essential oils (EOs) to drinking water for birds. EOs are plant-derived products, which have shown capability to decrease *Salmonella* contamination both *in vitro* and *in vivo* (Calo et al., 2015; Kollanoor Johny and Venkitanarayanan, 2016; Nair et al., 2014; O'Bryan et al., 2015). Alali et al. (2013a) concluded that the combination of EOs and lactic acid in drinking water could reduce *S.* Heidelberg shedding in broilers at 14 and 42 days postinoculation. This coincides with findings from another study, where decreased mortality and increased weight gain at the end of 42 days was observed by the same group (Alali et al., 2013b). They detected reduction in *S.* Heidelberg colonization in crops, feed conversion ratio, and increased weight gain in a group of birds at the end of 42 days receiving EOs in their drinking water. These findings suggest antimicrobial capability of EOs against *S.* Heidelberg in poultry. Carvacrol has been shown to be most effective against *S.* Heidelberg; however, xylanases with EOs such as cinnamaldehyde and thymol have also proven to be effective antimicrobials (Amerah et al., 2012). Similar studies might lead to an optimal choice of compounds or combination to be added in drinking water of birds to generally reduce *Salmonella* contamination.

Vaccination is another way to control colonization of layer chickens by *Salmonella* and hence subsequent contamination of the eggs (Aehle and Curtiss III, 2016). Vaccines typically consist of the target protective antigen and an immunostimulatory adjuvant (Curtiss III et al., 2009). The protective antigen elicits protection in the host by generating a targeted strong primary immune response followed by a lasting memory to resist subsequent exposures (Kong et al., 2008; Konjufca et al., 2006). Jiang et al. (2010) studied the effects of *S.* Typhimurium–based vaccine on *S.* Heidelberg prevalence in chickens. This study evaluated the use of an attenuated *S.* Typhimurium vaccine vector for delivery of a heterologous antigen to chicken (Jiang et al., 2010). After challenging birds with *S.* Heidelberg, *S.* Enteritidis, *S.* Anatum, *S.* Kentucky, or *S.* Typhimurium, they observed a decrease in colonization by *S.* Heidelberg and a prevention of necrotic enteritis in vaccinated birds. In another study, Nandre et al. (2015) compared the effectiveness of a live attenuated *S.* Enteritidis vaccine on egg contamination. The researchers noted increases in plasma immunoglobulin (Ig) G, intestinal secretory IgA, egg yolk IgY antibodies, and cellular immune response in layers after administration of this vaccine. After a challenge of the vaccinated laying hens with a virulent *Salmonella* strain, there was a decrease in the production of malformed eggs compared with the unvaccinated birds. Deguchi et al. (2009) examined the effectiveness of trivalent inactivated vaccine of *S.* Enteritidis, *S.* Typhimurium, and *S.* Infantis after a challenge with *S.* Heidelberg. Decreased *Salmonella* shedding was observed in this study, indicating that this trivalent vaccine could be an effective tool to control *S.* Heidelberg infections in chickens. The authors also concluded that this vaccine could be effective against different strains having common O antigens (O4, O7, and O9) with those of the vaccine strains (Deguchi et al., 2009). This finding is important because it likely increases the spectrum of serovars covered by the vaccine. Overall the different vaccine-based studies demonstrate the potential promise of vaccine-based approaches to reduce egg contamination by *Salmonella*.

## Interventions for Postharvest Egg Processing

Commercial egg washing and minimizing temperature abuse during transportation and storage has been proposed as a means to provide consumers with safer eggs (Curtis, 2007; O'Bryan et al., 2016). The addition of disinfectants, such as chlorine- and iodine-based compounds, to water used to wash eggs has been widely administered for cleaning eggs. In addition, naturally occurring chemicals, such as EOs with antimicrobial activity, are receiving more attention as consumer demands are changing (Upadhyaya et al., 2016). Some consumers consider EOs as more natural compared with other chemical agents such as chlorine and iodine (Calo et al., 2015). Egg washing with EOs could be a potential way to decrease *Salmonella* populations on the external surface of the eggshell. The mode of actions of EOs on egg surfaces is not clear; the phenolic compounds present in the EO molecule are hypothesized to penetrate the bacterial cell membrane (O'Bryan et al., 2015). The compounds are also hydrophobic in nature, which may increase their ability to disrupt lipid components of cell membrane (Calo et al., 2015). The antimicrobial properties of EOs are not limited to *S.* Heidelberg, as they have also been shown to be effective against other pathogens including *S.* Typhimurium, *Staphylococcus aureus*, *Listeria monocytogenes*, and *Campylobacter* (Callaway et al., 2011).

Pulsed UV light has also been evaluated as a potential approach to decontaminate the surface of eggs (O'Bryan et al., 2016). Keklik et al. (2010) detected significant reduction in *S.* Enteritidis cell populations after pulsed UV light treatment. They also observed that egg characteristics such as albumin content and eggshell strength did not change postexposure; thus this lack of residual effect of this treatment makes pulsed UV light treatment a potential attractive candidate for egg sanitization. Synergistic approaches using disinfectants containing chemicals, such as chlorine or iodine, and EOs, with physiochemical methods such as pulsed UV light therapy could be more beneficial than any one method along with in-shell pasteurization (Friedman, 2015; O'Bryan et al., 2016). Thus the combination of different approaches may give maximum reduction of *Salmonella* in layer hens and eggs; however, additional studies are needed to validate the approaches.

## Conclusions

Although *Salmonella* contamination in eggs remains a problem, considerable progress has been made in not only understanding how infection occurs in the laying hen but devising improved means for limiting contamination throughout the egg production cycle. Further advances are becoming possible as more is learned of the avian gut microbiome using next-generation sequencing. Such approaches have helped to delineate how the gut microbiota interacts with the host and which external factors have the most impact on the gut microbial community. Applying these approaches to studies on the layer hen gut ecosystem should reveal specific responses of the bird gut microbiota to dietary changes as well as addition of additives such as prebiotics and antimicrobials. This in turn will potentially help to design more effective and targeted treatments against *Salmonella* that avoid disrupting the gut microbiota that are positive contributors to the health and well-being of the laying hen host. For postharvest egg processing the trend may be moving away from traditional chemical sanitizers toward more natural compounds such as the various botanical-based antimicrobials. Such sources appear to be promising but will require additional studies to determine their efficacy and standardize their commercial application.

## References

Aehle, S., Curtiss III, R., 2016. Current and future perspectives on development of *Salmonella* vaccine technologies. In: Ricke, S.C., Gast, R.K. (Eds.), Producing Safe Eggs-The Microbial Ecology of *Salmonella*. Elsevier, Inc., San Diego, CA, pp. 281–299 (Chapter 14).

Alali, W.O., Hofacre, C.L., Mathis, G.F., Faltys, G., Ricke, S.C., Doyle, M.P., 2013a. Effect of non-pharmaceutical compounds on shedding and colonization of *Salmonella enterica* serovar Heidelberg in broilers. Food Control 31, 125–128.

Alali, W.O., Hofacre, C.L., Mathis, G.F., Faltys, G., 2013b. Effect of essential oil compound on shedding and colonization of *Salmonella enterica* serovar Heidelberg in broilers. Poultry Science 92, 836–841.

Alali, W.Q., Hofacre, C.L., Mathis, G.F., Batal, A.B., 2011. Effect of plant- based protein meal use in poultry feed on colonization and shedding of *Salmonella* Heidelberg in broiler birds. Agriculture, Food and Analytical Bacteriology 1 (1), 45–53.

Amerah, A.M., Mathis, G., Hofacre, C.L., 2012. Effect of xylanase and a blend of essential oils on performance and *Salmonella* colonization of broiler chickens challenged with *Salmonella* Heidelberg. Poultry Science 91, 943–947.

Arthur, T.M., Brichta-Harhay, D.M., Bosilevac, J.M., Guerin, M.N., Kalchayanand, N., Wells, J.E., Shackelford, S.D., Wheeler, T.L., Koohmaraie, M., 2008. Prevalence and characterization of *Salmonella* in bovine lymph nodes potentially destined for use in ground beef. Journal of Food Protection 71, 1685–1688.

Bacon, R.T., Sofos, J.N., Belk, K.E., Hyatt, D.R., Smith, G.C., 2002. Prevalence and antibiotic susceptibility of *Salmonella* isolated from beef animal hides and carcasses. Journal of Food Protection 65, 284–290.

Barnhart, H.M., Dreesen, D.W., Bastien, R., Pancorbo, O.C., 1991. Prevalence of *Salmonella* Enteritidis and other serovars in ovaries of layer hens at time of slaughter. Journal of Food Protection 54 (7), 488–491.

Barrow, P.A., Lovell, M.A., 1991. Experimental infection of egg-laying hens with *Salmonella* Enteritidis phage type 4. Avian Pathology 20 (2), 335–348.

Bäumler, A.J., Hargis, B.M., Tsolis, R.M., 2000. Tracing the origins of *Salmonella* outbreaks. Science 287 (5450), 50–52.

Berchieri Jr., A., Wigley, P., Page, K., Murphy, C.K., Barrow, P.A., 2001. Further studies on vertical transmission and persistence of *Salmonella enterica* serovar Enteritidis phage type 4 in chickens. Avian Pathology 30 (4), 297–310.

Bichler, L.A., Nagaraja, K.V., Halvorson, D.A., 1996. *Salmonella enteritidis* in eggs, cloacal swab specimens, and internal organs of experimentally infected white leghorn chickens. American Journal of Veterinary Research 57 (4), 489–495.

Braden, C.R., 2006. *Salmonella enterica* serotype Enteritidis and eggs: a national epidemic in the United States. Clinical Infectious Diseases 43 (4), 512–517.

Byrd, J.A., Corrier, D.E., DeLoach, J.R., Nisbet, D.J., Stanker, L.H., 1998. Horizontal transmission of *Salmonella* Typhimurium in broiler chicks. Journal of Applied Poultry Research 7 (1), 75–80.

Callaway, T.R., Edrington, T.S., Byrd, J.A., Nisbet, D.J., Ricke, S.C., 2016. Use of direct fed microbials in layer hen production-Performance response and *Salmonella* control. In: Ricke, S.C., Gast, R.K. (Eds.), Producing Safe Eggs-The Microbial Ecology of *Salmonella*. Elsevier, Inc., San Diego, CA, pp. 301–322 (Chapter 15).

Callaway, T.R., Ricke, S.C. (Eds.), 2012. Direct Fed Microbials/Prebiotics for Animals: Science and Mechanisms of Action. Springer Science, New York, NY.

Callaway, T., Carroll, J.A., Arthington, J.D., Edrington, T.S., Anderson, R.C., Rossman, M.L., Carr, M.A., Genovese, K.J., Ricke, S.C., Crandall, P., Nisbet, D.J., 2011. Orange peel products can reduce *Salmonella* populations in ruminants. Foodborne Pathogens and Disease 10, 1071–1075.

Calo, J.R., Crandall, P.G., O'Bryan, C.A., Ricke, S.C., 2015. Essential oils as antimicrobials in food systems – a review. Food Control 54, 111–119.

Carrasco, E., Morales-Rueda, A., García-Gimeno, R.M., 2012. Cross-contamination and recontamination by *Salmonella* in foods: a review. Food Research International 45, 545–556.

Carrique-Mas, J.J., Davies, R.H., 2008. Sampling and bacteriological detection of *Salmonella* in poultry and poultry premises: a review. Revue scientifique et technique OIE 27 (3), 665–677.

Carrique-Mas, J.J., Breslin, M., Snow, L., McLaren, I., Sayers, A.R., Davies, R.H., 2009. Persistence and clearance of different *Salmonella* serovars in buildings housing laying hens. Epidemiology and Infection 137 (06), 837–846.

Centers for Disease Control, Prevention (CDC), 2000. Outbreaks of *Salmonella* serotype Enteritidis infection associated with eating raw or undercooked shell eggs–United States, 1996–1998. MMWR Morbidity and Mortality Weekly Report 49 (4), 73.

Chittick, P., Sulka, A., Tauxe, R., Fry, A., 2006. A summary of national reports if foodborne outbreaks of *Salmonella* Heidelberg infections in the United States: clues for disease prevention. Journal of Food Protection 69, 1150–1153.

Clavijo, R.I., Loui, C., Andersen, G.L., Riley, L.W., Lu, S., 2006. Identification of genes associated with survival of *Salmonella enterica* serovar Enteritidis in chicken egg albumen. Applied and Environmental Microbiology 72 (2), 1055–1064.

Cox, N.A., Cason, J.A., Richardson, L.J., 2011. Minimization of *Salmonella* contamination on raw poultry. Annual Review of Food Science and Technology 2, 75–95.

Curtis, P., 2007. Microbiological challenges of poultry egg production in the US. World's Poultry Science Journal 63, 301–307.

Curtiss III, R., Wanda, S.-Y., Gunn, B.M., Zhang, X., Tinge, S.A., Ananthnarayan, V., Mo, H., Wang, S., Kong, W., 2009. *Salmonella enterica* serovar Typhimurium strains with regulated delayed attenuation in vivo. Infection and Immunity 77, 1071–1082.

Davies, R., Breslin, M., 2003a. Observations on *Salmonella* contamination of commercial laying farms before and after cleaning and disinfection. The Veterinary Record 152 (10), 283–287.

Davies, R.H., Breslin, M., 2003b. Persistence of *Salmonella* Enteritidis Phage Type 4 in the environment and arthropod vectors on an empty free-range chicken farm. Environmental Microbiology 5 (2), 79–84.

Dawoud, T.M., Shi, Z., Kwon, Y.M., Ricke, S.C., 2016. Overview of salmonellosis and foodborne *Salmonella*: historical and current perspectives. In: Ricke, S.C., Gast, R.K. (Eds.), Producing Safe Eggs-The Microbial Ecology of *Salmonella*. Elsevier, Inc., San Diego, CA, pp. 113–138 (Chapter 7).

Dawoud, T., Herrera, P., Hanning, I., Kwon, Y.M., Ricke, S.C., 2011. *In vitro* invasion of laying hen ovarian follicles by *Salmonella* Enteritidis strains. Poultry Science 90, 1134–1137.

Deguchi, K., Yokoyama, E., Honda, T., Mizuno, K., 2009. Efficacy of a novel trivalent inactivated vaccine against the shedding of *Salmonella* in a chicken challenge model. Avian Diseases 53, 281–286.

De Vylder, J., Dewulf, J., Van Hoorebeke, S., Pasmans, F., Haesebrouck, F., Ducatelle, R., Van Immerseel, F., 2011. Horizontal transmission of *Salmonella* Enteritidis in groups of experimentally infected laying hens housed in different housing systems. Poultry Science 90 (7), 1391–1396.

De Vylder, J., Raspoet, R., Dewulf, J., Haesebrouck, F., Ducatelle, R., Van Immerseel, F., 2013. *Salmonella* Enteritidis is superior in egg white survival compared with other *Salmonella* serotypes. Poultry Science 92 (3), 842–845.

De Vylder, J., Van Hoorebeke, S., Ducatelle, R., Pasmans, F., Haesebrouck, F., Dewulf, J., Van Immerseel, F., 2009. Effect of the housing system on shedding and colonization of gut and internal organs of laying hens with *Salmonella* Enteritidis. Poultry Science 88 (12), 2491–2495.

Donalson, L.M., Kim, W.K., Chalova, V.I., Herrera, P., McReynolds, J.L., Gotcheva, V.G., Vidanović, D., Woodward, C.L., Kubena, L.F., Nisbet, D.J., Ricke, S.C., 2008a. In vitro fermentation response of laying hen cecal bacteria to combinations of fructooligosaccharide (FOS) prebiotic with alfalfa or a layer ration. Poultry Science 87, 1263–1275.

Donalson, L.M., McReynolds, J.L., Kim, W.K., Chalova, V.I., Woodward, C.L., Kubena, L.F., Nisbet, D.J., Ricke, S.C., 2008b. The influence of a fructooligosaccharide prebiotic combined with alfalfa molt diets on the gastrointestinal tract fermentation, *Salmonella* Enteritidis infection and intestinal shedding in laying hens. Poultry Science 87, 1253–1262.

Donalson, L.M., Kim, W.K., Chalova, V.I., Herrera, P., Woodward, C.L., McReynolds, J.L., Kubena, L.F., Nisbet, D.J., Ricke, S.C., 2007. *In vitro* anaerobic incubation of *Salmonella enterica* serotype Typhimurum and laying hen cecal bacteria in poultry feed substrates and a fructooligosacharide prebiotic. Anaerobe 13, 208–214.

Dunkley, K.D., McReynolds, J.L., Hume, M.E., Dunkley, C.S., Callaway, T.R., Kubena, L.F., Nisbet, D.J., Ricke, S.C., 2007. Molting in *Salmonella* Enteritidis-challenged laying hens fed alfalfa crumbles. I. *Salmonella* Enteritidis colonization and virulence gene *hilA* response. Poultry Science 86 (8), 1633–1639.

Dunkley, K.D., Callaway, T.R., Chalova, V.I., McReynolds, J.L., Hume, M.E., Dunkley, C.S., Kubena, L.F., Nisbet, D.J., Ricke, S.C., 2009. Foodborne *Salmonella* ecology in the avian gastrointestinal tract. Anaerobe 15 (1), 26–35.

Durant, J.A., Corrier, D.E., Byrd, J.A., Stanker, L.H., Ricke, S.C., 1999. Feed deprivation affects crop environment and modulates *Salmonella* Enteritidis colonization and invasion of leghorn hens. Applied and Environmental Microbiology 65 (5), 1919–1923.

Erickson, M.C., 2010. Microbial risks associated with cabbage, carrots, celery, onions, and deli salads made with these produce items. Comprehensive Reviews in Food Science and Food Safety 9, 602–619.

Farnell, M.B., Donoghue, A.M., Solis de los Santos, F., Blore, P.J., Hargis, B.M., Tellez, G., Donoghue, D.J., 2006. Upregulation of oxidative burst and degranulation in chicken heterophils stimulated with probiotic bacteria. Poultry Science 85, 1900–1906.

Finstad, S., O'Bryan, C.A., Marcy, J.A., Crandall, P.G., Ricke, S.C., 2012. *Salmonella* and broiler production in the United States: relationship to foodborne salmonellosis. Food Research International 45, 789–794.

Foley, S.L., Johnson, T.J., Ricke, S.C., Nayak, R., Danzeisen, J., 2013. *Salmonella* pathogenicity and host adaptation in chicken-associated serovars. Microbiology and Molecular Biology Reviews: MMBR 77, 582–607.

Foley, S.L., Nayak, R., Hanning, I.B., Johnson, T.J., Han, J., Ricke, S.C., 2011. Population dynamics of *Salmonella enterica* serotypes in commercial egg and poultry production. Applied and Environmental Microbiology 77 (13), 4273–4279.

Foley, S.L., Lynne, A.M., Nayak, R., 2008. *Salmonella* challenges: prevalence in swine and poultry and potential pathogenicity of such isolates. Journal of Animal Science 86, E149–E162.

Friedman, M., 2015. Antibiotic-resistant bacteria: prevalence in food and inactivation byfood-compatible compounds and plant extracts. Journal of Agricultural and Food Chemistry 63, 3805–3822.

Fukata, T.K., Sasai, Y., Miyamoto, T., ad Baba, E., 1999. Inhibitory effects of competitive exclusion and fructooligosaccharide, singly or in combination, on *Salmonella* colonization of chicks. Journal of Food Protection 62, 229–233.

Fuller, R., 1989. Probiotics in man and animals. Journal of Applied Bacteriology 66, 365–378.

Galiş, A.M., Marcq, C., Marlier, D., Portetelle, D., Van, I., Beckers, Y., Théwis, A., 2013. Control of *Salmonella* contamination of shell eggs—preharvest and postharvest methods: a review. Comprehensive Reviews in Food Science and Food Safety 12 (2), 155–182.

Gantois, I., Ducatelle, R., Pasmans, F., Haesebrouck, F., Van Immerseel, F., 2008. *Salmonella enterica* serovar Enteritidis genes induced during oviduct colonization and egg contamination in laying hens. Applied and Environmental Microbiology 74 (21), 6616–6622.

Gantois, I., Ducatelle, R., Pasmans, F., Haesebrouck, F., Gast, R., Humphrey, T.J., Van Immerseel, F., 2009. Mechanisms of egg contamination by *Salmonella* Enteritidis. FEMS Microbiology Reviews 33 (4), 718–738.

Garber, L., Smeltzer, M., Fedorka-Cray, P., Ladely, S., Ferris, K., 2003. *Salmonella enterica* serotype Enteritidis in table egg layer house environments and in mice in U.S. layer houses and associated risk factors. Avian Diseases 47 (1), 134–142.

Gast, R.K., Guard-Petter, J., Holt, P.S., 2002. Characteristics of *Salmonella* Enteritidis contamination in eggs after oral, aerosol, and intravenous inoculation of laying hens. Avian Diseases 46 (3), 629–635.

Gast, R.K., Guard-Petter, J., Holt, P.S., 2003. Effect of prior serial in vivo passage on the frequency of *Salmonella* Enteritidis contamination in eggs from experimentally infected laying hens. Avian Diseases 47 (3), 633–639.

Gast, R.K., Guard-Bouldin, J., Holt, P.S., 2004. Colonization of reproductive organs and internal contamination of eggs after experimental infection of laying hens with *Salmonella* Heidelberg and *Salmonella* Enteritidis. Avian Diseases 48 (4), 863–869.

Gast, R.K., Guard-Bouldin, J., Holt, P.S., 2005. The relationship between the duration of fecal shedding and the production of contaminated eggs by laying hens infected with strains of *Salmonella* Enteritidis and *Salmonella* Heidelberg. Avian Diseases 49 (3), 382–386.

Gast, R.K., Guard-Bouldin, J., Guraya, R., Holt, P.S., 2009. Effect of prior passage through laying hens on invasion of reproductive organs by *Salmonella* Enteritidis. International Journal of Poultry Science 8, 116–212.

Gibson, G.R., Roberfroid, M.B., 1995. Dietary modulation of the human colonic microbiota: introducing the concept of prebiotics. Journal of Nutrition 125, 1401–1412.

Golden, N.J., Marks, H.H., Coleman, M.E., Schroeder, C.M., Bauer Jr., N.E., Schlosser, W.D., 2008. Review of induced molting by feed removal and contamination of eggs with *Salmonella enterica* serovar Enteritidis. Veterinary Microbiology 131 (3), 215–228.

Guard, J., Gast, R.K., Guraya, R., 2010. Colonization of avian reproductive-tract tissues by variant subpopulations of *Salmonella* Enteritidis. Avian Diseases 54 (2), 857–861.

Guard-Bouldin, J., Buhr, R.J., 2006. Evaluation of eggshell quality of hens infected with *Salmonella* Enteritidis by application of compression. Poultry Science 85 (1), 129–135.

Guard-Bouldin, J., Gast, R.K., Humphrey, T.J., Henzler, D.J., Morales, C., Coles, K., 2004. Subpopulation characteristics of egg-contaminating *Salmonella enterica* serovar Enteritidis as defined by the lipopolysaccharide O chain. Applied and Environmental Microbiology 70 (5), 2756–2763.

Guard-Petter, J., 2001. The chicken, the egg and *Salmonella enteritidis*. Environmental Microbiology 3 (7), 421–430.

Handley, J.A., Park, S.H., Shi, Z., Dawoud, T.M., Kwon, Y.M., Ricke, S.C., 2015. *Salmonella* and the potential role for microbial process indicators on chicken carcasses. In: Ricke, S.C., Donaldson, J.R., Phillips, C.A. (Eds.), Food Safety: Emerging Issues, Technologies and Systems. Elsevier, Oxford, UK, pp. 81–104 (Chapter 6).

Hanning, I., Nutt, J.D., Ricke, S.C., 2009. Salmonellosis outbreaks due to fresh produce: sources and potential intervention measures. Foodborne Pathogens and Disease 6, 635–648.

Hitchner, S.B., 2004. History of biological control of poultry diseases in the U.S.A. Avian Diseases 48, 1–8.

Holt, P.S., 1995. Horizontal transmission of *Salmonella* enteritidis in molted and unmolted laying chickens. Avian Diseases 39 (2), 239–249.

Holt, P.S., 2003. Molting and *Salmonella enterica* serovar enteritidis infection: the problem and some solutions. Poultry Science 82 (6), 1008–1010.

Holt, P.S., Macri, N.P., Porter Jr., R.E., 1995. Microbiological analysis of the early Salmonella enteritidis infection in molted and unmolted hens. Avian Diseases 39 (1), 55–63.

Holt, P.S., Mitchell, B.W., Gast, R.K., 1998. Airborne horizontal transmission of Salmonella enteritidis in molted laying chickens. Avian Diseases 42 (1), 45–52.

Howard, Z.R., Moore, R.W., Zabala-Diaz, I.B., Landers, K.L., Byrd, J.A., Kubena, L.F., Nisbet, D.J., Birkhold, S.G., Ricke, S.C., 2005. Ovarian laying hen follicular maturation and in vitro *Salmonella* internalization. Veterinary Microbiology 108 (1), 95–100.

Howard, Z.R., O'Bryan, C.A., Crandall, P.G., Ricke, S.C., 2012. *Salmonella* Enteritidis in shell eggs: current issues and prospects for control. Food Research International 45 (2), 755–764.

Humphrey, T.J., Baskerville, A., Whitehead, A., Rowe, B., Henley, A., 1993. Influence of feeding patterns on the artificial infection of laying hens with *Salmonella enteritidis* phage type 4. The Veterinary Record 132 (16), 407–409.

Humphrey, T.J., Whitehead, A., 1993. Egg age and the growth of *Salmonella enteritidis* PT4 in egg contents. Epidemiology and Infection 111 (02), 209–220.

Humphrey, T.J., Williams, A., McAlpine, K., Lever, M.S., Guard-Petter, J., Cox, J.M., 1996. Isolates of *Salmonella enterica* Enteritidis PT4 with enhanced heat and acid tolerance are more virulent in mice and more invasive in chickens. Epidemiology and Infection 117 (01), 79–88.

Jarvis, N.A., O'Bryan, C.A., Dawoud, T.M., Park, S.H., Kwon, Y.M., Crandall, P.G., Ricke, S.C., 2016. An overview of *Salmonella* thermal destruction during food processing and preparation. Food Control 68, 280–290.

Jiang, Y., Kulkarni, R.R., Parreira, V.R., Poppe, C., Roland, K.L., Prescott, J.F., 2010. Assessment of 2 *Salmonella enterica* serovar Typhimurium-based vaccines against necrotic enteritis in reducing colonization of chickens by *Salmonella* serovars of different serogroups. Canadian Journal of Veterinary Research 74, 264–270.

Kaldhone, P.R., Foley, S.L., Ricke, S.C., 2016. Chapter 12. *Salmonella* Heidelberg in layer hens and egg production: incidence and potential issues. In: Ricke, S.C., Gast, R.K. (Eds.), Producing Safe Eggs-The Microbial Ecology of *Salmonella*. Elsevier, Inc., San Diego, CA, pp. 235–256.

Kang, H., Loui, C., Clavijo, R.I., Riley, L.W., Lu, S., 2006. Survival characteristics of *Salmonella enterica* serovar Enteritidis in chicken egg albumen. Epidemiology and Infection 134 (05), 967–976.

Keklik, N.M., Demirci, A., Patterson, P.H., Puri, V.M., 2010. Pulsed UV light inactivation of *Salmonella* Enteritidis on eggshells and its effects on egg quality. Journal of Food Protection 73 (8), 1408–1415.

Keller, L.H., Benson, C.E., Krotec, K., Eckroade, R.J., 1995. *Salmonella enteritidis* colonization of the reproductive tract and forming and freshly laid eggs of chickens. Infection and Immunity 63 (7), 2443–2449.

Klein-Hessling, H., 2007. Managing breeder hens for more eggs. World Poultry 23 (5), 16–17. Retrieved from: http://www.worldpoultry.net/PageFiles/31569/001_boerderij-download-WP6827D01.pdf.

Kollanoor Johny, Venkitanarayanan, K., 2016. Preharvest food safety-Potential use of plant-derived compounds in layer chickens. In: Ricke, S.C., Gast, R.K. (Eds.), Producing Safe Eggs-The Microbial Ecology of *Salmonella*. Elsevier, Inc., San Diego, CA, pp. 347–372 (Chapter 17).

Kong, W., Wanda, S.-Y., Zhang, X., Bollen, W., Tinge, S.A., Roland, K.L., Curtiss III, R., 2008. Regulated programmed lysis of recombinant *Salmonella* in host tissue to release protective antigens and confer biological containment. Proceedings of the National Academy of Sciences of the United States of America 105, 9361–9366.

Konjufca, V., Wanda, S.-Y., Jenkins, M.C., Curtiss III, R., 2006. A recombinant attenuated *Salmonella enterica* serivar Typhimurium vaccine encoding *Eimeria acervulina* antigen offers protection against *E. acervulina* challenge. Infection and Immunity 74, 6785–6796.

Koo, O.K., Sirsat, S.A., Crandall, P.G., Ricke, S.C., 2012. Physical and chemical control of *Salmonella* in ready-to-eat products. Agriculture, Food and Analytical Bacteriology 2, 56–68.

Kubena, L.F., Byrd, J.A., Moore, R.W., Ricke, S.C., Nisbet, D.J., 2005. Effects of drinking water treatment on susceptibility of laying hens to *Salmonella* Enteritidis during forced molt. Poultry Science 84 (2), 204–211.

Latimer, H.K., Marks, H.M., Coleman, M.E., Schlosser, W.D., Golden, N.J., Ebel, E.D., Kause, J., Schroeder, C.M., 2008. Evaluating the effectiveness of pasteurization for reducing human illnesses from *Salmonella* spp. in egg products: results of a quantitative risk assessment. Foodborne Pathogens and Disease 5, 59–68.

Liljebjelke, K.A., Hofacre, C.L., Liu, T., White, D.G., Ayers, S., Young, S., Maurer, J.J., 2005. Vertical and horizontal transmission of *Salmonella* within integrated broiler production system. Foodborne Pathogens and Disease 2 (1), 90–102.

Lovell, T.M., Gladwell, R.T., Groome, N.P., Knight, P.G., 2003. Ovarian follicle development in the laying hen is accompanied by divergent changes in inhibin A, inhibin B, activin A and follistatin production in granulosa and theca layers. The Journal of Endocrinology 177 (1), 45–55.

Lu, S., Killoran, P.B., Riley, L.W., 2003. Association of *Salmonella enterica* serovar Enteritidis YafD with resistance to chicken egg albumen. Infection and Immunity 71 (12), 6734–6741.

Lutful Kabir, S.M., 2010. Avian colibacillosis and salmonellosis: a closer look at epidemiology, pathogenesis, diagnosis, control and public health concerns. International Journal of Environmental Research and Public Health 7 (1), 89–114.

Maciorowski, K.G., Jones, F.T., Pillai, S.D., Ricke, S.C., 2004. Incidence, sources, and control of foodborne *Salmonella* spp. in poultry feeds. World's Poultry Science Journal 60, 446–457.

Maciorowski, K.G., Herrera, P., Kundinger, M.M., Ricke, S.C., 2006. Animal feed production and contamination by foodborne *Salmonella*. Journal of Consumer Protection and Food Safety 1, 197–209.

Maciorowski, K.G., Herrera, P., Jones, F.T., Pillai, S.D., Ricke, S.C., 2007. Effects on poultry and livestock of feed contamination with bacteria and fungi. Animal Feed Science and Technology 133 (1), 109–136.

Martelli, F., Davies, R.H., 2012. *Salmonella* serovars from table eggs: an overview. Food Research International 45, 745–754.

Mazzuco, H., Avila, V.S., Coldebella, A., Mores, R., Jaenisch, F.R., Lopes, L.S., 2011. Comparison of the effect of different methods of molt: production and welfare evaluation. Poultry Science 90 (12), 2913–2920.

Menconi, A., Wolfenden, A.D., Shivaramiah, S., Terraes, J.C., Urbano, T., Kuttel, J., Kremer, C., Hargis, B.M., Tellez, G., 2011. Effect of lactic acid bacteria probiotic culture for the treatment of *Salmonella enterica* serovar Heidelberg in neonatal broiler chickens and turkey poults. Poultry Science 90, 561–565.

Miyamoto, T., Baba, E., Tanaka, T., Sasai, K., Fukata, T., Arakawa, A., 1997. Salmonella enteritidis contamination of eggs from hens inoculated by vaginal, cloacal, and intravenous routes. Avian Diseases 296–303.

Mollenhorst, H., van Woudenbergh, C.J., Bokkers, E.G.M., de Boer, I.J.M., 2005. Risk factors for *Salmonella* Enteritidis infections in laying hens. Poultry Science 84 (8), 1308–1313.

Moore, R.W., Holt, P.S., 2006. The effect of feed deprivation on tissue invasion by *Salmonella* Enteritidis. Poultry Science 85 (8), 1333–1337.

Nair, D.V.T., Nannapaneni, R., Kiess, A., Schilling, W., Sharma, S.C., 2014. Reduction of *Salmonella* on turkey breast cutlets by plant-derived compounds. Foodborne Pathogens and Disease 11, 981–987.

Nandre, R.M., Eo, S.K., Park, S.Y., Lee, J.H., 2015. Comparison of a live attenuated *Salmonella* Enteritidis vaccine candidate secreting *Escherichia coli* heat-labile enterotoxin B subunit with a commercial vaccine for efficacy of protection against internal egg contamination by *Salmonella* in hens. Canadian Journal of Veterinary Research 79, 235–240.

Nisbet, D., 2002. Defined competitive exclusion cultures in the prevention of enteropathogen colonization in poultry and swine. Antonie van Leeuwenhoek 81, 481–486.

O'Bryan, C.A., Ricke, S.C., Crandall, P.G., 2016. Chemical and physical sanitation and pasteurization methods for intact eggs. In: Ricke, S.C., Gast, R.K. (Eds.), Producing Safe Eggs-The Microbial Ecology of *Salmonella*. Elsevier, Inc., San Diego, CA, pp. 373–390 (Chapter 18).

O'Bryan, C.A., Pendleton, S.J., Crandall, P.G., Ricke, S.C., 2015. Potential of plant essential oils and their components in animal agriculture – *in vitro* studies on antibacterial mode of action. Frontiers in Veterinary Science 2 (Article 35), 1–8. http://dx.doi.org/10.3389/fvets.2015.00035.

Olsen, S.J., MacKinnon, L.C., Goulding, J.S., Bean, N.H., Slutsker, L., 2000. Surveillance for foodborne-disease outbreaks–United States, 1993–1997. MMWR CDC Surveillance Summaries 49 (1), 1–62.

Olsen, S.J., Bishop, R., Brenner, F.W., Roels, T.H., Bean, N., Tauxe, R.V., Slutsker, L., 2001. The changing epidemiology of *Salmonella*: trends in serotypes isolated from humans in the United States, 1987–1997. The Journal of Infectious Diseases 183 (5), 753–761.

Park, S.Y., Kim, W.K., Birkhold, S.G., Kubena, L.F., Nisbet, D.J., Ricke, S.C., 2004. Induced moulting issues and alternative dietary strategies for the egg industry in the United States. World's Poultry Science Journal 60, 196–209.

Parker, C., Asokan, K., Guard-Petter, J., 2001. Egg contamination by *Salmonella* serovar *enteritidis* following vaccination with Δ-*aroA Salmonella* serovar *typhimurium*. FEMS Microbiology Letters 195 (1), 73–78.

Parker, C.T., Harmon, B., Guard-Petter, J., 2002. Mitigation of avian reproductive tract function by *Salmonella* Enteritidis producing high-molecular-mass lipopolysaccharide. Environmental Microbiology 4 (9), 538–545.

Patwardhan, D.S., King, A.J., Mireles, A., 2011. Tomato pomace and safflower meal as ingredients in non-feed-removal molt diets. Journal of Applied Poultry Research 20 (3), 291–302.

Porwollik, S., Santiviago, C.A., Cheng, P., Florea, L., McClelland, M., 2005. Differences in gene content between *Salmonella enterica* serovar Enteritidis isolates and comparison to closely related serovars Gallinarum and Dublin. Journal of Bacteriology 187 (18), 6545–6555.

Rabsch, W., Hargis, B.M., Tsolis, R.M., Kingsley, R.A., Hinz, K.H., Tschäpe, H., Bäumler, A.J., 2000. Competitive exclusion of *Salmonella* Enteritidis by *Salmonella* Gallinarum in poultry. Emerging Infectious Diseases 6 (5), 443–448.

Rabsch, W., Tschäpe, H., Bäumler, A.J., 2001. Non-typhoidal salmonellosis: emerging problems. Microbes and Infection 3 (3), 237–247.

Rahimi, S., Grimes, J.L., Fletcher, O., Oviedo, E., Sheldon, B.W., 2009. Effect of a direct-feed microbial (Primalac) on structure and ultrastructure of small intestine in turkey poults. Poultry Science 88, 491–503.

Raspoet, R., Gantois, I., Devloo, R., Martel, A., Haesebrouck, F., Pasmans, F., Ducatelle, R., Van Immerseel, F., 2011. *Salmonella* Enteritidis universal stress protein (usp) gene expression is stimulated by egg white and supports oviduct colonization and egg contamination in laying hens. Veterinary Microbiology 153 (1), 186–190.

Rasschaert, G., Houf, K., De Zutter, L., 2007. Impact of the slaughter line contamination on the presence of *Salmonella* on broiler carcasses. Journal of Applied Microbiology 103 (2), 333–341.

Recheis, B., Osanger, A., Haubenwallner, S., Schneider, W.J., Nimpf, J., 2000. Chicken coagulation factor XIIIA is produced by the theca externa and stabilizes the ovarian follicular wall. The Journal of Biological Chemistry 275 (45), 35320–35327.

Revolledo, L., Ferreira, A.J.P., 2012. Current perspectives in avian salmonellosis:vaccines and immune mechanisms of protection. Journal of Applied Poultry Research 21, 418–431.

Revolledo, L., Ferreira, A.J.P., Mead, G.C., 2006. Prospects in *Salmonella* control: competitive exclusion, probiotics, and enhancement of avian intestinal immunity. Journal of Applied Poultry Research 15, 341–351.

Ricke, S.C., 2016. Gastrointestinal ecology of *Salmonella* Enteritidis in laying hens and intervention by prebiotic and non-digestible carbohydrate dietary supplementation. In: Ricke, S.C., Gast, R.K. (Eds.), Producing Safe Eggs-The Microbial Ecology of *Salmonella*. Elsevier, Inc., San Diego, CA, pp. 323–345 (Chapter 16).

Ricke, S.C., 2015. Potential of fructooligosaccharide prebiotics in alternative and nonconventional poultry production systems. Poultry Science 94, 1411–1418.

Ricke, S.C., 2014. Application of molecular approaches for understanding foodborne *Salmonella* Establishment in poultry production. Advances in Biology 2014. http://dx.doi.org/10.1155/2014/813275.

Ricke, S.C., 2010. Future prospects for advancing food - safety research in food animals. In: Ricke, S.C., Jones, F.T. (Eds.), Perspectives on Food Safety Issues of Food Animal Derived Foods. University of Arkansas Press, Fayetteville, AR, pp. 335–350.

Ricke, S.C., 2003a. The gastrointestinal tract ecology of *Salmonella* Enteritidis colonization in molting hens. Poultry Science 82, 1003–1007.

Ricke, S.C., 2003b. Perspectives on the use of organic acids and short chain fatty acids as antimicrobials. Poultry Science 82 (4), 632–639.

Ricke, S.C., Dawoud, T.M., Kwon, Y.M., 2015. Application of molecular methods for traceability of foodborne pathogens in food safety systems. In: Ricke, S.C., Donaldson, J.R., Phillips, C.A. (Eds.), Food Safety: Emerging Issues, Technologies and Systems. Elsevier, Oxford, UK, pp. 37–63 (Chapter 4).

Ricke, S.C., Saengkerdsub, S., 2015. *Bacillus* probiotics and biologicals for improving animal and human health: current applications and future prospects. In: Rai, R.R., Bai, J.A. (Eds.), Beneficial Microbes in Fermented and Functional Foods. CRC Press, Boca Raton, FL, pp. 341–353 (Chapter 19).

Ricke, S.C., Gast, R.K., 2014. *Salmonella/Salmonella* enteritidis. In: Batt, C., Tortello, M.L. (Eds.), Encyclopedia of Food Microbiology, vol. 3. second ed. Elsevier, Ltd., Oxford, UK, pp. 343–348.

Ricke, S.C., Dunkley, C.S., Durant, J.A., 2013. A review on development of novel strategies for controlling *Salmonella* Enteritidis colonization in laying hens: fiber – based molt diets. Poultry Science 92, 502–525.

Ricke, S.C., Pillai, S.D., 1999. Conventional and molecular methods for understanding probiotic bacteria functionality in gastrointestinal tracts. CRC Critical Reviews in Microbiology 25, 19–38.

Ricke, S.C., Kundinger, M.M., Miller, D.R., Keeton, J.T., 2005. Alternatives to antibiotics: chemical and physical antimicrobial interventions and foodborne pathogen response. Poultry Science 84, 667–675.

Roto, S.M., Rubinelli, P.M., Ricke, S.C., 2015. An introduction to the avian gut microbiota and the effects of yeast- based prebiotic-type compounds as potential feed additives. Frontiers in Veterinary Science 2 (Article 28), 1–18. http://dx.doi.org/10.3389/fvets.2015.00028.

Saeed, A.M., Gast, R.K., Potter, M.E., Wall, P.G. (Eds.), 1999. *Salmonella enterica* serovar Enteritidis in Humans and Animals – Epidemiology, Pathogenesis, and Control. Iowa State Press, Ames, IA.

Salminen, S.J., Nyborn, S., Meriluoto, J., Collado, M.S., Vertswerlung, S., El-Nezami, H., 2010. Interaction of probiotics and pathogens – benefits to human health? Current Opinion in Biotechnology 21, 157–167.

Scallan, E., Hoekstra, R.M., Angulo, F.J., Tauxe, R.V., Widdowson, M.A., Roy, S.L., Jones, J.L., Griffin, P.M., 2011. Foodborne illness acquired in the United States–major pathogens. Emerging Infectious Diseases 17, 7–15.

Shah, D.H., 2014. RNA sequencing reveals differences between the global transcriptomes of *Salmonella enterica* serovar Enteritidis strains with high and low pathogenicities. Applied and Environmental Microbiology 80 (3), 896–906.

Shah, D.H., Elder, J.R., Chiok, K.L., Paul, N.C., 2016. Genetic basis of *Salmonella* Enteritidis pathogenesis in chickens. In: Ricke, S.C., Gast, R.K. (Eds.), Producing Safe Eggs-The Microbial Ecology of *Salmonella*. Elsevier, Inc., San Diego, CA, pp. 187–208 (Chapter 10).

Shah, D.H., Casavant, C., Hawley, Q., Addwebi, T., Call, D.R., Guard, J., 2012a. *Salmonella* Enteritidis strains from poultry exhibit differential responses to acid stress, oxidative stress, and survival in the egg albumen. Foodborne Pathogens and Disease 9 (3), 258–264.

Shah, D.H., Zhou, X., Kim, H.Y., Call, D.R., Guard, J., 2012b. Transposon mutagenesis of *Salmonella enterica* serovar Enteritidis identifies genes that contribute to invasiveness in human and chicken cells and survival in egg albumen. Infection and Immunity 80 (12), 4203–4215.

Silva, C.A., Blondel, C.J., Quezada, C.P., Porwollik, S., Andrews-Polymenis, H.L., Toro, C.S., Zaldívar, M., Contreras, I., McClelland, M., Santiviago, C.A., 2012. Infection of mice by *Salmonella enterica* serovar Enteritidis involves additional genes that are absent in the genome of serovar Typhimurium. Infection and Immunity 80 (2), 839–849.

Siragusa, G.E., Ricke, S.C., 2012. Probiotics as pathogen control agents for organic meat production. In: Ricke, S.C., Van Loo, E.J., Johnson, M.G., O'Bryan, C.A. (Eds.), Organic Meat Production and Processing. John Wiley and Sons Inc., Ames, IA, pp. 331–349 (Chapter 20).

Sofos, J.N., Flick, G., Nychas, G.J., O'Bryan, C.A., Ricke, S.C., Crandall, P.G., 2013. Meat, poultry, and seafood. In: Doyle, M.P., Buchanan, R.L. (Eds.), Food Microbiology-Fundamentals and Frontiers, fourth ed. American Society for Microbiology, Washington, D.C., pp. 111–167 (Chapter 6).

St. Louis, M.E., Morse, D.L., Potter, M.E., DeMelfi, T.M., Guzewich, J.J., Tauxe, R.V., Blake, P.A., The Salmonella enteritidis Working group, 1988. The emergence of grade A eggs as a major source of *Salmonella enteritidis* infections: new implications for the control of salmonellosis. Journal of the American Medical Association 259, 2103–2107.

Thakur, S., Brake, J., Keelara, S., Zou, M., Susick, E., 2013. Farm and environmental distribution of *Campylobacter* and *Salmonella* in broiler flocks. Research in Veterinary Science 94 (1), 33–42.

Thiagarajan, D., Saeed, A.M., Asem, E.K., 1994. Mechanism of transovarian transmission of *Salmonella* Enteritidis in laying hens. Poultry Science 73 (1), 89–98.

Thomson, N.R., Clayton, D.J., Windhorst, D., Vernikos, G., Davidson, S., Churcher, C., Quail, M.A., Stevens, M., Jones, M.A., Watson, M., Barron, A., Layton, A., Pickard, D., Kingsley, R.A., Bignell, A., Clark, L., Harris, B., Ormond, D., Abdellah, Z., Brooks, K., Cherevach, I., Chillingworth, T., Woodward, J., Norberczak, H., Lord, A., Arrowsmith, C., Jagels, K., Moule, S., Mungall, K., Sanders, M., Whitehead, S., Chabalgoity, J.A., Maskell, D., Humphrey, T., Roberts, M., Barrow, P.A., Dougan, G., Parkhill, J., 2008. Comparative genome analysis of *Salmonella* Enteritidis PT4 and *Salmonella* Gallinarum 287/91 provides insights into evolutionary and host adaptation pathways. Genome Research 18 (10), 1624–1637.

Threlfall, E.J., Ward, L.R., Frost, J.A., Willshaw, G.A., 2000. The emergence and spread of antibiotic resistance in food-borne bacteria. International Journal of Food Microbiology 62 (1), 1–5.

Upadhyaya, I., Yin, H.-B., Nair, M.S., Venkitanarayanan, 2016. Natural approaches for improving postharvest safety of egg and egg products. In: Ricke, S.C., Gast, R.K. (Eds.), Producing Safe Eggs-The Microbial Ecology of *Salmonella*. Elsevier, Inc., San Diego, CA, pp. 391–420 (Chapter 19).

Upadhyaya, I., Upadhyay, A., Kollanoor-Johny, A., Darre, M.J., Venkitanarayanan, K., 2013. Effect of plant derived antimicrobials on *Salmonella* Enteritidis adhesion to and invasion of primary chicken oviduct epithelial cells in vitro and virulence gene expression. International Journal of Molecular Sciences 14 (5), 10608–10625.

Uzzau, S., Brown, D.J., Wallis, T., Rubino, S., Leori, G., Bernard, S., Casadesús, J., Platt, D.J., Olsen, J.E., 2000. Host adapted serotypes of *Salmonella enterica*. Epidemiology and Infection 125 (02), 229–255.

Vandeplas, S., Dauphin, R.D., Backers, Y., Thonart, P., Thewis, A., 2010. *Salmonella* in chicken: current and developing strategies to reduce contamination at farm level. Journal of Food Protection 72, 774–785.

Van Immerseel, F., 2010. Stress-induced survival strategies enable *Salmonella* Enteritidis to persistently colonize the chicken oviduct tissue and cope with antimicrobial factors in egg white: a hypothesis to explain a pandemic. Gut Pathogens 2, 23.

Van Immerseel, F., De Zutter, L., Houf, K., Pasmans, F., Haesebrouck, F., Ducatelle, R., 2009. Strategies to control *Salmonella* in the broiler production chain. World's Poultry Science Journal 65 (03), 367–392.

Xu, Z.R., Hu, C.H., Xia, M.S., Zhan, X.A., Wang, M.Q., 2003. Effects of dietary fructooligosaccharide on digestive enzyme activities, intestinal microflora and morphology of male broilers. Poultry Science 82, 1030–1036.

CHAPTER 10

# Cereal Grain Fractions as Potential Sources of Prebiotics: Current Status, Opportunities, and Potential Applications

**Xuhui Zhuang[1], Chen Zhao[1], Keshun Liu[2], Peter M. Rubinelli[3], Steven C. Ricke[3], Griffiths G. Atungulu[3]**

[1]*Academy of State Administration of Grain, Beijing, People's Republic of China;* [2]*United States Department of Agriculture, Agricultural Research Service, Aberdeen, ID, United States;* [3]*University of Arkansas, Fayetteville, AR, United States*

## Introduction

Approximately one-fourth of the world's children younger than 5 years have stunted growth, and it is believed to be associated with the intestinal bacteria (Owino et al., 2016). The intestinal bacteria also have been proven to be an important factor in the initiation and progression of colorectal cancers (Li et al., 2016), obesity and obesity-related metabolic disorders (Tilg and Kaser, 2011), blood diseases (Cheng et al., 2017), and other chronic diseases (Burcelin, 2016; Capurso et al., 2016). These recent insights suggest that the diversity of microorganisms inhabiting the gastrointestinal tract (GIT) of humans plays important roles in the host immune system and health status (Quigley, 2013). Therefore, the role of the GIT microbiota in these aspects has attracted considerable interest. The GIT appears to be an important organ that has been neglected for a long time, but now attention on how to maintain a healthy state of bacteria in human intestines is becoming the focus of more research.

Among the intestinal bacteria, probiotics are bacteria that would be considered beneficial to the host, such as *Bifidobacteria* and *Lactobacilli*, because they are viable nonpathogenic organisms that deliver beneficial effects on hosts (Hanning et al., 2015; Pandey et al., 2015). They are considered to be an effective alternative to the use of chemicals in nutrition and clinical medicine. Prebiotics serve as specific substrates for these bacteria and are defined as "a selectively fermented ingredient that allows specific changes, both in the composition or activity or both in the gastrointestinal microbiota that confers benefits upon host well-being and health" (Roberfroid, 2007). Chemically, they are believed to be the

Food and Feed Safety Systems and Analysis. http://dx.doi.org/10.1016/B978-0-12-811835-1.00010-5

nondigestible carbohydrates, such as the polysaccharides and oligosaccharides. At present, the most commonly used prebiotics in the market are oligosaccharides comprising 2 to 10 sugar moieties, such as lactulose, galactooligosaccharides, fructooligosaccharides, xylooligosaccharide, maltooligosaccharides, inulin, and its corresponding hydrolysates (Panesar, 2014; Ricke, 2015; Roto et al., 2015). These compounds have been shown to cause various physiological functions in the digestive system and therefore possess significant health implications for humans. Because of their remarkable properties of water dispersibility, viscosity, bulk, and fermentability into short-chain fatty acids, they are widely used in food, cosmetic, and medical industries (Ricke, 2003, 2015, 2016; Ricke et al., 2017). For example, after feeding chicory fructans to broiler chickens, the weight gain, feed conversion, and carcass weight increased while the serum cholesterol decreased (Yusrizal and Chen, 2003a). Altodextrins have been shown to decrease the number of *Escherichia coli* colonizing the GIT when used as the addition to piglet feeding, whereas the addition of the fructooligosaccharides can significantly increase the counts of *Lactobacillus* spp., *Bifidobacterium* spp., total anaerobes, and total aerobes compared with the control group (Bomba et al., 2002). More recently, prebiotics have been proposed to play an important role for food safety applications. According to Bjerrum et al. (2005), the carbohydrates in whole wheat can significantly reduce the number of *Salmonella* when fed to chickens. Yusrizal and Chen (2003b) observed decreases in *Campylobacter* and *Salmonella* and increases in *Lactobacilli* counts in the GIT after feeding chicory fructans to the broilers. Kleessen et al. (2003) reported a decrease in *Clostridium perfringens* and a decline in bacterial endotoxin levels by adding 0.5% of fructan-rich Jerusalem artichokes syrup into the water of broilers. In a series of in vitro and in vivo studies, Donalson et al. (2007, 2008a,b) demonstrated that fructooligosaccharides were potentially effective in reducing laying hen cecal levels of *Salmonella* Enteritidis and modifying short-chain volatile fatty acid profiles.

There are abundant prebiotic candidates in several cereal grains including rice, wheat, barley, and oats. Most of them are polysaccharides, oligosaccharides, and carbohydrates. In this chapter, these carbohydrate components from cereal grains are discussed with respect to their presence in grains, extraction, and isolation; health benefits as prebiotics based on animal studies; current applications of grain fractions enriched with prebiotic components; and economic and market opportunities. This is followed by suggestions for future research directions.

## Brief Overview of Cereal Grain Production and Processing

### Global Cereal Grain Production

According to the World Bank, the total global grain production in the crop year of 2008/2009 was estimated at 2.24 billion metric tons (Fig. 10.1). This number increased to 2.53 billion metric tons in the crop year of 2015/2016. However, the output of the cereal varies greatly

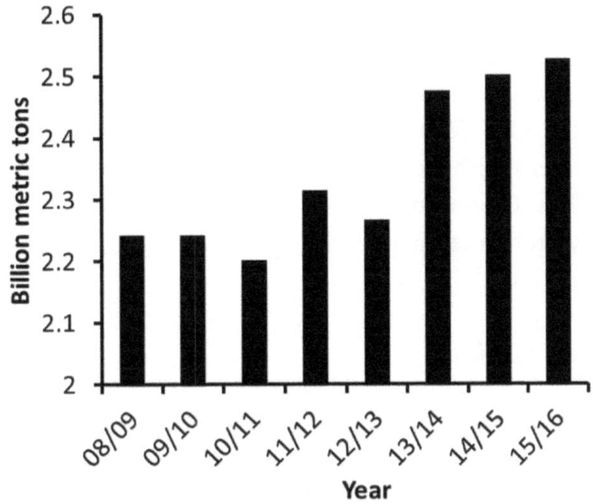


**Figure 10.1**
The total cereal grain production from 08/09 to 15/16 in the world. *The total cereal grain production from 08/09 to 14/15 is from* http://data.worldbank.org/indicator/AG.PRD.CREL.MT?. *The production of 15/16 is from* https://www.statista.com/statistics/271943/ *in 2017.*

with countries. For example, in 2014/2015, cereal production in China, United States, and India was 557.4, 442.9, and 294.0 million metric tons (MMT), respectively. The sum production of the three countries accounted for 52.3% of the total world cereal production (http://data.worldbank.org/indicator/AG.PRD.CREL.MT?).

Cereal grains include corn (maize), rice, wheat, barley, and oats along with a few others. Among the cereals, corn production was the greatest, reaching 1021.62 MMT in the crop year of 2014/2015, followed by rice (740.96 MMT), wheat (728.97 MMT), barley (144.33 MMT), and millet (27.82 MMT), respectively, based on the statistical data of "https://www.statista.com/statistics/263977/" in 2017.

Because of the climate and the corresponding types of vegetation, the staple foods among different countries exhibit a diverse geographical distribution (Table 10.1).

**Table 10.1: Geographical distribution of different cereals.**

| Cereal | First | Second | Third | Fourth | Fifth |
|---|---|---|---|---|---|
| Corn (maize) | USA | China | Brazil | Argentina | Ukraine |
| Rice | China | India | Indonesia | Bangladesh | Vietnam |
| Wheat | China | India | USA | Russia | France |
| Barley | Russia | Germany | France | Canada | Turkey |
| Millet | India | Niger | China | Mali | Burkina Faso |
| Oat | Russia | Canada | Finland | Poland | Australia |

Corn grows in a warm climate and needs more sunlight, so is primarily produced in the temperate climate areas, such as the United States, China, and Brazil. Rice is mostly grown in areas with higher temperature and abundant water. Therefore, rice is predominantly in China, Southeast Asia, and South Asia. Wheat grows well in cool, moist climates and ripens in a warm, dry climate. The areas with alternate cool winters and hot summers are better suited for wheat production. Thus, China, India, and United States are the three biggest wheat-producing countries. Barley prefers temperate climates, so it is mostly grown in Russia, Europe, United States, and Central Asia.

## *The Processing of Cereal Grains*

Most grains are rich in carbohydrates and may contain considerable levels of oil and protein. Generally, grains form the bulk of human staple food and domestic animal feed. In addition, grains can also be used in the chemical industry and to manufacture polymer materials and biofuels (Limayen and Ricke, 2012). For most grains, processing mainly consists of cleaning, drying, dehulling, and milling. The oil extracted from rice bran and corn is rich in unsaturated fatty acids and is widely used in the food market. These oils are considered healthier than those from animals (Larqué et al., 2001).

Besides their use as food and feed, in the United States, a substantial portion of the annual corn production is used as a feedstock to produce ethanol by a dry grind process (Table 10.2), involving enzymatic conversion of starch to glucose and fermentation of glucose by yeast. The main coproduct of the process is distillers dried grains with solubles (DDGS), which are rich in protein, oil, and other carbohydrates. Since 2000, corn-based bioethanol production has increased dramatically in the United States; DDGS has become a global commodity for animal feed (Liu and Rosentrater, 2011).

Rice possesses a hull that envelops the grain, and it must be removed by a milling process (Muthayya et al., 2014). However, the rice milling system can either be a one- or two-step process, or a multistage process. In the one-step milling, the hull and bran are removed

**Table 10.2: Domestic corn use in the United States from 1980 to 2013 (billions of bushels).**

| Corn use | 1980 | 1990 | 2000 | 2010 | 2013 |
|---|---|---|---|---|---|
| Animal feed and residual | 4.16 | 4.75 | 5.73 | 4.79 | 5.00 |
| Food, seed, and industrial | 0.81 | 1.08 | 1.15 | 1.25 | 1.46 |
| Ethanol production | 0.02 | 0.42 | 0.83 | 5.00 | 5.00 |
| Total | 4.99 | 6.25 | 7.71 | 11.04 | 11.46 |

*Ranum, P., Peña-Rosas, J.P., Garcia-Casal, M.N., 2014. Global maize production, utilization, and consumption. Annals of the New York Academy of Sciences 1312, 105–112.*

together. The hull is initially removed followed by the bran in the two-step procedure. For the multistage milling system, after sorting, the white rice, broken rice, outer bran, and inner bran are also obtained (Eicher and Staatz, 1998).

Nonstarch polysaccharides in the rice bran exhibit highly antibacterial and antioxidant activities and can be developed as prebiotics (Kim et al., 2014; Liu et al., 2017). Rice bran protein has a caloric value of 416 kcal/100 g and an *in vitro* digestibility of 80.9%–84.5%. Furthermore, unique hypoallergenic and anticancer properties make the bran proteins superior to other cereal proteins. Therefore, they have been suggested as a potential nourishment source to develop functional food for preschoolers and populations with protein allergies (Fabian and Ju, 2011; Friedman, 1996). Rice bran has been used in the form of full-fat bran, defatted bran, bran oil, and other fractions (Prakash and Ramaswamy, 1996).

In many countries and areas such as India, Pakistan, Bangladesh, Nigeria, Europe, and North America, parboiled rice consumption has been primarily associated with cultural preferences (Leethanapanich et al., 2016). Parboiled rice has the advantage in the food industry of being ready to eat, heat stable, aid in retention of nutrients, and lead to decreased breakage of rice when milling (Leethanapanich et al., 2016). Generally, the parboiling process is composed of soaking, steaming, and drying (Leethanapanich et al., 2016).

The wheat-milling process involves removing germ and bran layers which account for approximately 8%–10% of the whole wheat followed by milling in a conventional roller mill to produce flour or semolina (Fistes and Rakić, 2015). Wheat flour is one of the most widely used plant foods available. In the United States, per capita consumption of wheat exceeds that of any other single food staple. Wheat is used not only as a food source but also as a component of glue, paper, and pet food, among other products. The dietary fiber and resistant starch fractions in wheat are important prebiotics (Shewry and Hey, 2015). However, most of the bran, which contains the majority of the fiber, phytochemicals, and mineral substances of the grain serve mainly for animal feeding almost all over the world (Hemery et al., 2007). The concept of "the whole grain consuming" has promoted the development of the bran and the whole wheat. Some studies have also proved that the whole wheat or wheat bran is effective for decreasing the risk of some chronic diseases such as type 2 diabetes (Aune et al., 2013) and chronic nonorganic constipation (Badiali et al., 1995).

## Components From Cereal Grains for Potential Prebiotic Use: Extraction and Fractionation

Besides the fruits and vegetables, prebiotics can also be produced from the cereal grains, such as corn, rice, wheat, and straw of these grains resulting from their production and processing. Obviously, the cereal by-products for prebiotics production are more economical.

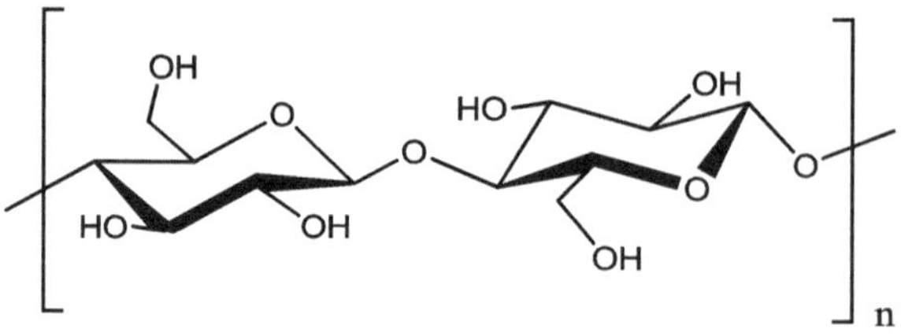


**Figure 10.2**
Structure of cellulose.

Prebiotics can be obtained as the nonstarch polysaccharides extracted from cereal grain. The major active ingredients are cellulose, β-glucan, α-glucan, arabinoxylan, and so forth. Their contents vary with types of grains (Englyst, 1989; Choct, 2002).

Cellulose is an indispensable component of plant cells and is the world's most abundant biopolymer (Limayen and Ricke, 2012). Cellulose is the dominant component of all cell walls in cereal grains. Its main structure is a linear homopolymer of β-(1–4)–linked glucose units (Fig. 10.2) with the molecular weights of over $10^6$ Da. Cellulose is highly insoluble in water but can swell in concentrated sodium hydroxide solutions. Through the use of chemicals or enzymes such as *N*-methylmorpholine *N*-oxide and cellulase, cellulose can be hydrolyzed into soluble oligosaccharides which can be used as a source of prebiotics (Kumar et al., 2012). After thermal and heat treatments, the corn bran can be transformed to cellulosic gel using alkaline peroxidation and shearing and then developed as a dietary fiber supplement (Kumar et al., 2012).

Besides cellulose, noncellulose polysaccharides are widely prevalent in cereals as well and include β-glucans, arabinoxylans, and α-glucans. The β-glucan fractions exist in most cereals, being particularly high in oats and barley. They are comprised of glucose units in a linear chain by both β-(1→3) and β-(1→4) linkages (Bengtsson et al., 1990). Barley β-glucans contain approximately 70% β-(1→4) linkages and 30% β-(1→3) linkages. The molecular weights of the water-soluble β-glucans are reported to range from 200,000 to 300,000, according to degrees of polymerization of 1200 to 1850 monomers (Woodward et al., 1983). Water-soluble β-glucans have become one of the most important dietary fibers for human health, as clinical studies have shown that β-glucans can reduce both serum cholesterol levels and postprandial blood glucose response (Tiwari and Cummins, 2011). On a dry-matter basis, β-glucan present in barley is roughly 3%–6%, and in oats 3%–5%. There is a growing interest in incorporating barley and oats into the human diet because they are naturally healthy, readily available, and relatively inexpensive.

In some cereal grains, the content of arabinoxylan is the highest among all polysaccharides except starch. Generally, the ubiquitous structural feature of cereal arabinoxylans consists of a chain linked with arabinose and xylose by the linear (1→4)-β-xylan backbone and xylosyl residues as substituents attached through 2-O and 3-O atoms (Annison et al., 1992). Although

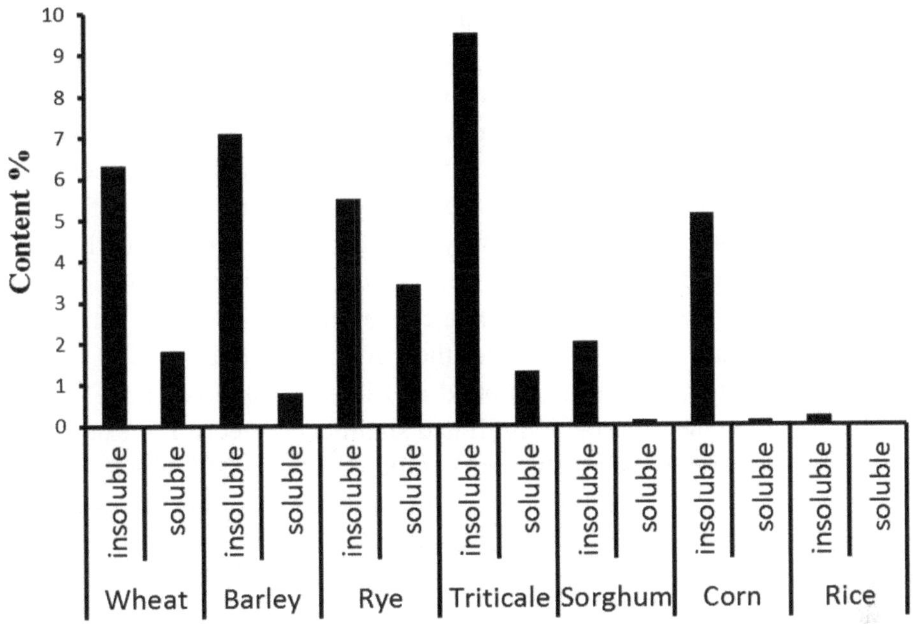


**Figure 10.3**

The levels of arabinoxylan in different cereal grains (% dry matter). *Englyst, H., 1989. Classification and measurement of plant polysaccharides. Animal Feed Science and Technology 23, 27–42 and Choct, M., 2002. Non-starch polysaccharides: effect on nutritive value. In: McNab, J.M., Boorman, K.N. (Eds.), Poultry Feedstuffs: Supply, Composition and Nutritive Value. CABI Publishing, Oxon, UK, pp. 221–235.*

the molecular weights of arabinoxylans are highly dependent on the source and extraction method, most of the ones derived from cereal grains are insoluble in water (Fig. 10.3).

In rice bran, the highest content of the polysaccharides is α-glucan. It is primarily composed of glucose units with α-(1→6) linkages. The average molecular weight of rice bran polysaccharides is approximately $10^6$ Da (Liu et al., 2017).

The preparation of the nonstarch polysaccharides from grains can be performed using wet extraction or dry fractionation. In the wet method, the soluble polysaccharides are extracted by a solvent, followed by the alcohol precipitation process. The solvent extraction may be assisted with heating, ultrasonic, microwave irradiation, or other physical methods. In general, the most widely used technology for polysaccharide extraction from cereal grains is hot-water extraction. Several studies have demonstrated that the ultrasonic- or microwave-assisted hot water extraction can accelerate the extraction process (Bonrath, 2004; Lianfu and Zelong, 2008). Ebringerová and Hromádková (2002) and Hromádková et al. (2008) compared the hot water extraction of hemicelluloses from corn bran and wheat bran at different conditions, such as with and without assistance of ultrasonic energy, including $H_2O_2$-containing alkaline media. They reported different yields and natural sugar composition with different extraction conditions.

Among all the reported studies on wet fractionation of barley and oats, almost all dealt with enrichment of one or two nutrients (such as β-glucan and/or protein) and neglected other potentially valuable components (such as starch or insoluble fiber). To increase production profitability, Liu (2014) reported a wet method that can recover all the major nutrients (β-glucan, protein, and starch) from oats simultaneously. The method is effective in recovering multiple nutrients simultaneously, but the concentration and recovery for some nutrients, particularly for β-glucan, is relatively low. Therefore, more recently, Liu and Barrows (2016) reported a further modified wet method, using barley as the experimental material (Fig. 10.4). In the same study, alkaline concentration, solvent to barley flour ratio, and extraction temperature were evaluated for their effects on concentration and recovery of β-glucan, protein, starch, and other carbohydrates in each fraction type. They found that the three parameters and their interactions all had significant effects. For example, for the β-glucan fractions, increasing the alkaline concentration decreased β-glucan content but increased its recovery significantly. With varying combinations of the three processing parameters, the β-glucan content in the resulting β-glucan fractions ranged from 21.5% to 87.0% while its recovery into the fractions ranged from 28.6% to 78.0%. Overall, the wet method was effective in processing barley into fractions concentrated in protein, beta-glucan, starch, or other carbohydrates, respectively.

Although the wet extraction is efficient and can optimally purify the polysaccharides, this fractionation technology is based on expensive chemical processes which may not retain the major components in intact form after separation. Furthermore, it generates high volumes of

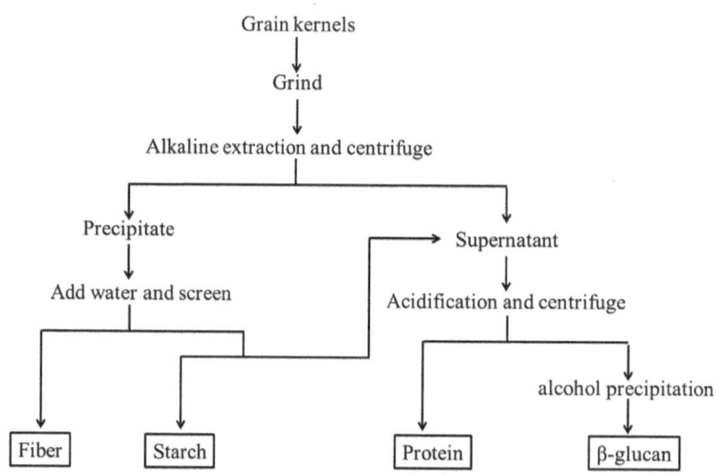


**Figure 10.4**
A flow diagram showing a modified method for fractionating barley into fractions enriched with several key nutrients, including β-glucan. *Liu, K., Barrows, F.T., 2016. Wet Processing of Barley Grains into Concentrates of Proteins,* B*-glucan, and Starch. Cereal Chemistry.* http://dx.doi.org/10.1094/CCHEM-09-14-0202-R.

effluents (Huber et al., 2006). Therefore, over the years, dry fractionation technologies have become important alternative means to wet extraction for polysaccharides preparation because they do not require chemicals and do not generate wastewater. This has been possible simply because of the heterogeneous distribution of various components throughout the kernel of a cereal grain. Several methods of dry fractionation of cereal grains for producing fractions enriched with a particular component, such as β-glucan fractions from barley or oats, have been used. These include: pearling, roller milling, milling followed by air classification, and milling followed by sieving. Pearling is a process to remove outer layers of cereal grains, typically by use of an abrasive dehulling device. Milling disintegrates grains into fine particles. Roller milling uses a special mill that allows grains to pass between rotating rollers for grinding and flaking and subsequently through shifters for fraction separation. Sieving separates milled flours on the basis of particle size, whereas air classification separates flours based on differences in density.

Liu et al. (2009) investigated the effect of several dry fractionation methods and their combinations on concentration and recovery rate of resulting fractions enriched respectively with β-glucan, protein, or starch and found that when the recovery rate was also taken into consideration, pearling alone was the most effective way to enrich protein in barley kernels. However, a combination of pearling with the method of milling and sieving was needed for maximizing β-glucan and starch contents in their respective fractions. Barakat and Rouau (2014) separated the cellulose from recalcitrant tissues enriched in lignin hemicelluloses using ultramilling and electrostatic fractionation from wheat straw.

Both the wet extraction and dry fractionation methods possess their own advantages and disadvantages. With wet extraction, the production is more pure, but when it is scaled up, the costs for chemical, enzyme, power, and wastewater treatment become too high. The dry fractionation route can avoid these problems. It is relatively energy efficient, with no solvent involvement, and thus requires low capital investments. Yet, resulting fractions tend to exhibit low purity, and the yield for higher purified fractions can also be low.

The polysaccharides extracted from cereal grains possess large molecular weights. Most of them have poor water solubility, which makes them difficult to be used as prebiotics. Therefore, some studies have focused on the modification of the polysaccharides using different means. Extrusion cooking, chemical reaction, and enzyme hydralyzation are considered to be feasible approaches. Björck et al. (1984) studied the effect of extrusion on the content of soluble fiber extracted from wheat flour and found that extrusion improved solubility of fiber extracted significantly (50%–75% soluble from extruded samples vs. 40% from raw wheat flour). Ning et al. (1991) compared the corn fiber treated with different methods including extrusion, acid hydrolysis, and alkaline hydrolysis, and they found that the acid-alkaline treatment was most effective in improving solubility of the fiber. By comparison, the enzyme treatment and the microbial fermentation are more advantageous. Liu et al. (2017) selected *Grifola frondosa* to ferment the defatted rice bran water extracts and obtained modified rice bran polysaccharides which had higher antioxidant activities and exhibited two-way effects on the immune system. Among the biomodified polysaccharides, MGN-3/biobran is the most widely studied product. MGN-3 is an arabinoxylan obtained from the bioreacting of rice bran hemicellulose with the multiple carbohydrate-hydrolyzing enzymes in shiitake mushrooms (Ghoneum, 1998).

## Animal Studies Involving Fractions for Prebiotics Application

The cereal fractions have been studied for their prebiotics effects in animals and human. Their potential effectiveness covers several areas, including human health response, feeding, reduction of foodborne pathogen, and regulating the balance of the gut microbiome.

### Human Health Responses

The polysaccharides applied in prebiotic development have been demonstrated to elicit activities contributing to the overall benefit of human health. In clinical settings, the gold standard includes typically double-blind, randomized and controlled trials. Animal studies and *in vitro* studies also are widely used to provide important clues for mechanisms for a relationship between disease and dietary exposure. In prebiotic studies, much of the information is based on *in vitro* models of fermentation. Fiber may play a beneficial role in reducing the incidence of heart disease by decreasing the levels of C-reactive protein, apolipoprotein, and blood pressure (Slavin, 2013). Water-soluble fibers were suggested to be effective for lowering serum low-density lipoprotein cholesterol concentrations (Slavin, 2013). Many large-scale cohort research efforts have demonstrated that dietary fiber consumption and development of type-II diabetes have a strong inverse relationship. People had significantly lower diabetes risk if they have more than 15 g of fiber per day (Hopping et al., 2010). Meyer et al. (2000) pointed out that the carbohydrate-rich foods such as whole grains, cereal fiber, and dietary magnesium can decrease the incidence of the type-II diabetes of older women. In more recent years, MGN-3 (the modified rice bran polysaccharides) has been considered to be a potential antitumor and immune-system stimulator. The antitumor mechanism of MGN-3 is linked to its important role in activating natural killer cells and inducing production of apoptosis and cytokine (Cholujova et al., 2013).

### Animal Feeding

A large part of the livestock animal diets are carbohydrates, and they supply carbon and energy from digestion and can also serve as sources of prebiotics. These dietary supplements can control the intestinal microbiota in food animals by two main modes (Halldor, 2012). The first one is to modulate the composition and number of commensal bacteria in the GIT of the animal, and this plays an important role to improve the health and productivity. The second one is to reduce asymptomatic intestinal colonization by pathogenic bacteria in the animals to reduce the possibility of foodborne transmission to humans. Although the information of what constitutes a healthy, balanced intestinal microbiota is still somewhat unknown, modulation by means of natural feed supplements such as prebiotics has been successfully practiced for a number of years. It is well recognized that cereal grain–based ingredients are viable for poultry and swine. Herfel et al. (2013) demonstrated improvement in the efficiency of nutrient utilization and increased intestinal *Bifidobacteria* concentrations in nursery pigs after being fed stabilized rice bran diets. The growth of the animals promoted by the cereal grain–based ingredients has evolved to the point it can be predicted accurately. Shi et al. (2015) generated eight equations to predict the digestible energy and metabolizable energy and generated results very close to the measured values.

From the nutritional standpoint, the viscosity of the grain-based ingredients should not be ignored, for the water-soluble fraction may increase the viscosity of the gut content in animal. Therefore, a method has been specially developed to measure the viscosity of the soluble ingredients from cereal grain in poultry diet formulations (Carré, 2002). This method can be performed to measure the nutrient digestibility. Bai et al. (1992) demonstrated that feeding laying hens with a diet containing 25% wheat middling achieved similar production and feed intake to other diets tested.

### Foodborne Pathogen Reduction

Foodborne disease is one of the primary public health concerns all over the world (Ricke et al., 2015). The microbiota of the chicken's GIT is very important to foodborne illness for humans. A balance of microorganisms in the intestine of the healthy poultry supports an efficient digestion, and optimal absorption of nutrients increases the body's resistance to infectious diseases (McCann et al., 1998). The imbalance of intestinal bacteria such as the increases in potential pathogens (e.g., *E. coli*, *Salmonella* spp., and *C. perfringens*) may bring about clinically apparent gastrointestinal disorders or subclinical manifestations (Ohtani and Shimizu, 2016; Ricke et al., 2013).

Prebiotics as substrates in the intestinal tract of food animals for host beneficial bacteria can be considered a natural resource to be harnessed to exclude pathogenic bacteria such as *Salmonella* (Ricke, 2015, 2016; Ricke et al., 2017). By feeding prebiotics in the food animals, it is believed that improving bacterial diversity can be targeted, and that in conjunction with this, exclusion of foodborne pathogenic bacterial colonization can be achieved.

Being rich in polysaccharides and low cost, cereal grains can be considered a promising resource to recover prebiotic candidates and thus be used in functional foods. Several studies have demonstrated the ability of prebiotic type compounds to prevent pathogen colonization of livestock animals. Other studies focused on the prevention of *Salmonella* Enteritidis infection in laying hens using cereal grain byproducts. Seo et al. (2001) examined the efficiency of wheat middlings as an alternative molt diet to reduce *Salmonella* Enteritidis shedding. They observed that although both the hens fed with and without wheat middlings stopped laying eggs within 7 day after the beginning of the molt, the infection rate of *Salmonella* Enteritidis of the hens fed with wheat middling was significantly less than the feed withdrawal hens. They also found that the levels of *Salmonella* Enteritidis in some organs (liver, spleens, and ovaries) in the hens fed with the diet without wheat middlings were significantly higher than the hens fed with wheat middlings.

### Gut Microbiome

The grain-based carbohydrates may provide several health benefits by altering the composition of the intestinal microbiota. They are considered to have the ability to increase the level of beneficial bacteria, such as *Bifidobacteria* and *Lactobacilli*. Resulting benefits from increasing these types of bacteria include improvement in host immunity and reduction of potentially pathogenic bacteria.

The ability to favorably alter the intestinal microbiota has been demonstrated in several studies. Polydextrose consumption resulted in a dose-dependent decrease in *Bacteroides*, as well as an increase in *Lactobacilli* and *Bifidobacteria* (Forssten et al., 2015; Magro et al., 2014). An *in vitro* fermentation of the fiber from wheat dextrin, partially hydrolyzed guar gum, and inulin was performed by Noack et al. (2013) and the result demonstrated that fermented fiber led to a higher total short-chain fatty acid concentrations compared with the negative control, and the level of the colony-forming units (CFUs) for *Bifidobacteria* (6.12 lgCFU/μL) and *Lactobacillus* (7.15) were much higher than the control (4.92 lgCFU/μL and 6.35 lgCFU/μL, respectively). Angelis et al. (2015) compared the fecal microbiota and metabolomes of 26 healthy subjects before and after 2 months of diet with durum wheat flour and whole-grain barley pasta containing β-glucans. The analysis of the metabolically active bacteria revealed that the levels of Clostridiaceae (*Clostridium orbiscindens* and *Clostridium* sp.), *Roseburia hominis*, and *Ruminococcus* sp. increased, while *Firmicutes* and *Fusobacteria* decreased. Furthermore, marked increases in short-chain fatty acids, such as 2-methyl-propanoic, acetic, butyric, and propionic acids, were detected in the samples.

## Economics for Application—Human and Animal Health Markets

The global animal medication market is projected to be up to US $42.9 billion by 2018, primarily driven by the current unmet animal health demands and growing need to ensure the safety of the food supply. Over the past few decades, animal health, especially farm animals, has received considerable interest all over the world due to their important roles in the human food safety and public health. Currently, animal health care market is categorized into three broad segments, namely feed additives, pharmaceuticals, and vaccines. The prebiotics are considered feed additives and are used to maintain the animal health and performance applications.

Cereal grain production may benefit from potential additional economic revenue generated in the form of prebiotics. The prices of some grains in February 2017 in the United States are shown in Table 10.3. In amazon.com, the price of a prebiotic fiber is approximately 30 dollars per bottle (240 g). Therefore, an appreciation for further development of the prebiotics using cereal grain as an economically viable resource should be of interest.

## Future Directions

Recently, most scientific research conducted on cereal grains focused primarily on toxin removal, grain and food preservation, grain processing, by-product utilization, and genetic modification along with other research efforts. In the next decades, there will continue to be interests that are also significantly associated with food safety and food security (Ricke, 2016; Ricke et al., 2017).

### Mycotoxin Removal

Mycotoxins are the toxic secondary metabolites of fungi. These fungi are parasitic in crops and can produce mycotoxins associated with diseases, such as cancer, feed refusal, vomiting, diarrhea, hepatotoxicity, and other health problems (Karlovsky et al., 2016). The major mycotoxins existing in cereal grains are listed in Table 10.4.

The key principle for methods to remove the mycotoxins is to destroy their chemical structures through de-epoxidation, acetylation, oxidation, epimerization, deamination, glucosylation, hydrolysis, lactone cleavage (hydrolysis), hydroxylation, peptide cleavage, sulfonation, reduction, and ether cleavage (Vanhoutte et al., 2016). Drying of grain to safe storage moisture content is recommended to inhibit the production of mycotoxins from fungi. Typical drying methods involve using heated air, solar drying, microwave irradiation,

**Table 10.3: The prices (US dollars per metric ton) of grains in February 2017 in the United States.**

| Product | Price |
|---|---|
| Rice | 368.10 |
| Wheat | 181.00 |
| Corn | 162.88 |
| Barley | 136.23 |
| Sorghum | 140.62 |

*indexmundi.com.*

**Table 10.4: Mycotoxins occurring in cereal grain.**

| Mycotoxin | Major producing fungi | Affected crops | Principal adverse effects |
|---|---|---|---|
| Aflatoxins | *Aspergillus parasiticus, Aspergillus flavus* | Almost every crops | Carcinogens |
| Ochratoxin A (OTA) | *Aspergillus alutaceus, Aspergillus carbonarius, Penicillium verrucosum* | Almost every crops | Nephrotoxic, carcinogens |
| Fumonisins | *Fusarium verticillioides, Fusarium proliferatum, Aspergillus niger* | Maize and grape | Carcinogens |
| Deoxynivalenol (DON) | *Fusarium graminearum, Fusarium culmorum* | Many crops | Feed refusal, vomiting, diarrhea |
| Zearalenone (ZEN) | *Fusarium* spp. | In all types of grains | Infertility |

*Karlovsky, P., Suman, M., Berthiller, F., De Meester, J., Eisenbrand, G., Perrin, I., Oswald, I.P., Speijers, G., Chiodini, A., Recker, T., Dussort, P.P., 2016. Impact of food processing and detoxification treatments on mycotoxin contamination. Mycotoxin Research 32, 179–205.*

and infrared drying. Solar drying is a widely used traditional method. With further technological development, new methods such as microwave heating (Atungulu et al., 2016; Olatunde et al., 2017) and infrared drying (Wilson et al., 2017; OKeyo et al., 2016) are gaining popularity.

The control of the mycotoxins should involve the criteria of a hazard analysis and critical control point approach. To achieve this from field to table, considerable work remains to be done: including but not being limited to studying the actions of toxigenic fungi in affected crops, developing on-farm production and harvest methods and technologies to mitigate aflatoxigenic mold proliferation, and developing safe methods for feeding animals with grains.

### *Grain Processing and By-product Utilization*

On processing, the endosperm of grain is typically used as the staple food. Grain byproducts, such as bran, are considered as animal feed simply without any further treatment. According to some studies, most of the nutritionally and active chemical components are in the bran, while the content of the endosperm is primarily starch (Gul et al., 2015). For the past few decades, development and efficient utilization of these products have gained considerable attention in the scientific community. Several studies have been conducted, including grain bran stabilization, use of bran in baked food, extraction and preparation for valuable chemical components, and other applications. To stabilize the brans, almost every method has been tested, including acid treatment, extrusion (Randall et al., 1985), heating (Dhingra et al., 2012), infrared (Tuncel et al., 2014), microwaving (Patil et al., 2016), and enzyme addition (Vallabha et al., 2015). MGN-3 is a good example of a preparation from rice bran fermentation that is widely used as a natural killer cell activator in the market. Other ingredients, such as oil, protein, and some active chemicals are also produced that can benefit human health.

The oil from cereal grain primarily consists of unsaturated fatty acid esters which are believed to be much healthier than animal fat.

In the future, the usage of the cereal grain by-products is anticipated to become more important. The extraction and purification of certain nutrients and their use in the preparation of functional food for certain segments of the human populations such as hypertensive patients, obese patients, immunodeficiency patients, and other applications have considerable potential in future markets.

### Genetic Modifications

Genome modification is a rapidly advancing technology that is widely applied in food crops and has become a globally important issue. Genome modification possesses considerable advantages, as they can be an effective tool to improve cereal grains with respects to both agronomic and end use traits. For instance, Li et al. (2015) obtained two distinct mutants (*GH9B* and *XAT*) genes among the selected groups of rice cell wall mutants. These two mutants have been shown to enhance plant lodging resistance, high biomass yield, and effective lignocellulose enzymatic digestibility. Zhang et al. (2015) built a systems biology approach and confirmed that silica plays a significant role in the biomass process and bioenergy rice breeding, and enhances plant mechanical strength and biomass digestibility. There is also an research effort focusing on nutritional improvements via genetic modification. For example La Frano et al. (2014) described approaches for increasing the content of vitamins and minerals, such as vitamin A, iron, zinc, and folate in food staples.

## Conclusions

Cereal grain, one of the most economical food resources in the world, can also be a rich source of prebiotics. These prebiotics have the roles of promoting beneficial bacteria, inhibiting growth of pernicious bacteria, and thus regulating the overall balance of intestinal bacteria. They also possess the unique advantages of abundant availability and low preparation costs, and thus have a considerable potential to be developed into value-added and healthy diets to benefit both humans and animals.

## References

Angelis, M.D., Montemurno, E., Vannini, L., Cosola, C., Cavallo, N., Gozzi, G., Maranzano, V., Cagno, R.D., Gobbetti, M., Gesualdo, L., 2015. Effect of whole-grain barley on the human fecal microbiota and metabolome. Applied and Environmental Microbiology 81, 7945–7956.

Annison, G., Choct, M., Cheetham, N.W., 1992. Analysis of wheat arabinoxylans from a large-scale isolation. Carbohydrate Polymers 19, 151–159.

Atungulu, G.G., Smith, D., Wilson, S., Sadaka, S., Rogers, S., 2016. Assessment of one-pass drying of rough rice with an industrial microwave system on milling quality. American Society of Biological and Agricultural Engineers. Applied Engineering Journal 32, 417–429.

Aune, D., Norat, T., Romundstad, P., Vatten, L.J., 2013. Whole grain and refined grain consumption and the risk of type 2 diabetes: a systematic review and dose–response meta-analysis of cohort studies. European Journal of Epidemiology 28, 845–858.

Badiali, D., Corazziari, E., Habib, F., Tomei, E., Bausano, G., Magrini, P., Anzini, F., Torsoli, A., 1995. Effect of wheat bran in treatment of chronic nonorganic constipation. Digestive Diseases and Sciences 40, 349–356.

Bai, Y., Sunde, M.L., Cook, M.E., 1992. Wheat middlings as an alternate feedstuff for laying hens. Poultry Science 71, 1007–1014.

Barakat, A., Rouau, X., 2014. New dry technology of environmentally friendly biomass refinery: glucose yield and energy efficiency. Biotechnology for Biofuels 7 (138), 1–11.

Bengtsson, S., Åman, P., Graham, H., Newman, C.W., Newman, R.K., 1990. Chemical studies on mixed-linked β-glucans in hul-lless barley cultivars giving different hypocholesterolaemic responses in chickens. Journal of the Science of Food and Agriculture 52, 435–445.

Bjerrum, L., Pedersen, K., Engberg, R.M., 2005. The influence of whole wheat feeding on *Salmonella* infection and gut flora composition in broilers. Avian Diseases 49, 9–15.

Björck, I., Nyman, M., Asp, N.-G., 1984. Extrusion cooking and dietary fiber: effects on dietary fiber content and on degradation in the rat intestinal tract. Cereal Chemistry 61, 174–179.

Bomba, A., Nemcová, R., Gancarcíková, S., Herich, R., Guba, P., Mudronová, D., 2002. Improvement of the probiotic effect of micro-organisms by their combination with maltodextrins, fructo-oligosaccharides and polyunsaturated fatty acids. British Journal of Nutrition 88, S95–S99.

Bonrath, W., 2004. Chemical reactions under "non-classical conditions", microwaves and ultrasound in the synthesis of vitamins. Ultrasonics Sonochemistry 11, 1–4.

Burcelin, R., 2016. Gut microbiota and immune crosstalk in metabolic disease. Molecular Metabolism 5, 771–781.

Capurso, G., Signoretti, M., Archibugi, L., Stigliano, S., Delle Fave, G., 2016. Systematic review and meta-analysis: small intestinal bacterial overgrowth in chronic pancreatitis. United European Gastroenterology Journal 4, 697–705.

Carré, B., 2002. Carbohydrate chemistry of the feedstuffs used for poultry. In: McNab, J.M., Boorman, K.N. (Eds.), Poultry Feedstuffs: Supply, Composition and Nutritive Value. CABI Publishing, Oxon, UK, pp. 39–56.

Cheng, X., Zhang, L., Xie, N.-C., Xu, H.-L., Lian, Y.-J., 2017. Association between small intestinal bacterial overgrowth and deep vein thrombosis in patients with spinal cord injuries. Journal of Thrombosis and Haemostasis 15, 304–311.

Choct, M., 2002. Non-starch polysaccharides: effect on nutritive value. In: McNab, J.M., Boorman, K.N. (Eds.), Poultry Feedstuffs: Supply, Composition and Nutritive Value. CABI Publishing, Oxon, UK, pp. 221–235.

Cholujova, D., Jakubikova, J., Czako, B., Martisova, M., Hunakova, L., Duraj, J., Mistrik, M., Sedlak, J., 2013. MGN-3 arabinoxylan rice bran modulates innate immunity in multiple myeloma patients. Cancer Immunology, Immunotherapy 62, 437–445.

Dhingra, D., Chopra, S., Rai, D.R., 2012. Stabilization of raw rice bran using ohmic heating. Agricultural Research 1, 392–398.

Donalson, L.M., Kim, W.K., Chalova, V.I., Herrera, P., Woodward, C.L., McReynolds, J.L., Kubena, L.F., Nisbet, D.J., Ricke, S.C., 2007. *In vitro* anaerobic incubation of *Salmonella enterica* serotype Typhimurum and laying hen cecal bacteria in poultry feed substrates and a fructooligosacharide prebiotic. Anaerobe 13, 208–214.

Donalson, L.M., Kim, W.K., Chalova, V.I., Herrera, P., McReynolds, J.L., Gotcheva, V.G., Vidanović, D., Woodward, C.L., Kubena, L.F., Nisbet, D.J., Ricke, S.C., 2008a. In vitro fermentation response of laying hen cecal bacteria to combinations of fructooligosaccharide (FOS) prebiotic with alfalfa or a layer ration. Poultry Science 87, 1263–1275.

Donalson, L.M., McReynolds, J.L., Kim, W.K., Chalova, V.I., Woodward, C.L., Kubena, L.F., Nisbet, D.J., Ricke, S.C., 2008b. The influence of a fructooligosaccharide prebiotic combined with alfalfa molt diets on the gastrointestinal tract fermentation, *Salmonella* Enteritidis infection and intestinal shedding in laying hens. Poultry Science 87, 1253–1262.

Ebringerová, A., Hromádková, Z., 2002. Effect of ultrasound on the extractibility of corn bran hemicelluloses. Ultrasonics Sonochemistry 9, 225–229.

Eicher, C.K., Staatz, J.M., 1998. International agricultural development. International Agricultural Development 81, 747–749.

Englyst, H., 1989. Classification and measurement of plant polysaccharides. Animal Feed Science and Technology 23, 27–42.

Fabian, C., Ju, Y.-H., 2011. A review on rice bran protein: its properties and extraction methods. Critical Reviews in Food Science and Nutrition 51, 816–827.

Fistes, A., Rakić, D., 2015. Using the eight-roller mill in the purifier-less mill flow. Journal of Food Science and Technology 52, 4661–4668.

Forssten, S.D., Röytiö, H., Hibberd, A.A., Ouwehand, A.C., 2015. The effect of polydextrose and probiotic *Lactobacilli* in a *Clostridium difficile*–infected human colonic model. Microbial Ecology in Health and Disease 26 (27988), 1–10.

Friedman, M., 1996. Nutritional value of proteins from different food sources. A review. Journal of Agricultural and Food Chemistry 44, 6–29.

Ghoneum, M., 1998. Anti-HIV activity in vitro of MGN-3, an activated arabinoxylane from rice bran. Biochemical and Biophysical Research Communications 243, 25–29.

Gul, K., Yousuf, B., Singh, A.K., Singh, P., Wani, A.A., 2015. Rice bran: nutritional values and its emerging potential for development of functional food—a review. Bioactive Carbohydrates and Dietary Fibre 6, 24–30.

Halldor, T., 2012. Patented non-antibiotic agents as animal feed additives. Recent Patents on Food. Nutrition & Agriculture 4, 155–168.

Hanning, I., Lingbeck, J., Ricke, S.C., 2015. Cardiovascular health and disease prevention: association with foodborne pathogens and potential benefits of probiotics. In: Watson, R.R., Preedy, V.R. (Eds.), Bioactive Foods in Promoting Health: Probiotics-Prebiotics, second ed. Elsevier, San Diego, CA, pp. 793–806.

Hemery, Y., Rouau, X., Lullienpellerin, V., Barron, C., Abecassis, J., 2007. Dry processes to develop wheat fractions and products with enhanced nutritional quality. Journal of Cereal Science 46, 327–347.

Herfel, T., Jacobi, S., Lin, X., Van Heugten, E., Fellner, V., Odle, J., 2013. Stabilized rice bran improves weaning pig performance via a prebiotic mechanism. Journal of Animal Science 91, 907–913.

Hopping, B.N., Erber, E., Grandinetti, A., Verheus, M., Kolonel, L.N., Maskarinec, G., 2010. Dietary fiber, magnesium, and glycemic load alter risk of Type 2 diabetes in a multiethnic cohort in Hawaii. The Journal of Nutrition 140, 68–74.

Hromádková, Z., Košt'álová, Z., Ebringerová, A., 2008. Comparison of conventional and ultrasound-assisted extraction of phenolics-rich heteroxylans from wheat bran. Ultrasonics Sonochemistry 15, 1062–1068.

Huber, G.W., Iborra, S., Corma, A., 2006. Synthesis of transportation fuels from biomass: chemistry, catalysts, and engineering. Chemical Reviews 106, 4044–4098.

Karlovsky, P., Suman, M., Berthiller, F., De Meester, J., Eisenbrand, G., Perrin, I., Oswald, I.P., Speijers, G., Chiodini, A., Recker, T., Dussort, P., 2016. Impact of food processing and detoxification treatments on mycotoxin contamination. Mycotoxin Research 32, 179–205.

Kim, S.P., Park, S.O., Lee, S.J., Nam, S.H., Friedman, M., 2014. A polysaccharide isolated from the liquid culture of *Lentinus edodes* (Shiitake) mushroom mycelia containing black rice bran protects mice against *Salmonellosis* through upregulation of the Th1 immune reaction. Journal of Agricultural and Food Chemistry 62, 2384–2391.

Kleessen, B., Elsayed, N.A.A.E., Loehren, U., Schroedl, W., Krueger, M., 2003. Jerusalem artichokes stimulate growth of broiler chickens and protect them against endotoxins and potential cecal pathogens. Journal of Food Protection 66, 2171–2175.

Kumar, V., Sinha, A.K., Makkar, H.P.S., de Boeck, G., Becker, K., 2012. Dietary roles of non-starch polysaccharides in human nutrition: a review. Critical Reviews in Food Science and Nutrition 52, 899–935.

La Frano, M.R., de Moura, F.F., Boy, E., Lönnerdal, B., Burri, B.J., 2014. Bioavailability of iron, zinc, and provitamin A carotenoids in biofortified staple crops. Nutrition Reviews 72, 289–307.

Larqué, E., Zamora, S., Gil, A., 2001. Dietary trans fatty acids in early life: a review. Early Human Development 65, S31–S41.

Leethanapanich, K., Mauromoustakos, A., Wang, Y.-J., 2016. Impacts of parboiling conditions on quality characteristics of parboiled commingled rice. Journal of Cereal Science 69, 283–289.

Li, F., Zhang, M., Guo, K., Hu, Z., Zhang, R., Feng, Y., Yi, X., Zou, W., Wang, L., Wu, C., Tian, J., Lu, T., Xie, G., Peng, L., 2015. High-level hemicellulosic arabinose predominately affects lignocellulose crystallinity for genetically enhancing both plant lodging resistance and biomass enzymatic digestibility in rice mutants. Plant Biotechnology Journal 13, 514–525.

Li, S., Konstantinov, S.R., Smits, R., Peppelenbosch, M.P., 2016. Bacterial biofilms in colorectal cancer initiation and progression. Trends in Molecular Medicine 23, 18–30.

Lianfu, Z., Zelong, L., 2008. Optimization and comparison of ultrasound/microwave assisted extraction (UMAE) and ultrasonic assisted extraction (UAE) of lycopene from tomatoes. Ultrasonics Sonochemistry 15, 731–737.

Limayen, A., Ricke, S.C., 2012. Lignocellulosic biomass for bioethanol production: current perspectives, potential issues and future prospects. Progress in Energy and Combustion Science 38, 449–467.

Liu, K., 2014. Fractionation of oats into products enriched with protein, beta-glucan, starch, or other carbohydrates. Journal of Cereal Science 60, 317–322.

Liu, K., Barrows, F.T., 2016. Wet processing of barley grains into concentrates of proteins, B-glucan, and starch.Cereal Chemistry 94, 161–169.

Liu, K., Rosentrater, K.A., 2011. Distillers Grains: Production, Properties, and Utilization. CRC Press & AOCS Press, Boca Raton, USA.

Liu, K., Barrows, F.T., Obert, D., 2009. Dry fractionation methods to produce barley meals varying in protein, beta-glucan, and starch contents. Journal of Food Science 74, C487–C499.

Liu, Q., Cao, X., Zhuang, X., Han, W., Guo, W., Xiong, J., Zhang, X., 2017. Rice bran polysaccharides and oligosaccharides modified by *Grifola frondosa* fermentation: antioxidant activities and effects on the production of NO. Food Chemistry 223, 49–53.

Magro, D.O., de Oliveira, L.M.R., Bernasconi, I., de Souza Ruela, M., Credidio, L., Barcelos, I.K., Leal, R.F., Ayrizono, M.D.L.S., Fagundes, J.J., Teixeira, L.D.B., Ouwehand, A.C., Coy, C.S.R., 2014. Effect of yogurt containing polydextrose, *Lactobacillus acidophilus* NCFM and *Bifidobacterium lactis* HN019: a randomized, double-blind, controlled study in chronic constipation. Nutrition Journal 13 (75), 1–5.

McCann, M.P., Solimeo, H.T., Cusick, F., Panunti, B., McCullen, C., 1998. Developmentally regulated protein synthesis during intraperiplasmic growth of *Bdellovibrio bacteriovorus*. Canadian Journal of Microbiology 44, 50–55.

Meyer, K.A., Kushi, L.H., Jacobs, D.R., Slavin, J., Sellers, T.A., Folsom, A.R., 2000. Carbohydrates, dietary fiber, and incident type 2 diabetes in older women. The American Journal of Clinical Nutrition 71, 921–930.

Muthayya, S., Sugimoto, J.D., Montgomery, S., Maberly, G.F., 2014. An overview of global rice production, supply, trade, and consumption. Annals of the New York Academy of Sciences 1324, 7–14.

Ning, L., Villota, R., Artz, W.E., 1991. Modification of corn fiber through chemical treatments in combination with twin-screw extrusion. Cereal Chemistry 68, 632–636.

Noack, J., Timm, D., Hospattankar, A., Slavin, J., 2013. Fermentation profiles of wheat dextrin, inulin and partially hydrolyzed guar gum using an in vitro digestion pretreatment and in vitro batch fermentation system model. Nutrients 5, 1500–1510.

Ohtani, K., Shimizu, T., 2016. Regulation of toxin production in *Clostridium perfringens*. Toxins 8 (207), 1–14.

OKeyo, A.A., Olatunde, G., Atungulu, G.G., Sadaka, S., McKay, T., 2016. Infrared drying characteristics of long-grain hybrid, long-grain pureline, and medium-grain rice cultivars. Cereal Chemistry Journal 94, 251–261.

Olatunde, G.A., Atungulu, G.G., Smith, D.L., 2017. One-pass drying of rough rice with an industrial 915 MHz microwave dryer: quality and energy use consideration. Biosystems Engineering 155, 33–43.

Owino, V., Ahmed, T., Freemark, M., Kelly, P., Loy, A., Manary, M., Loechl, C., 2016. Environmental enteric dysfunction and growth failure/stunting in global child health. Pediatrics 138, 1–10 e20160641.

Pandey, K.R., Naik, S.R., Vakil, B.V., 2015. Probiotics, prebiotics and synbiotics—a review. Journal of Food Science and Technology 52, 7577–7587.

Panesar, P.S., 2014. Prebiotics. In: Brar, S.K., Dhillon, G.S., Soccol, C.R. (Eds.), Biotransformation of Waste Biomass into High Value Biochemicals. Springer, pp. 237–259.

Patil, S.S., Kar, A., Mohapatra, D., 2016. Stabilization of rice bran using microwave: process optimization and storage studies. Food and Bioproducts Processing 99, 204–211.

Prakash, J., Ramaswamy, H.S., 1996. Rice bran proteins: properties and food uses. Critical Reviews in Food Science and Nutrition 36, 537–552.

Quigley, E.M.M., 2013. Gut bacteria in health and disease. Gastroenterology & Hepatology 9, 560–569.

Randall, J.M., Sayre, R.N., Schultz, W.G., Fong, R.Y., Mossman, A.P., Tribelhorn, R.E., Saunders, R.M., 1985. Rice bran stabilization by extrusion cooking for extraction of edible oil. Journal of Food Science 50, 361–364.

Ranum, P., Peña-Rosas, J.P., Garcia-Casal, M.N., 2014. Global maize production, utilization, and consumption. Annals of the New York Academy of Sciences 1312, 105–112.

Ricke, S.C., 2003. Perspectives on the use of organic acids and short chain fatty acids as antimicrobials. Poultry Science 82, 632–639.

Ricke, S.C., 2015. Potential of fructooligosaccharide prebiotics in alternative and nonconventional poultry production systems. Poultry Science 94, 1411–1418.

Ricke, S.C., 2016. Gastrointestinal ecology of *Salmonella* Enteritidis in laying hens and intervention by prebiotic and non-digestible carbohydrate dietary supplementation. In: Ricke, S.C., Gast, R.K. (Eds.), Producing Safe Eggs-The Microbial Ecology of *Salmonella*. Elsevier, Inc., San Diego, CA, pp. 323–345 (Chapter 16).

Ricke, S.C., Dunkley, C.S., Durant, J.A., 2013. A review on development of novel strategies for controlling *Salmonella* Enteritidis colonization in laying hens: fiber-based molt diets. Poultry Science 92, 502–525.

Ricke, S.C., Donaldson, J.R., Phillips, C.A. (Eds.), 2015. Food Safety: Emerging Issues, Technologies and Systems. Academic Press, Elsevier, Inc., Oxford, UK. 449 p.

Ricke, S.C., Perumalla, A.V.S., Hettiarachchy, N.S., 2017. Alternatives to antibiotics in preventing zoonoses and other pathogens in poultry: prebiotics and related compounds. In: Ricke, S.C. (Ed.), Achieving Sustainable Production of Poultry Meat, vol. 1 Safety, Quality and Sustainability. Burleigh Dodd Publishing, Cambridge, UK, pp. 87–108 (Chapter 5).

Roberfroid, M., 2007. Prebiotics: the concept revisited. The Journal of Nutrition 137, 830S–837S.

Roto, S.M., Rubinelli, P.M., Ricke, S.C., 2015. An introduction to the avian gut microbiota and the effects of yeast-based prebiotic-type compounds as potential feed additives. Frontiers in Veterinary Science 2, 28.

Seo, K., Holt, P., Gast, R., 2001. Comparison of *Salmonella* Enteritidis infection in hens molted via long-term feed withdrawal versus full-fed wheat middling. Journal of Food Protection 64, 1917–1921.

Shewry, P.R., Hey, S.J., 2015. The contribution of wheat to human diet and health. Food and Energy Security 4, 178–202.

Shi, C.X., Liu, Z.Y., Shi, M., Li, P., Zeng, Z.K., Liu, L., Huang, C.F., Zhu, Z.P., Li, D.F., 2015. Prediction of digestible and metabolizable energy content of rice bran fed to growing pigs. Asian-Australasian Journal of Animal Sciences 28, 654–661.

Slavin, J., 2013. Fiber and prebiotics: mechanisms and health benefits. Nutrients 5, 1417–1435.

Tilg, H., Kaser, A., 2011. Gut microbiome, obesity, and metabolic dysfunction. The Journal of Clinical Investigation 121, 2126–2132.

Tiwari, U., Cummins, E., 2011. Meta-analysis of the effect of β-glucan intake on blood cholesterol and glucose levels. Nutrition 27, 1008–1016.

Tuncel, N.B., Yılmaz, N., Kocabıyık, H., Uygur, A., 2014. The effect of infrared stabilized rice bran substitution on physicochemical and sensory properties of pan breads: part I. Journal of Cereal Science 59, 155–161.

Vallabha, S.V., Indira, T.N., Lakshmi, J.A., Radha, C., Tiku, P.K., 2015. Enzymatic process of rice bran: a stabilized functional food with nutraceuticals and nutrients. Journal of Food Science and Technology 52, 8252–8259.

Vanhoutte, I., Audenaert, K., Gelder, L.D., 2016. Biodegradation of mycotoxins: tales from known and unexplored worlds. Frontiers in Microbiology 7 (561), 1–20.

Wilson, S.A., Okeyo, A.A., Olatunde, G.A., Atungulu, G.G., 2017. Radiant heat treatments for corn drying and decontamination. Journal of Food Processing and Preservation. http://dx.doi.org/10.1111/jfpp.13193.

Woodward, J.R., Phillips, D.R., Fincher, G.B., 1983. Water-soluble (1→3), (1→4)-β-d-glucans from barley (*Hordeum vulgare*) endosperm. I. Physicochemical properties. Carbohydrate Polymers 3, 143–156.

Yusrizal, Chen, T.C., 2003a. Effect of adding chicory fructans in feed on broiler growth performance, serum cholesterol and intestinal length. International Journal of Poultry Science 2, 214–219.

Yusrizal, Chen, T.C., 2003b. Effect of adding chicory fructans in feed on fecal and intestinal microflora and excreta volatile ammonia. International Journal of Poultry Science 2, 188–194.

Zhang, J., Zou, W., Li, Y., Feng, Y., Zhang, H., Wu, Z., Tu, Y., Wang, Y., Cai, X., Peng, L., 2015. Silica distinctively affects cell wall features and lignocellulosic saccharification with large enhancement on biomass production in rice. Plant Science 239, 84–91.

SECTION 3

# *Postharvest Food Safety Systems – Incidence and Control of Foodborne Pathogens*

CHAPTER 11

# *Interventions to Reduce Shiga Toxin–Producing* Escherichia coli *on Beef Carcasses at Slaughter*

**Corliss A. O'Bryan, Sean J. Pendleton, Steven C. Ricke, Philip G. Crandall**
*University of Arkansas, Fayetteville, AR, United States*

## *Introduction*

The slaughter process for any meat-producing animal involves extracting bacteria-free meat from two contaminated surfaces, the hide and the gastrointestinal tract. No matter how carefully this process is performed, it is inevitable that some of the meat becomes contaminated with bacteria, some of which can cause foodborne illnesses in humans. Thus the primary food safety goal of the slaughter industry is to minimize bacterial contamination of the carcass and remove any contamination that has occurred.

*Escherichia coli* O157:H7 was recognized as a foodborne pathogen in 1982, and since then, it has become a major concern in the food industry, and specifically in the cattle industry (Nataro and Kaper, 1998; Wells et al., 1983). Shiga toxin–producing *E. coli* (STEC), including O157 and many non-O157 serotypes, cause illnesses ranging from mild intestinal discomfort to life-threatening hemolytic uremic syndrome (HUS) (Tarr et al., 1996), end-stage renal disease (ESRD), and death (Gould et al., 2009). Majowicz et al. (2014) assessed the global annual number of STEC infections and resulting cases of HUS, ESRD, and deaths. They estimated that STEC causes 2,801,000 acute illnesses annually leading to 3890 cases of HUS, 270 cases of ESRD, and 230 deaths; a sensitivity analysis indicated it is likely that these estimates are very conservative (Majowicz et al., 2014).

Cattle have been identified as a main reservoir of *E. coli* O157:H7 (Chapman et al., 1993). In addition, serological evidence suggests that all herds of cattle have some exposure to this pathogen (Laegreid et al., 1999). Non-O157 STEC serotypes are also often associated with cattle, and other ruminants and surveys have demonstrated the presence of non-O157 STEC in samples from cattle carcasses, retail beef, and raw milk (Cobbold et al., 2008; Hussein and Bollinger, 2005a,b; Hussein and Sakuma, 2005). Cattle often harbor multiple STEC serotypes, some of which appear to be less of a health risk to humans because they lack one or

Food and Feed Safety Systems and Analysis. http://dx.doi.org/10.1016/B978-0-12-811835-1.00011-7

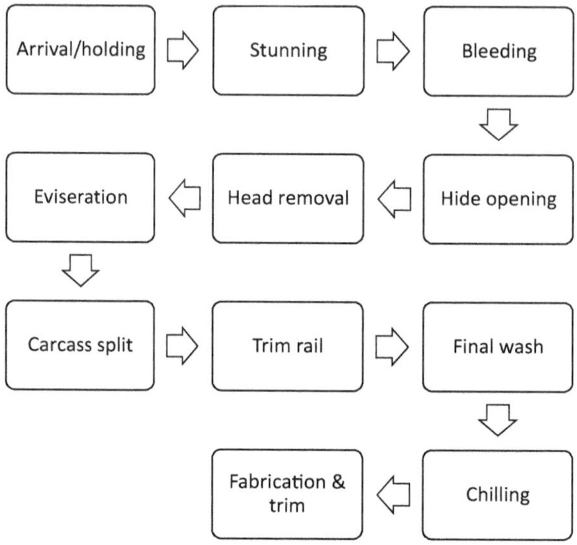


**Figure 11.1**
Flow diagram showing typical stages in beef processing; *arrows* point to areas where interventions to reduce or eliminate pathogens are most often applied.

more important virulence factors. Nevertheless, because microorganisms can readily exchange genetic information (Imamovic et al., 2009), the presence of any STEC in a food production animal is of concern. Barkocy-Gallagher et al. (2003) reported that the prevalence of O157:H7 in beef processing plants was 6.1%, 61%, 27.1%, and 1.3% in feces, hides, preevisceration carcasses, and postintervention carcasses, respectively. Non-O157 STEC had a higher prevalence than O157 for all samples other than hides (19.4%, 56.3%, 58.1%, and 9% in feces, hides, preevisceration carcasses, and posteevisceration carcasses, respectively) (Barkocy-Gallagher et al., 2003). A further review of STEC prevalence in the beef production chain can be found in the review by Rhoades et al. (2009).

The cattle industry has sought to eliminate *E. coli* O157 and other pathogens by multiple hurdles in both the preharvest and postharvest environments. There are several areas in the beef slaughterhouse where interventions can be applied (Fig. 11.1). This review seeks to draw together information from the scientific literature and government publications to discuss the effectiveness of interventions at the abattoir for reducing or eliminating STEC on beef.

## Risk Categorization of Incoming Cattle

The modern approach to meat inspection is risk based; to use this approach at the abattoir, batches of cattle intended for slaughter would need to be categorized based on risk posed to public health from, for instance, foodborne pathogens. Incoming cattle are currently categorized for risk at the antemortem inspection using factors such as farm management practices, movement of animals, and epidemiology (Buncic et al., 2014). However, this information really reveals little about the current public health hazard posed by the current herd of animals at slaughter. Some researchers are attempting to fill this knowledge gap. Blagojevic et al. (2011a) evaluated the usefulness of serum haptoglobin (Hp) testing for risk categorization. When animals are stressed from infection or other stress, the serum level of Hp increases (Murata et al., 2004). Blagojevic et al. (2011a) found no direct correlation between Hp concentration and specific postmortem abnormalities for individual cattle, although for groups of cattle the mean Hp level was statistically significantly higher for groups of cattle which were found to have abnormalities. More research is required on Hp to establish ranges that would allow effective risk categorization of incoming cattle using this screening tool. Blagojevic et al. (2012) also assessed the utility of visual hide cleanliness scoring in risk categorization of incoming cattle. They visually evaluated the cleanliness of the hides of 100 cattle and correlated the visual inspection with follow-up microbiological testing of the hides and the corresponding dressed carcasses. Although visibly dirtier hides were associated with higher levels of total viable microorganisms and *Enterobacteriaceae* on hides and carcasses, no correlation was found to exist between hide cleanliness and the presence of O157 (Blagojevic et al., 2012). Cleaner animals may be given a routine dressing process as opposed to slower more careful work of the dirtier animals to minimize cross-contamination from the hides, as well as slaughtering the cleaner animals first to reduce potential cross-contamination. However it remains to be seen whether this screening and prioritization of slaughter will reduce the public health risk from STEC.

## Hide-on Interventions

Cattle hides have long been known to be sources of microbial contamination on carcasses (Bell, 1997; Blagojevic et al., 2011b; Gill et al., 1996; McEvoy et al., 2000). Barkocy-Gallagher et al. (2003) found a much higher incidence of O157 and non-O157 STEC on cattle hides than in the feces of cattle. It has been noted that a significant number of carcasses are contaminated with O157 when tested immediately after hide removal, indicating the importance of hide decontamination and minimizing hide-to-meat contamination (Barkocy-Gallagher et al., 2003).

### Chemical Dehairing

Bowling and Clayton (1992) developed a chemical dehairing process designed to remove hair and contaminants from the hide before removal of the hide from the carcass. Chemical dehairing consists of washing hide-on carcass with a sodium sulfide solution, neutralizing with hydrogen peroxide, followed by a water rinse. This process was tested by Castillo et al. (1998a) by contaminating pieces of hide with fecal material containing *E. coli*. After chemical dehairing, *E. coli* O157:H7 populations were reduced from 5.0 log cfu/cm$^2$ inoculation levels to levels below the detection limit. Nou et al. (2003) also investigated the effects of chemical dehairing on prevalence of *E. coli* O157 on carcasses by sampling control (conventionally processed) and treated (chemically dehaired before hide removal) carcasses immediately after stunning but before treatment and immediately after hide removal in a commercial beef processing plant. The prevalence of *E. coli* O157:H7 was found to be lower on treated (1%) than on control carcasses (50%) immediately after hide removal. These data indicate that chemical dehairing of cattle hides is an effective intervention to significantly reduce the incidence of hide-to-carcass contamination with pathogens. The data also imply that any effective hide intervention process incorporated into beef processing procedures would significantly reduce carcass contamination by *E. coli* O157:H7.

### Chemical Washes

Bosilevac et al. (2004) tested the use of a combined water wash and cetylpyridinium chloride (CPC) treatment before stunning on cattle in the holding pens of a processing plant. Control cattle were processed in the normal manner, whereas the treatment group was prewashed with water the day before harvest. Immediately before stunning, the treatment group was sprayed twice with 1% CPC, first for 3 min, then for 1 min. Hides and preevisceration carcasses were sampled to determine *E. coli* O157 prevalence. Application of the treatment reduced the prevalence of *E. coli* O157 on hides from 56% positive controls to 34% and the prevalence on preevisceration carcasses from 23% to 3%, and did not result in any detectable CPC contamination of the chilled carcasses.

Bosilevac et al. (2005a) performed in vitro tests on cattle hides to compare the efficacy of several different types of washes followed by a rinse step using either water or acidified (pH 7.0) chlorine at 200 or 500 ppm. They determined that a 1.6% sodium hydroxide wash and a 1 ppm chlorinated water rinse were effective and therefore tested this treatment on hides in an on-line hide-washing cabinet. The prevalence of *E. coli* O157 on hides was reduced from 44% to 17% with the use of the wash. Bosilevac et al. (2005b) compared the efficacy of ozonated and electrolyzed oxidizing waters (EOW) (both acidic and basic) to washing with plain water for reduction of *E. coli* O157:H7 on cattle hides. The prevalence of *E. coli* O157 on hides was reduced from 89% to 31% after treatment with ozonated water and from 82% to 35% after EOW treatment, while control (plain water) wash treatments had little effect on the prevalence of *E. coli* O157:H7.

In 2011, the United States Department of Agriculture (USDA) approved the use of hypobromous acid (HOBr) for various applications in meat processing. Schmidt et al. (2012) studied the efficacy of HOBr as a hide antimicrobial. They used two different concentrations, 220 or 500 ppm, of HOBr. The 220-ppm treatment of HOBr reduced the prevalence of *E. coli* O157:H7 on hides from 25.3% to 10.1%, whereas treatment with 500 ppm of HOBr reduced the prevalence from 21.2 to, about the same level, 10.1%. Enumeration of *E. coli* O157:H7 was not possible in the experiment, but generic *E. coli* counts were found to be reduced by 2.2 (for 220 ppm) and 3.8 logs (for

500 ppm), respectively, which indicates the possibility of a similar reduction in *E. coli* O157:H7. However in a later study by Schmidt et al. (2014), immersion of beef cheek meat into 300 ppm of HOBr for 1 min only resulted in reductions of 0.2–0.3 logs of *E. coli* O157:H7 and non-O157:H7, not significantly different from immersion in plain tap water. As there was not a water-only control in the Schmidt et al. (2012), it is difficult to compare their results with those of Schmidt et al. (2014); however it does indicate the need for further research into this compound as an effective treatment for the control of STEC on cattle hides. Another thing to keep in mind, there may be environmental considerations in the use of some of these chemical hide treatments and this may be an advantage in rotating the use of hide treatments.

### *Steam Treatments*

McEvoy et al. (2001) used a laboratory-scale apparatus to apply steam at subatmospheric pressures to pieces of hide inoculated with *E. coli* O157:H7. Steam treatment at 80°C for 20 s, which included a heat-up time of 10 s and cool-down time of 13 s, gave approximately a 2-log reduction from 7.0 log cfu/g of *E. coli* O157:H7 to 5.99 log cfu/g. Treatment for 10 s (heat-up and cool-down time) resulted in significantly smaller reductions. This study demonstrates that steam condensing at or below 80°C can reduce *E. coli* O157:H7 on cattle hides, reducing the risk of cross-contamination to the carcass during slaughter and dressing. Steam hide decontamination may have a smaller environmental footprint.

### *Biological Treatments*

Bacteriophages have shown some promise as an alternative treatment for the decontamination of cattle hides, and in 2007, the Food Safety and Inspection Service (FSIS) of the USDA approved the use of an *E. coli* O157:H7 bacteriophage for the treatment of cattle hides. O'Flynn et al. (2004) isolated and identified two lytic *E. coli* O157:H7–specific bacteriophage, e11/2 and e4/1c, from bovine slurry. The genomes of these bacteriophages have been fully sequenced, and neither phage possesses genes that encode for virulence factors or other undesirable qualities. A cocktail of both phages was used to assess reduction of *E. coli* O157:H7 inoculated onto cattle hide pieces; control treatments were water only or no treatment (Coffey et al., 2011). Swab samples taken immediately after phage treatment showed no significant reduction of *E. coli* O157:H7 numbers compared with the water-treated or untreated samples, but after an exposure time of 1 h, a reduction of 1.5 log cfu/cm$^2$ was observed. However this treatment must still be tested in actual plant conditions.

### *Shellac*

The goal of all hide treatments is to prevent transfer of bacteria from the hide to the carcass meat surface. Researchers in the European Union (EU) developed a novel treatment using shellac, a food grade resin (Antic et al., 2010, 2011). Regulations in the EU do not specifically mention hide interventions. In the laboratory model system, Antic et al. (2010) sprayed hides with the shellac solution in ethanol and found that levels of aerobic bacteria and *E. coli* O157:H7 recoverable by swabbing were greatly reduced. In the laboratory, hide-to-meat contamination of aerobic bacteria was reduced by approximately 4 log cfu/cm$^2$ (Antic et al., 2010). When tested in a commercial abattoir, levels of aerobic bacteria found on skinned beef carcasses were reduced by approximately 2 log cfu/cm$^2$ (Antic et al., 2011). Further research is warranted to optimize the method and solution for use in commercial slaughter operations.

### Summary of Hide Treatments

The data reported in these research studies indicate that hide decontamination is an effective method of reducing the prevalence of *E. coli* O157. Because these methods, excepting the host-specific bacteriophage, also work to reduce prevalence of other pathogens as well as spoilage organisms, it seems logical to conclude that they would work equally well on the non-O157 STEC. Bacteriophage specific to each of the non-O157 STEC could also be isolated and used. The use of shellac or another compound to immobilize bacteria on the hide before skinning is also worthy of further research.

## Carcass Interventions

Hide treatments can minimize the level of bacterial contamination, but there is still a residual amount that may be transferred to the carcass during the hide removal process. In addition, there is the potential for cross-contamination from removal of the viscera, equipment, hands, aerosols, and many other vectors. As a consequence, most processors apply additional treatments to the hide-off carcass.

### Physical Interventions

#### Water Washes

The use of warm or hot water washes to remove bacteria from carcasses has been proven to be an effective form of intervention. Water interventions using temperatures near 35°C are considered warm water washes, whereas interventions using temperatures from 70°C and above are considered hot water washes (Dorsa et al., 1997; Phebus et al., 1997). Water washes work via physical removal of the pathogen from the surface and, in the case of hot water washes, the thermal inactivation of the bacteria. Barkate et al. (1993) examined the effect of hot water on the reduction of bacteria on beef carcasses and found that a 95°C hot water wash was able to reduce bacteria by approximately 1 log. Sprays of 95°C for 5 s at 24 psi from 12.5 cm gave reductions of up to 3 log in *E. coli* O157:H7 (Acuff et al., 1996). Dorsa et al. (1996a,b) examined the effects of warm or hot water on the decontamination of beef carcasses and determined that for the removal of *E. coli* from beef, warm water (30°C) applied at higher pressure, 125 psi, was just as effective as hot water (72°C) applied at 20 psi. Both treatments were found to reduce *E. coli* contamination by approximately 2.6 logs. Dorsa et al. (1997) demonstrated that hot water at

70°C and 40 psi produced a 2.6-log reduction in *E. coli* O157:H7, the same reduction seen at 2 degrees higher and 20 psi lower in their prior study. Cutter et al. (1997) investigated the parameters that affect spray washes on beef and determined that there was no significant difference in the reduction of their target bacteria, *E. coli* O157:H7, by water sprays at 30, 40, 50, 60, or 70°C. Although they did not find any significant difference between reduction of *E. coli* O157:H7 when sprayed with different temperatures, they did find that all water sprays reduced the pathogen by approximately 2 logs (Cutter et al., 1997). Two studies by Castillo et al. (1998b,c) found that a warm water carcass rinse at 35°C reduced *E. coli* O157:H7 and *Salmonella* Typhimurium by approximately 2 logs when applied at a pressure starting at 250 psi and increasing to 400 psi during treatment.

In the United States, the USDA has a "zero-tolerance" policy for fecal contamination on beef carcasses (USDA, 1996) and hot water treatments have been shown to remove fecal material and improve visual appearance. The effectiveness of the water washes depends somewhat on where in the processing the wash is applied; washing carcasses immediately after dehiding may prevent freshly contaminating STEC from attaching to the carcass (Dickson, 1995). Barkate et al. (1993) demonstrated that applying a hot water wash before any other treatment provides an average reduction of 1.3 log total bacteria compared to an average reduction of only 0.8 log when a cold water wash is applied first.

### *Trimming*

The use of knife trimming to remove visible contaminants such as ingesta, feces, milk, or urine is permitted by the EU and is recommended by the USDA (USDA, 1996). Some researchers have shown that trimming reduces *S.* Typhimurium and *E. coli* O157:H7 by approximately 3 logs (Hardin et al., 1995; Castillo et al., 1998c; Phebus et al., 1997). However, Gill et al. (1996) concluded that carcass trimming as an overall decontamination method is not effective. Reagan et al. (1996) demonstrated higher reductions by use of a combination of trimming and washing than with trimming alone.

### *Steam Treatment*

Because steam has a higher heat capacity than water at the same temperature, steam will raise the temperature of a surface faster than water. In laboratory trials, steam pasteurization has significantly reduced *E. coli* O157 levels on artificially inoculated samples, but few studies have examined the effects on naturally contaminated carcasses in a commercial environment. Nutsch et al. (1997, 1998) found significant reductions in total aerobic plate counts (APCs) and generic *E. coli* counts when steam pasteurization was used on naturally contaminated beef carcasses. Effectiveness of the steam pasteurization treatment was examined for reducing bacterial populations at several anatomical locations (inside round, loin, midline, brisket, or neck) on commercially slaughtered carcasses. Pasteurization treatment was performed at 82.2°C with a 6.5-s exposure time, and sterile sponges were used to sample a 300-$cm^2$ area at each location. Before treatment,

APCs at the midline were 4.5 log cfu/100 $cm^2$, approximately 3.8 log cfu/100 $cm^2$ at the inside round, brisket, and neck, and 3.4 log cfu/100 $cm^2$ at the loin. After treatment, APCs at all locations were reduced significantly to approximately 2.6 cfu/100 $cm^2$ for the inside round, loin, and brisket and 3.1 and 3.3 log cfu/100 $cm^2$, respectively, for the midline and neck. Generic *E. coli* populations were below the detection limit at 32% of the locations before treatment and below the detection limit on a majority of the carcasses (85%) after treatment. Another commercial trial showed significant reductions in generic *E. coli* and *Enterobacteriaceae* at sites where initial numbers were high but contamination was not totally eliminated (Minihan et al., 2003). A total of 30 carcass halves were steam-pasteurized at 90°C with a 10-s exposure time, whereas the 30 corresponding carcass halves were retained as untreated controls, with the neck, midline, and rump sampled on each carcass half. Significant reductions in *E. coli* counts (0.5 cfu/1000 $cm^2$) were seen only on the rump sites, but significant reductions of *Enterobacteriaceae* (more than 0.8 cfu/1000 $cm^2$) were observed at all carcass sites sampled, while total APC reductions were inconsistent. Although steam pasteurization can serve as an aid to hygienic beef processing, it has not been shown to be sufficient to eliminate STEC contamination.

### *Steam Vacuum*

USDA has approved the use of handheld steam-vacuum systems (for spots less than 2.5 cm in diameter) in place of trimming to remove visible contamination on beef carcasses. The process uses steam to loosen soil and kill bacteria, followed by application of a vacuum to remove contaminants (Dorsa et al., 1996a,b; Kochevar et al., 1997). Visible contaminants and bacterial counts have been reduced using commercial steam-vacuuming systems to at least those levels achieved by knife trimming (Dorsa et al., 1996a,b; Kochevar et al., 1997). Steam-vacuum treatments were able to lower levels of *E. coli* O157:H7 to 2.1 log cfu/$cm^2$ on beef carcasses artificially inoculated with 7.6 log cfu/$cm^2$ (Dorsa et al., 1996a,b). In commercial use, the effectiveness of steam vacuuming depends on the skill of the employees using the equipment as well as the equipment itself and this process, like knife trimming, is used only on areas that are visibly contaminated (Sofos and Smith, 1998).

### *Summary of Physical Interventions*

Reductions from water-based treatments depended on conditions such as application temperatures, exposure times, application pressures, and original contamination levels. Hot water and steam were most effective, but potential negative effects on appearance and quality of the beef must also be considered.

## Chemical Interventions

### *Cetylpyridinium Chloride*

Cutter et al. (2000) determined the effectiveness of CPC against *E. coli* O157:H7 and *S.* Typhimurium on lean and adipose beef surfaces. They performed several spray-washing experiments (125 psi, 15 s, 35°C) with 1% (wt/vol) CPC. On lean beef surfaces, CPC immediately reduced 5 to 6 log cfu/cm$^2$ of both pathogens to virtually undetectable levels and the bacteria did not increase during 35 days of refrigerated storage. On adipose beef surfaces, *E. coli* O157:H7 was reduced immediately to less than 2 log cfu/cm$^2$, and by day 35, the reduction was still 1.3 log cfu/cm$^2$ when compared to untreated controls. However the residual CPC found on the carcasses was found to be in excess of the daily allowed intake of CPC (Cutter et al., 2000).

The use of CPC has also been found to be very effective when used in a simulated spray-chilling process for beef carcasses, resulting in nearly 5-log reductions in *E. coli* O157:H7 after 24 h (Stopforth et al., 2004). Although the United States Food and Drug Administration has approved the use of CPC for the decontamination of poultry carcasses, it has not been approved for use on beef.

#### *Organic Acid Washes*

The most common method used in the beef processing industry in the United States to reduce STEC is an organic acid wash of the entire carcass after removing the hide and evisceration and before chilling. Currently, lactic acid and acetic acid are the most commonly used organic acids; however, formic, propionic, citric, fumaric, and L-ascorbic acid have all been researched either separately or in mixtures for use in chemical washes. These organic acids are generally delivered via a spray washer, at a concentration between 1% and 5% (Prasai et al., 1991; Cutter and Siragusa, 1994; Dorsa et al., 1998a,b) at various temperatures (Brackett et al., 1994; Castillo et al., 2001). Organic acids are most effective when used warm (50°C–55°C), but there are also reports that they are more corrosive at these warmer temperatures (Acuff, 2005). Lactic acid (2%) was shown to reduce *E. coli* O157:H7 on beef carcass tissue by 3.3 log, and 2% acetic acid reduced it by 1.6 log (Ransom et al., 2003). However, Gill and Landers (2003) found no efficacy for lactic acid when used on beef tissue under commercial conditions.

Castillo et al. (2001) examined the use of a 30-s postchill spray, using lactic acid heated to 55°C at a 4% concentration and found that the intervention provided a 2-log reduction of *E. coli* O157:H7. The bactericidal activity of lactic acid, levulinic acid (LV), and sodium dodecyl sulfate (SDS) was determined individually and in combination against STEC in pure culture. The use of 3% lactic acid for 2 min in pure cultures reduced *E. coli* O26:H11, O45:H2, O111:H8, O103:H2, O121:H2, O145:NM, and O157:H7 populations by 2.1, 0.4, 0.3, 1.4, 0.3, 2.1, and 1.7 log cfu/mL, respectively (Zhao et al., 2014). Thus there appears to be a significant difference in reduction of various STEC strains to these interventions. Treatment with 0.5% LV plus 0.05% SDS for less than 1 min reduced the populations of all STEC strains to undetectable levels. Treating cold (4°C) beef trim with LV-SDS at 21, 62, or 81°C for 30 s reduced *E. coli* O157:H7 by 1.0, 1.1, or 1.4 log cfu/cm$^2$, respectively, whereas treating beef trim at 8°C with LV-SDS at 12°C for 0.1, 1, 3, or 5 min reduced *E. coli* O157:H7 by 1.4, 2.4, 2.5, or 3.3 log cfu/cm$^2$, respectively. Spray treatment of beef trim at 4°C with 5% lactic acid only reduced the *E. coli* O157:H7 population by 1.3 log cfu/cm$^2$. This variance in the amount of reduction among the different STEC serotypes may be an indication that, at least for organic acids, all treatments may not be equally effective against all serotypes.

In the United States, organic acids (lactic, acetic, and citric) are applied as part of a carcass wash prechill and can be applied at levels up to 2.5% of a solution (FSIS, 2014). Recently the EU approved the use of lactic acid to reduce surface contamination on beef carcasses, cuts, and trimmings (Anonymous, 2013). Lactic acid is to be applied either by spraying or misting using a 2%–5% solution in potable water at temperatures of up to a maximum of 55°C.

#### *Peroxyacetic Acid*

Peroxyacetic acid is an oxidizer that can be used as a carcass wash in commercial beef processing plants. Researchers have achieved reductions of approximately 1.4 log for *E. coli* O157:H7 under in vitro conditions (Ransom et al., 2003). Kalchayanand et al. (2012) found that application of peroxyacetic acid at 200 ppm on beef flank tissue using a model spray wash cabinet reduced numbers of non-O157 STEC and O157 by 0.9–1.5 log. Gill and Badoni (2004) used 0.02% peroxyacetic acid in a trial in commercial beef slaughter plants and observed little effect on total bacterial numbers or generic *E. coli* on chilled beef quarters. King et al. (2005) determined that using even 600 ppm (three times the allowable level) did not reduce *E. coli* O157:H7 numbers on inoculated beef carcasses.

#### *Electrolyzed Water*

Electrolyzed water (EW) is emerging as an environmentally friendly antimicrobial treatment (Huang et al., 2007). It is produced by electrolysis of a dilute salt solution, and the reaction products include sodium hydroxide (NaOH) and hypochlorous acid (Huang et al., 2007). Three forms of the solution can be produced, an acidic form, a neutral pH form, and an alkaline form. Acidic EW exhibits an acid pH, a high oxidation reduction potential, and high free chlorine concentrations which makes it effective as an antimicrobial agent (Kim et al., 2000a,b; Hsu, 2003). Researchers have found EW able to control *E. coli* O157:H7 on various vegetables (Hung et al., 2010; Pangloli et al., 2009; Venkitanarayanan et al., 1999). Jadeja et al. (2013) used in vitro testing to determine that *E. coli* O26, O45, O103, O121, O111, and O145 were as susceptible to EW as was *E. coli* O157:H7. They also determined that EW was more active against the STEC than sodium hypochlorite at the same free chlorine concentration. In the United States, EW is allowed on beef carcasses applied as a spray at a level not to exceed 50 ppm calculated as free available chlorine (USDA/FSIS, 2014).

### Spray Chilling

Chilling slows the growth of most bacteria, and temperatures just above the freezing point can kill or also injure bacteria (Jay, 2000). Ice crystals may form within the bacteria and rupture their cell membrane, or chemical changes may occur which kill the organism (Jay, 2000). Chilling is the most widely used method for the preservation of meat (Ingram and MacKey, 1976). Spray chilling and blast chilling (ultralow temperature) are the most common methods of chilling used in the US beef industry. Spray chilling is normally performed with water or water mixed with chlorine. Heitter (1975) studied the use of the Chlor-chil system, a spray chill system with added chlorine, to reduce carcass shrinkage and bacteria, but although the system did reduce shrinkage, the reduction in bacteria was not significant. Dickson (1991) observed that incorporation of 2% acetic acid into the spray chill treatment could lead to reductions of up to 3 logs when applied to beef adipose tissue, and reductions of up to 1 log on lean tissue for *E. coli* O157:H7.

#### *Summary of Chemical Treatments*

A variety of chemical compounds have been tested for the decontamination of beef carcasses. The bactericidal activity of these chemicals is mainly based on the disruption of cellular membranes, other cellular constituents, and physiological cellular processes. For use in the beef slaughterhouse, it must also be considered that the antimicrobial activities of some are counteracted by organic matter, concentrated substances might constitute a health hazard or ecological threat, some agents show corrosive properties, or their stability is limited in solution. Carcass decontamination interventions other than potable water are not actually banned in the EU (Regulation [EC] no. 853/2004), but approval for an intervention can only be authorized after the European Food Safety Authority has provided a risk assessment (Hugas and Tsigarida, 2008). As previously noted, approval for the use of lactic acid sprays has been granted in the EU.

## *Emerging Technologies*

### *Electron Beam Irradiation*

Electron beam (E-beam) irradiation uses machine-generated high-energy electrons to inactivate microorganisms (Zhu et al., 2005). The mode of action of irradiation involves damaging the bacterial cells' genetic material and disrupting their normal functions. Doses as low as 0.4–0.6 kGy have been shown to give a 1-log reduction in *Listeria monocytogenes* (Radomyski et al., 1994). For beef carcasses, only the meat surface to a depth of approximately 15 mm receives a significant radiation dose (Koohmaraie et al., 2005). Researchers have demonstrated that a 1-kGy dose of E-beam radiation applied to chilled beef primals reduced *E. coli* O157:H7 numbers by 4 logs, and no off flavors or off odors were detected by a trained taste panel (Arthur et al., 2005). The American Meat Institute (AMI) submitted a petition for E-beam irradiation to be approved as a processing aid in beef carcasses, which would exempt it from labeling requirements (AMI, 2005). The petition was denied by FSIS because of a belief that the dose could not be applied evenly over an uneven shape such as a carcass so that the total absorbed dose might be higher than the maximum absorbed dose allowed, which is 4.5 kGy (USDA, 2011). Recently, Maxim et al. (2014) developed a method to ensure a uniform dose when E-beam is applied to irregular surfaces. Using the device, they were able to achieve a greater than 5-log reduction in *E. coli* O157:H7 artificially inoculated on rabbit carcasses. E-beam irradiation is allowed on chilled beef carcasses as long as the product meets the requirements for total absorbed dose and labeling as described in 9 CFR 424.

### Essential Oils

Essential oils (EOs) have a long history of use in the fragrance and flavor industry, but for several years, they have been studied as possible natural antimicrobials (Cowan, 1999; Holley and Patel, 2005; O'Bryan et al., 2015; Rivera Calo et al., 2015). The EOs of a wide range of plants including thyme, cinnamon, sage, rosemary, sassafras, clove, gardenia, and jasmine have been shown to exert antimicrobial effects against microorganisms including *E. coli* (Cosentino et al., 1999; Friedman et al., 2002; Moreira et al., 2005; Valero and Salmerón, 2003). The EOs of citrus compounds have also shown effective antimicrobial activity in vitro against both Gram-positive and Gram-negative foodborne pathogens (Fisher and Phillips, 2006, 2009; Friedly et al., 2009; Nannapaneni et al., 2008, 2009; O'Bryan et al., 2008; Pendleton et al., 2012). There have been some studies that indicate that citrus EOs and their components can exert their antimicrobial effect on foodstuffs even at refrigeration temperatures. For example, orange oil and terpineol at 1000 μL/L significantly reduced spoilage bacteria and extended the shelf life of pasteurized milk to more than 56 days when stored at 4°C (Dabbah et al., 1970). It was also determined that milk with orange oil at 1000 μL/L had an acceptable flavor, but terpineol gave the milk a medicinal flavor that was unacceptable (Dabbah et al., 1970). Citral and citron EO were found to increase the shelf life of fruit salads stored at 9°C (Belletti et al., 2008). Pendleton et al. (2012) reported that cold-pressed terpeneless Valencia orange oil suppressed the growth in vitro at 4°C and 10°C of 3 strains of *E. coli* O157:H7 isolated from beef. Pittman et al. (2011) reported that concentrations of 3% and 6% citrus EOs sprayed onto simulated beef subprimals artificially inoculated with *E. coli* O157:H7 significantly reduced the numbers of *E. coli* as well as total aerobic bacterial counts compared with inoculated, no spray, or water-sprayed controls, over a period of 90 days of refrigerated storage. These results give promise for new antimicrobial treatments to be used in the chilling and postchill stages of beef processing.

## Conclusions

Postharvest interventions have greatly improved the safety of beef. Numerous antimicrobial intervention compounds have been researched and implemented for carcasses, as well as physical strategies such as steam-vacuuming, hot water washing, steam pasteurization, and hide-on carcass washing systems. A multihurdle approach is commonly used to maximize the reduction in risk of pathogen contamination. In addition, it is necessary to proactively prevent cross-contamination and foodborne illness by educating employees, retailers, and consumers on the appropriate handling and storage of meat. Further studies are needed to provide more systematic data to fuel the development of novel approaches for control of STEC in foods, including beef.

## Acknowledgments

Preparation of this review was partially funded by a grant from the National Cattlemen's Association.

## References

Acuff, G.R., 2005. Chemical decontamination strategies for meat. In: Sofos, J.N. (Ed.), Improving the Safety of Fresh Meat. Woodhead Publishing Limited, CRC Press, New York, pp. 351–363.

Acuff, G.R., Castillo, A., Savell, J.W., 1996. Microbiological intervention methods: hot water rinses. In: Proceedings 49th Reciprocal Meat Conference, Provo, Utah, pp. 125–128 Chicago, IL.

AMI, 2005. Citizens Petition to Recognize the Use of E-Beam on Carcasses as a Processing Aid. Available at: http://www.fsis.usda.gov/wps/wcm/connect/31f4f6dc-19a1-41d0-bdcf-030b509e5fcd/Petition_Carcass_Surface_Irrad.pdf?MOD=AJPERES.

Anonymous, 4/2/2013. Commission Regulation (EC) No 101/2013 of 4 February 2013 concerning the use of lactic acid to reduce microbiological surface contamination on bovine carcasses. Official Journal of the European Union L34, 1–3.

Antic, D., Blagojevic, B., Buncic, S., 2011. Treatment of cattle hides with Shellac solution to reduce hide-to-beef microbial transfer. Meat Science 88, 498–502.

Antic, D., Blagojevic, B., Ducic, M., Mitrovic, R., Nastasijevic, I., Buncic, S., 2010. Treatment of cattle hides with Shellac-in-ethanol solution to reduce bacterial transferability—a preliminary study. Meat Science 85, 77–81.

Arthur, T.M., Wheeler, T.L., Shackelford, S.D., Bosilevac, J.M., Nou, X., Koohmaraie, M., 2005. Effects of low-dose, low-penetration electron beam irradiation of chilled beef carcass surface cuts on *Escherichia coli* O157:H7 and meat quality. Journal of Food Protection 68, 666–672.

Barkocy-Gallagher, G.A., Arthur, T.M., Rivera-Betancourt, M., Nou, X.W., Shackelford, S.D., Wheeler, T.L., Koohmaraie, M., 2003. Seasonal prevalence of Shiga toxin-producing *Escherichia coli*, including O157:H7 and non-O157 serotypes, and *Salmonella* in commercial beef processing plants. Journal of Food Protection 66, 1978–1986.

Barkate, M.L., Acuff, G.R., Lucia, L.M., Hale, D.S., 1993. Hot water decontamination of beef carcasses for reduction of initial bacterial numbers. Meat Science 35, 397–401.

Bell, R.G., 1997. Distribution and sources of microbial contamination on beef carcasses. Journal of Applied Microbiology 82, 292–300.

Belletti, N., Lancioti, R., Patrignani, F., Gardini, F., 2008. Antimicrobial efficacy of citron essential oil on spoilage and pathogenic microorganisms in fruit-based salads. Journal of Food Science 73, M331–M338.

Blagojevic, B., Antic, D., Ducic, M., Buncic, S., 2011a. A study of haptoglobin levels in groups of cattle and pigs with and without abnormalities at meat inspection. Foodborne Pathogens and Disease 8, 1119–1124.

Blagojevic, B., Antic, D., Ducic, M., Buncic, S., 2011b. Ratio between carcass- and skin microflora as an abattoir process hygiene indicator. Food Control 22, 186–190.

Blagojevic, B., Antic, D., Ducic, M., Buncic, S., 2012. Visual cleanliness scores of cattle at slaughter and microbial loads on the hides and the carcasses. Veterinary Record 170, 563. http://dx.doi.org/10.1136/vr.100477.

Bosilevac, J.M., Arthur, T.M., Wheeler, T.L., Shackelford, S.D., Rossman, M., Reagan, J.O., Koohmaraie, M., 2004. Prevalence of *Escherichia coli* O157 and levels of aerobic bacteria and enterobacteriaceae are reduced when hides are washed and treated with cetylpyridinium chloride at a commercial beef processing plant. Journal of Food Protection 67, 646–650.

Bosilevac, J.M., Nou, X., Osborn, M.S., Allen, D.M., Koohmaraie, M., 2005a. Development and evaluation of an on-line hide decontamination procedure for use in a commercial beef processing plant. Journal of Food Protection 68, 265–272.

Bosilevac, J.M., Shackelford, S.D., Brichta, D.M., Koohmaraie, M., 2005b. Efficacy of ozonated and electrolyzed oxidative waters to decontaminate hides of cattle before slaughter. Journal of Food Protection 68, 1393–1398.

Bowling, R.A., Clayton, R.P., 1992. Method for Dehairing Animals. U.S. Patent 5,149,295.

Brackett, R.E., Hao, Y.Y., Doyle, M.P., 1994. Ineffectiveness of hot acid sprays to decontaminate *Escherichia coli* O157:H7 on beef. Journal of Food Protection 57, 198–203.

Buncic, S., Nychas, G.J., Lee, M.R.F., Koutsoumanis, K., Hebraud, M., Desvaux, M., Chorianopoulos, N., Bolton, D., Blagojevic, B., Antic, D., 2014. Microbial pathogen control in the beef chain: recent research advances. Meat Science 97 (SI), 288–297.

Castillo, A., Dickson, J.S., Clayton, R.P., Lucia, L.M., Acuff, G.R., 1998a. Chemical dehairing of bovine skin to reduce pathogenic bacteria and bacteria of fecal origin. Journal of Food Protection 61, 623–625.

Castillo, A., Lucia, L.M., Goodson, K.J., Savell, J.W., Acuff, G.R., 1998b. Use of hot water for beef carcass decontamination. Journal of Food Protection 61, 19–25.

Castillo, A., Lucia, L.M., Goodson, K.J., Savell, J.W., Acuff, G.R., 1998c. Comparison of water wash, trimming, and combined hot water and lactic acid treatments for reducing bacteria of fecal origin on beef carcasses. Journal of Food Protection 61, 823–828.

Castillo, A., Lucia, L.M., Roberson, D.B., Stevenson, T.H., Mercado, I., Acuff, G.R., 2001. Lactic acid sprays reduce bacterial pathogens on cold beef carcass surfaces and in subsequently produced ground beef. Journal of Food Protection 64, 58–62.

Chapman, P.A., Siddons, C.A., Wright, D.J., Norman, P., Fox, J., Crick, E., 1993. Cattle as a possible source of verocytotoxin-producing *Escherichia coli* O157 infections in man. Epidemiology and Infections 111, 439–448.

Cobbold, R.N., Davis, M.A., Rice, D.H., Szymanski, M., Tarr, P.I., Besser, T.E., Hancock, D.D., 2008. Associations between bovine, human, and raw milk and beef isolates of non-O157 shiga toxigenic *Escherichia coli* within a restricted geographic area of the United States. Journal of Food Protection 71, 1023–1027.

Coffey, B., Rivas, L., Duffy, G., Coffey, A., Ross, R.P., McAuliffe, O., 2011. Assessment of *Escherichia coli* O157:H7-specific bacteriophages e11/2 and e4/1c in model broth and hide environments. International Journal of Food Microbiology 147, 188–194.

Cosentino, S., Tuberoso, C.I.G., Pisano, B., Sata, M., Mascia, V., Arzedi, E., Palmas, F., 1999. *In vitro* antimicrobial activity and composition of Sardinian *Thymus* essential oils. Letters in Applied Microbiology 29, 130–135.

Cowan, M.M., 1999. Plant products as antimicrobial agents. Clinical Microbiology Reviews 12, 564–582.

Cutter, C.N., Dorsa, W.J., Siragusa, G.R., 1997. Parameters affecting the efficacy of spray washes against *Escherichia coli* O157:H7 and fecal contamination on beef. Journal of Food Protection 60, 614–618.

Cutter, C.N., Dorsa, W.J., Handie, A., Rodriguez-Morales, S., Zhou, X., Breen, P.J., Compadre, C.M., 2000. Antimicrobial activity of cetylpyridinium chloride washes against pathogenic bacteria on beef surfaces. Journal of Food Protection 63, 593–600.

Cutter, C.N., Siragusa, G.R., 1994. Efficacy of organic acids against *Escherichia coli* O157:H7 attached to beef carcass tissue using a pilot scale model carcass washer. Journal of Food Protection 57, 97–103.

Dabbah, R., Edwards, V.M., Moats, W.A., 1970. Antimicrobial action of some citrus fruit oils on selected food-borne bacteria. Advances in Applied Microbiology 19, 27–31.

Dickson, J.S., 1991. Control of *Salmonella typhimurium*, *Listeria monocytogenes*, and *Escherichia coli* O157:H7 on beef in a model spray chilling system. Journal of Food Science 56, 191–193.

Dickson, J.S., 1995. Susceptibility of preevisceration washed beef carcasses to contamination by *Escherichia coli* O157:H7 and salmonellae. Journal of Food Protection 58, 514–518.

Dorsa, W.J., Cutter, C.N., Siragusa, G.R., 1996a. Effectiveness of a steam-vacuum sanitizer for reducing *Escherichia coli* O157:H7 inoculated to beef carcass surface tissue. Letters in Applied Microbiology 23, 61–63.

Dorsa, W.J., Cutter, C.N., Siragusa, G.R., Koohmaraie, M., 1996b. Microbial decontamination of beef and sheep carcasses by steam, hot water spray washes, and a steam-vacuum sanitizer. Journal of Food Protection 59, 127–135.

Dorsa, W.J., Cutter, C.N., Siragusa, G.R., 1997. Effects of steam-vacuuming and hot water spray wash on the microflora of refrigerated beef carcass surface tissue inoculated with *Escherichia coli* O157:H7, *Listeria innocua*, and *Clostridium sporogenes*. Journal of Food Protection 60, 114–119.

Dorsa, W.J., Cutter, C.N., Siragusa, G.R., 1998a. Long-term effect of alkaline, organic acid, or hot water washes on the microbial profile of refrigerated beef contaminated with bacterial pathogens after washing. Journal of Food Protection 61, 300–306.

Dorsa, W.J., Cutter, C.N., Siragusa, G.R., 1998b. Long-term bacterial profile of refrigerated ground beef made from carcass tissue, experimentally contaminated with pathogens and spoilage bacteria after hot water, alkaline, or organic acid washes. Journal of Food Protection 61, 1615–1622.

Fisher, K., Phillips, C., 2006. The effect of lemon, orange and bergamot essential oils and their components on the survival of *Campylobacter jejuni, Escherichia coli, Listeria monocytogenes, Bacillus cereus* and *Staphylococcus aureus in vitro* and in food systems. Journal of Applied Microbiology 101, 1232–1240.

Fisher, K., Phillips, C., 2009. The mechanism of action of a citrus oil blend against *Enterococcus faecium* and *Enterococcus faecalis*. Journal of Applied Microbiology 106, 1343–1349.

Friedly, E.C., Crandall, P.G., Ricke, S.C., Roman, M., O'Bryan, C., Chalova, V.I., 2009. *In vitro* antilisterial effects of citrus oil fractions in combination with organic acids. Journal of Food Science 74, M67–M72.

Friedman, M., Henika, P.R., Mandrell, R.E., 2002. Bacteriocidal activities of plant essential oils and some of their constituents against *Campylobacter jejuni, E. coli, Listeria monocytogenes*, and *Salmonella typhimurium*. Journal of Food Protection 65, 1545–1560.

FSIS, 2014. Safe and Suitable Ingredients Used in the Production of Meat, Poultry and Egg Products, FSIS Directive 7120.1 Rev. 23. Available at: http://www.fsis.usda.gov/wps/portal/fsis/topics/regulations/directives/7000-series.

Gill, C.O., Badoni, M., 2004. Effects of peroxyacetic acid, acidified sodium chlorite or lactic acid solutions on the microflora of chilled beef carcasses. International Journal of Food Microbiology 91, 43–50.

Gill, C.O., Landers, C., 2003. Effects of spray-cooling processes on the microbiological conditions of decontaminated beef carcasses. Journal of Food Protection 66, 1247–1252.

Gill, C.O., McGinnis, J.C., Badoni, M., 1996. Use of total or *Escherichia coli* counts to assess the hygienic characteristics of a beef carcass dressing process. International Journal of Food Microbiology 31, 181–196.

Gould, L.H., Demma, L., Jones, T.F., Hurd, S., Vugia, D.J., Smith, K., Shiferaw, B., Segler, S., Palmer, A., Zansky, S., Griffin, P., 2009. Hemolytic uremic syndrome and death in persons with *Escherichia coli* O157:H7 infection, foodborne diseases active surveillance network sites, 2000–2006. Clinical Infectious Diseases 49, 1480–1485.

Hardin, M.D., Acuff, G.R., Lucia, L.M., Oman, J.S., Savell, J.W., 1995. Comparison of methods for decontamination from beef carcass surfaces. Journal of Food Protection 58, 368–374.

Heitter, E.F., 1975. Chlor-chil. In: Proceedings of the Meat Industry Research Conference, pp. 31–32, 20–21 March, 1975, Arlington, VA.

Holley, R.A., Patel, D., 2005. Improvement in shelf-life and safety of perishable foods by plant essential oils and smoke antimicrobials. Food Microbiology 22, 273–292.

Hsu, S.Y., 2003. Effects of water flow rate, salt concentration and water temperature on efficiency of an electrolyzed oxidizing water generator. Journal of Food Engineering 60, 469–473.

Hugas, M., Tsigarida, E., 2008. Pros and cons of carcass decontamination: the role of the European Food Safety Authority. Meat Science 78, 43–52.

Hung, Y.C., Tilly, P., Kim, C., 2010. Efficacy of electrolyzed oxidizing (EO) water and chlorinated water for inactivation of *Escherichia coli* 0157:H7 on strawberries and broccoli. Journal of Food Quality 33, 559–577.

Hussein, H.S., Bollinger, L.M., 2005a. Prevalence of Shiga toxin-producing *Escherichia coli* in beef. Meat Science 71, 676–689.

Hussein, H.S., Bollinger, L.M., 2005b. Prevalence of Shiga toxin-producing *Escherichia coli* in beef cattle. Journal of Food Protection 68, 2224–2241.

Hussein, H.S., Sakuma, T., 2005. Prevalence of Shiga toxin-producing *Escherichia coli* in dairy cattle and their products. Journal of Dairy Science 88, 450–465.

Huang, Y.R., Hung, Y.C., Hsu, S., Huang, Y.W., Hwang, D.F., 2007. Application of electrolyzed oxidizing water in the food industry. Food Control 19, 329–334.

Imamovic, L., Jofre, J., Schmidt, H., Serra-Moreno, R., Muniesa, M., 2009. Phage-mediated shiga toxin 2 gene transfer in food and water. Applied and Environmental Microbiology 75, 1764–1768.

Ingram, M., MacKey, M.B., 1976. Inactivation by cold. In: Skinner, F.A., Hugo, W.B. (Eds.), Inhibition and Inactivation of Vegetative Microbes. Academic Press, New York, pp. 111–146.

Jadeja, R., Hung, Y.C., Bosilevac, J.M., 2013. Resistance of various shiga toxin-producing *Escherichia coli* to electrolyzed oxidizing water. Food Control 30, 580–584.

Jay, J.M., 2000. Low-temperature food preservation and characteristics of psychrotrophic microorganisms. In: Modern Food Microbiology, sixth ed. Aspen Publishers, Gaithersburg, MD.

Kalchayanand, N., Arthur, T.M., Bosilevac, J.M., Schmidt, J.W., Wang, R., Shackelford, S.D., Wheeler, T.L., 2012. Evaluation of commonly used antimicrobial interventions for fresh beef inoculated with Shiga toxin-producing *Escherichia coli* serotypes O26, O45, O103, O111, O121, O145, and O157:H7. Journal of Food Protection 75, 1207–1212.

Kim, C., Hung, Y.C., Brackett, R.E., 2000a. Efficacy of electrolyzed oxidizing (EO) and chemically modified water on different types of foodborne pathogens. International Journal of Food Microbiology 61, 199–207.

Kim, C., Hung, Y.C., Brackett, R.E., 2000b. Roles of oxidation-reduction potential in electrolyzed oxidizing and chemically modified water for the inactivation of food-related pathogens. Journal of Food Protection 63, 19–24.

King, D.A., Lucia, L.M., Castillo, A., Acuff, G.R., Harris, K.B., Savell, J.W., 2005. Evaluation of peroxyacetic acid as a post-chilling intervention for control of *Escherichia coli* O157:H7 and *Salmonella typhimurium* on beef carcass surfaces. Meat Science 69, 401–407.

Kochevar, S.L., Sofos, J.N., Bolin, R.R., Reagan, J.O., Smith, G.C., 1997. Steam vacuuming as a pre-evisceration intervention to decontaminate beef carcasses. Journal of Food Protection 60, 107–113.

Koohmaraie, M., Arthur, T.M., Bosilevac, J.M., Guerini, M.N., Shackelford, S.D., Wheeler, T.L., 2005. Post-harvest interventions to reduce/eliminate pathogens in beef. Meat Science 71, 79–91.

Laegreid, W.W., Elder, R.O., Keen, J.E., 1999. Prevalence of *Escherichia coli* O157:H7 in range beef calves at weaning. Epidemiology and Infections 123, 291–298.

Majowicz, S.E., Scallan, E., Jones-Bitton, A., Sargeant, J.M., Stapleton, J., Angulo, F.J., Yeung, D.H., Kirk, M.D., 2014. Global incidence of human Shiga toxin-producing *Escherichia coli* infections and deaths: a systematic review and knowledge synthesis. Foodborne Pathogens and Disease 11, 447–455.

Maxim, J.E., Neal, J.A., Castillo, A., 2014. Development of a novel device for applying uniform doses of electron beam irradiation on carcasses. Meat Science 96, 373–378.

McEvoy, J.M., Doherty, A.M., Finnerty, M., Sheridan, J.J., McGuire, L., Blair, I.S., McDowell, D.A., Harrington, D., 2000. The relationship between hide cleanliness and bacterial numbers on beef carcasses at a commercial abattoir. Letters in Applied Microbiology 30, 390–395.

McEvoy, J.M., Doherty, A.M., Sheridan, J.J., Blair, I.S., McDowell, D.A., 2001. Use of steam condensing at subatmospheric pressures to reduce *Escherichia coli* O157:H7 numbers on bovine hide. Journal of Food Protection 64, 1655–1660.

Minihan, D., Whyte, P., O'Mahony, M., Clegg, T., Collins, J.D., 2003. *Escherichia coli* O157 in Irish feedlot cattle: a longitudinal study involving preharvest and harvest phases of the food chain. Journal of Food Safety 23, 167–178.

Moreira, M.R., Ponce, A.G., Del Valle, C.E., Roura, S.I., 2005. Inhibitory parameters of essential oils to reduce a food borne pathogen. LWT Food Science and Technology 38, 565–570.

Murata, H., Shimada, N., Yoshioka, M., 2004. Current research on acute phase proteins in veterinary diagnosis: an overview. Veterinary Journal 168, 28–40.

Nannapaneni, R., Muthaiyan, A., Crandall, P.G., Johnson, M.G., O'Bryan, C.A., Chalova, V.I., Callaway, T.R., Arthington, J.D., Nisbet, D.J., Ricke, S.C., 2008. Antimicrobial activity of commercial citrus-based natural extracts against *Escherichia coli* O157:H7 isolates and mutant strains. Foodborne Pathogens and Disease 5, 695–699.

Nannapaneni, R., Chalova, V.I., Crandall, P.G., Ricke, S.C., 2009. *Campylobacter* and *Arcobacter* species sensitivity to commercial orange oil fractions. International Journal of Food Microbiology 129, 43–49.

Nataro, J.P., Kaper, J.B., 1998. Diarrheagenic *Escherichia coli*. Clinical Microbiology Reviews 11, 142–201.

Nou, X.W., Rivera-Betancourt, M., Bosilevac, J.M., Wheeler, T.L., Shackelford, S.D., Gwartney, B.L., Reagan, J.O., Koohmaraie, M., 2003. Effect of chemical dehairing on the prevalence of *Escherichia coli* O157:H7 and the levels of aerobic bacteria and *Enterobacteriaceae* on carcasses in a commercial beef processing plant. Journal of Food Protection 66, 2005–2009.

Nutsch, A.L., Phebus, R.K., Reimann, M.J., Schafer, D.E., Boyer, J.E., Wilson, R.C., Leising, J.D., Kastner, C.L., 1997. Evaluation of a steam pasteurization process in a commercial beef processing facility. Journal of Food Protection 60, 485–492.

Nutsch, A.L., Phebus, R.K., Reimann, M.J., Kotrola, J.S., Wilson, R.C., Boyer, J.E., Brown, T.L., 1998. Steam pasteurization of commercially slaughtered beef carcasses: evaluation of bacterial populations at five anatomical locations. Journal of Food Protection 61, 571–577.

O'Bryan, C.A., Crandall, P.G., Chalova, V.I., Ricke, S.C., 2008. Orange essential oils antimicrobial activities against *Salmonella* spp. Journal of Food Science 73, M264–M267.

O'Bryan, C.A., Pendleton, S.J., Crandall, P.G., Ricke, S.C., 2015. Potential of plant essential oils and their components in animal agriculture—in vitro studies on antibacterial mode of action. Frontiers in Veterinary Science 2, 35. http://dx.doi.org/10.3389/fvets.2015.00035.

O'Flynn, G., Ross, R.P., Fitzgerald, G.F., Coffey, A., 2004. Evaluation of a cocktail of three bacteriophages for biocontrol of *Escherichia coli* O157:H7. Applied and Environmental Microbiology 70, 3417–3424.

Pangloli, P., Hung, Y.C., Beuchat, L.R., King, C.H., Zhao, Z.-H., 2009. Reduction of *Escherichia coli* O157:H7 on produce by use of electrolyzed water under simulated food service operation conditions. Journal of Food Protection 72, 1854–1861.

Pendleton, S.J., Crandall, P.G., Ricke, S.C., Goodridge, L., O'Bryan, C.A., 2012. Inhibition of beef isolates of *E. coli* O157:H7 by orange oil at various temperatures. Journal of Food Science 77, M308–M311.

Phebus, R.K., Nutsch, A.L., Schafer, D.E., Wilson, R.C., Reimann, M.J., Leising, J.D., Kastner, C.L., Wolf, J.R., Prasai, R.K., 1997. Comparison of steam pasteurization and other methods for reduction of pathogens on surfaces of freshly slaughtered beef. Journal of Food Protection 60, 476–484.

Pittman, C.I., Pendleton, S., Bisha, B., O'Bryan, C.A., Belk, K.E., Goodridge, L., Crandall, P.G., Ricke, S.C., 2011. Activity of citrus essential oils against *Escherichia coli* O157:H7 and *Salmonella* spp. and effects on beef subprimal cuts under refrigeration. Journal of Food Science 76, M433–M438.

Prasai, R.K., Acuff, G.R., Lucia, L.M., Hale, D.S., Savell, J.W., Morgan, J.B., 1991. Microbiological effects of acid decontamination of beef carcasses at various locations in processing. Journal of Food Protection 54, 868–872.

Radomyski, T., Murano, E.A., Olson, D.G., 1994. Elimination of pathogens of significance in food by low dose irradiation: a review. Journal of Food Protection 57, 73–86.

Ransom, J.R., Belk, K.E., Sofos, J.N., Stopforth, J.D., Scanga, J.A., Smith, G.C., 2003. Comparison of intervention technologies for reducing *Escherichia coli* O157:H7 on beef cuts and trimmings. Food Protection Trends 23, 24–34.

Reagan, J.O., Acuff, G.R., Buege, D.R., Buyck, M.J., Dickson, J.S., Kastner, C.L., Marsden, J.L., Morgan, J.B., Nickelson II, R., Smith, G.C., Sofos, J.N., 1996. Trimming and washing of beef carcasses as a method of improving the microbiological quality of meat. Journal of Food Protection 59, 751–756.

Rhoades, J.R., Duffy, G., Koutsoumanis, K., 2009. Prevalence and concentration of verocytotoxigenic *Escherichia coli*, *Salmonella* Enterica and *Listeria monocytogenes* in the beef production chain: a review. Food Microbiology 26, 357–376.

Rivera Calo, J., Crandall, P.G., O'Bryan, C.A., Ricke, S.C., 2015. Essential oils as antimicrobials in food systems – A review. Food Control 54, 111–118.

Schmidt, J.W., Wang, R., Kalchayanand, N., Wheeler, T.L., Koohmaraie, M., 2012. Efficacy of hypobromous acid as a hide-on carcass antimicrobial intervention. Journal of Food Protection 75, 955–958.

Schmidt, J.W., Bosilevac, J.M., Kalchayanand, N., Wang, R., Wheeler, T.L., Koohmaraie, M., 2014. Efficacy of hypobromous acid as a hide-on carcass antimicrobial intervention. Journal of Food Protection 77, 538–548.

Sofos, J.N., Smith, G.C., 1998. Nonacid meat decontamination technologies: model studies and commercial applications. International Journal of Food Microbiology 44, 171–188.

Stopforth, J.D., Yoon, Y., Belk, K.E., Scanga, J.A., Kendall, P.A., Smith, G.C., Sofos, J.N., 2004. Effect of simulated spray chilling with chemical solutions on acid-habituated and non-acid-habituated *Escherichia coli* O157:H7 cells attached to beef carcass tissue. Journal of Food Protection 67, 2099–2106.

Tarr, P.I., Fouser, L.S., Stapleton, A.E., Wilson, R.A., Kim, H.H., Vary, J.C., Clausen, C.R., 1996. Hemolytic-uremic syndrome in a six-year-old girl after a urinary tract infection with Shiga-toxin-producing *Escherichia coli* O103:H2. New England Journal of Medicine 335, 635–638.

USDA, 1996. Notice of policy change; achieving the zero tolerance performance standard for beef carcasses by knife trimming and vacuuming with hot water or steam; use of acceptable carcass interventions for reducing carcass contamination without prior agency approval. United States Department of Agriculture, Food Safety and Inspection Service. Federal Register 61, 15024–15027.

USDA, 2011. Response to Petition Submitted by AMI. Available at: http://www.fsis.usda.gov/wps/wcm/connect/a185f497-cd3c-4ae9-b975-bba844a04fd2/Petition_Response_AMI_070805.pdf?MOD=AJPERES.

USDA/FSIS, 2014. Safe and Suitable Ingredients Used in the Production of Meat and Poultry Products. FSIS Directive 7120.1 Amendment 6, USDA-FSIS. http://www.fsis.usda.gov/wps/portal/fsis/topics/regulations/directives/7000-series/safe-suitable-ingredients-related-document.

Valero, M., Salmerón, M.C., 2003. Antibacterial activity of 11 essential oils against *Bacillus cereus* in tyndallized carrot broth. International Journal of Food Microbiology 85, 73–81.

Venkitanarayanan, K.S., Ezeike, G., Hung, Y.-C., Doyle, M.P., 1999. Efficacy of electrolyzed oxidizing water for inactivation of *Escherichia coli* 0157:H7, *Salmonella enteritidis*, and *Listeria monocytogenes*. Applied and Environmental Microbiology 65, 4276–4279.

Wells, J.G., Davis, B.R., Wachsmuth, I.K., Riley, L.W., Remis, R.S., Sokolow, R., Morris, G.K., 1983. Laboratory investigation of hemorrhagic colitis outbreak associated with a rare *Escherichia coli* serotype. Journal of Clinical Microbiology 18, 512–520.

Zhao, T., Zhao, P., Chen, D., Jadeja, R., Hung, Y.C., Doyle, M.P., 2014. Reductions of Shiga toxin-producing *Escherichia coli* and *Salmonella typhimurium* on beef trim by lactic acid, levulinic acid, and sodium dodecyl sulfate treatments. Journal of Food Protection 77, 528–537.

Zhu, M., Du, M., Cordray, J., Ahn, D.U., 2005. Control of *Listeria monocytogenes* contamination in ready-to-eat meat products. Comprehensive Reviews in Food Science and Food Safety 4, 34–42.

CHAPTER 12

# Microbiological Safety of Unpasteurized Fruit and Vegetable Juices Sold in Juice Bars and Small Retail Outlets

**Armitra Jackson-Davis[1], Aubrey Mendonca[2], Shakaree Hale[1], Jealae Jackson[1], Alescia King[1], Joy Jackson[3]**
[1]*Alabama Agricultural and Mechanical University, Huntsville, AL, United States;* [2]*Iowa State University, Ames, IA, United States;* [3]*University of Texas-Arlington, Arlington, TX, United States*

## Introduction

Unpasteurized fruit juice has gained considerable popularity mainly due to the demand for more freshlike foods that have not been subjected to processing methods (such as heat) that may lead to loss of color, flavor, vitamins, and minerals (Elez-Martínez et al., 2006). However, these products may be vehicles for pathogenic microorganisms due to the lack of processing and antimicrobials typically added to pasteurized juice (See Table 12.1). Unpasteurized juices present additional concerns with pathogenic bacteria contamination, acid tolerant microorganisms, osmophilic microbiota, and a shorter shelf life due to microbial and enzymatic spoilage (Aneja et al., 2014; Mosqueda-Melgar et al., 2008). Despite the popularity of these juices, they are still subject to foodborne outbreaks. Table 12.1 lists outbreaks of human foodborne disease from various microorganisms associated with juices during the period of 1922–2014. The majority of the outbreaks relate to the consumption of unpasteurized juice.

According to IBISWorld, as of May 2015, there are 1639 juice and smoothie bars in the United States. Although this number represents a very conservative estimation, the popularity of these establishments is clear. With this increased popularity, there is also an increased risk of foodborne illness. Some juice establishments use these raw fruits and vegetables to produce unpasteurized juice products. In addition, combining different fruits and vegetables into one juice product has also gained popularity. One example of a juice blend is apples, kale, spinach, cucumber, parsley, and lemon. Another example is beet, carrot, apple, ginger, and lemon. Because raw fruits and commodities from different sources can possess different microbial loads based on different agricultural practices, the chance of contamination increases substantially. In addition to the leisurely consumption of unpasteurized juices, many

Food and Feed Safety Systems and Analysis. http://dx.doi.org/10.1016/B978-0-12-811835-1.00012-9

**Table 12.1: Outbreak of human foodborne disease from various microorganisms associated with juices during the period of 1922–2014.**

| Type | Product | Pathogen[a] | Year | Location | Venue | Cases (deaths)[b] | References[c] |
|---|---|---|---|---|---|---|---|
| Açaí | Unspecified | *Trypanosoma cruzi*[d] | 2004 | Brazil | Retail | 27 | SVS (2005b) and Pereira et al. (2009) |
| | Unspecified | *T. cruzi* | 2007 | Brazil | Retail | 25 | SVS (2007b) and Pereira et al. (2009) |
| Apple | Unpasteurized | *S.* Typhi | 1922 | France | NR[e] | 23 (0) | Paquet (1923) |
| | Unpasteurized | *S.* Typhimurium | 1974 | USA (NJ) | Farm, small retail outlets | 296 (0) | CDC (1975) |
| | Unpasteurized | *E. coli* O157:H7 (suspected) | 1980 | Canada (ON) | Local market | 14 (1) | Steele et al. (1982) |
| | Unpasteurized | *E. coli* O157:H7 | 1991 | USA (MA) | Small cider mill | 23 (0) | Besser et al. (1993) |
| | Unpasteurized | *Cryptosporidium* | 1993 | USA (ME) | School | 213 (0) | Millard et al. (1994) |
| | Unpasteurized | *C. parvum* | 1996 | USA (NY) | Small cider mill | 31 (0) | CDC (1997) |
| | Unpasteurized | *E. coli* O157:H7 | 1996 | USA (CT) | Small cider mill | 14 (0) | CDC (1997) |
| | Unpasteurized | *E. coli* O157:H7 | 1996 | USA (WA) | Small cider mill | 6 (0) | FDA (2001) |
| | Unpasteurized | *E. coli* O157:H7 | 1996 | Canada (BC), USA (CA, CO, WA) | Retail | 70 (1) | CDC (1996) and Cody et al. (1999) |
| | Unpasteurized | *E. coli* O157:H7 | 1997 | USA (IN) | Farm | 6 | INS DOH (1997) |
| | Unpasteurized | *E. coli* O157:H7 | 1998 | Canada (ON) | Farm/Home | 14 (0) | Tamblyn et al. (1999) |
| | Unpasteurized | *E. coli* O157:H7 | 1999 | USA (OK) | NR | 25 | CDC (2011) |
| | Unpasteurized (ozonated) | *C. parvum* | 2003 | USA (OH) | Farm/Retail | 144 | Blackburn et al. (2006) and Vojdani et al. (2008) |
| | Unpasteurized | *E. coli* O111 and *C. parvum* | 2004 | USA (NY) | Farm/Home | 213 | Vojdani et al. (2008) and Schaffzin et al. (2012) |
| | Unpasteurized | *E. coli* O157:H7 | 2005 | Canada (ON) | NR | 4 | LSDEPC (2005) |
| | Unpasteurized | *E. coli* O157:H7 | 2007 | USA (MA) | NR | 9 | CDC (2011) |
| | Unpasteurized | *E. coli* O157:H7 | 2008 | USA (IA) | Fair | 7 | CDC (2011) |
| | Unpasteurized | *E. coli* O157:H7 | 2010 | USA (MD) | Retail | 7 | FDA (2010) |
| | Unpasteurized | *Cryptosporidium*, *E. coli* O111:NM | 2011 | USA (MN) | Farm | 14 | MDH (2014) |
| | Unpasteurized | *Cryptosporidium* | 2011 | USA (OH) | Private home | 4 (0) | CDC (2015) |
| | Unpasteurized | *E. coli* O157:H7 | 2012 | USA (MI) | Farm | 4 | Clark (2015) |
| | Unpasteurized | *Cryptosporidium* | 2013 | USA (IA) | Private home | 10 | IDPH (2014) |
| | Unpasteurized | *Cryptosporidium* | 2013 | USA (OH) | Private home | 8 (0) | CDC (2015) |

| | | | | | | | |
|---|---|---|---|---|---|---|---|
| | Unpasteurized | *S.* Typhimurium | 2013 | USA (PA) | Fair, festival; private home | 10 (0) | CDC (2015) |
| | Unpasteurized | *E. coli* O157:H7 | 2014 | Canada (ON) | Farm, cider mill, local farmer's market | 3 | CFIA (2014) |
| Carrot | Homemade | *C. botulinum* | 1993 | USA (WA) | Home | 1 (0) | Buzby and Crutchfield (1999) |
| | Pasteurized | *C. botulinum* | 2006 | USA | Retail | 4 (1) | CDC (2006) |
| Coconut | Milk[f] | *Vibrio cholerae* | 1991 | USA (MD) | Home/picnic | 4 | CDC (1991) and Taylor et al. (1993) |
| Guava | Unspecified | *T. cruzi* | 2007 | Venezuela | School | 103 (1) | de Noya et al. (2010) |
| Mamey | Frozen Puree | *S.* Typhi | 1999 | USA | NR | 19 | Katz et al. (2002) |
| | Frozen Pulp | *S.* Typhi | 2010 | USA | Retail | 12 | CDC (2010) |
| Mixed Fruit | Unspecified | *Shigella sonnei* | 2002 | Canada, USA, UK, British West Indies | Resort | 78 | CDC (2011) |
| | Açaí, sugarcane | *T. cruzi* | 2006 | Brazil | NR | 94 (6) | SVS (2007a) and Pereira et al. (2009) |
| | Açaí, banana, strawberry, sugarcane | Hepatitis A | 2007 | USA (FL) | Food Service | 3 | CDC (2011) |
| Mixed Fruit/ Vegetable | Clover, sweet potato leaf, apple | *Angiostrongylus cantonensis*[g] | 2001 | Taiwan | Home | 5 | Tsai et al. (2004) |
| Orange | Unpasteurized | Enterotoxigenic *E. coli* | 1992 | India | Roadside vendor | 6 (0) | Singh et al. (1995) |
| | Unpasteurized | *Salmonella* Gaminara, Hartford, and Rubislaw | 1995 | USA (FL) | Retail | 63 (0) | CDC (1995), Cook et al. (1998) and Parish (1998) |
| | Unpasteurized | *Shigella flexneri* | 1995 | South Africa | Restaurant | 14 | Thurston et al. (1998) |
| | Unpasteurized | Virus suspected | 1996 | USA | Food Service | 2 | Parish (2000) |
| | Unpasteurized | *S.* Muenchen | 1999 | Canada and USA | Restaurant | 423 (1) | CDC (1999) |
| | Unpasteurized | *S.* Anatum | 1999 | USA (FL) | Roadside stand | 6 (0) | Krause et al. (2001) |
| | Unpasteurized | *S.* Typhimurium | 1999 | Australia | Retail | 405 (0) | National Centre for Disease Control (1999) |
| | Unpasteurized | *S.* Enteritidis | 2000 | USA (6 states) | Retail and Food Service | 88 | Butler (2000) |
| | Unpasteurized | Hepatitis A | 2004 | Egypt | Hotel | 351 | Frank et al. (2005) |
| | Unpasteurized | *S.* Typhimurium and *S.* Saintpaul | 2005 | USA (23 states) | Retail and Food Service | 152 | Jain et al. (2009) |

*Continued*

**Table 12.1: Outbreak of human foodborne disease from various microorganisms associated with juices during the period of 1922–2014.—cont'd**

| Type | Product | Pathogen[a] | Year | Location | Venue | Cases (deaths)[b] | References[c] |
|---|---|---|---|---|---|---|---|
| | Unpasteurized | *S.* Panama | 2008 | The Netherlands | Retail | 33 | Noël et al. (2010) |
| | Reconstituted | *S.* Typhi | 1944 | USA (OH) | Hotel | 18 (1) | Duncan et al. (1946) |
| | Reconstituted | Hepatitis A | 1962 | USA (MO) | Hospital | 24 | Eisenstein et al. (1963) |
| | Reconstituted | Unknown | 1965 | USA (CA) | Football game | 563 | Tabershaw et al. (1967) |
| | Reconstituted | *S.* Typhi | 1989 | USA (NY) | Hotel | 69 | Birkhead et al. (1993) |
| | Unspecified | Norwalk-like virus | 1991 | Australia | Airline | 3053 | Lester et al. (1991) |
| Sugarcane | Unspecified | *T. cruzi* | 2005 | Brazil | Roadside kiosk | 25 (3) | SVS (2005a) and Pereira et al. (2009) |
| Watermelon | Homemade | *Salmonella* spp. | 1993 | USA (FL) | Home | 18 (0) | FDA (1998) |

[a]Pathogens abbreviated and associated with outbreaks include *S.*, *Salmonella*; *E.*, *Escherichia*; *C. parvum*, *Cryptosporidium parvum*; *C. botulinum*, *Clostridium botulinum*.
[b]The number in parenthesis represents the number of deaths, if reported.
[c]References for each outbreak appear in the following reference list.
[d]*Trypanosoma cruzi* is the causative agent of Chagas disease.
[e]*NR*, not reported.
[f]Coconut milk is the liquid that is squeezed from the coconut meat; coconut juice or coconut water is the liquid obtained from a whole coconut when one breaks the shell.
[g]*Angiostrongylus cantonensis*, also known as rat lungworm, is the major cause of eosinophilic meningitis in the Pacific Islands and southeast Asia.
*Reproduced from Danyluk, M.D., Goodrich-Schneider, R.M., Schneider, K.R., Harris, L.J., Worobo, R.W., 2012. Outbreaks of Foodborne Disease Associated with Fruit and Vegetable Juices, 1922–2010, EDIS Publication FSHN12-04. Available from:http://ucfoodsafety.ucdavis.edu/files/223883.pdf and revised with permission.*

consumers are now using unpasteurized juices for detoxification in the form of juice cleansing. To accomplish this, raw juice producers now sell unpasteurized juices in packages that can include 1 day (6 juices) to 5 days (30 juices) worth of juices. These juices are comprised of different combinations of different fruits and vegetables. Because of this, the chances of foodborne illnesses increase.

## Raw Juice Characteristics: Influence on Pathogen Survival

Juices derived from raw agricultural products including coconut, carrots, and wheatgrass are all susceptible to contamination with human enteric pathogens from environmental sources (Doyle and Erickson, 2012). These unpasteurized juices are sold in juice bars or retail stores. Intrinsic factors are those that are inherent to the juice itself or those that become part of the juice. The pH of many fruit juices ranges from up to 5.0–5.5 for juices such as some apple and pear, to below pH of 2.0 for lemon juice. Vegetable juices such as carrot, beet, and celery have somewhat higher pH values (pH = 6 and above). Furthermore, foodborne illness outbreaks have been associated with unpasteurized fruit juices, indicating that juice pH itself is not sufficient to ensure microbial control and food safety (Table 12.1). In addition, research has been conducted that demonstrates the ability of foodborne pathogens to survive and grow in low pH and high acid environments (Duan and Zahao, 2009; Mosqueda-Melgar et al., 2008). This had been recognized by early researchers (Goverd et al., 1979; Murdock and Hatcher, 1975). Therefore, it is important to recognize that many unpasteurized juices may not fall into the high acid dogma of the more commonly known juices that are considered traditional juices. Therefore, they lack the built-in protections that are provided for traditional juices.

In the United States and abroad, unpasteurized juice derived from fruits and vegetables is sold in many establishments directly to the consumer. In addition to the more common fruit and vegetable juices (such as apple juice and carrot juice), other juice types are now available. For example, wheatgrass, derived from the wheat plant *Triticum aestivum*, is widely recognized for its variety of nutrients and vitamins and is commonly sold as a juice product (Bodla, 2011). This product is also sold in more local juice bars as well. Many of these juice bars market their products as "healthier," which makes these establishments popular to consumers who desire more natural and minimally processed food products. The minimally processed form of these juices places the product at more risk as it relates to foodborne pathogens such as *Listeria monocytogenes* and *Salmonella* (Mritunjay and Kumar, 2015). *L. monocytogenes* and *Salmonella* are both enteric pathogens that have been shown to survive in a variety of traditional juice products. *L. monocytogenes* is ubiquitous in the environment, is common on fruits and vegetables, and has been associated with human disease due to the consumption of raw vegetables (Beuchat, 1996a,b). Because carrot juice is derived from a root vegetable and wheatgrass juice is derived from a product that grows close to the ground in soil, the minimally processed juice is at risk for microbial contamination (Santamarıa and Toranzos, 2003).

The microbiological concerns of juice processors and packagers typically involve prevention of spoilage and control of microorganisms in the processing and distribution systems. As a consequence of *Salmonella*, *Escherichia coli* O157:H7, and *Cryptosporidium parvum* outbreaks in 1990s associated with raw juices (primarily apple and orange) processed at commercial facilities, the U.S. Food and Drug Administration (FDA) introduced regulation (21 Code of Federal Regulations 120) mandating that 100% juice be produced under a Hazard Analysis and Critical Control Points (HACCP) plan having supporting Good Manufacturing Practices (GMPs) and Sanitation Standard Operating Procedures (FDA, 2001). This regulation further requires that juice processors demonstrate at least a 5-log reduction of the "pertinent microorganism," which is defined as "the most resistant microorganism of public health significance that is likely to occur in the juice."

Determination of the public health microorganism that is most pertinent for any particular juice may be based on disease outbreak data and any other appropriate information available. Currently, *Salmonella* is generally accepted as the pertinent pathogen in citrus juices, whereas *E. coli* O157:H7 and *C. parvum* need to be taken into consideration for apple juice (FDA, 2001). For many nontraditional juice products, where outbreaks have not previously occurred, there is limited information available for producers to determine the appropriate pathogen of concern. There have been very few surveys of retail juices for the presence of pathogens in part due to the very low probability of finding pathogens in these products (Harris et al., 2003).

## Nonthermal Technologies to Control Pathogens in Juice

To circumvent the adverse affects of heat-treated juices, research has been conducted that evaluates nonthermal interventions. Tomadoni et al. (2016) evaluated the effects of ultrasound, vanillin, and pomegranate extract for shelf-stable unpasteurized strawberry juice. Research findings suggest that the combined effects decreased the natural microbiota and the inoculated pathogens without adversely affecting sensory attributes. Kim and Rhee (2015) investigated the synergistic antimicrobial effect of caprylic acid and citric acid on *E. coli* O157:H7 and indigenous microbiota in carrot juice. Results revealed that caprylic acid and citric acid were more effective on controlling microorganisms compared with individual use. Specifically, reduction in *E. coli* O157:H7 and indigenous microbiota was 7.46 and 3.07 log CFU/ml, respectively. Tomadoni et al. (2015) evaluated the ability of vanillin and geraniol to reduce *E. coli* O157:H7 on strawberry juice over a 14-day period. On day 0 of the study, *E. coli* counts for samples treated with vanillin and geraniol were 3.0–3.2 log compared with 5.4 log in untreated samples. From day 3 to day 14 of the study, *E. coli* O157:H7 populations were below the detectable limit (<2 log).

High pressure processing (HPP) involves the application of high pressures ranging from 100 to 1000 MPa. The antimicrobial mode of action involves protein denaturation and disruption of cell membrane integrity, which results in cell death. Because of the low acid nature of vegetable juices, *Clostridium botulinum* is the most resistant pertinent pathogen of public health concern. Because the pressure resistance of *C. botulinum* spores has been proven to be much greater than that of vegetative cells, higher temperatures and pressures (600 MPa, 220 s, 121°C) to inactivate *C. botulinum* spores (Juliano et al., 2009) are necessary. Picouet et al. (2015) evaluated the changes that are induced by thermal and nonthermal treatments on both acidified and nonacidified carrot juice. Both treatments resulted in the same level of microbial reduction, and compared with the unprocessed samples, both HPP and mild heating processes resulted in modifications in the carotene content, color, acidity and Brix. In this same study, these changes remained unchanged throughout the storage period. HPP processing provides carrot juice with better sensory properties than mild heating. Acidification by adding glucono delta-lactone did not significantly extend the shelf life, but exhibited a detrimental effect on the sensory attributes.

Mohideen et al. (2015) conducted a study that investigated the effect of continuous ultrasonication processing on the microbiological, chemical, and physical properties of unpasteurized blueberry juice. Continuous flow sonication was successful in reducing microbial populations. In addition, sonication at higher intensities significantly ($P<.05$) reduced microbial populations. The highest log reduction in total aerobic microorganisms (1.36 log cycles) was achieved at high flow rate (93.5 min/mL) with high-intensity (100 amplitude) sonication condition compared with untreated and control juice samples. Continuous flow sonication processing preserved the anthocyanins and the color of the blueberry juice.

Unluturk and Atilgan (2015) evaluated the effects of UV-C irradiation on the inactivation of *E. coli* K-12 and on the shelf life of freshly squeezed turbid white grape juice. The shelf life of grape juice was determined as day 14 for UV-C–treated product. Untreated white grape juice samples spoiled on the seventh day of storage. Therefore, the microbial shelf life of product was increased by twofold with the treatment. Although the microbial shelf life of product doubled, the quality of juice was adversely affected. Taze et al. (2015) evaluated the effect of UV-C irradiation on inactivation of spoilage microorganisms and color of freshly squeezed orange juice. Resistance of yeast to UV light and existence of suspended particles limited the effectiveness of the process. Time and UV dose of first decimal reduction were obtained as 5.7 min and 31 $mJ/cm^2$, respectively. The data suggest that biodosimetric studies performed by using inoculated microorganisms for assessment of the efficiency of UV irradiation treatment in the shelf life extension of juices must be carefully evaluated. UV-C irradiation had no influence on the color of orange juice. Santhirasegaram et al. (2015) compared the effect of UV-C light and thermal pasteurization on microbial inactivation, physiochemical properties, antioxidant activities, and other quality parameters of freshly squeezed Chokanan mango juice. After UV-C treatment and thermal pasteurization, no significant changes occurred in physicochemical properties. A significant increase in extractability of carotenoids (6%), polyphenols (31%), and flavonoids (3%) were observed in juice exposed to UV-C light for 15 and 30 min, when compared with freshly squeezed juice.

In addition, enhancement of antioxidant activity was observed after UV-C treatment. Thermal pasteurization and UV-C treatment exhibited significant reduction in microbial load and prolonged shelf life of juice. The results obtained support the use of UV-C light to improve the quality of Chokanan mango juice along with safety standards as an alternative to thermal pasteurization.

Yin et al. (2015) studied the inactivation and reactivation potential of pathogenic *E. coli* O157:H7 in pasteurized apple juice after exposure to UV light at three monochromatic wavelengths: far UV ($\lambda = 222$ nm), far UV+ ($\lambda$ 32 = 282 nm), and UV-C light ($\lambda = 254$ nm). The results showed that *E. coli* O157:H7 is acid resistant, and up to 99.50% of cells survived in apple juice when incubated at 20°C for 24 h. Inactivation of *E. coli* O157:H7 after exposure to far UV light (2.81 log reduction) was higher ($P < .05$) than the inactivation caused by UV-C light (1.95 log reduction) and far UV + light (1.83 log reduction) at similar levels of UV fluence of 75 mJ/cm$^2$. No reactivation potential was observed for *E. coli* O157:H7 in dark incubation phases after exposure to UV light as determined by the regular plating method. In addition, the exposure to far UV light at 222 nm followed by incubating at 37°C significantly decreased ($P < .05$) the survival of *E. coli* O157:H7 during dark incubation phase compared with that of UV-C and far UV + light. Yoo et al. (2015) investigated the effect of titanium dioxide–UV-C photocatalysis (TUVP), a nonthermal technique capable of microbial inactivation via generation of hydroxyl radicals to decontaminate orange surfaces, and high hydrostatic pressure (HHP). Results showed that levels of spot-inoculated *E. coli* O157:H7 (initial level of 7.0 log CFU/cm$^2$) on oranges (12 cm$^2$) were reduced by 4.3 log CFU/mL when treated with TUVP (17.2 mW/cm$^2$). Reductions of 1.5, 3.9, and 3.6 log CFU/mL were achieved using tap water, chlorine (200 ppm), and UV-C alone (23.7 mW/cm$^2$), respectively. *E. coli* O157:H7 in juice from TUVP (17.2 mW/cm$^2$)–treated oranges was reduced by 1.7 log CFU/mL. After orange juice was treated with HHP at 400 MPa for 1 min without any prior fruit surface disinfection, the level of *E. coli* O157:H7 was reduced by 2.4 log CFU/mL. However, the *E. coli* O157:H7 level in juice was reduced by 4.7 log CFU/mL (to lower than the detection limit) when TUVP treatment of oranges was followed by HHP treatment of juice, indicating a synergistic inactivation effect. HHP treatment did not affect the pH, Brix, or color of juice. However, the ascorbic acid concentration and pectinmethylesterase activity were reduced by 35.1% and 34.7%, respectively.

## Regulatory Perspective of Raw Juice

The Fresh Produce Subcommittee of the National Advisory Committee on Microbiological Criteria for Foods met on December 18, 1996, to consider the safety of all fruit juices. The committee concluded that the safety concerns associated with unpasteurized juices was greater and there was a need for safety interventions. The committee also recommended mandatory HACCP for all juice products and that further research be conducted on the "efficacy of new technologies and intervention strategies for safety" of juices. On April 24, 1998, the Federal Register proposed the HACCP approach to establish requirements related to the processing of juice and juice products. The proposal would require processors and importers to implement HACCP principles to ensure the safety of the juice produced. The occurrence of foodborne illness outbreaks associated with the consumption of unpasteurized juices resulted in the FDA approving regulations to improve the safety of juices in January 2001 (FDA, 21 CRF Part 120). Under Juice HACCP regulations, juice processors are required to identify chemical, physical, and microbiological hazards that could possibly contaminate the product. Juice processors must prevent, reduce, or eliminate these hazards and achieve a 5-log (or 100,000-fold) reduction in the number of potential pathogens in the final juice product (FDA, 2001). Therefore, failure of a juice processor to establish and adequately implement a HACCP system renders the juice produced adulterated (FDA, 21 CRF Part 120). Juices that have not been processed to achieve a 5-log reduction in the number of potential pathogens must include a warning label stating that "this product has not been pasteurized and, therefore, may contain harmful bacteria which can cause serious illness in children, the elderly, and persons with weakened immune systems." Juice retailers that sell directly to consumers are not covered by Juice HACCP regulations (FDA, 21 CRF Part 120).

## Conclusions

Although the consumption of unpasteurized juices has increased, research continues to confirm the ability of foodborne pathogens to survive in these food systems. Given the popularity of unpasteurized juices and the establishments that sell this product directly to consumers, additional research is needed to ensure the safety of these products. Current research involves the use of nonthermal technologies to control potential pathogens. Although Juice HACCP is designed to control pathogens of concern, juice produced in juice bar establishments do not fall under this regulation. Therefore, those juices are at risk for contamination.

## References

Aneja, K.R., Dhiman, R., Aggarwal, N.K., Aneja, A., 2014. Emerging preservation techniques for controlling spoilage and pathogenic microorganisms in fruit juices. International Journal of Food Microbiology 2014, 14 pp.

Besser, R.E., Lett, S.M., Weber, J.T., Doyle, M.P., Barrett, T.J., Wells, J.G., Griffin, M.P., 1993. An outbreak of diarrhea and hemolytic uremic syndrome from *Escherichia coli* O157:H7 in fresh-pressed apple cider. JAMA 269, 2217–2220.

Beuchat, L.R., 1996a. Pathogenic microorganisms associated fresh produce. Journal of Food Protection 59, 204–216.

Beuchat, L.R., 1996b. *Listeria monocytogenes*: incidence on vegetables. Food Control 7, 223–228.

Birkhead, G.S., Morse, D.L., Levine, W.C., Fudala, J.K., Kondracki, S.F., Chang, H.G., Shaydgani, M., Novick, L., Blake, P.A., 1993. Typhoid fever at a resort hotel in New York: a large outbreak with an unusual vehicle. The Journal of Infectious Diseases 167, 1228–1232.

Blackburn, B.G., Mazurek, J.M., Hlavsa, M., Park, J., Tillapaw, M., Parrish, M., Salehi, E., Franks, W., Koch, E., Smith, F., Xiao, L., Arrowood, M., Hill, V., da Silva, A., Johnston, S., Jones, J.L., 2006. Cryptosporidiosis associated with ozonated apple cider. Emerging Infectious Diseases 12, 684–686.

Bodla, R., 2011. A study on wheat grass and its nutritional value. Food Science and Quality Management. Available at: http://www.forrestgrasses.com.au/documents/Wheat-Grass-Study.pdf.

Butler, M.E., 2000. *Salmonella* outbreak leads to juice recall in Western states. Food Chemical News 42, 19–20.

Buzby, J.C., Crutchfield, S.R., 1999. New juice regulations underway. Food Review. http://www.ers.usda.gov/publications/foodreview/may1999/frmay99f.pdf.

CDC (Centers for Disease Control and Prevention), 1975. *Salmonella typhimurium* outbreak traced to a commercial apple cider—New Jersey. Morbidity and Mortality Weekly Report 24, 87–88.

CDC, 1991. Cholera associated with imported frozen coconut milk—Maryland. MMWR. Morbidity and Mortality Weekly Report 40, 844–845.

CDC, 1995. Outbreak of *Salmonella* Hartford infections among travelers to Orlando, Florida. EPI-AID Trip Report 95–62.

CDC, 1996. Outbreak of *Escherichia coli* O157:H7 infections associated with drinking unpasteurized commercial apple juice—British Columbia, California, Colorado, and Washington, October 1996. MMWR. Morbidity and Mortality Weekly Report 45, 975.

CDC, 1997. Outbreaks of *Escherichia coli* O157:H7 infection and cryptosporidiosis associated with drinking unpasteurized apple cider—Connecticut and New York, October 1996. MMWR. Morbidity and Mortality Weekly Report 46, 4–8.

CDC, 1999. Outbreak of *Salmonella* serotype Muenchen infections associated with unpasteurized orange juice—United States and Canada, June 1999. MMWR. Morbidity and Mortality Weekly Report 48, 582–585.

CDC, 2006. Botulism associated with commercial carrot juice—Georgia and Florida. MMWR. Morbidity and Mortality Weekly Report 55, 1098–1099.

CDC, 2010. Investigation Update: Multistate Outbreak of Human Typhoid Fever Infections Associated with Frozen Mamey Pulp. http://www.cdc.gov/salmonella/typhoidfever/index.html.

CDC, 2011. Foodborne Outbreak Online Database (FOOD). http://wwwn.cdc.gov/foodborneoutbreaks/Default.aspx.

CDC, 2015. Foodborne Outbreak Online Database (FOOD). http://wwwn.cdc.gov/foodborneoutbreaks/Default.aspx.

CFIA (Canadian Food Inspection Agency), 2014. Food Recall Warning – Unpasteurized Apple Cider Processed by Rolling Acres Cider Mill Recalled Due to *E. coli* O157:H7. http://www.inspection.gc.ca/about-the-cfia/newsroom/food-recall-warnings/completelisting/2014-10-30/eng/1414720185030/1414720197088.

Clark, M., 2015. Foodborne Illness Outbreak Database. Available at: http://www.outbreakdatabase.com.

Cody, S.H., Glynn, K., Farrar, J.A., Cairns, K.L., Griffin, P.M., Kobayashi, J., Fyfe, M., Hoffman, R., King, A.S., Lewis, J.H., Swaminathan, B., Bryant, R.G., Vugia, D.J., 1999. An outbreak of *Escherichia coli* O157:H7 infection from unpasteurized commercial apple juice. Annals of Internal Medicine 130, 202–209.

Cook, K.A., Dobbs, T.E., Hlady, W.G., Wells, J.G., Barrett, T.J., Puhr, N.D., Lancette, G.A., Bodager, D.W., Toth, W., Genese, C.A., Highsmith, A.K., Pilot, K.E., Finelli, L., Swerdlow, D.L., 1998. Outbreak of *Salmonella* Hartford infections associated with unpasteurized juice. JAMA 280, 1504–1509.

Danyluk, M.D., Goodrich-Schneider, R.M., Schneider, K.R., Harris, L.J., Worobo, R.W., 2012. Outbreaks of Foodborne Disease Associated with Fruit and Vegetable Juices, 1922–2010. EDIS Publication FSHN12-04. Available from: http://ucfoodsafety.ucdavis.edu/files/223883.pdf.

de Noya, B., Díaz-Bello, Z., Colmenares, C., Ruiz-Guevara, R., Mauriello, L., Zavala-Jaspe, R., Suarez, J.A., Abate, T., Naranjo, L., Paiva, M., Rivas, L., Castro, J., Márques, J., Mendoza, I., Acquatella, H., Torres, J., Noya, O., 2010. Large urban outbreak of orally acquired acute Chagas disease at a school in Caracas, Venezuela. The Journal of Infectious Diseases 201, 1308–1315.

Doyle, M., Erickson, M., 2012. Opportunities for mitigating pathogen contamination during on-farm food production. International Journal of Food Microbiology 152, 54–74.

Duan, J., Zahao, Y., 2009. Antimicrobial efficiency of essential oil and freeze-thaw treatments against *Escherichia coli* O157:H7 and *Salmonella enterica* ser. Enteritidis in strawberry juice. Food Microbiology and Safety 74, 131–137.

Duncan, T.G., Coull, J.A., Miller, E.R., Bancroft, H., 1946. Outbreak of typhoid fever with orange juice as the vehicle illustrating the value of immunization. American Journal of Public Health 36, 34–36.

Eisenstein, A.B., Aach, R.D., Jacobson, W., Goldman, A., 1963. An epidemic of infectious hepatitis in a general hospital. JAMA 185, 171–174.

Elez-Martínez, P., Soliva-Fortuny, R., Martín-Belloso, O., 2006. Comparative study on shelf life of orange juice processed by high intensity pulsed electric fields or heat treatment. European Food Research and Technology 222, 321–329.

FDA (U.S. Food and Drug Administration), April 24, 1998. Federal Register proposed rules – 63 FR 20449 April 24, 1998 – HACCP; procedures for the safe and sanitary processing and importing of juice; food labeling: warning notice statements; labeling of juice products. Federal Register. 63, 20449–20486. http://www.fda.gov/Food/FoodSafety/HazardAnalysisCriticalControlPointsHACCP/JuiceHACCP/ucm082031.htm.

FDA (U.S. Food and Drug Administration), 2010. DHMH Issues Consumer Alert Regarding Recall of Baugher's Apple Cider. Available at: http://www.fda.gov/Safety/Recalls/ucm232878.htm.

Food and Drug Administration, 2001. Hazard analysis and critical control point (HACCP): procedures for the safe and sanitary processing and importing of juice. Federal Register 66, 6138–6202.

Frank, C., Walter, J., Muehlen, M., Jansen, A., van Treeck, U., Hauri, A.M., Zoellner, I., Schreier, E., Hamouda, O., Stark, K., 2005. Large outbreak of hepatitis A in tourists staying at a hotel in Hurghada, Egypt, 2004 – orange juice implicated. Eurosurveillance 10, 2720.

Goverd, K.A., Beech, F.W., Hobbs, R.P., Shannon, R., 1979. The occurrence and survival of coliforms and Salmonellosis in apple juice and cider. Journal of Applied Microbiology 46, 521–530.

Harris, L.J., Farber, J.N., Beuchat, L.R., Parish, M.E., Suslow, T.V., Garrett, E.H., Busta, F.F., 2003. Outbreaks associated with fresh produce: incidence, growth and survival of pathogens in fresh and fresh-cut produce. Comprehensive Reviews in Food Science and Food Safety 2S, 78–141.

IDPH (Iowa Department of Public Health), 2014. Iowa Surveillance of Notifiable and Other Diseases: Annual Report 2013. Available at: http://www.idph.state.ia.us/IDPHChannelsService/file.ashx?file=0FB37AF6-665A-4C2F-BAF5-7E6CB33C3816.

INS DOH (Indiana State Department of Health), 1997. Summary of Special Disease Outbreak Investigations—1997, Appendix E. Available at: http://www.in.gov/isdh/21186.htm.

Jain, S., Bidol, S.A., Austin, J.L., Berl, E., Elson, F., Williams, M.L., Deassy III, M., Moll, M.E., Rea, V., Vojdani, J.D., Yu, P.A., Hoekstra, R.M., Braden, C.R., Lynch, M.F., 2009. Multistate outbreak of *Salmonella typhimurium* and Saintpaul infections associated with unpasteurized orange juice—United States, 2005. Clinical Infectious Diseases 48, 1065–1071.

Juliano, P., Knoerzer, K., Fryer, P.J., Versteeg, C., 2009. *C. botulism* inactivation kinetics implemented in a computational model of a high-pressure sterilization process. Biotechnology Progress 25, 163–175.

Katz, D.J., Cruz, M.A., Trepka, M.J., Sarez, J.A., Filrella, P.D., Hammond, R.M., 2002. An outbreak of typhoid fever in Florida associated with an imported fruit. The Journal of Infectious Diseases 186, 234–239.

Kim, S.A., Rhee, M.S., 2015. Synergistic antimicrobial activity of caprylic acid in combination with citric acid against both *Escherichia coli* O157:H7 and indigenous microflora in carrot juice. Food Microbiology 49, 166–172.

Krause, G., Terzagian, R., Hammond, R., 2001. Outbreak of *Salmonella* serotype Anatum infection associated with unpasteurized orange juice. Southern Medical Journal 94, 1168–1172.

Lester, R., Stewart, T., Carnie, J., Ng, S., Taylor, R., 1991. Air travel-associated gastroenteritis outbreak, August 1991. Communicable Disease Intelligence 15, 292–293.

LSDEPC (Laboratory Surveillance Data for Enteric Pathogens in Canada), 2005. Annual Summary. Available at: http://www.nml-lnm.gc.ca/N.

MDH (Minnesota Department of Health), 2014. *Escherichia coli* O157 Infection and Hemolytic Uremic Syndrome (HUS), 2011. Available at: http://www.health.state.mn.us/divs/idepc/newsletters/dcn/sum11/ecoli.html.

Millard, P.S., Gensheimer, K.F., Addiss, D.G., Sosin, D.M., Beckett, G.A., Houck-Jankoski, A., Hudson, A., 1994. An outbreak of cryptosporidiosis from fresh-pressed apple cider. JAMA 272, 1592–1596.

Mohideen, F.W., Solval, K.M., Li, J., Zhang, J., Chouljenk, A., Chotiko, A., Sathivel, S., 2015. Effect of continuous ultra-sonication on microbial counts and physico-chemical properties of blueberry (*Vaccinium corymbosum*) juice. LWT-Journal of Food Science and Technology 60 (1), 563–570.

Mosqueda-Melgar, J., Raybaudi-Massilia, R.M., Martín- Belloso, O., 2008. Combination of high-intensity pulsed electric fields with natural antimicrobials to inactivate pathogenic microorganisms and extend the shelf-life of melon and watermelon juices. Food Microbiology 25, 479–491.

Mritunjay, S., Kumar, V., 2015. Fresh farm produce as a source of pathogens: a review. RJET 9, 59–70.

Murdock, D.I., Hatcher, W.S., 1975. Growth of microorganisms in chilled orange juice. Journal of Milk and Food Technology 38, 393–396.

National Centre for Disease Control/Communicable Diseases Network Australia New Zealand, Australian Department of Health and Aged Care, 1999. Salmonellosis outbreak, South Australia. Communicable Diseases Intelligence 23 (3), 73.

Noël, H., Hofhuis, A., De Jonge, R., Heuvelink, A.E., De Jong, A., Heck, M.E., De Jager, C., van Pelt, W., 2010. Consumption of fresh fruit juice: how a healthy food practice caused a national outbreak of *Salmonella panama* gastroenteritis. Journal of Foodborne Pathogens and Disease 7, 375–381.

Paquet, P., 1923. Épidémie de fièvre typhoi¨de: Déterminée par la consummation de petit citre. Revue d'Hygine 45, 165–169.

Parish, M.E., 1998. Coliforms, *Escherichia coli*, and *Salmonella serovars* associated with a citrus-processing facility implicated in a Salmonellosis outbreak. Journal of Food Protection 61, 280–284.

Parish, M., 2000. Relevancy of *Salmonella* and pathogenic *E. coli* to fruit juices. Proceedings IFU-Workshop Microbiology. Fruit Processing 10, 246–250.

Pereira, K.S., Schmidt, F.L., Guaraldo, A.M.A., Franco, R.M.B., Dias, V.L., Passos, L.A.C., 2009. Chagas' disease as food borne illness. Journal of Food Protection 72, 441–446.

Picouet, P.A., Sárraga, C., Cofán, S., Belletti, N., Guàrdia, M.D., 2015. Effects of thermal and high-pressure treatments on the carotene content, microbiological safety and sensory properties of acidified and of non-acidified carrot juice. Journal of Food Science and Technology 62 (1), 920–926.

Santamarıa, J., Toranzos, G.A., 2003. Enteric Pathogens and Soil: A Short Review. Available at: http://revistes.iec.cat/index.php/IM/article/viewFile/9390/9387.

Santhirasegaram, V., Razali, Z., George, D.S., Somasundram, C., 2015. Comparison of UV-C treatment and thermal pasteurization on quality of Chokanan mango (*Mangifera indica* L.) juice. Journal of Food and Bioproducts Processing 94, 313–321.

Schaffzin, J.K., Coronado, F., Dumas, N.B., Root, T.P., Halse, T.A., Schoonmaker-Bopp, D.J., Lurie, M.M., Nicholas, D., Gerzonich, B., Johnson, G.S., Wallace, B.J., Musser, K.A., 2012. Public health approach to detection of non-O157 Shiga toxin-producing *Escherichia coli*: summary of two outbreaks and laboratory procedures. Emerging Infectious Diseases 140, 283–289.

Singh, B.R., Kulshreshtha, S.B., Kapoor, K.N., 1995. An orange juice borne outbreak due to enterotoxi genic *Escherichia coli*. Journal of Food Science and Technology 3, 504–506.

Steele, B.T., Murphy, N., Rance, C.P., 1982. An outbreak of hemolytic uremic syndrome associated with ingestion of fresh apple juice. The Journal of Pediatrics 101, 963–966.

SVS (Secretaria de Vigilância em Saúde do Ministério da Saúde, Brasil), 2005a. Doença de Chagas Aguda relacionada à ingestão de caldo de cana em Santa Catarina. In: Nota Técnica, 4 de abril de 2005. http://portal.saude.gov.br/portal/saude/visualizar_texto.cfm?idtxt=21270.

SVS, 2005b. Doença de Chagas Aguda no municípiode Santana—Amapá. In: Nota Técnica, 4 de abril de 2005. Available at: http://portal.saude.gov.br/portal/saude/visualizar_texto.cfm?idtxt=21310.

SVS, 2007a. Doença a de Chagas Aguda. In: Nota Técnica, 9 de outubro de 2007. http://portal.saude.gov.br/portal/arquivos/pdf/nota_chagas_091007.pdf.

SVS, 2007b. Surto de doença de Chagas agudo (DCA) em Coari/Amazonas—abril de 2007. In: Nota Técnica, 11 de maio de 2007. http://portal.saude.gov.br/portal/saude/visuali-zar_texto.cfm?idtxt=26087.

Tabershaw, I.R., Schmelzer, L.L., Bruhn, H.B., 1967. Gastroenteritis from an orange juice preparation. Archives of Environmental Health 15, 72–77.

Tamblyn, S., de Grosbois, J., Taylor, D., Stratton, J., 1999. An outbreak of *Escherichia coli* O157:H7 infection associated with unpasteurized non-commercial, custom-pressed apple cider – Ontario, 1998. Canada Communicable Disease Report 25, 113–117.

Taylor, J.L., Tuttle, J., Pramukul, T., O'Brien, K., Barrett, T.J., Jolbaito, B., Lim, Y.L., Vugia, D.J., Morris Jr., J.G., Tauxe, R.V., Dwyer, D.M., 1993. An outbreak of cholera in Maryland associated with imported commercial frozen fresh coconut milk. The Journal of Infectious Diseases 167, 1330–1335.

Taze, B.H., Unluturk, S., Buzrul, S., Alpas, H., 2015. The impact of UV-C irradiation on spoilage microorganisms and colour of orange juice. Journal of Food Science and Technology 52 (2), 1000–1007.

Thurston, H., Stuart, J., McDonnell, B., Nicholas, S., Cheasty, T., 1998. Fresh orange juice implicated in an outbreak of *Shigella flexneri* among visitors to a South African game reserve. The Journal of Infectious Diseases 36, 350.

Tomadoni, B., Cassani, L., Moreira, M.R., Ponce, A., 2015. Efficacy of vanillin and geraniol in reducing *Escherichia coli* O157:H7 on strawberry juice. Food Science and Technology 64, 554–557.

Tomadoni, B., Cassani, L., Ponce, A., Moreira, M.R., Agu¨ero, M.V., 2016. Optimization of ultrasound. LWT – Food Science and Technology 72, 475–484.

Tsai, H.-C., Lee, S., Huang, C.-K., Yen, C.-M., Chen, E.-R., Liu, Y.-C., 2004. Outbreak of eosinophilic meningitis associated with drinking raw vegetable juice in southern Taiwan. The American Journal of Tropical Medicine and Hygiene 71, 222–226.

Unluturk, S., Atilgan, M.R., 2015. Microbial safety and shelf life of UV-C treated freshly squeezed white grape juice. Journal of Food Science 80 (8), M1831–M1841.

Vojdani, J., Beuchat, L., Tauxe, R., 2008. Juice-associated outbreaks of human illness in the United States, 1995 through 2005. Journal of Food Protection 71 (2), 356–364.

Yin, F., Zhu, Y., Koutchma, T., Gong, J., 2015. Inactivation and potential reactivation of pathogenic *Escherichia coli* O157:H7 in apple juice following ultraviolet light exposure at three monochromatic wavelengths. Journal of Food Microbiology 46, 329–335.

Yoo, S., Ghafoor, K., Kim, J.U., Kim, S., Jung, B., Lee, D.U., Park, J., 2015. Inactivation of *Escherichia coli* O157:H7 on orange fruit surfaces and in juice using photocatalysis and high hydrostatic pressure. Journal of Food Protection 78 (6), 1098–1105.

CHAPTER 13

# Listeria monocytogenes *Adaptation and Growth at Low Temperatures: Mechanisms and Implications for Foodborne Disease*

**Joshua C. Saldivar[1], Morgan L. Davis[2], Michael G. Johnson[2], Steven C. Ricke[2]**
*[1]Stanford University School of Medicine, Stanford, CA, United States; [2]University of Arkansas, Fayetteville, AR, United States*

## *Introduction*

*Listeria monocytogenes* is a Gram-positive, rod-shaped bacterium that occurs throughout the environment. In these environments, *L. monocytogenes* encounters several stresses such as low and high temperatures, nutrient starvation, low water content, anaerobic atmospheres, oxidative stresses, and osmotic pressures (Lado and Yousef, 2007; Milillo et al., 2012a). This foodborne pathogen is noted for its ability to survive or grow in several different stressful environments and can be transmitted to humans via a wide range of foods. Preservation methods to limit the growth of viable pathogenic and spoilage organisms on contaminated foods often rely on freezing or refrigerating foods, modified atmosphere packaging, chemical preservatives, acidic and osmotic action, high pressure, and high temperatures (Ricke, 2003; Ricke et al., 2005; Lado and Yousef, 2007; Lungu et al., 2009; Jarvis et al., 2016a). Alternatively, some foods are only minimally processed to provide a "fresh, ready-to-eat (RTE)" product. In many cases, the shelf life of a food is dependent on its intrinsic properties such as being acidic, having a low water content, and increased osmolarity that prevent bacterial growth.

The ability of *L. monocytogenes* to survive and even grow under several of the preservative stresses used to preserve foods plays a crucial role in concerns on food safety associated with this foodborne pathogen (Lado and Yousef, 2007; Lungu et al., 2009). Additionally, studies have shown *L. monocytogenes* to become more resistant to different processing treatments after preexposure to particular types of stress (Lou and Yousef, 1996, 1997; Hill and Gahan, 2000). This is important because cells with stronger resistance to preservative stresses pose a difficult problem for food industries to solve. Ingestion of contaminated foods (RTE meats, ice cream, produce, cheeses, and related products) is considered one of the leading cause of listeriosis cases (Zhu et al., 2005; Milillo et al., 2012a; Rodrigues et al., 2017).

Food and Feed Safety Systems and Analysis. http://dx.doi.org/10.1016/B978-0-12-811835-1.00013-0

Many of these types of food products at certain stages of their production and/or retail also involve some form of low temperature exposure. This becomes problematic with *L. monocytogenes* contamination, because not only can it grow over a wide range of temperatures, but it also possesses the ability to adapt and grow, albeit more slowly at low temperatures (Barria et al., 2013; Cordero et al., 2016). Consequently, it is not surprising that in the past decade several studies have focused on various aspects of the ability of *L. monocytogenes* to adapt to low temperatures (Tasara and Stephan, 2006; Chan et al., 2007; Chan and Wiedmann, 2009; Cacace et al., 2010; Soni et al., 2011; Cabrita et al., 2013; Pittman et al., 2014; Shen et al., 2014; Cordero et al., 2016; Hingston et al., 2017). Given the importance of *L. monocytogenes* as a foodborne pathogen and its relationship with food products, particularly those that are considered RTE, it remains imperative that research should be continued to further characterize and understand the mechanisms utilized by *L. monocytogenes* to counter cold stress. Relatively recent comprehensive reviews have been published that detail the different aspects of *L. monocytogenes* cold shock response and its role in food production, processing, and retail (Zhu et al., 2005; Chan and Wiedmann, 2009; Ryser and Buchanan, 2013; Donaldson et al., 2015; NicAogáin and O'Byrne, 2016; Rodrigues et al., 2017). Therefore, the purpose of this chapter is to provide an overview on general virulence response, followed by mechanisms utilized by *L. monocytogenes* to counter cold stress and how that influences responses to other stresses associated with food production, and concluding with opportunities for further study.

## Listeriosis *Epidemiology*

Listeriosis is a foodborne illness of serious public health concern due to the severity of the symptoms caused by the disease. The pathogen is responsible for nearly 20% of all deaths in the United States caused by known foodborne bacterial pathogens (Painter and Slutsker, 2007; Swaminathan et al., 2007; Scallan et al., 2011). Listeriosis poses the highest risk for individuals with a predisposing condition impairing their T cell–mediated immunity, including pregnant women, neonates, and immunocompromised adults; however, less susceptible persons have occasionally contracted listeriosis, causing a concern for all individuals (Painter and Slutsker, 2007). The serotypes of *L. monocytogenes* responsible for 95% of human listeriosis reported have been identified as three main serovars, 1/2a, 1/2b, and 4b (Seeliger and Hohne, 1979; Kathariou, 2002; Nelson et al., 2004; Graves et al., 2007; Milillo et al., 2012a). Among these three serotypes, serotype 4b strains are responsible for most of the outbreaks and sporadic cases of listeriosis (Nelson et al., 2004; Swaminathan et al., 2007; Donaldson et al., 2015).

Several human cases and outbreaks of listeriosis have been reported over the years (Scallan et al., 2011). *L. monocytogenes* continues to receive considerable attention due in part to the high mortality rate observed among those afflicted with the disease, making it one of the more lethal foodborne pathogens (Donaldson, 2010; Fuchs et al., 2012; Cartwright et al., 2013; Rodrigues et al., 2017). For example, in 1999, it was estimated that human listeriosis in the

United States alone caused 2500 illnesses, 2300 hospitalizations, and 500 deaths each year (Mead et al., 1999). From reported outbreaks of listeriosis, it is evident that this pathogen is quite capable of causing very costly economic losses. In 1994 the United Sates Department of Agriculture (USDA) reported that the total medical cost of illness for the year 1993 from acute listeriosis was between $61.7 and $64.8 million, and the estimated productivity loss was between $125.8 and $154.4 million (Buzby et al., 1994). More recently, it has been implicated in the cause of meningitis, septicemia, and abortion, with a high case fatality rate of approximately 19% (Scallan et al., 2011). In the following section, a discussion of *L. monocytogenes* and the characteristics that contribute to its pathogenesis is presented.

## Listeria monocytogenes *Pathogenesis Mechanisms*

*Listeria* spp. are fairly widespread in nature and capable of residing in a multitude of environments including soil, plants, sewage, and water among other ecosystems (Sauders and Weidmann, 2007). Under these conditions, based on gene profiles and transcriptional analyses, it possesses the metabolic and survival capabilities to exist in diverse ecosystems (Gray et al., 2006; Toledo-Arana et al., 2009). This widespread dissemination in the environment also ensures ample opportunity for *Listeria* to be not only encountered by animals through ingestion of contaminated water, feed, and silage among other sources, but also be present in the environments where human food is being processed (Sauders and Weidmann, 2007; Wesley, 2007). Once pathogenic *Listeria* are ingested, animals can exhibit the characteristic symptoms of listeriosis infection including septicemia and organ colonization followed by abortion in pregnant animals (Wesley, 2007). In humans, certain demographics of the population are considered at highest risk, such as pregnant women, neonates, and immunocompromised adults, and severe infections can lead to bacteremia, meningoencephalitis, abortion, or neonatal sepsis (Painter and Slutsker, 2007; Fuchs et al., 2012).

There has been considerable interest in the means by which free-living *L. monocytogenes* go from a nonpathogenic physiological state in the environment to becoming a fairly lethal foodborne pathogen in a host. Gray et al. (2006) has suggested that *L. monocytogenes* may require these pathogenic mechanisms in the environment to ward off a myriad of potential eukaryotic predators. Once ingested in the mammalian host, these virulence genes are expressed and the classic pathogenesis route occurs. The general mechanism of pathogenicity of *L. monocytogenes* involves the invasion of viable host cells, the escape from the phagocytic vacuole, replication in the host cytosol, intracytoplasmic movement and cell-to-cell spread, and the start of a new infection cycle (Kuhn and Goebel, 2007; Fuchs et al., 2012). Several factors impact pathogenicity response as *L. monocytogenes* cells enter the mammalian intestinal tract and subsequently become invasive. Exposure and resistance to stresses, such as the presence of bile salts, osmotic stress, and low stomach pH, encountered in the host can play a role (Donaldson, 2010; NicAogáin and O'Bryne, 2016). Availability of nutrients, such as iron, are influential factors as well (Gray et al., 2006). Expanding on the nutritional aspect,

Fuchs et al. (2012) have suggested that *L. monocytogenes* metabolic pathways should be considered an important component of the virulence response.

As expected, numerous genes are involved in *L. monocytogenes* invasion of host target cells. These genes will not be covered in detail here as several comprehensive reviews discuss these and other aspects of *Listeria* pathogenesis much more extensively (Kathariou, 2002; Kuhn and Goebel, 2007; Swaminathan et al., 2007; Fuchs et al., 2012; Ryser, and Buchanan, 2013; Donaldson et al., 2015). Instead a brief discussion of some of the more familiar virulence genes will be discussed in the current review. The most widely understood of these genes include *inlA* and *inlB*, both part of the internalin multigene family and coding for the proteins internalin A and internalin B, respectively (Gailliard et al., 1991). The listerial cell surface protein internalin A (InlA) initiates contact with E-cadherin, a eukaryotic cell receptor (Mengaud et al., 1996a,b). InlA has been shown to be involved in the intestinal phase of infection (Temm-Grove et al., 1994; Cepek et al., 1994), and InlB has been implicated in the hepatic phase of infection (Dramsi et al., 1995). InlA contains an N-terminal leucine-rich repeat (LRR) domain and a C-terminal sorting signal, including an LPXTG motif. The transpeptidase sortase A (SrtA) cleaves the LPXTG motif and covalently links it to the peptidoglycan anchoring the InlA to the bacterial surface (Sabet et al., 2005). The LRR region and the LPXTG motif are necessary and sufficient to promote bacterial entry into cells expressing the receptor E-cadherin (Sabet et al., 2005; Swaminathan et al., 2007).

Other virulence factors involved in *L. monocytogenes* invasion of eukaryotic cells are protein p60, the listerial surface protein autolysin amidase (Ami), and a *Listeria* adhesion protein (LAP). Protein p60 has been shown to be involved in the invasion of fibroblasts by *L. monocytogenes* (Kuhn and Goebel, 1989) and is encoded by the gene termed invasion-associated protein (*iap*) (Kohler et al., 1990). Ami is involved in bacterial adhesion to eukaryotic cells and LAP has been shown to be responsible for adherence to numerous intestinal culture cell lines and has been suggested to be involved in the intestinal phase of infection (Wampler et al., 2004).

By inducing its own uptake into host cells, *L. monocytogenes* becomes localized inside host cell vacuoles. Once internalized in host cell vacuoles, *L. monocytogenes* is able to escape the phagocytic vacuoles, enter into the cytoplasm, and replicate. The pore-forming toxin listeriolysin (LLO) and the phosphatidylinositol-specific phospholipase C (PC-PLC) are key virulence factors that mediate the rapid escape of *L. monocytogenes* into the cytosol of the host cells. LLO binds to the surface of the cell membrane and is believed to form pores in the vacuoles, allowing the organism to escape from both the primary vacuole and the double-membrane vacuole that encases the bacteria during cell-to-cell spread (Swaminathan et al., 2007).

In the cytoplasm of host cells, *L. monocytogenes* is capable of movement due to continuously polymerizing the host cell actin filaments propelling the bacterium through the cytoplasm

(Swaminathan et al., 2007). The surface protein ActA, encoded by the gene *actA*, provides *L. monocytogenes* with this actin-based motility. Bacteria moving through the assembly of the actin comet tails are thus capable of forming outward extending protrusions from the host cells that can become engulfed by neighboring cells, resulting in a double-membrane vacuole surrounding the *L. monocytogenes*. Once in the double-membrane vacuole, LLO and PC-PLC mediate the escape of *L. monocytogenes* from the vacuole into the cytoplasm by producing pores in the vacuole cell membrane, and consequently the bacterium is able to begin a new infectious cycle during its cell-to-cell spread (Dussurget et al., 2004).

Most of the virulence genes in *L. monocytogenes* that have been identified are regulated by the transcriptional activator PrfA, a member of the Crp/Fnr family of regulatory proteins. The list of virulence genes located within the PrfA regulon are *prfA*, *plcA*, *hly*, *mpl*, *actA*, *plcB*, *inlA*, *inlB*, *inlC*, and *hpt* (Dussurget et al., 2004). It is likely that the full activity of PrfA requires the binding of a cofactor or else some modification after it has been translated (Mueller and Freitag, 2005). Several environmental signals affect the activity of PrfA and consequently affect the virulence gene expression of *L. monocytogenes*. Distinguishing between the external environment and the environment within a host is essential for intracellular pathogens that are also ubiquitous in nature, such as *L. monocytogenes*. The expression of many PrfA-regulated genes is greatly increased within the cytosol of infected host cells, suggesting that PrfA activation is induced in the cytosol (Mueller and Freitag, 2005). In addition, PrfA expression is thermoregulated by an RNA thermosensor (Johansson et al., 2002) and has increased activity at 37°C and relatively no activity below 20°C, which suggests expression of the virulence genes in *L. monocytogenes* also has increased activity at 37°C. Other environmental signals affecting the virulence gene expression of *L. monocytogenes* include both physicochemical signals (iron, glucose, cellobiose, salt, and pH) and stress conditions (heat shock, oxidative stress, nutritional stress, and intracellular growth) (Kuhn and Goebel, 2007).

It has been determined that the alternative sigma factor $\sigma^B$ (SigB), the general stress response regulator, contributes to the *L. monocytogenes* invasion of Caco-2 cells by controlling the expression of *inlA* and *inlB* (Kim et al., 2004, 2005). Furthermore, Rauch et al. (2005) observed *prfA* transcription, in vitro, to be SigB dependent, as well as other newly identified genes regulated by PrfA. Garner et al. (2006a) reported SigB contributing to *L. monocytogenes* gastrointestinal infection using a guinea pig model. Stress conditions in the gastrointestinal tract of a host may be what initiates the SigB-dependent regulation of *prfA* and the PrfA regulon in *L. monocytogenes*. More recently, Guildimann et al. (2017) demonstrated the importance of PrfA and $\sigma^B$ under cold stress and other various environment stress conditions such as thermal and salt stress.

Over the past few years, several additional factors involved in the virulence of *L. monocytogenes* have been identified, including a second key virulence regulator and regulon in addition to the PrfA regulon. Mandin et al. (2005) identified a response regulator of the two-component system termed VirR, named for virulence regulator. Another novel virulence factor, Vip, involved in the entry into some mammalian cells was discovered by comparative genomics. From whole genomic comparisons, the Vip protein of *L. monocytogenes* was identified as having an LPXTG surface protein motif not present in avirulent *Listeria* species and it was determined to be regulated by PrfA (Cabanes et al., 2005). A study using DNA microarray analysis, real-time PCR, and transposon mutagenesis revealed that several genes in *L. monocytogenes* exhibited increased transcription for bacterial cells located in the cytoplasm of Caco-2 cells, including genes encoding transporter proteins for carbon and nitrogen sources, factors involved in anabolic pathways, stress proteins, transcriptional regulators, and proteins of unknown function (Joseph et al., 2006). These findings provide strong evidence on the importance of metabolism, stress adaptation, gene expression regulation, and environmental sensing of *L. monocytogenes* during the cytosolic stage of infection (Joseph et al., 2006). These findings also suggest the possibility of additional unidentified genes involved in the mechanisms of pathogenesis of *L. monocytogenes*. In the following sections, the occurrence of *L. monocytogenes* on foods and the association with refrigeration and cold stress are discussed.

## Listeria *and Food Sources*

Listeriosis became known to be associated with food sources during the past 25 years, after an outbreak occurred in 1981 in Nova Scotia involving coleslaw made from contaminated cabbage (Schlech et al., 1983; Gellin and Broome, 1989; Rodrigues et al., 2017). Although *L. monocytogenes* infections may have been as prominent prior to the 1981 outbreak, only after that event did it receive recognition as a serious foodborne disease (Swaminathan et al., 2007; Rodrigues et al., 2017). Since 1981 there have been numerous listeriosis reports every year, originating from a wide range of food product sources (Hanning et al., 2008; Cartwright et al., 2013). *L. monocytogenes* infections in humans have been most commonly associated with contaminated foods such as unpasteurized milk, soft cheeses, frankfurters, delicatessen meats, poultry products, and some seafoods (Pradhan et al., 2010; Carpentier and Cerf, 2011; Gambarin et al., 2012; Ryser and Buchanan, 2013; Cartwright et al., 2013; Rodrigues et al., 2017). Particularly because of the risk associated with *L. monocytogenes*–contaminated RTE meats and the corresponding connection with human cases of listeriosis, the USDA developed a zero-tolerance policy for *L. monocytogenes* in RTE meat products (USDA-FSIS, 2003; Zhu et al., 2005; Lianou and Sofos, 2007; Pradhan et al., 2010).

Although RTE meat products have garnered most of the attention, more awareness of non–RTE meat contaminated food sources has emerged as important contributors to human cases of listeriosis. Non–RTE meats and other animal products are known to harbor *L. monocytogenes*. For example, Mendonça et al. (2016) isolated several different strains of *L. monocytogenes* from packaged refrigerated chicken that had been collected from Brazilian supermarkets and subsequently identified them by serotyping and pulsed field gel electrophoresis profiling. While it is not clear where *L. monocytogenes* may have been in contact with poultry during the poultry production cycle, other research indicates that live birds in certain environmental surroundings can encounter *Listeria* (Rothrock et al., 2017). Milillo et al. (2012b) recovered *L. monocytogenes* and hemolytic *Listeria innocua* from cecal contents of broilers raised in pasture flocks as well as environmental samples.

Pasteurization of bulk milk is an effective control measure for eliminating *L. monocytogenes* but unpasteurized raw milk, which has become popular in the past few years, has also been identified as a potential risk vehicle for listeriosis (Frye and Donnelly, 2005; Lejune and Rajala-Schultz, 2009; Latorre et al., 2011). In a comprehensive risk assessment study on raw milk and listeriosis risk, Latorre et al. (2011) concluded that risk could be reduced if the milk was refrigerated at proper refrigeration temperatures and consumed as soon after purchase as possible by the customer. However, refrigeration of postpasteurized milk, even if pasteurized at higher temperatures, still poses a threat if it becomes contaminated with *L. monocytogenes* after pasteurization (Stasiewicz et al., 2014).

While a diverse array of fresh produce products has been associated with listeriosis (Hanning et al., 2008; Garner and Kathariou, 2016), the major listeriosis 2011 outbreak originating from whole cantaloupe brought additional scrutiny and in-depth research on the ecology, transmission routes, and causative factors related to *L. monocytogenes* and its association with this particular fresh produce product. Martinez et al. (2016) pointed out that the outbreak originating from cantaloupe revealed that environmental and equipment contamination in the processing plant was an important source, thus in some ways similar with other RTE foods. Garner and Kathariou (2016) concluded in a review of produce outbreaks that although *L. monocytogenes* is likely to be present in the agricultural fields where produce is grown, field sources have not been identified with human listeriosis but rather can be traced to the packing plant. After surveying retail environments, Hoelzer et al. (2011) found that within small commercial establishments or those with a history of failed inspections, *L. monocytogenes* was more prevalent on nonfood contact surfaces than food contact surfaces. Ferreira et al. (2014) suggested that persistence of *L. monocytogenes* in food processing plants over several years is an important factor not only in transmission but also as a primary source for listeriosis outbreaks. Storage conditions of produce after processing may also be a factor as indicated by listeriosis cases linked to prepackaged lettuce (Jackson et al., 2016). Hoelzer et al. (2012) reviewed the data on *L. monocytogenes* growth responses to generate predictive modeling for *L. monocytogenes* growth on produce. They concluded that listeriosis outbreaks from processed produce appeared to often be associated with suboptimal storage conditions such as refrigeration for several months. Francis and O'Beirne (2001) noted that growth responses and survival of *L. monocytogenes* were influenced by vegetable type, package atmosphere, and temperature, but viable organisms could still be recovered at the end of 4°C storage.

As more has become known, it is apparent *L. monocytogenes* has the ability to survive cold temperatures such as those encountered during refrigeration. It also appears that it can survive over lengthy periods and therefore represents a risk for certain food products that are stored under such conditions. This coupled with potential cross-contamination occurring at the food processing plants offers the opportunity for *L. monocytogenes* to become established on food products and remain there potentially for several months. Clearly, *L. monocytogenes* has mechanisms that enable it to counteract environmental stress associated with cold temperatures. The following sections address the mechanisms possessed by *L. monocytogenes* and how these relate to other stress responses and virulence expression by this organism.

## Listeria monocytogenes *and Cold Stress*

Many of the foods classified as high-risk vehicles for *L. monocytogenes* contamination are RTE, minimally processed foods (Zhu et al., 2005; Rodrigues et al., 2017). This is supported by the recent outbreaks that have been associated with RTE foods that do not require heat treatment before consumption (Rodrigues et al., 2017). A major concern with minimally RTE processed foods is that the means of bacterial inhibition and preservation is quite limited and primarily depends on storage at refrigerated temperatures (Cacace et al., 2010). Because refrigeration may be used for some of these food products, pathogens that are also psychrotrophs, such as *L. monocytogenes*, are able to grow at cold temperatures and consequently cause foodborne disease (Walker et al., 1990). The ability of *L. monocytogenes* to grow at refrigerated temperatures coupled with the long-term preservation of minimally processed foods by refrigeration offers this pathogen an optimal environment to survive and remain viable for extended periods (Swaminathan et al., 2007).

*L. monocytogenes* has been observed to initiate growth at temperatures as low as −1.5° C to as high as 45° C, although it multiplies more slowly at the lower temperatures (Hudson et al., 1994; Barbosa et al., 1994). In addition, although it does not grow at temperatures below −1.5°C, it can still survive at even lower temperatures (Lado and Yousef, 2007). Given this propensity to grow at lower temperatures, cold enrichment historically was used for isolation of *L. monocytogenes* from samples before more selective media became available (Jarvis et al., 2016b). In general, when *L. monocytogenes* is exposed to a lowered temperature, it responds by a stoppage in growth followed by adaptation and subsequent slower growth at the decreased environmental temperature (Barria et al., 2013; Cordero et al., 2016). The lag phase of the organism at 4°C varies between 69.8 and 270 h, and at 10°C it varies from 36.5 to 68.9 h, before it enters the log growth phase. Once in the log phase, its generation time averages 43 h at 4°C and 6.6 h at 10°C (Barbosa et al., 1994). However, there is strain variation in both lag phase and growth rate responses among *L. monocytogenes* isolates when exposed to different temperatures and variability on adaptation behavior to different temperatures (Lianou and Koutsoumanis, 2013; Cordero et al., 2016). Therefore, when attempting to frame a generalized cold stress response by *L. monocytogenes*, strain variation should certainly be considered as well as the culture media the isolates are being grown in for physiological and genetic characterization (Jarvis et al., 2016b).

The mechanisms used by *L. monocytogenes* for growth at low temperatures have been extensively studied as well as many of the molecular adaptations generated by the bacterial cells described (Lado and Yousef, 2007). Most of the mechanisms used by *L. monocytogenes* to grow at cold temperatures are similar to the mechanisms used by other psychrotrophic bacteria; however, some of the survival and growth strategies employed by *L. monocytogenes* at cold temperatures are unique to this bacterium. Adaptation mechanisms to low temperatures of *L. monocytogenes* involves several responses including maintenance of the cellular membrane fluidity, the stabilization of macromolecule structures such as ribosomes that lose stability under low temperatures, and the uptake or synthesis of compatible solutes among others (Zhu et al., 2005b; Tasara and Stephan, 2006; Lado and Yousef, 2007; Garmyn et al., 2012; Rantsiou et al., 2012; Durack et al., 2013; Cordero et al., 2016). Some of these adaptation mechanisms have also been identified in other psychrotrophic microorganisms.

In addition, *L. monocytogenes* is thought to have some mechanisms that are not yet understood. Liu et al. (2002) identified several genes upregulated after exposure to low temperatures having no similarity to previously identified gene sequences in other bacteria. Liu et al. (2002) proposed that these identified genes might in fact be considered unique to this particular species. The proteins encoded by these genes have not yet been identified and the likely functions of these proteins remain to be determined (Liu et al., 2002). Detailed examination of individual genes and designing approaches for identifying their functionality would be needed to sort out their role in cold stress response. For example, Liu et al. (2006) further examined a *pgpH* mutant generated by using a Tn*917* insertional gene inactivation approach (Tomich et al., 1979, 1980; Camilli et al., 1990) as described in their previous study (Zhu et al., 2005a). The inactivated *pgpH* gene encoded the protein PgpH, a component of the metal-dependent phosphohydrolase family, and as a result the mutant exhibited accumulated levels of highly phosphorylated guanosine nucleotide ((p)ppGpp) at greater concentrations compared with the levels detected in its wild-type counterpart. The slower growth and longer lag phase of the mutant at low temperature and restoration of normal growth response when the mutant gene was complemented with wild-type *pgpH* gene demonstrated that the *pgpH* gene may have a role in cold stress response. Based on these results, Liu et al. (2006) speculated that the PgpH system may modulate cellular (p)ppGpp levels and serve as mediator of cold sensing.

The adaptation of *L. monocytogenes* membranes to low temperatures is also achieved by altering the branching in the methyl end of the fatty acid from iso to anteiso and by shortening the length of fatty acid chains (Annous et al., 1997; Zhu et al., 2005b; Lado and Yousef, 2007). These changes result in an increase of anteiso-$C_{15:0}$ fatty acids in the bacterial cell membrane. Low amounts of straight-chain saturated fatty acids as well as increased amounts of anteiso-$C_{15:0}$ are critical to maintaining an appropriate degree of membrane fluidity at low temperatures (Annous et al., 1997; Edgcomb et al., 2000; Jones et al., 2002; Rogiers et al., 2017). Further evidence supporting these findings resulted from comparing *Bacillus subtilis* and *Staphylococcus aureus* with *L. monocytogenes*, both being closely related to *L. monocytogenes*. They (*S. aureus* and *B. subtilis*) possess large quantities of iso and anteiso fatty acids (de Mendoza and Cronan, 1983; Wilkinson, 1997); however, these organisms have substantially greater amounts of straight-chain fatty acids than *L. monocytogenes*. In addition, both organisms are more cold temperature sensitive than *L. monocytogenes*, indicating the importance of lowering the straight-chain saturated fatty acids for low temperature growth of this organism (Annous et al., 1997).

Stabilizing temperature-sensitive macromolecules at low temperatures is achieved mainly by means of proteases and protein chaperones. Several cold-shock proteins (Csps), cold-acclimation proteins (Caps), heat-shock proteins, and caseinolytic proteins (Clps) may act as peptide chaperones under several stressful conditions including cold stress adaptation (Gottesman and Maurizi, 1992). The caseinolytic proteins ClpP and ClpB act as proteases to degrade proteins damaged by cold stress and providing additional sources of amino acids for protein synthesis. The general stress response protein GroEL and proteins from the main cold shock family CspA are important chaperones for stabilizing proteins damaged or misfolded by cold stress (Tasara and Stephan, 2006).

## Cellular Metabolism During Cold Stress

The accumulation of certain osmolytes, mainly glycine betaine and carnitine, also contribute to the survival and growth of *L. monocytogenes* in low-temperature environments (Verheul et al., 1997; Ko et al., 1994). Osmolytes can stimulate the growth of *L. monocytogenes* at temperatures as low as 4°C. Osmolytes are taken up by *L. monocytogenes* through the activities of three protein transporters: BetL, Gbu, and OpuC. BetL and Gbu are the main transporters for glycine betaine, while OpuC is the main transporter for carnitine (Angelidis et al., 2002; Sleator et al., 1999; Wemekamp-Kamphuis et al., 2002b). Because of the relative abundance of carnitine in animal tissues, the uptake of carnitine by *L. monocytogenes* allows the bacteria to grow effectively both in foods of animal origin when stored at refrigerated temperatures and under osmotic stress during subsequent infection while in the small intestines (Sleator et al., 2001). Because carnitine is relatively abundant in foods of animal origin, it is believed to contribute more than other osmolytes to the growth of *L. monocytogenes* on contaminated foods kept at refrigerated temperatures.

Lado and Yousef (2007) summarized a list of genes from several research studies (Borezee et al., 2000; Liu et al., 2002; Brondsted et al., 2003) that were believed to be involved in listerial regulatory adaptive responses, namely, *rpoN*, *lhkA*, *yycJ*, *bglG*, *adaB*, and *psr*;

general microbial stress response genes, *groEL*, *clpP*, *clpB*, *flp*, and *trxB*; genes likely needed for amino acid metabolism at cold temperatures, *hisJ*, *trpG*, *cysS*, and *aroA*; and genes required for cell surface alterations, *fbp*, *psr*, and *flaA*, along with an oligopeptide permease (OppA) and a high-affinity potassium uptake system. How these all contribute and fit together in an overall cold stress response by *L. monocytogenes* at the cellular level is still a work in progress but is becoming better understood as methodologies that can be used for global analyses at the transcriptional and proteomics level have been developed and applied. For example, Cacace et al. (2010) used two-dimensional protein gel electrophoresis to demonstrate that *L. monocytogenes* grown at 4°C expressed increased levels of chaperones, folding catalysts, osmolyte and oligopeptide transporters, and bacterial stress proteins along with proteins associated with energy production pathways.

Cabrita et al. (2013) more specifically profiled the extracellular proteome of *L. monocytogenes* using a combination of sodium dodecyl sulfate–polyacrylamide gel electrophoresis, glycoprotein detection, and liquid chromatography–tandem mass spectrometry. Based on this approach, Cabrita et al. (2013) suggested that the cold induction of OppA may be related to the need for more efficient oligopeptide uptake at lower temperatures. They also noted that *L. monocytogenes* was motile and flagellated below 30°C. Cordero et al. (2016) isolated RNA from fast- and slow-growing *L. monocytogenes* to conduct microarray global transcriptomic analyses of cells grown at 8°C and integrating the results into a metabolic network. They observed induction of peptidoglycan synthesis pathways while repressing synthesis of polar- and pyruvate-based amino acid pathways which they noted serve as precursors for iso-branched-chain fatty acids that function to decrease membrane fluidity. They also observed induction of energy metabolism, which could serve as the energy source for flagella functioning and was consistent with the increased expression of flagella genes albeit less expressed in the faster growing *L. monocytogenes* strains.

## Potential for Cross-Protection and Cold Stress

In addition to the ability of *L. monocytogenes* to grow at low temperatures and the difficulties low temperature growth represents for food industries to maintain safe foods, low temperature exposure of *L. monocytogenes* may provide cross-protection and increase stress resistance of the bacterium. Wemekamp-Kamphuis et al. (2002a) showed Csps from the major CspA family were not only involved in cold stress adaptation, but also provided cross-protection against hydrostatic pressure or freezing stress. Future studies identifying additional cross-protections or increased resistance to stresses developed by cold-temperature exposed cells should provide a more complete understanding of *L. monocytogenes* persistence in foods. Determining the extent of cross-protection and degree of increased stress resistance should also provide information to the corresponding food industries as to the extent of processing and preserving measures needed to affectively eliminate *L. monocytogenes* from food products. In addition, prior exposure to low temperatures may "harden" *L. monocytogenes* cells to different stresses, resulting in *L. monocytogenes* cells more resilient to preservative stresses and/or host stresses. For example, Stephens et al. (1991) observed *L. monocytogenes* cells grown at low temperatures exhibiting increased survival during infection using a mouse model.

This concept of *L. monocytogenes* stress cross-protection response is supported by the genetic and functional analyses of the genes being identified with cold stress response. What is interesting about several of the genes identified by Liu et al. (2002) is that they are not only expressed in response to exposure to low temperatures but are also important for survival under other unrelated stresses. *trxB* expression was increased on exposure of cells to 10°C in the study by Liu et al. (2002). *trxB* encodes thioredoxin reductase and is used to alleviate oxidative stress by reducing the disulfide bonds of proteins in the cytoplasm of bacteria growing aerobically (Prinz et al., 1997). *flp* expression was also increased at low temperatures (Liu et al., 2002), and this gene is known to provide protection from oxidative stress in *Campylobacter jejuni* (Wai et al., 1996). Flp, the major Csp in *L. monocytogenes*, is a ferritin-like protein that alleviates oxidative stress associated with cold stress (Hebraud and Guzzo, 2000; Liu et al., 2002). The increased expression of *trxB* and *flp* at low temperatures indicate that cold temperatures impose oxidative stress on bacterial cells (Liu et al., 2002). Another example is the *L. monocytogenes rpoN* gene, which is expressed at 10°C and codes for the sigma factor $\sigma^{54}$ (Liu et al., 2002). In addition to cold stress acclimation, *rpoN* is known to play a role in resistance to the antibacterial peptide mesentericin Y105 (Robichon et al., 1997). Likewise, sigma factor $\sigma^{B}$ has been demonstrated to provide *L. monocytogenes* multi-stress resistance to acid stress, high hydrostatic pressure, and freezing (Wemekamp-Kamphuis et al., 2004).

However, multistress protection does not always hold true and can depend on the sequence at which the stressors are introduced. For example, acclimation to cold temperatures of 4°C followed by exposure to sublethal acid (lactic acid) did not induce acid adaptation and protect *L. monocytogenes* against lethal acid shock (Shen et al., 2014). It would be interesting to compare other organic acids to see if this is universally true for all acids or is inherent with a specific type of acid. An examination of multiple *L. monocytogenes* strains as well as other species such as *L. innocua* would also be of interest because strain and genetic differences in both acid and cold tolerance are known to exist in *L. monocytogenes* (Lianou and Koutsoumanis, 2013; Cordero et al., 2016; Hingston et al., 2017).

## Cold Stress and Virulence

Some reports have indicated that growth of *L. monocytogenes* at reduced temperatures increased the virulence of the organism (Durst, 1975; Wood and Woodbine, 1979; Stephens et al., 1991; Dons et al., 1992). Specifically, Czuprynski et al. (1989) showed that several strains of *L. monocytogenes*, when grown at 4°C significantly increased their resistance to killing by human neutrophils in vitro. Durst (1975) observed that weakly virulent strains of *L. monocytogenes* develop increased virulence in mice after storing cells at 4°C for several months. Additionally, Wood and Woodbine (1979) noted one *L. monocytogenes* strain to become increasingly virulent to chick embryos when grown at 4°C. Clearly the possibility exists that low temperature storage may increase the virulence of some strains of *L. monocytogenes*.

In addition to these findings, more recent studies have identified several genes playing roles in cold temperature growth and virulence of *L. monocytogenes*. Caseinolytic proteins (Clps) have both proteolytic and chaperone activities (Liu et al., 2002). Clps in *L. monocytogenes* are believed to degrade abnormal peptides produced at low temperatures and act as chaperones to adjust the structures of damaged proteins. In addition, Gaillot et al. (2000) demonstrated *clpP* to be necessary for survival in host macrophages. GroEL, which was identified by Liu et al. (2002) to be expressed at 10°C, has been shown to be needed for virulence, being expressed at elevated levels during infection (Gahan et al., 2001). However, it must be noted that a more recent study by Garner et al. (2006b) observed *L. monocytogenes* 10403S grown at 7°C to have reduced invasiveness in Caco-2 cells compared with *L. monocytogenes* grown at 37°C.

To date, the genes that are virulence determinants regulated by PrfA, which is not expressed at low temperatures, are likely not critical for low temperature growth. However, the observation that certain *L. monocytogenes* cells exposed to low temperatures develop increased virulence suggests that the bacterium possesses some genetic factors encoding proteins important for both low temperature exposure and virulence. Such genetic factors are yet to be identified if indeed they are present in *L. monocytogenes*. Further experiments utilizing the molecular techniques that have been more recently developed are needed to investigate the genome of *L. monocytogenes* in attempt to identify any genetic factors necessary for virulence and involved in low temperature growth.

## Genomic Approaches for Identifying Cold Stress Function

There have been numerous approaches used to further the understanding of the mechanisms employed by *L. monocytogenes*. Mutational analysis has identified several novel genes that have helped reveal the genetic mechanisms behind the adaptation of *L. monocytogenes* to cold temperatures (Liu et al., 2002, 2006; Zhu et al., 2005a) and the genetic mechanisms involved in pathogenesis of *L. monocytogenes* (Gailliard et al., 1991; Kim et al., 2004; Mandin et al., 2005). Transcriptional analysis using microarrays and other methods has brought further insight into gene regulation and other genetic mechanisms involved in low temperature growth (Liu et al., 2002; Cordero et al., 2016) and during growth while localized in the cytoplasm of a host cell (Joseph et al., 2006). While these contributions have significantly impacted the understanding of the cold temperature growth and adaptation of *L. monocytogenes*, the need to continue the investigation of this psychrotrophic pathogen remains critical, if novel methods of preventing its growth on contaminated, refrigerated foods are to be developed.

One approach that can be used to deepen the understanding of the cold temperature adaptation of *L. monocytogenes* is the use of whole genomic comparisons among many strains of *L. monocytogenes* as well as with other psychrotrophs and psychrophiles. The genomes of some *L. monocytogenes* strains and other common psychrotrophs and psychrophiles have not yet been completely sequenced. However, the introduction of whole genome next generation–based sequencing methods has enabled more total genomic sequences for bacteria to become cataloged on a regular basis (Metzker, 2010; Glenn, 2011; Quail et al., 2012). Knowing the entire genomic sequence will allow whole genomic comparisons among *Listeria* species and strains for tracking origins and sources of *Listeria*. Whole genome sequencing (WGS) has successfully been used to identify and characterize isolates from several *L. monocytogenes* foodborne listeriosis outbreaks (Law et al., 2015; Jackson et al., 2016). Comparison of *L. monocytogenes* from different environmental sources is also possible. Holch et al. (2013) used an Illumina Genome Analyzer II to identify a persistent *L. monocytogenes* in two different fish processing plants and demonstrate that they were nearly identical although they were isolated 6 years apart. Based on this, they suggested that perhaps *L. monocytogenes* subtypes with specific traits were being selected by these food processing environments, hence their WGS similarities.

Along these lines the employment of WGS can also be used to identify differences among isolates and serotypes that relate to their environmental prevalence and identify previously unknown genes. Fox et al. (2016) used WGS to demonstrate that a subgroup of Australian *L. monocytogenes* isolates possessed a highly conserved core genome with most of the genetic variability associated with the mobile genetic components. Based on phenotypic characterization and WGS, Hingston et al. (2017) demonstrated that minor genetic differences may influence stress responses among *L. monocytogenes* isolates that possess impaired or enhanced tolerances to cold, salt, acid, or desiccation. As more WGS work is done, it is anticipated that stress cross-protection genetic circuitry will become more elucidated and potentially offer targets for more focused intervention approaches that can effectively limit *L. monocytogenes* in environments where it is capable of survival for extended periods such as during cold storage.

## Conclusions

Given the importance of refrigeration to the food industry, understanding the ecology of foodborne pathogenic *L. monocytogenes* is a critical research area. It appears that *L. monocytogenes* is present in the environment of food processing plants and can contaminate food as it is being processed. Once food becomes contaminated and is subsequently stored at refrigeration temperatures for extended periods, it can retain viable *L. monocytogenes* that in turn become a foodborne illness threat. While RTE meat products have received most of the attention, fresh produce is also a food vector that can lead to foodborne illness in humans. Foodborne listeriosis, once contracted in susceptible individuals, can be lethal depending on the physiological status of the victim and there are certain demographic groups such as pregnant women and immunocompromised individuals who are particularly vulnerable.

Considerable research has been directed toward understanding the mechanisms that *L. monocytogenes* uses to tolerate cold temperatures and withstand cold shock. It appears that membrane composition along with other metabolic changes is engaged as *L. monocytogenes* shifts into a physiological state that allows it to survive cold temperatures. Further research utilizing molecular approaches such as WGS can be used to investigate the genome of *L. monocytogenes* in an attempt to identify any unknown genetic factors involved in low temperature growth and survival. While most WGS efforts on *L. monocytogenes* have been directed toward tracking and identifying listeriosis outbreak isolates and sources, the potential to use these approaches to probe the genome and the transcriptome to uncover global metabolic network responses to cold stress are intriguing. As these types of studies are conducted, it is anticipated that a better understanding of the cold shock mechanisms and their sensor and regulatory components can be achieved. In addition, as these molecular components are characterized and better understood, this will create opportunities to develop intervention strategies that more effectively target *L. monocytogenes* while it exists in a physiological state to survive cold temperatures. This in turn could lead to development of novel compounds and biological elements that attack vulnerable essential pathways utilized by *L. monocytogenes* under these conditions.

## References

Angelidis, A.S., Smith, L.T., Hoffman, L.M., Smith, G.M., 2002. Identification of OpuC as a chill-activated and osmotically activated carnitine transporter in *Listeria monocytogenes*. Applied and Environmental Microbiology 68, 2644–2650.

Annous, B., Becker, L.A., Bayles, D.O., Labeda, D.P., Wilkinson, B.J., 1997. Critical role of anteiso-C15:0 fatty acid in the growth of *L. monocytogenes* at low temperatures. Applied and Environmental Microbiology 63, 3887–3894.

Barbosa, W.B., Cabedo, L., Wederquist, H.J., Sofos, J.N., Schmidt, F.R., 1994. Growth variation among species and strains of *Listeria monocytogenes*. Journal of Food Protection 57 765–769, 775.

Barria, C., Malecki, M., Arraiano, C.M., 2013. Bacterial adaptation to cold. Microbiology 159, 2437–2443.

Borezee, E., Pellegrini, F., Berche, P., 2000. OppA of *Listeria monocytogenes*, an oligopeptide binding protein required for bacterial growth in low temperature and involved in intracellular survival. Infection and Immunity 68, 7069–7077.

Brondsted, L., Kallipolitis, B.H., Igmer, H., Knochel, S., 2003. KdpE and a putative RsBQ homologue contribute to growth of *Listeria monocytogenes* at high osmolarity and low temperature. FEMS Microbiology Letters 219, 233–239.

Buzby, J.C., Roberts, T., Lin, C.T.J., MacDonald, J.M., 1994. Bacterial Foodborne Disease: Medical Costs and Productivity Losses. Report from Economic Research Service/USDA. AER-741.

Cabanes, D., Sousa, S., Cebria, A., Lecuit, M., Garcia-del Portillo, F., Cossart, P., 2005. Gp96 is a receptor for a novel *Listeria monocytogenes* virulence factor, Vip, a surface protein. The EMBO Journal 24, 2827–2838.

Cabrita, P., Batista, S., Mchado, H., Moes, S., Jeno, P., Manadas, B., Trigo, M.J., Monterio, S., Ferreira, R.B., Brito, L., 2013. Comparative analysis of the exoproteomes of *Listeria monocytogenes* strains grown at low temperatures. Foodborne Pathogens and Disease 10, 428–434.

Cacace, G., Mazzeo, M.F., Sorrention, A., Spada, V., Malorni, A., Sicilano, R.A., 2010. Proteomics for the elucidation of cold adaptation mechanisms in *Listeria monocytogenes*. Journal of Proteomics 73, 2021–2030.

Camilli, A., Portnoy, D.A., Youngman, P., 1990. Insertional mutagenesis of *Listeria monocytogenes* with a novel Tn*917* derivative that allows direct cloning of DNA flanking transposon insertions. Journal of Bacteriology 172, 3738–3744.

Carpentier, B., Cerf, O., 2011. Review – persistence of *Listeria monocytogenes* isolates of food industry equipment and premises. International Journal of Food Microbiology 145, 1–8.

Cartwright, E.J., Jackson, K.A., Johnson, S.D., Graves, L.M., Silk, B.J., Mahon, B.E., 2013. Listeriosis outbreaks and associated food vehicles, United States, 1998-2008. Emerging Infectious Diseases 19, 1–9.

Cepek, K.L., Shaw, S.K., Parker, C.M., Russel, G.J., Morrow, J.S., Rimm, D.L., Brenner, M.B., 1994. Adhesion between epithelial cells and T lymphocytes mediated by E-cadherin and the $\alpha^E\beta_7$ integrin. Nature 372, 190–193.

Chan, Y.C., Wiedmann, M., 2009. Physiology and genetics of *Listeria monocytogenes* survival and growth at cold temperatures. Critical Reviews in Food Science and Nutrition 49, 237–253.

Chan, Y.C., Raengpradub, S., Boor, K.J., Wiedmann, M., 2007. Microarray-based characterization of the *Listeria monocytogenes* cold regulon in log- and stationary phase cells. Applied and Environmental Microbiology 73, 6484–6498.

Cordero, N., Maza, F., Navea-Perez, H., Aravena, A., Marquez-Fontt, B., Navarrete, P., Fiuero, G., Gonzãlez, M., Latorre, M., Reyes-Jara, A., 2016. Different transcriptional responses from slow and fast growth rate strains of *Listeria monocytogenes* adapted to low temperature. Frontiers in Microbiology 7 (229), 1–9. http://dx.doi.org/10.3389/fmicb.2016.00229.

Czuprynski, C., Brown, J.F., Roll, J.T., 1989. Growth at reduced temperature increases the virulence of *Listeria monocytogenes* for intravenously but not intragastrically inoculated mice. Microbial Pathogenesis 7, 213–223.

Donaldson, J.R., 2010. The gastrointestinal tract: a friend or foe to *Listeria monocytogenes*? Archives of Clinical Microbiology 1, 3. http://dx.doi.org/10.3823/202.

Donaldson, J.R., Hercik, K., Rai, A.N., Reddy, S., Lawrence, M.L., Nanduri, B., Edelmann, M., 2015. Chapter 8. *Listeria* and -omics approaches for understanding its biology. pp. 135–158. In: Ricke, S.C., Donaldson, J.R., Phillips, C.A. (Eds.), Food Safety: Emerging Issues, Technologies and Systems. Academic Press, Elsevier, Inc., Oxford, UK.

Dons, L., Rasmussen, O.F., Olsen, J.E., 1992. Cloning and characterization of a gene encoding flagellin of *Listeria monocytogenes*. Molecular Microbiology 6, 2919–2929.

Dramsi, S., Biswas, I., Maguin, E., Braun, L., Mastroeni, P., Cossart, P., 1995. Entry of *Listeria monocytogenes* into hepacytes requires expression of InlB, a surface protein of the internalin multigene family. Molecular Microbiology 16, 251–261.

Durack, J., Ross, T., Bowman, J.P., 2013. Characterisation of the transcriptomes of genetically diverse *Listeria monocytogenes* exposed to hyperosmotic and low temperature conditions reveal global stress adaptation mechanisms. PLoS One 8, e73603. http://dx.doi.org/10.1371/journal.pone.0073603.

Durst, J., 1975. The role of temperature factors in the epidemiology of listeriosis. Zentralblatt für Bakteriologie, Parasitenkunde, Infektionskrankheiten und Hygiene. Erste Abteilung Originale. Reihe A: Medizinische Mikrobiologie und Parasitologie 233, 72–74.

Dussurget, O., Pizarro-Cerda, J., Cossart, P., 2004. Molecular determinants of *Listeria monocytogenes* virulence. Annual Review of Microbiology 58, 587–610.

Edgcomb, M.R., Sirimanne, S., Wilkinson, B.J., Drouin, P., Morse II, R.P.D., 2000. Electron paramagnetic resonance studies of the membrane fluidity of the foodborne pathogenic psychrotroph *Listeria monocytogenes*. Biochimica et Biophysica Acta 1463, 31–42.

Ferreira, V., Wiedmann, M., Teixeira, P., Stasiewicz, M.J., 2014. *Listeria monocytogenes* persistence in food-associated environments: epidemiology, strain characteristics, and implications for public health. Journal of Food Protection 77, 150–170.

Fox, E.M., Allnutt, T., Bradbury, M.J., Fanning, S., Chandry, P.S., 2016. Comparative genomics of the *Listeria monocytogenes* ST204 subgroup. Frontiers in Microbiology 7 (2057), 1–12. http://dx.doi.org/10.3389/fmicb.2016.02057.

Francis, G.A., O'Beirne, D., 2001. Effects of vegetable type. Package atmosphere and storage temperature on growth and survival of *Escherichia coli* O157:H7 and *Listeria monocytogenes*. Journal of Industrial Microbiology and Biotechnology 27, 111–116.

Frye, C., Donnelly, C.W., 2005. Comprehensive survey of pasteurized fluid milk produced in the United States reveals a low prevalence of *Listeria monocytogenes*. Journal of Food Protection 68, 973–979.

Fuchs, T.M., Eisenreich, W., Kern, T., Dandekar, T., 2012. Toward a systemic understanding of *Listeria monocytogenes* metabolism during infection. Frontiers in Microbiology 3 (23), 1–12 DOI:10.3389fmicrob.2012 00023.

Gahan, C.G., O'Mahony, J., Hill, C., 2001. Characterization of the groESL operon in *Listeria monocytogenes*: utilization of two reporter systems (*gfp* and *hly*) for evaluating in vivo expression. Infection and Immunity 69, 3924–3932.

Gailliard, J.L., Berche, P., Frehel, C., Gouin, E., Cossart, P., 1991. Entry of *Listeria monocytogenes* into cells is mediated by internalin, a repeat protein reminiscent of surface antigens from Gram-positive cocci. Cell 65, 1127–1141.

Gaillot, O., Pellegrini, E., Bregenholt, S., Nair, S., Berche, P., 2000. The ClpP serine protease is essential for the intracellular parasitism and virulence of *Listeria monocytogenes*. Molecular Microbiology 35, 1286–1294.

Gambarin, P., Magnabosco, C., Losio, M.N., Pavoni, E., Gattuso, A., Arcangeli, G., Favretti, M., 2012. *Listeria monocytogenes* in ready-to-eat seafood and potential hazards for consumers. International Journal of Microbiology 212, 1–10. http://dx.doi.org/10.1155/2012/497635. Article ID:497635.

Garmyn, D., Augagneur, Y., Gal, L., Vivant, A.L., Piveteau, P., 2012. *Listeria monocytogenes* differential transcriptome analysis reveals temperature dependent Agr regulation and suggests overlaps with other regulons. PLoS One 7, e43154. http://dx.doi.org/10.1371/journal.pone.0043154.

Garner, D., Kathariou, S., 2016. Fresh produce-associated listeriosis outbreaks, sources of concern, teachable moments, and insights. Journal of Food Protection 79, 337–344.

Garner, M.R., Njaa, B.L., Wiedmann, M., Boor, K.J., 2006a. Sigma B contributes to *Listeria monocytogenes* gastrointestinal infection but not to systemic spread in the Guinea pig infection model. Infection and Immunity 74, 876–886.

Garner, M.R., James, K.E., Callahan, M.C., Wiedmann, M., Boor, K.J., 2006b. Exposure to salt and organic acids increases the ability of *Listeria monocytogenes* to invade Caco-2 cells but decreases its ability to survive gastric stress. Applied and Environmental Microbiology 72, 5384–5395.

Gellin, B.G., Broome, C.V., 1989. Listeriosis. JAMA 261, 1313–1320.

Glenn, T.C., 2011. Field guide to next-generation sequencing. Molecular Ecology Resources 11, 759–769.

Gottesman, S., Maurizi, M.R., 1992. Regulation by proteolysis: energy-dependent proteases and their targets. Microbiological Reviews 56, 592–621.

Graves, L.M., Swaminathan, B., Hunter, S., 2007. Subtyping *Listeria monocytogenes*, Chapter 9, p. 283–304. In: Ryser, E.T., Marth, E.H. (Eds.), *Listeria*, Listeriosis, and Food Safety, third ed. CRC Press, Boca Raton, FL.

Gray, M.J., Freitag, N.E., Boor, K.J., 2006. How the bacterial pathogen *Listeria monocytogenes* mediates the switch from environmental Dr. Jekyll to pathogenic Mr. Hyde. Infection and Immunity 74, 2505–2512.

Guildimann, C., Guariglia-Oropeza, V., Harrand, S., Kent, D., Boor, K.J., Wiedmann, M., 2017. Stochastic and differential activation of $\sigma^B$ and PrfA in *Listeria monocytogenes* at the single cell level under different environmental stress conditions. Frontiers in Microbiology 8 (348), 1–14. http://dx.doi.org/10.3389/fmmicb.2017.00348.

Hanning, I., Johnson, M.G., Ricke, S.C., 2008. Pre-cut lettuce pre-packaged lettuce: a risk for listeriosis? Foodborne Pathogens and Disease 5, 731–746.

Hebraud, M., Guzzo, J., 2000. The main cold shock protein of *Listeria monocytogenes* belongs to the family of ferritin-like proteins. FEMS Microbiology Letters 190, 29–34.

Hill, C., Gahan, C., 2000. *Listeria monocytogenes*: role of stress in virulence and survival in food. Irish Journal of Agricultural and Food Research 39, 195–201.

Hingston, P., Chan, J., Dhillon, B.K., Laing, C., Bertelli, C., Hansen, L.T., Wang, S., 2017. Genotypes associated with *Listeria monocytogenes* isolates displaying impaired or enhanced tolerances to cold, salt, acod, or dessication stress. Frontiers in Microbiology 8 (369), 1–20. http://dx.doi.org/10.3389/fmmicb.2017.00369.

Hoelzer, K., Sauders, B.D., Sanchez, M.D., Olsen, P.T., Pickett, M.M., Mangione, K.J., Rice, D.H., Corby, J., Stich, S., Fortes, E.D., Roof, S.E., Gröhn, Y.T., Wiedmann, M., Oliver, H.F., 2011. Prevalence, distribution, and diversity of *Listeria monocytogenes* in retail environments, focusing on small establishments and establishments with a history of failed inspections. Journal of Food Protection 74, 1083–1095.

Hoelzer, K., Pouillot, R., Dennis, S., 2012. *Listeria monocytogenes* growth dynamics on produce: a review of the available data for predictive modeling. Foodborne Pathogens and Disease 9, 661–673.

Holch, A., Webb, K., Lukancenko, O., Ussery, D., Rosenthal, B.M., Gram, L., 2013. Genome sequencing identifies two nearly unchanged strains of persistent *Listeria monocytogenes* isolated at two different fish processing plants sampled 6 years apart. Applied and Environmental Microbiology 79, 2944–2951.

Hudson, J.A., Mott, S.J., Penney, N., 1994. Growth of *Listeria monocytogenes*, *Aeromonas hydrophila*, and *Yersinia enterocolitica* on vacuum and saturated carbon dioxide controlled atmosphere-packaged sliced roast beef. Journal of Food Protection 57, 204–208.

Jackson, K.A., Stroika, S., Katz, L.S., Beal, J., Brandt, E., Nadon, C., Reimer, A., Major, B., Conrad, A., Tarr, C., Jackson, B.R., Mody, R.K., 2016. Use of whole genome sequencing and patient interviews to link a case of sporadic listeriosis to consumption of prepackage lettuce. Journal of Food Protection 79, 806–809.

Jarvis, N.A., O'Bryan, C.A., Dawoud, T.M., Park, S.H., Kwon, Y.M., Crandall, P.G., Ricke, S.C., 2016a. An overview of *Salmonella* thermal destruction during food processing and preparation. Food Control 68, 280–290.

Jarvis, N.A., O'Bryan, C.A., Ricke, S.C., Johnson, M.G., Crandall, P.G., 2016b. A review of minimal and defined media for growth of *Listeria monocytogenes*. Food Control 66, 256–269.

Johansson, J., Mandin, P., Renzoni, A., Chiaruttini, C., Springer, M., Cossart, P., 2002. An RNA thermosensor controls expression of virulence genes in *Listeria monocytogenes*. Cell 110, 551–561.

Jones, S.L., Drouin, P., Wilkinson, B.J., Morse II, P.D., 2002. Correlation of long-range membrane order with temperature-dependent growth characteristics of parent and a cold-sensitive, branched-chain-fatty acid deficient mutant of *Listeria monocytogenes*. Archives of Microbiology 177, 217–222.

Joseph, B., Przybilla, K., Stuhler, C., Schauer, K., Slaghuis, J., Fuchs, T., Goebel, W., 2006. Identification of *Listeria monocytogenes* genes contributing to intracellular replication by expression profiling and mutant screening. Journal of Bacteriology 188, 556–568.

Kathariou, S., 2002. *Listeria monocytogenes* virulence and pathogenicity, a food safety perspective. Journal of Food Protection 65, 1811–1829.

Kim, H., Boor, K.J., Marquis, H., 2004. *Listeria monocytogenes* $\sigma^B$ contributes to invasion of human intestinal epithelial cells. Infection and Immunity 72, 7374–7378.

Kim, H., Marquis, H., Boor, K.J., 2005. $\sigma^B$ contributes to *Listeria monocytogenes* invasion by controlling expression of *inlA* and *inlB*. Microbiology 151, 3215–3222.

Ko, R., Smith, L.T., Smith, G.M., 1994. Glycine betaine confers enhanced osmotolerance and cryotolerance on *Listeria monocytogenes*. Journal of Bacteriology 176, 426–431.

Kohler, S., Leimeister-Wachter, M., Chakraborty, T., Lottspeich, F., Goebel, W., 1990. The gene coding for protein p60 of *Listeria monocytogenes* and its use as a species specific probe for *Listeria monocytogenes*. Infection and Immunity 58, 1943–1950.

Kuhn, M., Goebel, W., 1989. Identification of an extracellular protein of *Listeria monocytogenes* possibly involved in the intracellular uptake by mammalian cells. Infection and Immunity 57, 55–61.

Kuhn, M., Goebel, W., 2007. Molecular virulence determinants of *Listeria monocytogenes*, Chapter 5. In: Ryser, E.T., Marth, E.H. (Eds.), *Listeria*, Listeriosis, and Food Safety, third ed. CRC Press, Boca Raton, FL, pp. 111–155.

Lado, B.H., Yousef, A.E., 2007. Characteristics of *Listeria monocytogenes* important to food processors, Chapter 6. In: Ryser, E.T., Marth, E.H. (Eds.), *Listeria*, Listeriosis, and Food Safety, third ed. CRC Press, Boca Raton, FL, pp. 157–213.

Latorre, A.A., Prahan, A.K., Van Kessel, J.A., Karns, J.S., Boor, K.J., Rice, D.H., Mangione, K.J., Gröhn, Y.T., Schukken, Y.H., 2011. Quantitative risk assessment of listeriosis due to consumption of raw milk. Journal of Food Protection 74, 1268–1281.

Law, J.W.-F., Mutalib, N.-S.A., Chan, K.-G., Lee, L.-H., 2015. An insight into the isolation, enumeration, and molecular detection of *Listeria monocytogenes* in food. Frontiers in Microbiology 6 (1227), 1–15. http://dx.doi.org/10.3389/fmicb.2015.01227.

Lejune, J.T., Rajala-Schultz, P.J., 2009. Unpasteurized milk: a public health risk. Clinical Infectious Diseases 48, 93–100.

Lianou, A., Koutsoumanis, K.P., 2013. Strain variability of the behavior of foodborne bacterial pathogens: a review. International Journal of Food Microbiology 167, 310–321.

Lianou, A., Sofos, J.N., 2007. A review of the incidence and transmission of *Listeria monocytogenes* in ready-to-eat products in retail and food service environments. Journal of Food Protection 70, 2172–2198.

Liu, S., Graham, J.E., Bigelow, L., Morse II, P.D., Wilkinson, B.J., 2002. Identification of *Listeria monocytogenes* genes expressed in response to growth at low temperature. Applied and Environmental Microbiology 68, 1697–1705.

Liu, S., Bayles, D.O., Mason, T.M., Wilkinson, B.J., 2006. A cold-sensitive *Listeria monocytogenes* mutant has a transposon insertion in a gene encoding a putative membrane protein and shows altered (p)ppGpp levels. Applied and Environmental Microbiology 72, 3955–3959.

Lou, Y., Yousef, A.E., 1996. Resistance of *Listeria monocytogenes* to heat after adaptation to environmental stresses. Journal of Food Protection 59, 465–471.

Lou, Y., Yousef, A.E., 1997. Adaptation to sublethal environmental stresses protects *Listeria monocytogenes* against lethal preservation factors. Applied and Environmental Microbiology 63, 1252–1255.

Lungu, B., Ricke, S.C., Johnson, M.G., 2009. Growth, survival, proliferation and pathogenesis of *L. monocytogenes* under low oxygen or anaerobic conditions: a review. Anaerobe 15, 7–17.

Mandin, P., Fsihi, H., Dussurget, O., Vergassola, M., Milohanic, E., Toledo-Arana, A., Lasa, I., Johansson, J., Cossart, P., 2005. VirR, a response regulator critical for *Listeria monocytogenes* virulence. Molecular Microbiology 57 (5), 1367–1380.

Martinez, M.R., Osborne, J., Jayeola, V.O., Katic, V., Kathariou, S., 2016. Capacity of *Listeria monocytogenes* strains from the 2011 cantaloupe outbreak to adhere, survive, and grow on cantaloupe. Journal of Food Protection 79, 757–763.

Mead, P.S., Slutsker, L., Dietz, V., McCaig, L.F., Bresee, J., Shapiro, C., Griffin, P.M., Tauxe, R.V., 1999. Food-related illness and death in the United States. Emerging Infectious Diseases 5, 607–625.

Mendonça, K.S., Michael, G.B., Nalério, E.S., Mendonça, M., Cardosa, M.R.I., de Silva, W.P., 2016. Genotypic profile of *Listeria monocytogenes* isolated in refrigerated chickens in southern Rio Grande do Sul, Brazil. Ciência Rural Santa Maria 46, 132–137.

de Mendoza, D., Cronan Jr., J.E., 1983. Thermal regulation of membrane lipid fluidity in bacteria. Trends in Biochemical Sciences 8, 49–52.

Mengaud, J., Lecuit, M., Lebrun, M., Nato, F., Mazie, J.-C., Cossart, P., 1996a. Antibodies to the leucine-rich repeat region of internalin block entry of *Listeria monocytogenes* into cells expression E-cadherin. Infection and Immunity 64, 5430–5433.

Mengaud, J., Ohayon, H., Gounon, P., Mege, R.-M., Cossart, P., 1996b. E-cadherin is the receptor for internalin, a surface protein required for entry of *Listeria monocytogenes* into epithelial cells. Cell 84, 923–932.

Metzker, M.L., 2010. Sequencing technologies – the next generation. Nature Reviews 11, 31–46.

Milillo, S.R., Friedly, E.C., Saldivar, J.C., Muthaiyan, A., O'Bryan, C.A., Crandall, P.G., Johnson, M.G., Ricke, S.C., 2012a. A review of the ecology, genomics and stress response of *Listeria innocua* and *Listeria monocytogenes*. Critical Reviews in Food Science and Nutrition 52, 712–725.

Milillo, S.R., Stout, J.C., Hanning, I.B., Clement, A., Fortes, E.D., den Bakker, H.C., Wiedemann, M., Ricke, S.C., 2012b. Characteristics of *Listeria* from pasture-reared poultry reveals *L. monocytogenes* and hemolytic *L. innocua* in poultry ceca and the pasture environment. Poultry Science 91, 2158–2163.

Mueller, K.J., Freitag, N.E., 2005. Pleiotropic enhancement of bacterial pathogenesis resulting from the constitutive activation of the *Listeria monocytogenes* regulatory factor PrfA. Infection and Immunity 73, 1917–1926.

Nelson, K.E., Fouts, D.E., Mongodin, E.F., Ravel, J., DeBoy, R.T., Kolonay, J.F., Rasko, D.A., Angioli, S.V., Gill, S.R., Paulsen, I.T., Peterson, J., White, O., Nelson, W.C., Nierman, W., Beanan, M.J., Brinkac, L.M., Daughtery, S.C., Dodson, R.J., Durkin, A.S., Madupu, R., Haft, D.H., Selengut, J., Van Aken, S., Khouri, H., Federova, N., Forberger, H., Tan, B., Kothariou, S., Wonderling, L.D., Uhlich, G.A., Bayles, D.O., Luchansky, J.B., Frase, C.M., 2004. Whole genome comparisons of serotype 4b and 1/2a strains of the food-borne pathogen *Listeria monocytogenes* reveal new insights into the core genome components of this species. Nucleic Acids Research 32, 2386–2395.

NicAogáin, K., O'Bryne, C.P., 2016. The role of stress and stress adaptations in determining the fate of the bacterial pathogen *Listeria monocytogenes* in the food chain. Frontiers in Microbiology 7 (1865), 1–16. http://dx.doi.org/10.3389/fmicb. 2016. 01865.

Painter, J., Slutsker, L., 2007. Listeriosis in humans, Chapter 4, p85–109. In: Ryser, E.T., Marth, E.H. (Eds.), *Listeria*, Listeriosis, and Food Safety, third ed. CRC Press, Boca Raton, FL.

Pittman, J.R., Buntyn, J.O., Posadas, G., Nanduri, B., Pendarvis, K., Donaldson, J.R., 2014. Proteomic analysis of cross protection provided between cold and osmotic stress in *Listeria monocytogenes*. Journal of Proteome Research 13, 1896–1904.

Pradhan, A.K., Ivanek, R., Gröhn, Y.T., Bukowski, R., Geornaras, I., Sofos, J., Wiedmann, M., 2010. Quantitative risk assessment of listeriosis-associated deaths due to *Listeria monocytogenes* contamination of deli meats originating from manufacture and retail. Journal of Food Protection 73, 620–630.

Prinz, W.A., Aslund, F., Holmgren, A., Beckwith, J., 1997. The role of the thioredoxin and glutaredoxin pathways in reducing protein disulfide bonds in the *Escherichia coli* cytoplasm. Journal of Biological Chemistry 272, 15661–15667.

Quail, M.A., Smith, M., Coupland, P., Otto, T.D., Harris, T.D., Connor, T.R., Bertoni, A., Swerdlow, H.P., Gu, Y., 2012. A tale of three next generation sequencing platforms: comparison of Ion Torrent, Pacific Biosciences and Illumina MiSeq sequencers. BMC Genomics 12, 341.

Rantsiou, K., Greppi, A., Garosi, M., Acquadro, A., Mataragas, M., Cocolin, L., 2012. Strain dependent expression of stress response and virulence genes of *Listeria monocytogenes* in meat juices as determined by microarray. International Journal of Food Microbiology 152, 116–122.

Rauch, M., Luo, Q., Muller-Altrock, S., Goebel, W., 2005. SigB-dependent in vitro transcription of *prfA* and some newly identified genes of *Listeria monocytogenes* whose expression is affected by PrfA in vivo. Journal of Bacteriology 187, 800–804.

Ricke, S.C., 2003. Perspectives on the use of organic acids and short chain fatty acids as antimicrobials. Poultry Science 82, 632–639.

Ricke, S.C., Kundinger, M.M., Miller, D.R., Keeton, J.T., 2005. Alternatives to antibiotics: chemical and physical antimicrobial interventions and foodborne pathogen response. Poultry Science 84, 667–675.

Robichon, D., Gouin, E., Debarbouille, M., Cossart, P., Cenatiempo, Y., Hechard, Y., 1997. The rpoN ($\sigma^{54}$) gene from *Listeria monocytogenes* is involved in resistance to mesentericin Y105, an antibacterial peptide from *Leuconostoc mesenteroides*. Journal of Bacteriology 179, 7591–7594.

Rodrigues, C.S., de Sá, C.V.G., de Melo, C.B., 2017. An overview of *Listeria monocytogenes* in ready to eat meat, dairy, and fishery foods. Ciência Rural Santa Maria 47, 1–7.

Rogiers, G., Kebede, B.T., Loey, A.V., Michiels, C.W., 2017. Membrane fatty acid composition as a determinant of *Listeria monocytogenes* sensitivity to trans-cinnamaldehyde. Research in Microbiology 168, 536–546. http://dx.doi.org/10.1016/j.resmic.2017.03.001.

Rothrock Jr., M.J., Davis, M.L., Locatelli, A., Bodie, A., McIntosh, T.G., Ricke, S.C., 2017. *Listeria* occurrence in poultry flocks: detection and potential implications. Frontiers in Veterinary Science 4 (125), 1–7. http://dx.doi.org/10.3389/fvets.2017.00125.

Ryser, E.T., Buchanan, R.L., 2013. Chapter 20. *Listeria monocytogenes*. pp. 503–545. In: Doyle, M.P., Buchanan, R.L. (Eds.), Food Microbiology: Fundamantals and Frontiers, fourth ed. American Society for Microbiology Press, Washington D.C.

Sabet, C., Lecuit, M., Cabanes, D., Cossart, P., Bierne, H., 2005. LPXTG protein InlJ, a newly identified internalin involved in *Listeria monocytogenes* virulence. Infection and Immunity 73, 6912–6922.

Sauders, B.D., Weidmann, M., 2007. Ecology of *Listeria* species and *Listeria monocytogenes* in the natural environment, Chapter 2, pp 21–53. In: Ryser, E.T., Marth, E.H. (Eds.), *Listeria*, Listeriosis, and Food Safety, third ed. CRC Press, Boca Raton, FL.

Scallan, E., Hoekstra, R.M., Angulo, F.J., Tauxe, R.V., Widdowson, M.A., Jones, J.L., Griffin, P.M., 2011. Foodborne illness acquired in the United States—major pathogens. Emerging Infectious Diseases 17 (1), 7–15.

Schlech, W.F., Lavigne, P.M., Bortolussi, R.A., Allen, A.C., Haldane, E.V., Wort, A.J., Hightower, A.W., Johnson, S.E., King, S.H., Nicholls, E.S., Broome, C.V., 1983. Epidemic listeriosis—evidence for transmission by food. The New England Journal of Medicine 308, 203–206.

Seeliger, H.P.R., Hohne, K., 1979. Serotyping of *Listeria monocytogenes* and related species. Methods in Microbiology 13, 31–49.

Shen, Q., Soni, K.A., Nannapaneni, R., 2014. Influence of temperature on acid-stress adaptation in *Listeria monocytogenes*. Foodborne Pathogens and Disease 11, 43–49.

Sleator, R.D., Gahan, C.G.M., Abee, T., Hill, C., 1999. Identification and disruption of BetL, a secondary glycine betaine transport system linked to the salt tolerance of *Listeria monocytogenes* LO28. Applied and Environmental Microbiology 65, 2078–2083.

Sleator, R.D., Wouters, J., Gahan, C., Abee, T., Hill, C., 2001. Analysis of the role of OpuC, an osmolyte transport system, in salt tolerance and virulence potential of *Listeria monocytogenes*. Applied and Environmental Microbiology 67, 2692–2698.

Soni, K.A., Nannapeneni, R., Tasara, T., 2011. The contribution of transcriptomic and proteomic analysis in elucidating stress adaptation responses of *Listeria monocytogenes*. Foodborne Pathogens and Disease 8, 843–852.

Stasiewicz, M.J., Martin, N., Laue, S., Lane, S., Gröhn, Y.T., Boor, K.J., Wiedmann, M., 2014. Responding to bioterror concerns by increasing milk pasteurization temperature would increase estimated annual deaths from listeriosis. Journal of Food Protection 27, 696–705.

Stephens, J.C., Roberts, I.S., Jones, D., Andrew, P.W., 1991. Effect of growth temperature on virulence of strains of *Listeria monocytogenes* in the mouse: evidence for a dose dependence. Journal of Applied Bacteriology 70, 239–244.

Swaminathan, B., Cabanes, D., Zhang, W., Cossart, P., 2007. *Listeria monocytogenes*, Chapter 21, pp. 457–491. In: Doyle, M.P., Beuchat, L.R. (Eds.), Food Microbiology: Fundamentals and Frontiers, third ed. American Society for Microbiology Press, Washington D.C.

Tasara, T., Stephan, R., 2006. Cold stress tolerance of *Listeria monocytogenes*: a review of molecular adaptive mechanisms and food safety implications. Journal of Food Protection 69, 1473–1484.

Temm-Grove, C.T., Jokusch, B., Rohde, M., Niebuhr, K., Chakraborty, T., Wehland, J., 1994. Exploitation of microfilament proteins by *Listeria monocytogenes*: microvillus-like composition of the comet tails and vectorial spreading in polarized epithelial sheets. Journal of Cell Science 107, 2951–2960.

Toledo-Arana, A., Dussurget, O., Nikitas, G., Sesto, N., Guet-Revillet, H., Balestrino, D., Loh, E., Gripenland, J., Tiensuu, T., Vaitkevicius, K., Barthelemy, M., Vergassola, M., Nahori, M.-A., Soubigou, G., Régnault, B., Coppée, J.-Y., Lecuit, M., Johansson, J., Cossart, P., 2009. The *Listeria* transcriptional landscape from saprophytism to virulence. Nature 459, 950–956.

Tomich, P., An, F., Clewell, D.B., 1979. A transposon (Tn*917*) in *Streptococcus faecalis* that exhibits enhanced transposition during induction of drug resistance. Cold Spring Harbor Symposia on Quantitative Biology 43, 1217–1221.

Tomich, P.K., An, F.Y., Clewell, D.B., 1980. Properties of erythromycin-inducible Tn*917* in *Streptococcus faecalis*. Journal of Bacteriology 141, 1366–1374.

U.S. Department of Agriculture, Food Safety and Inspection Service, 2003. Control of *Listeria monocytogenes* in ready-to-eat meat and poultry products: final rule. Federal Register 68, 34208–34254.

Verheul, A., Glaasker, E., Poolman, B., Abee, T., 1997. Betaine and L-carnitine transport by *Listeria monocytogenes* ScottA in response to osmotic signals. Journal of Bacteriology 179, 6979–6985.

Wai, S.N., Nakayama, K., Umene, K., Moriya, T., Amako, K., 1996. Construction of a ferritin-deficient mutant of *Campylobacter jejuni*: contribution of ferritin to iron storage and protection against oxidative stress. Molecular Microbiology 20, 1127–1134.

Walker, S.J., Archer, P., Banks, J.G., 1990. Growth of *Listeria monocytogenes* at refrigeration temperatures. Journal of Applied Bacteriology 68, 157–162.

Wampler, J.L., Kim, K., Jaradat, Z., Bhunia, A.K., 2004. Heat shock protein 60 acts as a receptor for the *Listeria monocytogenes* adhesion protein in Caco-2 cells. Infection and Immunity 72, 931–936.

Wemekamp-Kamphuis, H.H., Karatzas, A.K., A Wouters, J., Abee, T., 2002a. Enhanced levels of cold shock proteins in *Listeria monocytogenes* LO28 upon exposure to low temperature and high hydrostatic pressure. Applied and Environmental Microbiology 68, 456–463.

Wemekamp-Kamphuis, H.H., Wouters, J.A., Sleator, R.D., Gahan, C.G., Hill, C., Abee, T., 2002b. Multiple deletions of the osmolyte transporters BetL, Gbu, and OpuC of *Listeria monocytogenes* affect virulence and growth at high osmolarity. Applied and Environmental Microbiology 68, 4710–4716.

Wemekamp-Kamphuis, H.H., Wouters, J.A., de Leeuw, P.P.L.A., Hain, T., Chakraborty, T., Abee, T., 2004. Identification of sigma factor $\sigma^B$ -controlled genes and their impact on acid stress, high hydrostatic pressure, and freeze survival in *Listeria monocytogenes* EGD-e. Applied and Environmental Microbiology 70, 3457–3466.

Wesley, I.V., 2007. Listeriosis in animals, Chapter 3, pp 55–84. In: Ryser, E.T., Marth, E.H. (Eds.), *Listeria*, Listeriosis, and Food Safety, third ed. CRC Press, Boca Raton, FL.

Wilkinson, B.J., 1997. The biology of staphylococci, pp. 1–38. In: Crossley, K.B., Archer, G.L. (Eds.), The Staphylococci in Human Disease. Churchill Livingstone, New York, NY.

Wood, L.V., Woodbine, M., 1979. Low temperature virulence of *Listeria monocytogenes* in the avian embryo. Zentralblatt für Bakteriologie, Parasitenkunde, Infektionskrankheiten und Hygiene. Erste Abteilung Originale. Reihe A: Medizinische Mikrobiologie und Parasitologie 243, 74–81.

Zhu, M., Du, M., Cordray, J., Ahn, D.U., 2005. Control of *Listeria monocytogenes* contamination in ready-to-eat meat products. Comprehensive Reviews in Food Science and Food Safety 4, 34–42.

Zhu, K., Bayles, D.O., Xiong, A., Jayaswal, R.K., Wilkinson, B.J., 2005a. Precursor and temperature modulation of fatty acid composition and growth of *Listeria monocytogenes* cold-sensitive mutants with transposon-interrupted branched-chain α-keto acid dehydrogenase. Microbiology 151, 615–623.

Zhu, K., Ding, X., Julotok, M., Wilkinson, B.J., 2005b. Exogenous isoleucine and fatty acid shortening ensure the high content of anteiso-$C_{15:0}$ fatty acid required for low temperature growth of *Listeria monocytogenes*. Applied and Environmental Microbiology 71, 8002–8007.

CHAPTER 14

# Use of Natural Antimicrobials of Plant Origin to Improve the Microbiological Safety of Foods

**Aubrey Mendonca[1], Armitra Jackson-Davis[2], Rkia Moutiq[1], Emalie Thomas-Popo[1]**

[1]*Iowa State University, Ames, IA, United States;* [2]*Alabama Agricultural and Mechanical University, Huntsville, AL, United States*

## Introduction

Increasing health consciousness of consumers and their preference for natural food additives are major contributory factors to food manufacturers' decisions to replace synthetic food additives with those from natural sources. Ensuring microbial food safety while satisfying consumer preference for natural food additives has fueled increased interest in the use of naturally derived antimicrobials. Natural antimicrobial compounds from plants including, but not limited to, cinnamon, thyme, rosemary, garlic, sage, oregano, basil, marjoram, savory, and clove are mainly in the form of essential oils (EOs) and their active components. The EOs are extracted from leaves (basil, oregano, thyme, marjoram, rosemary, and sage), flowers or buds (clove), seeds (caraway, fennel, parsley, and nutmeg), bulbs (garlic and onions), rhizomes (ginger), fruits (cardamom and pepper), bark (cinnamon), and other plant parts (Gutierrez et al., 2008b; Nychas et al., 2003). Extraction of EOs is facilitated by methods such as steam distillation and physical expression (cold pressing). The EOs are complex mixtures, which may contain about 20–80 individual components (ICs) each.

The major groups of ICs in EOs are terpenes, terpenoids, aldehydes, isoflavonoids, acids, and ketones (Burt, 2004; Dorman and Deans, 2000; Lopez-Malo et al., 2005). It is well established that EOs and several ICs have a broad spectrum of antimicrobial activity against foodborne pathogenic and spoilage microorganisms (Burt and Reinders, 2003; Chorianopoulos et al., 2008; Gutierrez et al., 2008a, 2009).

The antimicrobial activity of EOs is attributed to the chemical structure of their ICs, especially hydrophilic functional groups including the hydroxyl group of phenolics and/or hydrophobicity of several other ICs (Dorman and Deans, 2000). Typically, EOs with components that contain phenolic groups such as oils of thyme, sage, clove, oregano,

Food and Feed Safety Systems and Analysis. http://dx.doi.org/10.1016/B978-0-12-811835-1.00014-2

vanillin, and rosemary are very effective antimicrobials (Skandamis et al., 2002) that are generally more inhibitory to Gram-positive than Gram-negative bacteria (Chorianopoulos et al., 2004; Gutierrez et al., 2008a; Marino et al., 2001); however, some EOs such as cinnamon, clove, and citral are effective against both bacterial groups (Kim and Fung, 2004; Skandamis et al., 2002). In addition, some nonphenolic constituents of EOs such as allyl isothiocyanate (AIT), a main component of garlic oil or horseradish oil, exhibits a greater antimicrobial effect against Gram-negative bacteria (Ward et al., 1998) and fungi (Nielsen and Rios, 2000) with little or no effect on lactic acid producing bacteria (Muthukumarasamy et al., 2003).

Antimicrobials in plants are generally produced to protect plants from disease. Many of them are part of the plants' preinfection or postinfection defense systems against infectious or parasitic agents (Rauha et al., 2000). In this respect, edible plants that exhibit strong antimicrobial activity may be sources of antimicrobials for use against foodborne pathogens (Ibrahim et al., 2006). In addition, some antimicrobials are generated as a result of the plant's response to stress or from inactive precursors, which are activated by enzymes such as oxidases or hydrolases typically found in plant tissues (Holley and Patel, 2005). For example, vegetables such as horseradish, mustard, and wasabi from the *Brassicaceae* family contain glucosinolates that are inactive. When the plant tissue is damaged, the enzyme myrosinase is released and converts glucosinolates to thiocyanates including AIT, which is a potent antimicrobial (Delaquis and Mazza, 1995).

There is a rapidly growing list of published reports on the food use of natural antimicrobials of plant origin to control foodborne human enteric pathogens. As our knowledge on pathogen control by plant-derived antimicrobials increases, there is an urgent need for information on optimizing the effectiveness of these antimicrobials in foods. In this respect, the present chapter aims to address selected plant-derived antimicrobials (mainly EOs) with regard to (1) applications in selected food categories, (2) limitations of use in foods, (3) techniques for delivery in foods, (4) exploitation of synergistic actions, and (5) regulations for food applications.

## Food Applications for Pathogen Control

The growing interest in using plant EOs as antimicrobial agents in foods is fueled by the need to reduce the use of synthetic chemicals in food preservation. Accordingly, several EOs have been evaluated for controlling human enteric pathogens in certain foods namely fruit and vegetable juices, meat products, fish products, and dairy products.

### Fruit and Vegetable Juices

Based on FAO/WHO recommendations, a daily intake of five portions of fruit and vegetables is beneficial to human health (World Health Organization, 2003). Accordingly, consumption of these nutritious products has increased, as consumers become more aware of associated health benefits. Consequently, minimally processed fruit and vegetable products including fresh cut fruits and vegetables and raw juices are one of the fastest growing segments in terms of sales at food retail outlets. Raw fruit and vegetable juices are a rich source of vitamins, minerals, antioxidants, and dietary fiber; however, they can be contaminated with pathogenic microorganisms such as *Salmonella enterica* and pathogenic strains of *Escherichia coli*. To control these pathogens in juices, research efforts have focused on the use of natural antimicrobials as consumers reject the use of synthetic chemical food preservatives. In this respect, certain plant EOs have shown good potential to improve the microbial safety of a variety of juices (Table 14.1).

Friedman et al. (2004) evaluated the antimicrobial effects of EOs (0.00065%–0.67%) derived from several sources (apricot, orange, bergamot, clove, cinnamon, lime, grapefruit, lemon, lavender, lemon grass, Melissa, and oregano) against *E. coli* O157:H7 and *S. enterica* in clear and cloudy apple juice held at 4, 21, and 37°C. Antimicrobial efficacy of each EO was expressed as the EO concentration (%) capable of inactivating 50% of the bacterial population (BA50). The lowest BA50 values were evident when juices were held at higher temperatures (21 and 37°C), and antibacterial effects of the EOs were greater in clear juices compared with cloudy juices. Interestingly, thyme EO inactivated both pathogens in apple juice held at 4°C. In another study, thyme EO (0.5%) killed 0.71 log CFU/mL of *Listeria monocytogenes* in a mixed (apple–carrot) juice after 5 days at 4°C (Irkin and Korukluoglu, 2009).

The antibacterial effects of lemongrass, cinnamon, clove, and palmarose EOs at concentrations ranging from 2 to 10 μL/mL were evaluated against *Listeria innocua*, *Salmonella* Enteritidis , and *E. coli* in apple, pear, and melon juices (Raybaudi-Massilia et al., 2006). In pear and apple juices held at 35°C, lemongrass EO or cinnamon EO at 2 μL/mL inactivated >5.0 log CFU/mL of each pathogen after 24 h. Similar bactericidal effects were achieved in melon juices but only when higher concentrations such as 5 μL/mL (lemongrass EO) and 10 μL/mL (cinnamon EO) were added. The pH values for apple, pear, and melon juices were 4.2, 3.9, and 5.9, respectively. In this regard the low pH of apple and pear juices was a likely contributory factor to the bactericidal effects of lemongrass EO in those juices.

The addition of lemongrass EO to pineapple juice (4°C) at EO concentrations of 0.6, 1.2, 2.5, and 5 μL/mL drastically reduced populations of *L. monocytogenes*, *E. coli*, and *S.* Enteritidis. At the lowest EO concentration tested (0.6 μL/mL), ≥5.0 log reduction in *E. coli* and *L. monocytogenes* occurred after 1 h and 45 min, respectively. At that same EO concentration, similar reductions in *S.* Enteritidis were observed in pineapple juice after 12 h (Leite et al., 2016). The low pH of the pineapple juice (3.92) and the relatively high quantity of monoterpenes (77.9%) such as geraniol, citral, and neral in lemongrass EO are factors that likely

**Table 14.1: Control of human enteric pathogens in fruit juices by the use of essential oils.**

| Juice (pH) | Essential oil | Concentration (μL/mL) | Target pathogen | Storage temperature (°C), time | Effect: microbial reduction | References |
|---|---|---|---|---|---|---|
| Apple (3.7) | Apricot, orange, bergamot, clove, cinnamon, lime, grapefruit, lemon, lavender, lemon grass, Melissa, and tangerine | 0.0006 to 0.67 | *Escherichia coli* O157:H7 and *Salmonella enterica* | 37, 1 h | 50% of bacterial population | Friedman et al. (2004) |
| a) Apple (4.2) and pear (3.97)<br>b) Melon (5.91)<br>c) Pineapple (3.92) | Lemon grass oil | a) 0.2<br>b) 0.5<br>c) 2.5 or 5.0 | a, b) *Listeria innocua*, *Salmonella* Enteritidis, *E. coli*<br>c) *Listeria monocytogenes*, *E. coli*, and *S.* Enteritidis | a, b) 35, 24 h<br>c) 4, 0.25 h | a, b) > 5 log CFU/mL<br>c) ≥ 5 log CFU/mL | a, b) Raybaudi-Massilla et al. (2006)<br>c) Leite et al. (2016) |
| a) Apple (4.2) and pear (3.97)<br>b) Melon (5.91)<br>c) Strawberry (3.16), apple (4.46), orange (3.44), pear (4.40), and tomato (4.30)<br>d) Melon (6.11) and watermelon (5.73)<br>e) Carrot (6.2) | Cinnamon oil | a) 0.2<br>b) 0.8<br>c) 0.1 to 0.3<br>d) 0.3<br>e) 0.05 | a, b) *L. innocua*, *S.* Enteritidis, *E. coli*<br>c) *S.* Enteritidis, *E. coli* O157:H7<br>d) *S.* Enteritidis, *E. coli* O157:H7, *L. monocytogenes*<br>e) *Bacillus cereus* EPSO-35AS and *B. cereus* INRA TZ415 spores (2.0 log CFU/mL) | a, b) 35, 24 h<br>c) 22, 1 h<br>d) 22, 1 h<br>e) 8 and 12, 60 days | a, b, c) > 5 log CFU/mL<br>d) 3.1 to 3.9, 1.4 to 1.9, and 3.4 to 4.4 log CFU/mL of *S.* Enteritidis, *E. coli* O157:H7, and *L. monocytogenes*, respectively<br>e) Germination inhibited | a, b) Raybaudi-Massilla et al. (2006)<br>c) Mosqueda-Melgar et al. (2008a,b)<br>d) Mosqueda-Melgar et al. (2008c)<br>e) Hernandez-Herrero et al. (2008) |
| Apple/carrot (4.8) | Thyme oil | 0.5 | *L. monocytogenes* AUFE 39237 | 4, 5 days | 0.71 log CFU/mL | Irkin and Korukluoglu (2009) |
| Vegetable/fruit (3.9) | a) Clove oil<br>b) Cinnamon oil<br>c) Star anise oil | 0.1 | *E. coli* O157:H7 | 20, 7 days | 5 log CFU/mL from a) 24 h, b) 48 h, and c) 72 h | Pan et al. (2014) |
| Pineapple (3.77), cashew (4.27), guava (4.17), mango (5.68) | Peppermint oil | 10 | *E. coli*, *S.* Enteritidis, and *L. monocytogenes* | 4, 24 h | ≥5 log CFU/mL | De Sousa Guedes et al. (2016) |

contributed to the bactericidal effects of lemongrass EO against the bacteria tested (Leite et al., 2016). The increased sensitivity of bacteria to EOs as pH decreases is due to increased hydrophobicity of EOs at low pH that facilitates their interaction with lipids of the bacterial cell membrane (Burt, 2004).

### Meat Products

The antibacterial activity of several plant extracts including grapeseed extract, cranberry extract, and EOs has been tested against enteric pathogens in a variety of meat products. Ahn et al. (2007) investigated the effects of grapeseed extract (ActiVin), pine bark extract (Pycnogenol), and oleoresin rosemary (Herbalox) on growth of pathogens in cooked ground beef. In that study 1.0% ActiVin or Pycnogenol reduced initial populations of *E. coli* O157:H7 and *Salmonella* Typhimurium and slowed down growth of *L. monocytogenes* and *Aeromonas hydrophila*. Pycnogenol was most effective and reduced the viable count of *E. coli* O157:H7, *L. monocytogenes*, *S.* Typhimurium, and *A. hydrophila* by 1.7, 2.0, 0.8, and 0.4 log CFU/g, respectively, after 9 days of refrigerated storage. The behavior of *L. monocytogenes* in frankfurters formulated with 3% cranberry powder, or with 1% or 2% cranberry powder each with cherry powder (0.6%), lime powder (60 mg/kg), or a blend of cherry, lime, and vinegar (1.4%), was researched by Xi et al. (2012). Compared with control frankfurters, those with added cranberry powder at 3% significantly retarded growth of *L. monocytogenes* by 5.3 log CFU/g ($P < .05$). However, from a meat quality perspective, added cranberry at concentrations above 1% resulted in significant product pH decrease and negatively affected the color, texture, and sensory attributes of the frankfurters.

The effectiveness of EOs of thyme and balm to inhibit growth of *Salmonella* and *E. coli* in raw chicken slices stored aerobically at 4°C for 21 days was demonstrated by Fratianni et al. (2010). In that same study, thyme EO exhibited a marked growth inhibitory effect on *E. coli* compared with balm EO. After 21 days, the viable count *E. coli* in chicken samples surface-treated with balm and thyme were, respectively, about 0.30 and 1.46 log less than that of control. Interestingly, those authors reported that thyme EO was less effective against *Salmonella*. After 21 days, about 2.30 CFU of *Salmonella* were detected in thyme-treated chicken, whereas the pathogen was not detected in chicken treated with balm EO. The addition of thyme EO (0.6%) to ground beef resulted in growth inhibition of *E. coli* O157:H7 during storage of the beef at 10°C but not at 4°C. Burt and Reinders (2003) reported a similar growth inhibitory effect of thyme oil against *E. coli* O157:H7 in vitro. Additionally, those same authors observed that thyme oil at 0.12% and 0.25% caused a bacteriostatic and bactericidal effect, respectively, on *E. coli* O157:H7. In that study two major phenolic compounds of thyme and oregano oils, namely, carvacrol and thymol, were deemed responsible for their antibacterial effect on *E. coli* O157:H7 (Burt et al., 2005).

Govaris et al. (2010) evaluated the antibacterial effect of 0.6% or 0.9% oregano EO in ground sheep meat stored at 4°C or 10°C for 12 days. Based on results of that study, the viable count of the pathogen in meat treated with EO at 0.6% decreased by 2.43 log CFU/g on the second day of storage at 4°C and remained almost the same throughout 12 days. At 10°C, the viable count of *S.* Enteritidis in meat samples 0.6% EO decreased to 2.30 log CFU/g on the second day and remained at this level up to the sixth day before increasing to approximately 4.37 log CFU/g by the end of storage. Those same authors reported more effective antibacterial activity of 0.9% oregano EO in the ground sheep meat at both storage temperatures. Ismaiel and Pierson (1990) determined the antimicrobial effectiveness of oregano EO against *Clostridium botulinum* spores in vacuum packaged ground pork. Oregano EO (400 ppm) with sodium nitrite (50–100 ppm) significantly decreased spore germination and delayed the growth of *C. botulinum*; however, this effect was not observed when oregano EO was applied in the absence of nitrite.

EOs of coriander, clove, oregano, and thyme EOs were strongly inhibitory to *L. monocytogenes*, *A. hydrophila*, and indigenous microbiota in meat products at levels of 5–20 mL/g (Skandamis and Nychas, 2001; Tsigarida et al., 2000). Barbosa et al. (2009) evaluated minimum inhibitory concentration (MIC) of thyme, marjoram, oregano basil, ginger, lemongrass, and clove EOs against *E. coli*, *Staphylococcus aureus*, *L. monocytogenes*, and *S.* Enteritidis strains in vitro and in artificially inoculated ground meat. In vitro, all pathogens were sensitive to the EOs tested, with clove exhibiting the highest antimicrobial activity ($MIC_{90\%}=0.09\%$ v/v), followed by lemongrass ($MIC_{90\%}=0.25\%$ v/v) and thyme ($MIC_{90\%}=0.26\%$ v/v). However, the $MIC_{90\%}$ values observed in the ground meat were higher indicating decreased effectiveness of the EOs in that meat product. In addition, log reductions of 1.3 and 1.0 in numbers of viable pathogens were the highest obtained in ground meat. Menon and Garg (2001) examined the activity of clove oil against *L. monocytogenes* in fresh ground mutton held at 30°C and 7°C. At 0.5% and 1.0%, clove oil inhibited the growth of *L. monocytogenes* at both temperatures with 1% clove oil exhibiting the larger growth inhibitory effect. Viable counts of *L. monocytogenes* in treated meat samples were 1–3 $\log_{10}$CFU/g lower than those of controls at different times during storage. Busatta et al. (2008) demonstrated that addition of marjoram EO to fresh sausages at levels lower than the MIC (2.3 mg/g) for *E. coli* resulted in a bacteriostatic effect against that organism. It was suggested that use of marjoram EO in fresh sausages is a good approach to extend the shelf life of the product. Based on those results indicating growth cessation of *E. coli*, marjoram EO may also have good potential for improving the microbial safety of fresh sausage.

Basil EO has substantial antimicrobial effect against foodborne organisms including *Bacillus cereus*, *S. aureus*, *E. coli*, *Salmonella* spp., and *L. monocytogenes* (Suppakul et al., 2003). In addition, the antifungal efficacy of basil EO on the surface of an Italian-style sausage indicated the good potential of that EO for fungal control on sausages (Salarbashi et al., 2012). Gillet al. (2002) diluted cilantro EO in sterile canola oil or incorporated it into gelatin to obtain EO concentrations (v/v) of 0.1%, 0.5%, and 6% before applying those concentrations of the EO on the surface of ham slices artificially inoculated with a five-strain mixture of *L. monocytogenes*. The treated ham was then vacuum-packaged and held at 10°C for 4 weeks. The only treatment that retarded growth of the pathogen was 6% cilantro EO in gel. Numbers of viable *L. monocytogenes* in ham treated with 0.6% EO in gel were 1.3 log CFU/mL lower than controls after 1 week of storage; however, there was no difference among treatments from the second week to the end of storage. The authors suggested that immobilization of the antimicrobial in a gel might have enhanced the effect of cilantro EO treatment (6% v/v). Based on the overall results of that study, cilantro EO oil does not seem to be a suitable natural antimicrobial for the control of *L. monocytogenes* on ham (Gill et al., 2002).

### *Fish Products*

The antibacterial activity of oregano EO against *S.* Enteritidis has been demonstrated in vitro or in other foods such as traditional Greek salted fish roe appetizer called taramasalata (Koutsoumanis et al., 1999). Taramasalata, with added oregano EO (0.0, 0.5, 1.0, 2.0% v/w), was artificially inoculated with *S.* Enteritidis and stored at different temperatures (5, 10, 15, and 20°C). Loss of viability of *S.* Enteritidis was observed in all instances and its death rate of the pathogen was influenced by pH, temperature, and the EO concentration. Death responses as a function of those parameters were described using a quadratic function, which facilitated prediction of the death rate of *S.* Enteritidis in taramasalata of varying compositions. Frangos et al. (2010) investigated the effect of oregano EO (0.2%) on fresh rainbow trout fillets that were salted, packaged in a modified atmosphere (45% $CO_2$, 5% $O_2$, and 50% $N_2$), and refrigerated (4°C) for 21 days. During storage of the fish fillets, the population of lactic acid bacteria, $H_2S$-producing bacteria, Enterobacteriaceae, and *Pseudomonas* spp. significantly decreased. Because the family Enterobacteriaceae includes numerous genera that include pathogens such as *Salmonella*, *Yersinia*, *Shigella*, *Escherichia*, *Enterobacter*, and *Klebsiella*, substantial reduction in numbers of Enterobacteriaceae in fish by oregano EO suggests good potential of this EO for improving the microbiological safety of fish. The antibacterial effect of EOs against a four-strain mixture of *L. monocytogenes* on fresh raw catfish fillets was determined by Desai et al. (2012). Thyme and oregano EOs and carvacrol displayed both concentration and time-dependent inhibition of the pathogen, whereas lemon, orange, or tangerine EO exhibited a time-dependent mode of inhibition at 1% concentration. The antibacterial activity of thyme EO at 0.4%, 0.8%, and 1.2% was evaluated against *L. monocytogenes* in artificially inoculated ground fish stored aerobically at refrigeration temperature (4°C) for 12 days. The initial viable count of *L. monocytogenes* in the inoculated fish was 4.0 log CFU/g. Thyme EO at 0.8% and 1.2% reduced the viable count of the pathogen below 2 log CFU/g after 6 days (Abdollahzadeh et al., 2014).

Djenane (2015) reported in vitro MICs of citrus extracts ranging from 0.25 to 0.40 μL/mL for *S. aureus*. Based on those results, lemon and bergamot citrus peel EOs were added at concentrations of 1 × MIC and 4 × MIC to *Sardina pilchardus* artificially inoculated with *S. aureus* (3.5 $\log_{10}$CFU/g) and stored at 8 ± 1 °C. The 4 × MIC value of bergamot citrus peel EO completely inhibited the growth of *S. aureus* from day 2 until the end of storage. These results suggest that citrus EOs could potentially be used to improve the microbial safety of fish.

Lin et al. (2004) evaluated the antibacterial effect of oregano and cranberry at 75% and 25% (wt/wt), respectively, against *L. monocytogenes* in artificially inoculated slices of fish held at 37°C for 18 h and at 4°C for 8 days. The initial viable count of *L. monocytogenes* cells was ~$10^3$ CFU/g. The combined extracts at 75% oregano and 25% cranberry (wt/wt) produced an optimized phenolic content of 0.1 mg/mL. At pH 7.0 the fish samples treated with the extracts exhibited no significant difference in viable cell numbers compared with controls after 18 h. However, at pH 6.0, after pH adjustment of fish slices with lactic acid, the viable cell count of fish with the oregano–cranberry mixture was 2.3 log lower than control fish after 18 h. At 4°C complete inactivation of the pathogen occurred at 4 days in fish samples (pH 6.0) treated with the oregano–cranberry extract.

## Dairy Products

EOs (cinnamon, garlic, and clove), selected nanopowders (zinc oxide, silver–copper), and polyethylene (PE) glycol were incorporated in polylactide-based films via a solvent casting method. Films were evaluated for antibacterial activity against foodborne pathogens on experimentally inoculated cheese. Films with incorporated nanopowders alone had poor antibacterial activity, whereas EOs of cinnamon and clove incorporated into films displayed a significant antimicrobial activity against the pathogens; growth of *L. monocytogenes* and *S.* Typhimurium on cheese was inhibited for up to 11 days during storage at refrigeration temperature (Ahmed et al., 2017).

Selim (2011) evaluated the in vitro antibacterial activity of 11 EOs against vancomycin-resistant enterococci (VRE) and *E. coli* O157:H7. The most effective EOs were thyme, clove, eucalyptus, and juniper. The antibacterial activity of the EOs was also evaluated against those same organisms in experimentally inoculated ($10^3$ CFU/g) in Feta soft cheese stored at 7°C for 14 days. Clove and mint EOs were ineffective against VRE and *E. coli* O157:H7 in the cheese. In contrast, thyme oil at 0.5% and 1% produced a significant reduction in microbial growth. In the treated cheese, there was about a 1 $\log_{10}$CFU/g decrease in numbers of viable VRE compared with control cheese after storage at 7°C for 1, 2, and 3 days. However, on subsequent days, there was no significant difference in bacterial counts between the two groups. These results highlight the potential of thyme EO as a natural antimicrobial to enhance the microbial safety of Feta soft cheese.

Govaris et al. (2011) investigated the antibacterial activity of oregano EO and thyme EO added to Feta cheese at concentrations of 0.1 or 0.2 and 0.1 mL/100 g, respectively. The cheese was experimentally inoculated with *E. coli* O157:H7 or *L. monocytogenes* and stored in modified atmosphere packaging (50% $CO_2$ and 50% $N_2$) at 4°C. In control Feta cheese (without added EOs), *E. coli* O157:H7 and *L. monocytogenes* survived up to 32 and 28 days, respectively. In cheese treated with oregano EO (0.1 mL/100 g), *E. coli* O157:H7 and *L. monocytogenes* survived up to 22 and 18 days, respectively. At the higher concentration of oregano EO (0.2 mL/100 g), the pathogens survived up to 16 days (*E. coli* O157:H7) and 14 days (*L. monocytogenes*). Pathogen survival in Feta cheese with added thyme EO (0.1 mL/100 g) was not significantly different ($P > .05$) from that in cheese with added oregano EO (0.1 mL/100 g).

Pires et al. (2009) tested the effectiveness of antimicrobial sachets containing AIT against yeast and molds, *Staphylococcus* sp., and psychrotrophs in sliced mozzarella cheese stored at 12°C ± 2°C for 15 days. The AIT was added to a highly absorptive polymeric resin (Accurel XP200, HDPE, Accurel Systems International Corp., Sunnyvale, California, USA). The resin with AIT was inserted in nontissue sachets, which were heat-sealed and fixed inside bags of cheese via the use of double-face tape. Control cheese was packaged without the antimicrobial sachet. The packaged samples of cheese were stored at 12°C ± 2°C and analyzed at 0, 3, 6, 9, 12, and 15 days. The $\log_{10}$ reductions of fungi and *Staphylococcus* sp. in treated cheese were 3.6 (after 15 days) and 2.4 (after 12 days), respectively. Psychrotrophic bacteria were the most resistant to the antimicrobial treatment with no observed significant difference in microbial numbers at day 12 or day 15.

## Limitations in Food Applications

### Sensory Changes

The use of EOs as food preservatives has been limited by several factors. A major factor is change in sensory characteristics of foods due to relatively high EO concentrations needed to achieve microbial control. This is especially true when an EO is used as the sole hurdle. Most EOs have strong aroma and taste, and even at low concentrations they can have negative effects on the organoleptic properties of the food. Further lowering of the EO concentration to overcome undesirable sensory characteristics may in turn compromise the effectiveness of the EO antimicrobial activity. However, the use of low concentrations of EOs in combination with other compounds with antimicrobial properties may result in a synergistic effect without negative organoleptic effects (Nguefack et al., 2012). Encapsulation of the active compounds of plant antimicrobials instead of directly applying to the food (Fang and Bhandari, 2010) or encapsulation of EOs into nanoemulsions (Donsì et al., 2011) may be other ways of minimizing organoleptic effects. The EOs and their individual constituents may also be used in combination with other emerging preservation technologies such as high hydrostatic pressure, atmospheric cold plasma, ultrasound, and pulsed electric field to inhibit microbial growth while avoiding negative organoleptic effects.

### Miscibility Problems

Another major challenge experienced with the use of plant antimicrobials, particularly EOs in foods, is the issue of miscibility. The EOs are volatile, aromatic hydrophobic liquids obtained from various plant parts such as the leaves, flowers, fruits, and roots (Burt, 2004). Because they are oily, many of these EOs exhibit poor solubility in hydrophilic solutions (Samperio et al., 2010). Poor solubility results in the inability of the antimicrobial to be dispersed homogenously in the food resulting in it being unable to reach foodborne microorganisms. One approach to overcome this problem has been to use cyclodextrins. Cyclodextrins are compounds that are generally recognized as safe (GRAS) by the U.S. Food and Drug Administration (FDA) and form complexes with EOs resulting to aid in their dispersion (Samperio et al., 2010). The use of natural surfactants is another potential solution that needs to be further explored to exploit consumers' increasing demand for more natural food additives.

### Interactions With Food Matrix Components

For food applications, the antimicrobial efficacy of plant antimicrobials is highly dependent on their stability. Because of the complex nature and composition of food products, the effectiveness of EOs and their constituents is affected by interactions with food components. This poses a major challenge for the use of plant antimicrobials in foods. Intrinsic components of foods such as nutrients (starch, proteins, fats) can interact with plant antimicrobials and decrease their effectiveness. Plant antimicrobials have been reported to interact with fat (Cava-Roda et al., 2012) and a high fat content in foods can decrease the antimicrobial activity of phenolic compounds such as carvacrol and cymene (Rattanachaikunsopon and Phumkhachorn, 2010). The efficacy of EOs can also be altered by their interactions with other food components such as starch (Gutierrez et al., 2008a) and proteins (Kyung, 2012). It is hypothesized that the high fat and protein content in foods could protect pathogenic and spoilage microorganisms from the antimicrobial activity of EOs (Tassou et al., 1995). The EOs are hydrophobic and readily dissolve in the lipid phase of the food. In this regard, there would be limited amount of EO interaction with microorganisms in the aqueous phase of the food (Mejlholm and Dalgaard, 2002).

### pH Dependency

Another intrinsic property of the food that can also influence the effectiveness of plant antimicrobials is pH. The effectiveness of plant-derived natural antimicrobials has been shown to be pH dependent. At acidic pH, EOs are more effective in foods in spite of a high protein content (Gutierrez et al., 2009). At low pH, EOs become more hydrophobic and thus can dissolve easily in the lipid bilayer membrane of pathogenic and spoilage microbes, resulting in greater antimicrobial efficacy at low pH (Juven et al., 1994).

### Temperature Dependency

Extrinsic properties of foods such as temperature (Rattanachaikunsopon and Phumkhachorn, 2010) as well as extent of microbial contamination (Somolinos et al., 2010) and type of microorganism (Mosqueda-Melgar et al., 2008c) are also factors influencing the effectiveness of plant antimicrobials. Based on a study investigating the effectiveness of 17 EOs and 9 oil compounds against *E. coli* O157:H7 and *S. enterica* in apple juice, the EOs and their components were more effective at 37°C than at 4 or 21°C (Friedman et al., 2004). At higher temperatures, the bacterial cell membrane becomes more fluid and thus less ordered. This increased fluidity enhances the diffusion of an antimicrobial into the microbial cell, allowing it to reach its target site causing inhibition or death (Raybaudi-Massilia et al., 2009).

### Microbial Contamination and Type of Microbe

The level of microbial contamination in foods is another important factor that influences the antimicrobial potency of plant antimicrobials. Somolinos et al. (2010) confirmed this by demonstrating that the initial inoculum size affected the efficacy of citral, a component of citrus EO, against *E. coli* cells. Differences in inhibitory effects of plant antimicrobials against different types of microbes are also limitations in their use in foods. It is hypothesized that EOs are generally more effective against Gram-positive bacteria than Gram-negative bacteria. This is attributed to the lipopolysaccharide layer of the outer membrane of Gram-negative bacteria that is absent in Gram-positive bacteria and that acts as a barrier for some antimicrobials (Burt, 2004; Mosqueda-Melgar et al., 2008c). Although this may be generally true, researchers recognize that this may not be true for every EO and its constituents. ICs of EOs may have different degrees of effectiveness against Gram-negative and Gram-positive bacteria (Dorman and Deans, 2000). In a study evaluating the effect of an EO obtained from mint against *S.* Enteritidis and *L. monocytogenes* in Greek appetizers, the EO was more effective against *S.* Enteritidis than *L. monocytogenes* (Tassou et al., 1995). The effectiveness of an antimicrobial is not only dependent on the type of microbe (Gram-negative vs. Gram-positive) but may also differ due to microbial species and even strain. Researchers must therefore consider all the types of microorganisms that could potentially be present in a food system when evaluating the effectiveness of a plant antimicrobial in that food.

## Techniques for Delivery of Essential Oils and Their Components

Growing consumer preference for more natural preservatives has forced the food industry to utilize natural extracts, especially from plants as alternatives to synthetic antimicrobials. Because EOs are categorized as GRAS by U.S. FDA (Hyldgaard et al., 2012; Persico et al., 2009), researchers are extensively studying their application in food preservation, especially as antimicrobials. In this regard, the method of delivery of EOs or their components in foods is crucial for their effective applications in foods. Apart from direct addition to foods, antimicrobial EOs can be delivered in several ways, including incorporation in food packaging, encapsulation, and incorporation in biodegradable edible coatings. In choosing a method, factors such as the cost, stability, antimicrobial efficacy, transportation, and storage conditions need to be considered.

### Incorporation in Food Packaging

The interest in incorporating antimicrobial agents into food packaging has more than doubled in the last 5 years (Appendini and Hotchkiss, 2002). Antimicrobial plant extracts can be incorporated into paper, thermoplastics, and thermosets to produce active packaging (Manso et al., 2013). There are two methods of the application of these active films: (1) overlay diffusion method involving direct contact of the film on the food and (2) vapor phase methods that exploit EO volatiles for use as a kind of fumigant. In this regard, direct contact between the food and the polymer is not necessary (Du et al., 2011). Manso et al. (2013) tested the antifungal effect of vapors from cinnamon EO (incorporated in polyethylene terephthalate (PET) films) against *Aspergillus flavus*. The compounds released by the active packaging limited the development of that mold. Multilayer films consisting of control, matrix, and barrier layers (Martiñon et al., 2014; Sipahi et al., 2013) may allow minimum rate of antimicrobial release from the polymer to maintain the surface concentration at a critical inhibitory rate (Han, 2003; Sipahi et al., 2013). The inner layer controls the rate of diffusion, the matrix layer contains the active substance, and the barrier layer prevents the outward migration of the volatiles from the package. Those EOs that might be sensitive to the temperatures used in polymer processing can be adsorbed to coat films after forming or added to cast films such as PE and PET films (Appendini and Hotchkiss, 2002).

### Encapsulation

In some instances, EOs exhibit limited stability especially against physical degradation mainly by temperature. They also present solubility problems in aqueous food systems and undesirable strong taste and smell (Fang and Bhandari, 2010). Encapsulation systems, also known as "delivery systems," are used to overcome these limitations. For the delivery of different EOs or EO compounds in foods, many encapsulation systems of different formulations, size, and effects were tested to develop small solid particles, liquid droplets, or gases.

Encapsulation of antimicrobial plant extracts has been extensively studied and assessed in foods. The antimicrobial effects of the encapsulated antimicrobials are extended and contribute to prolong the shelf life of foods without affecting their quality and hygiene, yet decreasing the sensory impact of the antimicrobial volatiles (Nazzaro et al., 2012). Encapsulation of rosemary EO increased its antimicrobial activity against *L. monocytogenes* in pork liver sausage (Pandit and Shelef, 1994). Similarly, the nanoencapsulation of a mixture of terpenes extracted from *Melaleuca alternifolia* increased their activity against *Lactobacillus delbrueckii* in pear and orange juices (Donsì et al., 2011).

### Incorporation in Biodegradable Edible Coating

Research efforts continue to be focused on active food coating, especially incorporating EOs or their most active compounds into edible coating (Valdés et al., 2017). These films can limit surface microbial growth, serve as an oxygen barrier, and contribute to the protection of the environment by reducing the use of synthetic packaging materials. Incorporating antimicrobial compounds into edible films provides a novel way to improve the safety and shelf life of ready-to-eat foods. In the literature, different EOs or their compounds have been combined in different package coating films for meat, fish, fruits, and vegetables (Table 14.2) (Du et al., 2011).

## Exploiting Synergistic Actions of Essential Oils and Their Components

Exploiting synergistic actions among EO components without compromising their antimicrobial effectiveness may be the key to overcoming several limitations of using EOs in foods. However, synergism may not always be the result of using antimicrobials in combinations. Instead, there are three interactive outcomes that may result due to the combination of EO components. Synergism occurs if the combination of two antimicrobial components results in a much greater antimicrobial effect than the sum of the individual antimicrobial components. On the other hand, antagonism occurs if the combination results in a combined effect that is less than when the antimicrobials are applied individually. If the combined effect is equal to the sum of the antimicrobial effects of individual antimicrobial components, then the interactive outcome is considered additive (Burt, 2004; Davidson and Parish, 1989).

### Interactive Outcomes and Intra–Essential Oil Interactions

Plant EOs are very complex mixtures containing more than 60 individual constituents in various concentrations. Major constituents may be up to 85% of the EO fraction, whereas the minor constituents are typically found in trace amounts (Senatore, 1996). It has been reported that EOs are more effective antimicrobials compared with mixtures of just their major individual constituents (Gill et al., 2002; Mourey and Canillac, 2002). This suggests that minor constituents in EOs are likely interacting to produce additive and/or synergistic effects when combined with major constituents. On the contrary, combinations with minor constituents may also result in antagonistic outcomes. Carvacrol, a major constituent of oregano oil, was found to be more effective than the whole oregano extract (Rao et al., 2010).

Several studies evaluated the synergistic effects of EO constituents when applied in combinations (Nguefack et al., 2012). In one study evaluating the bactericidal action of thymol, carvacrol, and eugenol individually and in binary and ternary mixtures against nonpathogenic

**Table 14.2: Different types of antimicrobial coating with different essential oils (EOs) or EO compounds applied to a wide variety of foods.**

| Food application | Product | Matrix | Antimicrobial agent | References |
|---|---|---|---|---|
| Fish | Sliced fresh *Channa argus* | Chitosan and polyethyleneimine | Thyme EO | Yang et al. (2015) |
| | Fresh Indian salmon | Gelatin from waste of Nemipterus japonicus | Garlic (*Allium sativum*), lime (*Citrus aurantiifolia*) | Thaker et al. (2017) |
| | Gilthead seabream fillets | Methylcellulose | *Satureja thymbra* (L.) EO | Choulitoudi et al. (2016) |
| | Fresh silver carp fillets | Methylcellulose | *Pimpinella affinis* EO | Ariaii et al. (2015) |
| | Fish | Gelatin–chitosan | Clove (*Syzygium aromaticum* L.) | Gómez-Estaca et al. (2010) |
| Meat | Chicken breasts | κ-Carrageenan and chitosan | Mustard extract | Olaimat et al. (2014) |
| | | Sodium alginate | Thyme and propionic acid | Matiacevich et al. (2015) |
| | Pork meat | Chitosan | Clove oil and/or ethylenediaminetetraacetate | He et al. (2014) |
| | Beef muscle | Milk protein based | Oregano, pimento | Oussalah et al. (2004) |
| | Chicken breast | Whey protein Isolate | Oregano/clove EOs | Fernández-Pan et al. (2014) |
| Fruits | Watermelon (*Citrullus lanatus*) | Sodium alginate, pectin, and calcium lactate | *trans*-Cinnamaldehyde | Forgiarini et al. (2001) |
| | Cantaloupe melon | Chitosan and pectin | *trans*-Cinnamaldehyde | Ezhilarasi et al. (2013) |
| | Strawberries | Pectin and calcium chloride | Eugenol, citral, ascorbic acid | Guerreiro et al. (2015c) |
| | Raspberries | Pectin and calcium chloride | Eugenol, citral, ascorbic acid | Guerreiro et al. (2015a) |
| | *Arbutus unedo* L. fruit | Sodium alginate | Citral, eugenol | Guerreiro et al. (2015b) |
| | Pineapple | Sodium alginate | Lemongrass EO | Azarakhsh et al. (2014) |
| | *Fuji* apples | Sodium alginate | Lemongrass EO | Salvia-Trujillo et al. (2015) |
| | Blueberry | Chitosan | Carvacrol, cinnamaldehyde, *trans*-cinnamaldehyde | Sun et al. (2014) |
| | Avocado | Gum arabic, aloe vera, and chitosan | Thyme oil | Bill et al. (2014) |
| | Pepper | Chitosan | Lemongrass EO | Ali et al. (2015) |
| | | Pullulan | Leather bergenia leaves ethanolic extracts | Kraśniewska et al. (2015) |
| | | Pullulan | *Satureja hortensis* aqueous or ethanolic extracts | Kraśniewska et al. (2014) |
| Vegetables | Pumpkin | Starch | Carvacrol | Santos et al. (2016) |
| | Cherry tomatoes | Zein | Cinnamon, mustard EO, commercial wax | Yun et al. (2015) |
| | Fresh cut broccoli | Chitosan | Bioactive compounds and EOs | Alvarez et al. (2013) |
| | Cauliflower florets | Maltodextrins and methylcellulose | Lactic acid, citrus extract, lemongrass EO | Boumail et al. (2016) |
| | Green beans | Modified chitosan | Mandarin EO | Severino et al. (2014) |
| | Rucola | Modified chitosan | Lemon, mandarin, oregano, or clove EOs | Sessa et al. (2015) |

*Adapted from Valdés, A., Ramos, M., Beltrán, A., Jiménez, A., Garrigós, M.C., 2017. State of the art of antimicrobial edible coatings for food packaging applications. Coatings 7 (4), 56.*

*L. innocua*, the most synergistic effect was evident between carvacrol and thymol (García-García et al., 2011). Synergism was also reported between carvacrol and its precursor p-cymene, an otherwise weak antimicrobial, which appeared to swell cell membranes of vegetative cells of *Bacillus cereus* enabling carvacrol to easily enter the cells and act on its target (Ultee et al., 2002). Synergy was similarly evident when a study showed that the combination of menthol or linalool with eugenol was more effective than the individual constituents (Bassolé et al., 2010).

### Analysis of Combined Effects

The checkerboard method can be used to evaluate the efficacy of combinations of various EOs and forms the basis for calculating the fractional inhibitory concentration index (FIC Index). Analyzing the combined effects of a mixture of EOs and their components to determine synergy, additive effects, or antagonism requires calculating the FIC Index. FIC Index is calculated by finding the sum of the minimum concentrations that inhibited microbial growth for both EOs and their components A and B. FIC Index = FICA + FICB. Measurements of MIC for the ICs are required to calculate the FIC Index. In the following equations, FICA = MICA + B/MICA and FICB = MICB + A/MICB, the value for MICA + B is equal to the MIC of component A in the presence of component B and the value for MICB + A is equal to the MIC of component B in the presence of component A (Davidson and Parish, 1989). A FIC Index $<0.5$ is interpreted as synergistic, additive if $0.5 < \text{FIC Index} < 4$, or antagonistic if FIC Index $>4$ (Odds, 2003).

### Theoretical Mechanisms

Although EOs and their constituents in combination may cause additive, synergistic, or antagonistic outcomes, very little about the mechanisms of these interactions are known. There are four theoretical mechanisms that could explain synergism among antimicrobial components. First, an antimicrobial may interact with the cell wall or cell membrane of a microbial cell and enhance the uptake of other antimicrobials. Second, numerous antimicrobials may interact with the cell wall. A third theory proposed is that sequential inhibition may occur whereby many steps in a common biochemical pathway are inhibited, thus producing synergistic effects. The fourth theoretical mechanism hypothesized that synergy results from the inhibition of enzymes that degrade antimicrobials (Davidson and Parish, 1989; Eliopoulos et al., 1996). A more recent theory proposed that antimicrobials may have diverse mode of actions and that, when combined, each antimicrobial component acts on different target sites on a cell. Each component may be indirectly dependent on each other (Hyldgaard et al., 2012).

A few theoretical mechanisms were proposed to explain antagonism among antimicrobial components. It was hypothesized that if the antimicrobials in the combination have the same target sites on microbial cells, then antagonism may result. A second theory is that one antimicrobial may inhibit the growth or reproduction of microbes, whereas the other in the combination may kill the microbe. This combined effect may result in a less effective antimicrobial effect than if the antimicrobials were individually applied. It is also proposed that the interaction of antimicrobials with each other, rather than with the target sites of microbes, could produce antagonistic effects (Davidson and Parish, 1989; Roller, 2003).

## Regulations for Food Applications

Food safety is a global issue and several countries have agencies in place that evaluate the safety of food additives. Plant-derived antimicrobials are among food additives that are controlled by legislation globally. The major governmental regulations on food additives are the Codex Alimentarius, which makes up the FAO/WHO joint regulatory document, the U.S. Food and Drug Act, and the European Union Standards. Based on those regulations, natural antimicrobials are GRAS; however, their GRAS status depends on factors such as a history of safe consumption by humans, their origin in an edible or inedible product, and proven absence of toxicity in a concentrated form. Generally, provision of scientific data on the safety of new food additives is required for evaluation by competent authorities before permission is granted for food use.

### U.S. Food and Drug Administration

EOs are considered GRAS food additives by FDA. Food packaging materials containing antimicrobials are considered food additives and are regulated by the FDA. The FDA evaluates the safety of an additive by considering both short-term and long-term health effects (Hintz et al., 2015). The FDA is also responsible for evaluating data on the daily intake of a food additive before setting regulatory limits on their accepted daily intake (Hyldgaard et al., 2012). Additionally, the FDA sets regulations on the types of foods in which an approved food additive can be used and how it should be identified on food labels. Regulations pertaining to food and drugs within the United States must be published in Title 21 of the Code of Federal Regulations (Hintz et al., 2015). Some food additives including most plant-derived antimicrobials that are proven harmless to the health of consumers are placed on the GRAS list. Thyme, oregano, clove, cinnamon, and nutmeg are some of the EOs that are approved as GRAS by the FDA and are used at low levels as flavorings (Hyldgaard et al., 2012).

### European Food Safety Authority

In Europe, numerous plant-derived antimicrobials are currently added to foods to enhance their flavors. Several registered compounds, such as thymol, eugenol, cinnamaldehyde, citral, and vanillin, are considered safe and pose no risk to consumers (Hyldgaard et al., 2012).

However, only a small number of food additives, namely organic acids, are approved for use as food preservatives. The European Union sets very stringent guidelines for the approval of new compounds as flavorings. Very thorough toxicological studies must be completed before a food additive is approved. It must be clear that it poses no health risk to the consumer. Therefore, an applicant must submit scientific data on the safety of the new food additive to the European Commission when applying for its approval for food use. This information is then submitted to the European Food Safety Authority (EFSA) to evaluate the safety of the new additive. EFSA's responsibility is to advise the European Commission based on scientific data and review any new scientific data that may arise for already-existing approved food additives (Hintz et al., 2015).

### *China Food Additives Association*

In China, the China Food Additives Association (CFAA) is responsible for assessing the safety of new food additives before they are approved for human consumption (Hintz et al., 2015). According to the CFAA, it is the only registered food additive and ingredients industry organization permitted by the Chinese government to assist the government in creating regulations and policies associated with the food industry. The CFAA also plays an integral role in setting standards for the food industry.

## *Conclusions and Future Prospects*

Information in the present chapter indicates that EOs have good potential for replacing synthetic food additives and harsh food processing treatments to meet consumer expectations. Optimal applications of EOs still require circumventing certain hurdles such as high EO levels for microbial control, negative alteration of foods' sensory properties, and poor solubility in aqueous food systems. Despite the growing body of knowledge on food applications of EOs, in vivo studies to provide toxicological information on the vast array of EO components are lacking.

Food applications of natural antimicrobials are likely to increase in the future due to the food industry's efforts to meet consumer demands for fresh or freshlike food products with only natural food additives. Further research on the influence of intrinsic food factors and food storage factors, as well as on the antimicrobial efficacy of EOs combined with nonthermal processes, is warranted along with strategies to eliminate or mask undesirable sensory changes in foods with added EOs. Scientifically based evaluations of major and/or minor EO components to obtain synergistic antimicrobial effects remain a major focus area for future research. Considering the wide variety of food products and variations in stability and efficacy of natural antimicrobials, it is important that EOs be carefully selected and delivered to exhibit strong antimicrobial effects against foodborne pathogens while maintaining stability throughout the food's shelf life.

## References

Abdollahzadeh, E., Rezaei, M., Hosseini, H., 2014. Antibacterial activity of plant essential oils and extracts: the role of thyme essential oil, nisin, and their combination to control *Listeria monocytogenes* inoculated in minced fish meat. Food Control 35 (1), 177–183. http://dx.doi.org/10.1016/j.foodcont.2013.07.004.

Ahmed, J., Hiremath, N., Jacob, H., 2017. Antimicrobial efficacies of essential oils/nanoparticles incorporated polylactide films against *L. monocytogenes* and *S.* Typhimurium on contaminated cheese. International Journal of Food Properties 20 (1), 53–67.

Ahn, J., Grun, I.U., Mustapha, A., 2007. Effects of plant extracts on microbial growth, color change, and lipid oxidation in cooked beef. Food Microbiology 24 (1), 7–14. http://dx.doi.org/10.1016/j.fm.2006.04.006.

Ali, A., Noh, N.M., Mustafa, M.A., 2015. Antimicrobial activity of chitosan enriched with lemongrass oil against anthracnose of bell pepper. Food Packaging and Shelf Life 3, 56–61.

Alvarez, M.V., Ponce, A.G., Moreira, M. d. R., 2013. Antimicrobial efficiency of chitosan coating enriched with bioactive compounds to improve the safety of fresh cut broccoli. LWT – Food Science and Technology 50 (1), 78–87.

Appendini, P., Hotchkiss, J.H., 2002. Review of antimicrobial food packaging. Innovative Food Science & Emerging Technologies 3 (2), 113–126. http://dx.doi.org/10.1016/S1466-8564(02)00012-7.

Ariaii, P., Tavakolipour, H., Rezaei, M., Elhami Rad, A.H., Bahram, S., 2015. Effect of methylcellulose coating enriched with Pimpinella affinis oil on the quality of silver carp fillet during refrigerator storage condition. Journal of Food Processing and Preservation 39 (6), 1647–1655.

Azarakhsh, N., Osman, A., Ghazali, H.M., Tan, C.P., Adzahan, N.M., 2014. Lemongrass essential oil incorporated into alginate-based edible coating for shelf-life extension and quality retention of fresh-cut pineapple. Postharvest Biology and Technology 88, 1–7.

Barbosa, L.N., Rall, V.L., Fernandes, A.A., Ushimaru, P.I., da Silva Probst, I., Fernandes Jr., A., 2009. Essential oils against foodborne pathogens and spoilage bacteria in minced meat. Foodborne Pathogens and Disease 6 (6), 725–728. http://dx.doi.org/10.1089/fpd.2009.0282.

Bassolé, I.H., Lamien-Meda, A., Bayala, B., Tirogo, S., Franz, C., Novak, J., Dicko, M.H., 2010. Composition and antimicrobial activities of *Lippia multiflora* Moldenke, *Mentha x piperita* L. and *Ocimum basilicum* L. essential oils and their major monoterpene alcohols alone and in combination. Molecules: a Journal of Synthetic Chemistry and Natural Product Chemistry 15 (11), 7825–7839. http://dx.doi.org/10.3390/molecules15117825.

Bill, M., Sivakumar, D., Korsten, L., Thompson, A.K., 2014. The efficacy of combined application of edible coatings and thyme oil in inducing resistance components in avocado (*Persea americana* Mill.) against anthracnose during post-harvest storage. Crop Protection 64, 159–167.

Boumail, A., Salmieri, S., St-Yves, F., Lauzon, M., Lacroix, M., 2016. Effect of antimicrobial coatings on microbiological, sensorial and physico-chemical properties of pre-cut cauliflowers. Postharvest Biology and Technology 116, 1–7.

Burt, S., 2004. Essential oils: their antibacterial properties and potential applications in foods—a review. International Journal of Food Microbiology 94 (3), 223–253. http://dx.doi.org/10.1016/j.ijfoodmicro.2004.03.022.

Burt, S.A., Reinders, R.D., 2003. Antibacterial activity of selected plant essential oils against *Escherichia coli* O157:H7. Letters in Applied Microbiology 36 (3), 162–167.

Burt, S.A., Vlielander, R., Haagsman, H.P., Veldhuizen, E.J., 2005. Increase in activity of essential oil components carvacrol and thymol against *Escherichia coli* O157:H7 by addition of food stabilizers. Journal of Food Protection 68 (5), 919–926.

Busatta, C., Vidal, R.S., Popiolski, A.S., Mossi, A.J., Dariva, C., Rodrigues, M.R.A., Cansian, R.L., 2008. Application of *Origanum majorana* L. essential oil as an antimicrobial agent in sausage. Food Microbiology 25, 207–211.

Cava-Roda, R., Taboada-Rodriguez, A., Valverde-Franco, M., Marin-Iniesta, F., 2012. Antimicrobial activity of vanillin and mixtures with cinnamon and clove essential oils in controlling *Listeria monocytogenes* and *Escherichia coli* O157:H7 in milk. Food Bioprocess Technology 5 (6), 2120–2131. http://dx.doi.org/10.1007/s11947-010-0484-4.

Chorianopoulos, N., Kalpoutzakis, E., Aligiannis, N., Mitaku, S., Nychas, G.J., Haroutounian, S.A., 2004. Essential oils of *Satureja*, *Origanum*, and *Thymus* species: chemical composition and antibacterial activities against foodborne pathogens. Journal of Agricultural and Food Chemistry 52 (26), 8261–8267. http://dx.doi.org/10.1021/jf049113i.

Chorianopoulos, N.G., Giaouris, E.D., Skandamis, P.N., Haroutounian, S.A., Nychas, G.J., 2008. Disinfectant test against monoculture and mixed-culture biofilms composed of technological, spoilage and pathogenic bacteria: bactericidal effect of essential oil and hydrosol of *Satureja thymbra* and comparison with standard acid-base sanitizers. Journal of Applied Microbiology 104 (6), 1586–1596. http://dx.doi.org/10.1111/j.1365-2672.2007.03694.x.

Choulitoudi, E., Bravou, K., Bimpilas, A., Tsironi, T., Tsimogiannis, D., Taoukis, P., Oreopoulou, V., 2016. Antimicrobial and antioxidant activity of *Satureja thymbra* in gilthead seabream fillets edible coating. Food and Bioproducts Processing 100, 570–577.

Davidson, P.M., Parish, M.E., 1989. Methods for testing the efficacy of food antimicrobials. Food Technology 43 (1), 148–155.

de Sousa Guedes, J.P., da Costa Medeiros, J.A., de Souza e Silva, R.S., Batista de Sousa, J.M., Conceição, M.L., de Souza, E.L., 2016. The efficacy of *Mentha arvensis* L. and *M. piperita* L. essential oils in reducing pathogenic bacteria and maintaining quality characteristics in cashew, guava, mango, and pineapple juices. International Journal of Food Microbiology 238 (5), 183–219.

Delaquis, P.J., Mazza, G., 1995. Antimicrobial properties of isothiocyanates in food preservation. Food Technology 49 (11), 73–84.

Desai, M.A., Soni, K.A., Nannapaneni, R., Schilling, M.W., Silva, J.L., 2012. Reduction of *Listeria monocytogenes* in raw catfish fillets by essential oils and phenolic constituent carvacrol. Journal of Food Science 77 (9), M516–M522. http://dx.doi.org/10.1111/j.1750-3841.2012.02859.x.

Djenane, D., 2015. Chemical profile, antibacterial and antioxidant activity of algerian citrus essential oils and their application in *Sardina pilchardus*. Foods 4 (2), 208–228. http://dx.doi.org/10.3390/foods4020208.

Donsì, F., Annunziata, M., Sessa, M., Ferrari, G., 2011. Nanoencapsulation of essential oils to enhance their antimicrobial activity in foods. LWT – Food Science and Technology 44 (9), 1908–1914. http://dx.doi.org/10.1016/j.lwt.2011.03.003.

Dorman, H.J., Deans, S.G., 2000. Antimicrobial agents from plants: antibacterial activity of plant volatile oils. Journal of Applied Microbiology 88 (2), 308–316.

Du, W., Avena Bustillos, R.D., Hua, S.T., Mchugh, T.H., 2011. In: Mendez-Vilas, A. (Ed.), Antimicrobial Volatile Essential Oils in Edible Films for Food Safety. Formatex, Badajoh, Spain.

Eliopoulos, G.M., Moellering, R.C.J., Pillai, S.K., 1996. Antimicrobial combinations. In: Lorian, V. (Ed.), Antibiotics in Laboratory Medicine, fifth ed. Williams & Wilkins, Philadelphia, pp. 365–440.

Ezhilarasi, P.N., Karthik, P., Chhanwal, N., Anandharamakrishnan, C., 2013. Nanoencapsulation techniques for food bioactive components: a review. Food and Bioprocess Technology 6 (3), 628–647. http://dx.doi.org/10.1007/s11947-012-0944-0.

Fang, Z., Bhandari, B., 2010. Encapsulation of polyphenols – a review. Trends in Food Science & Technology 21 (10), 510–523. http://dx.doi.org/10.1016/j.tifs.2010.08.003.

Fernández-Pan, I., Carrión-Granda, X., Maté, J.I., 2014. Antimicrobial efficiency of edible coatings on the preservation of chicken breast fillets. Food Control 36 (1), 69–75.

Forgiarini, A., Esquena, J., González, C., Solans, C., 2001. Formation of nano-emulsions by low-energy emulsification methods at constant temperature. Langmuir: the ACS Journal of Surfaces and Colloids 17 (7), 2076–2083. http://dx.doi.org/10.1021/la001362n.

Frangos, L., Pyrgotou, N., Giatrakou, V., Ntzimani, A., Savvaidis, I.N., 2010. Combined effects of salting, oregano oil and vacuum-packaging on the shelf-life of refrigerated trout fillets. Food Microbiology 27 (1), 115–121. http://dx.doi.org/10.1016/j.fm.2009.09.002.

Fratianni, F., De Martino, L., Melone, A., De Feo, V., Coppola, R., Nazzaro, F., 2010. Preservation of chicken breast meat treated with thyme and balm essential oils. Journal of Food Science 75 (8), M528–M535. http://dx.doi.org/10.1111/j.1750-3841.2010.01791.x.

Friedman, M., Henika, P.R., Levin, C.E., Mandrell, R.E., 2004. Antibacterial activities of plant essential oils and their components against *Escherichia coli* O157:H7 and *Salmonella enterica* in apple juice. Journal of Agricultural and Food Chemistry 52 (19), 6042–6048. http://dx.doi.org/10.1021/jf0495340.

García-García, R., López-Malo, A., Palou, E., 2011. Bactericidal action of binary and ternary mixtures of carvacrol, thymol, and eugenol against *Listeria innocua*. Journal of Food Science 76 (2), M95–M100. http://dx.doi.org/10.1111/j.1750-3841.2010.02005.x.

Gill, A.O., Delaquis, P., Russo, P., Holley, R.A., 2002. Evaluation of antilisterial action of cilantro oil on vacuum packed ham. International Journal of Food Microbiology 73 (1), 83–92.

Govaris, A., Botsoglou, E., Sergelidis, D., Chatzopoulou, P.S., 2011. Antibacterial activity of oregano and thyme essential oils against *Listeria monocytogenes* and *Escherichia coli* O157:H7 in feta cheese packaged under modified atmosphere. LWT Food Science and Technology 44, 1240–1244.

Govaris, A., Solomakos, N., Pexara, A., Chatzopoulou, P.S., 2010. The antimicrobial effect of oregano essential oil, nisin and their combination against *Salmonella* Enteritidis in minced sheep meat during refrigerated storage. International Journal of Food Microbiology 137 (2–3), 175–180. http://dx.doi.org/10.1016/j.ijfoodmicro.2009.12.017.

Guerreiro, A.C., Gago, C.M., Faleiro, M.L., Miguel, M.G., Antunes, M.D., 2015a. Raspberry fresh fruit quality as affected by pectin-and alginate-based edible coatings enriched with essential oils. Scientia Horticulturae 194, 138–146.

Guerreiro, A.C., Gago, C.M., Faleiro, M.L., Miguel, M.G., Antunes, M.D., 2015b. The effect of alginate-based edible coatings enriched with essential oils constituents on *Arbutus unedo* L. fresh fruit storage. Postharvest Biology and Technology 100, 226–233.

Guerreiro, A.C., Gago, C.M., Faleiro, M.L., Miguel, M.G., Antunes, M.D., 2015c. The use of polysaccharide-based edible coatings enriched with essential oils to improve shelf-life of strawberries. Postharvest Biology and Technology 110, 51–60.

Gutierrez, J., Barry-Ryan, C., Bourke, P., 2008a. The antimicrobial efficacy of plant essential oil combinations and interactions with food ingredients. International Journal of Food Microbiology 124 (1), 91–97. http://dx.doi.org/10.1016/j.ijfoodmicro.2008.02.028.

Gutierrez, J., Barry-Ryan, C., Bourke, P., 2009. Antimicrobial activity of plant essential oils using food model media: efficacy, synergistic potential and interactions with food components. Food Microbiology 26 (2), 142–150. http://dx.doi.org/10.1016/j.fm.2008.10.008.

Gutierrez, J., Rodriguez, G., Barry-Ryan, C., Bourke, P., 2008b. Efficacy of plant essential oils against foodborne pathogens and spoilage bacteria associated with ready-to-eat vegetables: antimicrobial and sensory screening. Journal of Food Protection 71 (9), 1846–1854.

Gómez-Estaca, J., De Lacey, A.L., López-Caballero, M., Gómez-Guillén, M., Montero, P., 2010. Biodegradable gelatin–chitosan films incorporated with essential oils as antimicrobial agents for fish preservation. Food Microbiology 27 (7), 889–896.

Han, J.H., 2003. 4-Antimicrobial food packaging A2-Ahvenainen, Raija. In: Novel Food Packaging Techniques. Woodhead Publishing, pp. 50–70.

He, S., Yang, Q., Ren, X., Zi, J., Lu, S., Wang, S., Wang, Y., 2014. Antimicrobial efficiency of chitosan solutions and coatings incorporated with clove oil and/or ethylenediaminetetraacetate. Journal of Food Safety 34 (4), 345–352.

Hernandez-Herrero, L.A., Giner, M.J., Valero, M., 2008. Effective chemical control of psychrotrophic *Bacillus cereus* EPSO-35AS and INRA TZ415 spore outgrowth in carrot broth. Food Microbiology 25 (5), 714–721.

Hintz, T., Matthews, K.K., Di, R., 2015. The use of plant antimicrobial compounds for food preservation. Biomed Research International 2015, 246264. http://dx.doi.org/10.1155/2015/246264.

Holley, R.A., Patel, D., 2005. Improvement in shelf-life and safety of perishable foods by plant essential oils and smoke antimicrobials. Food Microbiology 22 (4), 273–292. http://dx.doi.org/10.1016/j.fm.2004.08.006.

Hyldgaard, M., Mygind, T., Meyer, R.L., 2012. Essential oils in food preservation: mode of action, synergies, and interactions with food matrix components. Front Microbiology 3, 12. http://dx.doi.org/10.3389/fmicb.2012.00012.

Ibrahim, S.A., Salameh, M.M., Phetsomphou, S., Yang, H., Seo, C.W., 2006. Application of caffeine, 1,3,7-trimethylxanthine, to control *Escherichia coli* O157:H7. Food Chemistry 99 (4), 645–650. http://dx.doi.org/10.1016/j.foodchem.2005.08.026.

Irkin, R., Korukluoglu, M., 2009. Growth inhibition of pathogenic bacteria and some yeasts by selected essential oils and survival of *L. monocytogenes* and *C. albicans* in apple-carrot juice. Foodborne Pathogens and Disease 6 (3), 387–394. http://dx.doi.org/10.1089/fpd.2008.0195.

Ismaiel, A.A., Pierson, M.D., 1990. Effect of sodium nitrite and *Origanum* oil on growth and toxin production of *Clostridium botulinum* in TYG broth and ground pork. Journal of Food Protection 53 (11), 958–960. http://dx.doi.org/10.4315/0362-028X-53.11.958.

Juven, B.J., Kanner, J., Schved, F., Weisslowicz, H., 1994. Factors that interact with the antibacterial action of thyme essential oil and its active constituents. Journal of Applied Bacteriology 76 (6), 626–631.

Kim, S., Fung, D.Y., 2004. Antibacterial effect of water-soluble arrowroot (*Puerariae radix*) tea extracts on foodborne pathogens in ground beef and mushroom soup. Journal of Food Protection 67 (9), 1953–1956.

Koutsoumanis, K., Lambropoulou, K., Nychas, G.J., 1999. A predictive model for the non-thermal inactivation of *Salmonella enteritidis* in a food model system supplemented with a natural antimicrobial. International Journal of Food Microbiology 49 (1–2), 63–74.

Kraśniewska, K., Gniewosz, M., Synowiec, A., Przybył, J.L., Bączek, K., Węglarz, Z., 2014. The use of pullulan coating enriched with plant extracts from *Satureja hortensis* L. to maintain pepper and apple quality and safety. Postharvest Biology and Technology 90, 63–72.

Kraśniewska, K., Gniewosz, M., Synowiec, A., Przybył, J.L., Bączek, K., Węglarz, Z., 2015. The application of pullulan coating enriched with extracts from Bergenia crassifolia to control the growth of food microorganisms and improve the quality of peppers and apples. Food and Bioproducts Processing 94, 422–433.

Kyung, K.H., 2012. Antimicrobial properties of allium species. Current Opinion in Biotechnology 23 (2), 142–147. http://dx.doi.org/10.1016/j.copbio.2011.08.004.

Leite, C.J., de Sousa, J.P., Medeiros, J.A., da Conceicao, M.L., dos Santos Falcao-Silva, V., de Souza, E.L., 2016. Inactivation of *Escherichia coli*, *Listeria monocytogenes*, and *Salmonella* Enteritidis by *Cymbopogon citratus* D.C. Stapf. Essential oil in pineapple juice. Journal of Food Protection 79 (2), 213–219. http://dx.doi.org/10.4315/0362-028x.jfp-15-245.

Lin, Y.T., Labbe, R.G., Shetty, K., 2004. Inhibition of *Listeria monocytogenes* in fish and meat systems by use of oregano and cranberry phytochemical synergies. Applied and Environmental Microbiology 70 (9), 5672–5678. http://dx.doi.org/10.1128/AEM.70.9.5672-5678.2004.

Lopez-Malo, A., Alzamora, S.M., Palou, E., 2005. In: Davidson, P.M., Sofos, J.N., Branen, A.L. (Eds.), Naturally Occurring Compounds: Plant Sources, third ed. CRC Press, New York.

Manso, S., Cacho-Nerin, F., Becerril, R., Nerin, C., 2013. Combined analytical and microbiological tools to study the effect on *Aspergillus flavus* of cinnamon essential oil contained in food packaging. Food Control 30, 370–378.

Marino, M., Bersani, C., Comi, G., 2001. Impedance measurements to study the antimicrobial activity of essential oils from Lamiaceae and Compositae. International Journal of Food Microbiology 67 (3), 187–195.

Martiñon, M.E., Moreira, R.G., Castell-Perez, M.E., Gomes, C., 2014. Development of a multilayered antimicrobial edible coating for shelf-life extension of fresh-cut cantaloupe (*Cucumis melo* L.) stored at 4°C. LWT – Food Science and Technology 56 (2), 341–350. http://dx.doi.org/10.1016/j.lwt.2013.11.043.

Matiacevich, S., Acevedo, N., López, D., 2015. Characterization of edible active coating based on alginate–thyme oil–propionic acid for the preservation of fresh chicken breast fillets. Journal of Food Processing and Preservation 39 (6), 2792–2801.

Mejlholm, O., Dalgaard, P., 2002. Antimicrobial effect of essential oils on the seafood spoilage micro-organism *Photobacterium phosphoreum* in liquid media and fish products. Letters in Applied Microbiology 34 (1), 27–31.

Menon, K.V., Garg, S.R., 2001. Inhibitory effect of clove oil on *Listeria monocytogenes* in meat and cheese. Food Microbiology 18, 647–650.

Mosqueda-Melgar, J., Raybaudi-Massilia, R.M., Martın-Belloso, O., 2008a. Non-thermal pasteurization of fruit juices by combining high-intensity pulsed electric fields with natural antimicrobials. Innovative Food Science and Emerging Technologies 9, 328–340.

Mosqueda-Melgar, J., Raybaudi-Massilia, R.M., Martın-Belloso, O., 2008b. Inactivation of *Salmonella enterica* ser. *Enteritidis* in tomato juice by combining of high-intensity pulsed electric fields with natural antimicrobials. Journal of Food Science 73, M47–M53.

Mosqueda-Melgar, J., Raybaudi-Massilia, R.M., Martín-Belloso, O., 2008c. Combination of high-intensity pulsed electric fields with natural antimicrobials to inactivate pathogenic microorganisms and extend the shelf-life of melon and watermelon juices. Food Microbiology 25 (3), 479–491. http://dx.doi.org/10.1016/j.fm.2008.01.002.

Mourey, A., Canillac, N., 2002. Anti-*Listeria monocytogenes* activity of essential oils components of conifers. Food Control 13 (4–5), 289–292. http://dx.doi.org/10.1016/S0956-7135(02)00026-9.

Muthukumarasamy, P., Han, J.H., Holley, R.A., 2003. Bactericidal effects of *Lactobacillus reuteri* and allyl isothiocyanate on *Escherichia coli* O157:H7 in refrigerated ground beef. Journal of Food Protection 66 (11), 2038–2044. http://dx.doi.org/10.4315/0362-028X-66.11.2038.

Nazzaro, F., Orlando, P., Fratianni, F., Coppola, R., 2012. Microencapsulation in food science and biotechnology. Current Opinion in Biotechnology 23 (2), 182–186. http://dx.doi.org/10.1016/j.copbio.2011.10.001.

Nguefack, J., Tamgue, O., Dongmo, J.B.L., Dakole, C.D., Leth, V., Vismer, H.F., Nkengfack, A.E., 2012. Synergistic action between fractions of essential oils from *Cymbopogon citratus*, *Ocimum gratissimum* and *Thymus vulgaris* against *Penicillium expansum*. Food Control 23 (2), 377–383. http://dx.doi.org/10.1016/j.foodcont.2011.08.002.

Nielsen, P.V., Rios, R., 2000. Inhibition of fungal growth on bread by volatile components from spices and herbs, and the possible application in active packaging, with special emphasis on mustard essential oil. International Journal of Food Microbiology 60 (2–3), 219–229.

Nychas, G.J.E., Skandamis, P.N., Tassou, C.C., 2003. 9-Antimicrobials from herbs and spices A2-Roller, Sibel. In: Natural Antimicrobials for the Minimal Processing of Foods. Woodhead Publishing, pp. 176–200.

Odds, F.C., 2003. Synergy, antagonism, and what the chequerboard puts between them. Journal of Antimicrobial Chemotherapy 52 (1), 1. http://dx.doi.org/10.1093/jac/dkg301.

Olaimat, A.N., Fang, Y., Holley, R.A., 2014. Inhibition of *Campylobacter jejuni* on fresh chicken breasts by κ-carrageenan/chitosan-based coatings containing allyl isothiocyanate or deodorized oriental mustard extract. International Journal of Food Microbiology 187, 77–82.

Oussalah, M., Caillet, S., Salmiéri, S., Saucier, L., Lacroix, M., 2004. Antimicrobial and antioxidant effects of milk protein-based film containing essential oils for the preservation of whole beef muscle. Journal of Agricultural and Food Chemistry 52 (18), 5598–5605.

Pan, X., Mak, L.-M., Nakano, H., 2014. Efficacy of essential oils on inactivation of *Escherichia coli* O157:H7 in vegetable juice. Food Science and Technology Research 20 (5), 1043–1049.

Pandit, V.A., Shelef, L.A., 1994. Sensitivity of *Listeria monocytogenes* to rosemary (*Rosmarinus officinalis* L.). Food Microbiology 11 (1), 57–63. http://dx.doi.org/10.1006/fmic.1994.1008.

Persico, P., Ambrogi, V., Carfagna, C., Cerruti, P., Ferrocino, I., Mauriello, G., 2009. Nanocomposite polymer films containing carvacrol for antimicrobial active packaging. Polymer Engineering and Science 49 (7), 1447–1455.

Pires, A.C.S., Soares, N.F.F., Andrade, N.J., Silva, L.H.M., Camilloto, G.P., Bernardes, P.C., 2009. Increased preservation of sliced mozzarella cheese by antimicrobial sachet incorporated with allyl isothiocyanate. Brazilian Journal of Microbiology 40, 1002–1008.

Rao, A., Zhang, Y., Muend, S., Rao, R., 2010. Mechanism of antifungal activity of terpenoid phenols resembles calcium stress and inhibition of the TOR pathway. Antimicrobial Agents and Chemotherapy 54 (12), 5062–5069. http://dx.doi.org/10.1128/AAC.01050-10.

Rattanachaikunsopon, P., Phumkhachorn, P., 2010. Assessment of factors influencing antimicrobial activity of carvacrol and cymene against *Vibrio cholerae* in food. Journal of Bioscience and Bioengineering 110 (5), 614–619. http://dx.doi.org/10.1016/j.jbiosc.2010.06.010.

Rauha, J.P., Remes, S., Heinonen, M., Hopia, A., Kahkonen, M., Kujala, T., Vuorela, P., 2000. Antimicrobial effects of Finnish plant extracts containing flavonoids and other phenolic compounds. International Journal of Food Microbiology 56 (1), 3–12.

Raybaudi-Massilia, R.M., Mosqueda-Melgar, J., Martin-Belloso, O., 2006. Antimicrobial activity of essential oils on *Salmonella* Enteritidis, *Escherichia coli*, and *Listeria innocua* in fruit juices. Journal of Food Protection 69 (7), 1579–1586.

Raybaudi-Massilia, R.M., Mosqueda-Melgar, J., Soliva-Fortuny, R., Martín-Belloso, O., 2009. Control of pathogenic and spoilage microorganisms in fresh-cut fruits and fruit juices by traditional and alternative natural antimicrobials. Comprehensive Reviews in Food Science and Food Safety 8 (3), 157–180. http://dx.doi.org/10.1111/j.1541-4337.2009.00076.x.

Roller, S., 2003. In: Roller, S. (Ed.), Natural Antimicrobials for the Minimal Processing of Foods. Woodhead Publishing, Limited, Cambridge.

Salarbashi, D., Tajik, S., Ghasemlou, M., Shojaee-Aliabadi, S., Shahidi Noghabi, M., Khaksar, R., 2012. Characterization of soluble soybean polysaccharide film incorporated essential oil intended for food packaging. Carbohydrate Polymers 98, 1127–1136.

Salvia-Trujillo, L., Rojas-Graü, M.A., Soliva-Fortuny, R., Martín-Belloso, O., 2015. Use of antimicrobial nanoemulsions as edible coatings: impact on safety and quality attributes of fresh-cut Fuji apples. Postharvest Biology and Technology 105, 8–16.

Samperio, C., Boyer, R., Eigel 3rd, W.N., Holland, K.W., McKinney, J.S., O'Keefe, S.F., Marcy, J.E., 2010. Enhancement of plant essential oils' aqueous solubility and stability using alpha and beta cyclodextrin. Journal of Agricultural and Food Chemistry 58 (24), 12950–12956. http://dx.doi.org/10.1021/jf103275a.

Santos, A.R., Silva, A.F., Amaral, V.C., Ribeiro, A.B., Abreu Filho, B.A., Mikcha, J.M., 2016. Application of edible coating with starch and carvacrol in minimally processed pumpkin. Journal of Food Science and Technology 53 (4), 1975–1983.

Selim, S., 2011. Antimicrobial activity of essential oils against vancomycin-resistant enterococci (vre) and *Escherichia coli* O157:H7 in feta soft cheese and minced beef meat. Brazilian Journal of Microbiology 42 (1), 187–196. http://dx.doi.org/10.1590/s1517-83822011000100023.

Senatore, F., 1996. Influence of harvesting time on yield and composition of the essential oil of a thyme (*Thymus pulegioides* L.) growing wild in Campania (Southern Italy). Journal of Agricultural and Food Chemistry 44 (5), 1327–1332. http://dx.doi.org/10.1021/jf950508z.

Sessa, M., Ferraria, G., Donsìb, F., 2015. Novel edible coating containing essential oil nanoemulsions to prolong the shelf life of vegetable products. Chemical Engineering Transactions 43, 55–60.

Severino, R., Vu, K.D., Donsì, F., Salmieri, S., Ferrari, G., Lacroix, M., 2014. Antibacterial and physical effects of modified chitosan based-coating containing nanoemulsion of mandarin essential oil and three non-thermal treatments against *Listeria innocua* in green beans. International Journal of Food Microbiology 191, 82–88.

Sipahi, R.E., Castell-Perez, M.E., Moreira, R.G., Gomes, C., Castillo, A., 2013. Improved multilayered antimicrobial alginate-based edible coating extends the shelf life of fresh-cut watermelon (*Citrullus lanatus*). LWT – Food Science and Technology 51, 9–15.

Skandamis, P., Tsigarida, E., Nychas, G.J.E., 2002. The effect of oregano essential oil on survival/death of *Salmonella* Typhimurium in meat stored at 5°C under aerobic, VP/MAP conditions. Food Microbiology 19 (1), 97–103. http://dx.doi.org/10.1006/fmic.2001.0447.

Skandamis, P.N., Nychas, G.J., 2001. Effect of oregano essential oil on microbiological and physico-chemical attributes of minced meat stored in air and modified atmospheres. Journal of Applied Microbiology 91 (6), 1011–1022.

Somolinos, M., Garcia, D., Condon, S., Mackey, B., Pagan, R., 2010. Inactivation of *Escherichia coli* by citral. Journal of Applied Microbiology 108 (6), 1928–1939. http://dx.doi.org/10.1111/j.1365-2672.2009.04597.x.

Sun, X., Narciso, J., Wang, Z., Ference, C., Bai, J., Zhou, K., 2014. Effects of chitosan-essential oil coatings on safety and quality of fresh Blueberries. Journal of Food Science 79 (5), M955–M960.

Suppakul, P., Miltz, J., Sonneveld, K., Bigger, S.W., 2003. Antimicrobial properties of basil and its possible application in food packaging. Journal of Agricultural and Food Chemistry 51 (11), 3197–3207. http://dx.doi.org/10.1021/jf021038t.

Tassou, C.C., Drosinos, E.H., Nychas, G.J., 1995. Effects of essential oil from mint (*Mentha piperita*) on *Salmonella enteritidis* and *Listeria monocytogenes* in model food systems at 4 degrees and 10°C. Journal of Applied Bacteriology 78 (6), 593–600.

Thaker, M., Hanjabam, M.D., Gudipati, V., Kannuchamy, N., 2017. Protective effect of fish gelatin-based natural antimicrobial coatings on quality of Indian salmon fillets during refrigerated storage. Journal of Food Process Engineering 40 (1).

Tsigarida, E., Skandamis, P., Nychas, G.J., 2000. Behaviour of *Listeria monocytogenes* and autochthonous flora on meat stored under aerobic, vacuum and modified atmosphere packaging conditions with or without the presence of oregano essential oil at 5°C. Journal of Applied Microbiology 89 (6), 901–909.

Ultee, A., Bennik, M.H., Moezelaar, R., 2002. The phenolic hydroxyl group of carvacrol is essential for action against the food-borne pathogen *Bacillus cereus*. Applied and Environmental Microbiology 68 (4), 1561–1568.

Valdés, A., Ramos, M., Beltrán, A., Jiménez, A., Garrigós, M.C., 2017. State of the art of antimicrobial edible coatings for food packaging applications. Coatings 7 (4), 56, https://doi.org/10.3390/coatings7040056.

Ward, S.M., Delaquis, P.J., Holley, R.A., Mazza, G., 1998. Inhibition of spoilage and pathogenic bacteria on agar and pre-cooked roasted beef by volatile horseradish distillates. Food Research International 31, 19–26.

World Health Organization, 2003. Diet, Nutrition and the Prevention of Chronic Diseases. Report of a Joint FAO/WHO Expert Consultation, Geneva, World Health Organization Technical Report Series, No. 916. Retrieved on 6-30-2017 from http://www.who.int/dietphysicalactivity/publications/trs916/download/en/.

Xi, Y., Sullivan, G.A., Jackson, A.L., Zhou, G.H., Sebranek, J.G., 2012. Effects of natural antimicrobials on inhibition of *Listeria monocytogenes* and on chemical, physical and sensory attributes of naturally-cured frankfurters. Meat Science 90 (1), 130–138. http://dx.doi.org/10.1016/j.meatsci.2011.06.013.

Yang, F., Hu, S., Lu, Y., Yang, H., Zhao, Y., Li, L., 2015. Effects of coatings of polyethyleneimine and thyme essential oil combined with chitosan on sliced fresh *Channa argus* during refrigerated storage. Journal of Food Process Engineering 38 (3), 225–233.

Yun, J., Fan, X., Li, X., Jin, T.Z., Jia, X., Mattheis, J.P., 2015. Natural surface coating to inactivate *Salmonella enterica* serovar Typhimurium and maintain quality of cherry tomatoes. International Journal of Food Microbiology 193, 59–67.

CHAPTER 15

# Characteristics of Bacteriocins and Use as Food Antimicrobials in the United States

**Corliss A. O'Bryan[1], Ok Kyung Koo[2], Michael L. Sostrin[1,3], Steven C. Ricke[1], Philip G. Crandall[1], Michael G. Johnson[1]**

[1]*University of Arkansas, Fayetteville, AR, United States;* [2]*Gyeongsang National University, Jinju, Republic of Korea;* [3]*Walmart Stores, Inc., Bentonville, AR, United States*

## Food Antimicrobials

Food antimicrobials are compounds used to inhibit the growth of microorganisms or in some instances to achieve complete cell death (Davidson et al., 2013; Davidson and Harrison, 2002). Most of the antimicrobials and sanitizing agents currently used in food manufacturing have been in use for 50 to 100 plus years; for instance, benzoic acid and its salts were one of the first groups of antimicrobials approved for use in foods (Davidson and Harrison, 2002). In the United States, many traditional antimicrobials are approved for use in food products including acetic acid in baked goods and condiments, benzoic acid in beverages and margarine, lactic acid in meats and fermented foods, nitrate and nitrite in cured meats, sorbic acid in wines, and sulfites in fruit and potato products (Davidson and Harrison, 2002). Although food antimicrobials have been used to inhibit spoilage organisms and prolong the shelf life and quality of foods for many years, only a select few are used solely for the purpose of controlling the growth of specific bacterial pathogens (Davidson and Harrison, 2002). Examples of this include the use of nitrite to inhibit the growth of *Clostridium botulinum* in cured meats, specific organic acid sprays used to reduce pathogens on beef carcass surfaces, nisin and lysozyme used to inhibit *C. botulinum* in pasteurized processed cheese, and lactate and diacetate used to inhibit *Listeria monocytogenes* in processed meats (Lindsay, 2008).

Consumers have voiced an increased interest in food products that are minimally processed and have reduced amounts of synthetic additives (Gould, 2000). Food manufacturers worldwide are striving to offer a wider selection of such food products (Lockie et al., 2002). The organic food industry has sales of $63 billion worldwide, and the United States represents the largest single market for organic food and beverages, with $31.5 billion in sales in 2011 (Anonymous, 2013). A wide range of effective natural antimicrobial systems can be derived from animals, plants, and microorganisms (Tiwari et al., 2009). Natural antimicrobials derived from animals include

Food and Feed Safety Systems and Analysis. http://dx.doi.org/10.1016/B978-0-12-811835-1.00015-4

enzymes such as lysozymes, proteins such as lactoferrin, and small peptides such as histatins (Davidson et al., 2013). However, natural antimicrobials derived from animals are still viewed as undesirable by the increasing population of consumers whose vegetarian or vegan diets restrict or prohibit the consumption of animal products (Jabs et al., 2000). Therefore, inhibition of foodborne pathogens by natural antimicrobials derived from plants (e.g., phenolics and essential oils) or microorganisms (e.g., bacteriocins) may be preferred. In this chapter, we will discuss the history and classification of bacteriocins, their various modes of action, and applications in food systems, especially as used in the United States.

## Bacteriocins

Although there is some confusion as to the definition of the term "bacteriocin," the definition proposed by Tagg et al. (1976) is most frequently quoted. This definition describes bacteriocins as bactericidal proteins with a narrow inhibitory spectrum, mainly for the same species as the bacteriocin producer. It was also proposed that a bacteriocin should be coded by plasmid-borne genetic determinants, adhere to specific cell receptors, and be produced by a lethal biosynthesis mechanism, meaning the producer cell dies (Tagg et al., 1976). These general points are largely accepted to be common to many bacteriocins but are by no means universal. In recent years the definition has broadened somewhat as the diversity of these compounds has become increasingly apparent, and one can consider a bacteriocin as being "a non–self-propagating protein or protein-based complex, displaying growth-inhibitory activity against a restricted range of bacteria" (Kaiser and Montville, 1993).

### Classification and Nomenclature of Bacteriocins

Bacteriocins, according to the classification procedure proposed by Klaenhammer (1993) and modified by Nes et al. (1996), are divided into four classes based on their molecular weight, physicochemical properties, and their amino acid sequence, with the majority of those produced by bacteria associated with food belonging to class I or II (Table 15.1).

**Table 15.1: Classification and characteristics of bacteriocins.**

| Class | Subclass | Mass | Characteristics | Bacteriocins |
|---|---|---|---|---|
| I | A | <5 kDa | Lantibiotics | Nisin, subtilin, lacticin |
| | B | | | Merscacidin |
| II | a | <10 kDa | Antilisterial | Pediocin, sakacin |
| | b | | Require two components | Lactococcin, lactacin |
| | c | | Signal peptide secretion dependent | Enterocin P |
| III | | >30 kDa | Heat-sensitive proteins | Helveticin J |
| IV | | >30 kDa | Require lipid or carbohydrate moiety | Glycocin |

Class I bacteriocins are made up of small membrane-active peptides (less than 5 kDa) containing the unusual amino acid lanthionine. The oldest and the most extensively studied class I bacteriocin is nisin A, which was discovered by Rogers and Whittier (1928). Nisin A is produced by many strains of *Lactococcus lactis*, a lactic acid bacterium widely used for cheese manufacture. Nisin A affects a wide range of Gram-positive genera, including staphylococci, streptococci, *Listeria* spp., bacilli, and enterococci. Nisin A has been used in the food industry as a biopreservative for more than 50 years without inducing microbial resistance (Cotter et al., 2005). Nisin binds to the precursor of peptidoglycan, lipid II, on the bacterial cell wall and the nisin–lipid II complex forms pores in the bacterial cell membrane allowing leakage of essential cellular materials which causes cell death (Delves-Broughton &Weber, 2011). Other examples of class I bacteriocins include lacticin 481, carnocin U149, and lactosin S (Stoyanova et al., 2012).

Class II bacteriocins are made up of small non–lanthionine-containing heat-stable, membrane-active peptides (Perez et al., 2014). Class II is further subdivided into three subgroups, IIa, IIb, and IIc. Subgroup IIa bacteriocins are active against *Listeria* and include pediocin PA-1, pediocin AcH, sakacin A, sakacin P, curvacin A, and leucocin A (Drider et al., 2006). Pediocin PA-1, produced by *Pediococcus acidilactici*, is the most well known of the class IIa bacteriocins. Pediocin PA-1 permeabilizes the cytoplasmic membrane of the receptor bacteria, resulting in a leakage of ions and small molecules and cell death (Sun et al., 2015).

Subgroup IIb bacteriocins include lactacin F, Lactococcin M, and Lactococcin G and are characterized by their poration complexes requiring two different peptides for activity (Balla et al., 2000). The best characterized of these two peptide systems are plantaricins EF/JK and plantaricin S produced by C11 and *Lactobacillus plantarum* LPCO10, respectively (Anderssen et al., 1998; Jiménez-Díaz et al., 1995). Subgroup IIc bacteriocins include lactococcin B and are thiol-activated peptides requiring reduced cysteine residues for activity (Kawai et al., 2004).

The class III and class IV bacteriocins are not well characterized. The class III bacteriocins include helveticin J, helveticin V-1829, lactacin A and B, and acidophilus A and are large heat-labile proteins (greater than 30 kDa) (Joerger and Klaenhammer, 1986; Nilsen et al., 2003). The class IV bacteriocins consist of complex bacteriocins that require carbohydrate or lipid moieties for activity, but information about this class is limited and contradictory (Franz et al., 2007).

### *Mode of Action of Bacteriocins*

Class I bacteriocins form a complex with the bacterial cell wall precursor lipid II, thereby inhibiting cell wall biosynthesis, after which the complexes aggregate, incorporate further peptides, and form an actual pore through the bacterial membrane (Bierbaum and Sahl, 2009).

The pore formation leads to loss of membrane integrity and causes loss of proton motive force (PMF) and intracellular ions and cell death. This loss results in a disruption of the chemical gradients across the membrane, leading to the loss of ability to transport protons and generate ATP (Bruno et al., 1992; Bruno and Montville, 1993). Similar losses of $K^+$ ions for sensitive cells were reported for the activities of the bacteriocin bifidocin B produced by *Bifidobacterium bifidum* NCFB by Yildirim and Johnson (1998) and Yildirim et al. (1999). Class Ib bacteriocins are mainly globular, either uncharged or negatively charged, and inhibit the synthesis of peptidoglycan, the essential sugar–peptide backbone of the bacterial cell wall (Brötz and Sahl, 2000). Class IIa bacteriocins permeabilize the cell membrane of sensitive cells, and recent research has indicated that the target of these molecules is a mannose permease protein on the target cell surface (Drider et al., 2006). Subclass IIb bacteriocins are composed of two peptides, and includes the alpha enterocins and lactococcin G peptides; they inhibit the growth of *Enterococcus* spp. and a few other Gram-positive bacteria, acting as pore-forming toxins that create cell membrane channels through a barrel-stave mechanism in which bacteriocin monomers bind, insert, and oligomerize the cytoplasmic membrane to form a pore with a hydrophobic residue facing outward and a hydrophilic residue facing inward, causing membrane leakage, and thus producing an ionic imbalance in the cell (Balla et al., 2000). Subclass IIc bacteriocins have various modes of action including membrane permeabilization, specific inhibition of septum formation, and pheromone activity (Héchard and Sahl, 2002). For a more in-depth discussion of the various modes of action of bacteriocins, see Snyder and Worobo (2014).

## Nisin

In the late 1920s and early 1930s, nisin was discovered and described as a toxic substance in milk which had adverse effects on the performance of cheese starter cultures (Rogers and Whittier, 1928). The first investigation into its potential as a food preservative was by Hirsch et al. (1951), and in the following year, research was performed on the use of nisin as a preservative in processed cheese (McClintock et al., 1952). Nisaplin, the first commercially developed nisin preparation, was produced by Aplin and Barrett in 1957 (Delves-Broughton, 2005). Currently nisin is approved for use in over 40 countries and is the only bacteriocin that has generally recognized as safe (GRAS) status in the United States (Hurst, 1981; De Vuyst and Vandamme, 1994). Table 15.2 lists the most common food uses of nisin and levels used.

Nisin is a peptide formed by 34 amino acids, with a molecular weight of less than 5 kDa (Gross and Morell, 1971). A fully active nisin molecule contains a five-ring structure composed of thio-ester bridges that form during the posttranslational dehydration of serine and threonine (Sahl et al., 1995). This structure was confirmed by proton nuclear magnetic resonance and illustrated in a study by Chan et al. (1989). Molecular biological studies

**Table 15.2: Most frequent use of nisin in foods.**

| Food | Purpose | Level of usage (mg/kg) |
|---|---|---|
| Processed cheese | Prevent growth of *Clostridia* | 5 to 15 |
| Ricotta cheese | *Listeria* control | 2.5 to 5.0 |
| Pasteurized milk products | Used when addition of flavorings increase bacterial count | 0.25 to 10.0 |
| Shredded cheese | Prevent spoilage | Less than 250 |
| Salad dressings | Prevent spoilage by lactic acid bacteria | 1.25 to 6.25 |
| Alcoholic beverages | Prevent spoilage by lactic acid bacteria | 0.25 to 1.25 |
| Canned foods | Prevent growth of *Clostridium botulinum* | 2.5 to 5.0 |
| Frankfurters/hot dogs, deli meats | *Listeria* control | 5.5 to 6.9 |
| Egg products | Increase shelf life | 2.5 to 5.0 |

have shown that nisin is derived from ribosomally synthesized precursors (Chan et al., 1989). The precursor of nisin is 57 amino acid residues long and is composed of a 23-residue leader sequence followed by a 34-residue sequence (Chan et al., 1989). This 34-residue sequence becomes an active nisin molecule when its serine, threonine, and cysteine precursors transform into lanthionine, 3-methylanthionine, and α, β-didehydroamino acid residues (Chan et al., 1989). Nisin acts on the cytoplasmic membrane of cells, forming transient pores which allow the efflux of compounds such as ATP, ADP, amino acids, and essential cations such as $K^+$. The formation of transient pores is initiated by an electrostatic attraction between the cationic nisin molecule and the anionic phospholipids of the cell (Abee et al., 1994). Nisin subsequently inserts into the cell membrane in a perpendicular orientation, and because contact with phospholipids is continuous in this formation, the phospholipids bend the bilayer resulting in the formation of a transient pore (Abee et al., 1994). This pore formation results in the dissipation of membrane potential and ionic gradient across the membrane and disrupts the PMF, the transport of amino acids, and the release of accumulated amino acids. Energy metabolism of the cell is eventually disrupted, resulting in cell death (Abee et al., 1994).

A wide range of Gram-positive organisms including germinating *Bacillus* and *Clostridium* spores are inhibited by nisin (Stevens et al., 1991), although little or no inhibition against Gram-negative bacteria, yeasts, or molds has been observed (Hurst, 1981). The outer membrane of Gram-negative bacteria acts as a permeability barrier for the cell and may explain the decreased activity of nisin against these organisms (Stevens et al., 1991). However, nisin-induced inhibition of Gram negative bacteria may be achieved if the target cells have been pretreated with the chelating agent ethylenediaminetetraacetic acid (ETDA) which disrupts the outer cell envelope, allowing nisin to reach and disrupt the underlying cell membrane (Stevens et al., 1991).

## Pediocins

*Pediococcus* is a genus of Gram-positive lactic acid bacteria, belonging to the family of Lactobacillacea. *P. acidilactici* and *Pediococcus pentosaceus* are mostly associated with food fermentations, either as indigenous microbiota or in starters, and both have been used in natural and controlled fermentations of vegetables and sausages (Wood, 1997; Knorr, 1998). The ability of these organisms to produce bacteriocins has attracted interest for the use of either the cultures or their products as protective cultures or biopreservatives in foods. The class II bacteriocins produced by pediococci are termed as "antilisterial" (Papagianni, 2003).

Pediocins are synthesized with a leader peptide attached which is cleaved and removed during extracellular exposure by a proteolytic enzyme, with the cleavage occurring usually after a double glycine residue (Ray, 1995). Studies with pediocin AcH revealed that at the translation level, it is synthesized as a biologically inactive peptide with 66 amino acid residues, which then undergoes a posttranslational modification, which includes the removal of a leader fragment of 18 amino acids from the N-terminal, to produce a 44 amino acid peptide that is biologically active (Motlagh et al., 1992). Posttranslational modification is dependent on enzymes that are activated at low pH (Johnson et al., 1992; Ray, 1995). The importance of the C-terminal region of pediocins in determining the target cell specificity was demonstrated in research with pediocin AcH (same as pediocin PA-1) in which cleaving the peptide region from residue 20 to residue 34 greatly diminished the bactericidal activity of the pediocin (Fimland et al., 1998).

Pediocins target the cytoplasmic membrane of Gram-positive bacteria and, similarly to other class IIa bacteriocins whose modes of action have been studied, permeabilize the cytoplasmic membrane through pore formation by insertion of the C-terminal regions into the membrane (Drider et al., 2006; Papagianni, 2003). Pediocin AcH/PA-1 was found to bind to receptors on the cell surface of both sensitive and resistant cells, but destabilization of the membrane occurred only in sensitive cells (Bhunia et al., 1991). Bauer et al. (2005) concluded that pediocin PD-1 acts on the cytoplasmic membrane of *Oenococcus oeni* by generating pores in the membrane. The ability of pediocin PD-1 to form pores in sensitive cells of *O. oeni*, as observed by $K^+$ loss was found to be pH dependent and increased when the extracellular pH was reduced from 7.0 to 5.0. In the presence of 10-mM gadolinium ($Gd3^+$), pediocin PD-1 did not affect cells of *O. oeni*, suggesting that the mode of action of the pediocin relies on a net negatively charged cell surface. Anastasiadou et al. (2008) observed a bactericidal mode of action in studies with purified pediocins SA-1 and SM-1 and indicator cells of *Micrococcus luteus* at midlogarithmic phase of growth. Colony-forming units of plated *M. luteus* decreased from $10^8$ to 0 within 4 h, but the turbidity was unchanged at OD 600 nm, suggesting that no cell lysis took place.

## Bacteriocins in Food Systems

### Regulatory Environment in the United States

The assessment of toxic effects of food additives is based on amount of exposure. Because most bacteriocins are composed largely of polypeptides, they thought to be largely inactivated *in situ* by proteolytic digestive enzymes. Therefore, bacteriocin exposure levels experienced by consumers should be very low (Kroes et al., 2005). The primary legislation regulating bacteriocins comes from the Food Additives Amendment of 1958, which requires premarket approval for food additives, except for substances that are GRAS. A substance is granted GRAS status based on empirical evidence or evidence of years of safe consumption and includes ingredients such as sugar and salt, as well as ingredients from plants or animals such as processed starches or enzyme preparations (Fields, 1996). The U.S. Food and Drug Administration (FDA) maintains a list of approved uses for given GRAS compounds, including specifications on concentration limits and what food applications are allowed (Gillor et al., 2007). Bacteriocins produced naturally by food-grade bacteria, dried in the production medium such as milk, and added to food products in concentrations that are no higher than what would be found in a fermented food can be considered GRAS. However, if the producing bacterium is not food grade, has been genetically modified, or if the bacteriocin is added in elevated concentrations, then it is not considered GRAS and is subject to premarket approval. In the United States, the FDA has approved nisin for use as a food additive in certain foods at levels of 2.9 mg (activity not less than 900 International Units per mg) per person per day (Federal Register, 1988).

Regulation of the use of bacteriocins in foods is under the authority of a multitude of agencies depending on food product. Although the FDA determines GRAS status for food ingredients including bacteriocins, use may be under the authority of the Center for Food Safety and Applied Nutrition of FDA, the Center for Veterinary Medicine of FDA, or the Food Safety Inspection Service in the USDA. Authority can even fall to the EPA under regulations for pesticides covered in the Federal Insecticide, Fungicide, and Rodenticide Act if a bacteriocin is to be used on whole fruits or vegetables (Fields, 1996; Post, 1996).

### Methods to Use Bacteriocins in Food Products

Bacteriocins can be integrated into a food product in a variety of ways: addition of the purified compound as a food additive, addition of a bacterial fermentate as a GRAS substance, addition of the producer bacterium to the product or as part of the starter culture, incorporation into an edible food coating or bound to food packaging. Each approach has specific advantages and limitations and may be more or less appropriate in particular foods.

Complete purification of a bacteriocin from the crude fermentate is expensive and not necessary if the producing strain is a food-grade bacterium (Garsa et al., 2014), and therefore commercial use of purified bacteriocins is rare (Deegan et al., 2006). The purification process involves several partial purification steps including salt precipitations, filtrations, and chromatography increasing the concentration several fold to almost total purity (De Vuyst and Leroy, 2007; Joshi et al., 2006).

Alternatively, bacteriocins can be added as freeze-dried end products of a bacterial fermentation. The producer bacterium is grown in a milk-based medium, and the end product will have the secondary metabolites produced by the bacteria, including bacteriocins and organic acids (Galvez et al., 2008). These types of preparations may be seen as food additives or ingredients from the point of view of the FDA because some of their components may have a function in the food, for instance increasing protein content or serving as a thickening agent (O'Sullivan et al., 2002). Concentrates currently on the market include ALTA 2351 (Rozum and Maurer, 1997) and Microgard (von Staszewski and Jagus, 2008). Other milk-based preparations have been described such as lacticin 3147 (Suda et al., 2012) and variacin (O'Mahony et al., 2001).

Addition of the bacteriocin-producing microorganism is especially useful in fermented foods where it can be included as part of the starter culture (Callewaert et al., 2000). The addition of a food-grade producer directly to the food product is less costly and meets regulatory requirements, but the producer must be compatible with the starter culture bacteria as well as having the capacity to express the bacteriocin under food storage conditions (De Vuyst and Leroy, 2007; O'Sullivan et al., 2002). It is also possible for the bacteriocin-producing organism to be inactivated in food by bacteriophage activity, and spontaneous loss of bacteriocin expression can also occur (Schillinger et al., 1996).

To avoid the problems posed by the food matrix, bacteriocins can also be applied in the form of coatings. The carrier-coating material protects the bacteriocin from inactivation due to interaction with food components and enzymatic inactivation. In addition, the application of bacteriocin molecules on the food surface requires much lower amounts of bacteriocin (compared to application in the whole food volume), decreasing the processing costs. Ming et al. (1997) placed pediocin powder onto the surface of plastic packaging of meat or poultry inoculated with *L. monocytogenes* and demonstrated a complete inhibition of growth through 12 weeks of storage at 4°C. Nisin was incorporated into a polyethylene-based plastic film and used to wrap beef carcass surface tissue sections surface inoculated with the psychrotrophic spoilage bacterium *Brocothrix thermosphacta* and held at 4°C. At the end of 20 days, the nisin-impregnated plastic-wrapped samples had significantly lower counts as compared to controls (Siragusa et al., 1999). Natrajan and Sheldon (2000) examined the use of protein- and polysaccharide-based films containing nisin to inhibit *Salmonella* Typhimurium on fresh broiler skin and were able to demonstrate a 2-log reduction after 96 h at 4°C. Lungu and Johnson (2005b) incorporated nisin into zein coatings for frankfurters surface inoculated with *L. monocytogenes*. Over 28-day storage at 4°C, the nisin films reduced *L. monocytogenes* populations on frankfurters by 6.6 log CFU/g. These studies demonstrate that the use of antimicrobial films containing bacteriocins can improve the quality and safety and prolong the shelf life of food products.

### Factors Affecting Use of Bacteriocins in Food

Using bacteriocins effectively in foods requires the consideration of a variety of factors including the type of food and the types and numbers of bacteria that require inhibition or killing. Data obtained in culture media often suggest much greater inhibition than what is obtained in foods and often as much as 10 times higher concentration of bacteriocin may be required in foods to achieve an equivalent effect (Schillinger et al., 1996). Bacteriocins can be partially or completely inactivated because of a variety of reasons, including adsorption to food components, the action of proteases in the food, deleterious effects of processing, or even the pH of the food matrix (Hugas et al., 1998; Stiles, 1994). Higher fat levels also decrease the effectiveness of nisin (Jung et al., 1992).

The food matrix also has an effect on the application of bacteriocins to food. For instance, nisin has been well studied and has been found to be most effective against Gram-positive bacteria, in foods with low fat and protein content and a pH of less than 6.0. One such example is the use of nisin in beer and brandy to control the growth of spoilage bacteria without affecting yeast growth (Henning et al., 1986; Ogden, 1986). Studies performed by Jones (1974) on controlling *Staphylococcus aureus* in milk found that nisin was more effective in skim milk as opposed to whole milk due to the butter fat content of the latter. Other studies by Henning et al. (1986) corroborate the diminished activity of nisin in foods containing fat. Jung et al. (1992) reported that a concentration of 1.25 mg/L nisin caused a four- to six-log reduction of *L. monocytogenes* in nonfat milk, whereas in half and half cream, less than 1-log reduction was achieved. A study by Davies et al. (1997) reported that addition of 2.5 mg/L nisin inhibited *L. monocytogenes* in ricotta cheese for up to 55 days at 6°C–8°C and that lowering the pH by adding acetic acid extended this inhibition to over 10 weeks. The solubility of nisin is also affected by pH, and it is most stable at pH 3.0 to 3.5 (Delves-Broughton, 2005). In addition, Daeschel et al. (1991) found that nisin exhibited 90% diminished activity in red wines as compared to white wines during the first 4 months of storage, possibly due to the inactivation of nisin by the higher content of phenolic compounds in red wine.

## Food Applications

*L. monocytogenes* is considered to be the Gram-positive pathogen of most interest in ready-to-eat foods because it can survive and grow over a wide range of foods and environmental conditions including at refrigerator temperatures. Nisin and class IIa bacteriocins are known to be active against this foodborne pathogen, but there has been an emergence of bacteriocin-resistant strains as well as cross-resistance observed between bacteriocins (Naghmouchi et al., 2007; Tessema et al., 2009). Good inhibitory activities against *L. monocytogenes* on some selected foods were demonstrated for pediocin on refrigerated chicken (Goff et al., 1996), bifidocin B on cooked chicken (Yildirim et al., 2007), and nisin with other selected GRAS chemicals in a corn protein zein film on cooked chicken or turkey (Janes et al., 2002; Lungu and Johnson, 2005a). More recently, the synergistic activity of the food flavoring agent diacetyl with nisin against starved cells of *L. monocytogenes* has also been demonstrated (O'Bryan et al., 2009). Inactivation of *S. aureus*, another Gram-positive foodborne pathogen, has had variable results in food systems such as sauces, desserts, meat, or milk when bacteriocins were added (Jofré et al., 2008; Martinez Viedma et al., 2009; Ananou et al., 2010). Inactivation of spores of *Bacillus cereus* and *C. botulinum* is still a concern in the food industry, and many bacteriocins, particularly nisin, have shown antimicrobial activity against endospore-forming bacterial cells and germinating spores in food systems (Galvez et al., 2008). Bacteriocins alone are not effective against Gram-negative bacteria but could be used in combination with other treatments to induce cell damage and partial disorganization of the outer cell membrane protective layer (Galvez et al., 2008).

## Conclusions

Although a wide range of bacteriocins exhibit significant antimicrobial and other technologically important qualities, the use of a bacteriocin alone in food will not ensure sufficient safety. Gram-negative bacteria do not represent target cells for bacteriocins as they are protected by an outer membrane. Bacteriocins should be used in combination with other preservation methods, such as other antimicrobials or organic acids in true multiple hurdle technology, and use genetic methods to optimize synergy in the choice of antimicrobials used as combinations. In addition, with the developments in molecular biology that have occurred in the past few years, there are unprecedented opportunities to not only potentially modify existing bacteriocins to enhance and/or expand their antimicrobial properties but use genetic information to develop high-throughput methods to hunt and screen for new bacteriocin-producing microbial species in the environment. Likewise, as more is understood about the genetic regulation of bacteriocin-producing bacteria, there may be potential to develop more precise delivery vehicles that can be used to target foodborne pathogens in a broader range of food products as well as food animals.

## References

Abee, T., Rombouts, F.M., Hugenholtz, J., Guihard, G., Letellier, L., 1994. Mode action of nisin Z against *Listeria monocytogenes* Scott A grown at high and low temperatures. Applied and Environmental Microbiology 60, 1962–1968.

Ananou, S., Garriga, M., Jofré, A., Aymerich, T., Gálvez, A., Maqueda, M., Martínez-Bueno, M., Valdivia, E., 2010. Combined effect of enterocin AS-48 and high hydrostatic pressure to control food-borne pathogens inoculated in low acid fermented sausages. Meat Science 84, 594–600.

Anastasiadou, S., Papagianni, M., Filiousis, G., Ambrosiadis, I., Koidis, P., 2008. Growth and metabolism of a meat isolated strain of *Pediococcus pentosaceus* in submerged fermentation: purification, characterization and properties of the produced pediocin SM-1. Enzyme and Microbial Technology 43, 448–454.

Anderssen, E.L., Diep, B.D., Nes, I.F., Eijsink, V.G.H., Nissen-Meyer, J., 1998. Antagonistic activity of *Lactobacillus plantarum* C11: two new two-peptide bacteriocins, plantaricins EF and JK, and the induction factor plantaricin A. Applied and Environmental Microbiology 64, 2269–2272.

Anonymous, 2013. Global Organic Sales Reach $63 Billion, U.S. is Largest Market. Available from: http://www.agprofessional.com/news/212753341.html.

Balla, E., Dicks, L.M., Du Toit, M., Van Der Merwe, M.J., Holzapfel, W.H., 2000. Characterization and cloning of the genes encoding enterocin 1071A and enterocin 1071B, two antimicrobial peptides produced by *Enterococcus faecalis* BFE 1071. Applied and Environmental Microbiology 66, 1298–1304.

Bauer, R., Chikindas, M.L., Dicks, L.M.T., 2005. Purification, partial amino acid sequence and mode of action of pediocin PD-1, a bacteriocin produced by *Pediococcus damnosus* NCFB1832. International Journal of Food Microbiology 101, 17–27.

Bhunia, A.K., Johnson, M.C., Ray, B., Kalchayanad, N., 1991. Mode of action of pediocin AcH from *Pediococcus acidilactici* H on sensitive bacterial strains. Journal of Applied Bacteriology 70, 23–25.

Bierbaum, G., Sahl, H.G., 2009. Lantibiotics: mode of action, biosynthesis and bioengineering. Current Pharmaceutical Biotechnology 10, 2–18.

Brötz, H., Sahl, H.-G., 2000. New insights into the mechanism of action of lantibiotics—diverse biological effects by binding to the same molecular target. Journal of Antimicrobials and Chemotherapy 46, 1–6.

Bruno, M.E.C., Kaiser, A., Montville, T.J., 1992. Depletion of proton motive force by nisin in *Listeria monocytogenes* cells. Applied and Environmental Microbiology 58, 2255–2259.

Bruno, M.E.C., Montville, T.J., 1993. Common mechanistic action of bacteriocins from lactic acid bacteria. Applied and Environmental Microbiology 59, 3003–3010.

Callewaert, R., Hugas, M., De Vuyst, L., 2000. Competitiveness and bacteriocin production of enterococci in the production of Spanish-style dry fermented sausages. International Journal of Food Microbiology 57, 33–42.

Chan, W.C., Lian, L.Y., Bycroft, B.W., Roberts, G.K.C., 1989. Confirmation of the structure of nisin by complete H-1-NMR resonance assignment in aqueous and dimethyl sulfoxide solutions. Journal of the Chemical Society-Perkin Transactions 1, 2359–2367.

Cotter, P.D., Hill, C., Ross, R.P., 2005. Bacteriocins: developing innate immunity for food. Nature Reviews Microbiology 3, 777–788.

Daeschel, M.A., Jung, D.S., Watson, B.T., 1991. Controlling wine malolactic fermentation with nisin and nisin-resistant strains of *Leuconostoc oenos*. Applied and Environmental Microbiology 57, 601–603.

Davidson, P.M., Critzer, F.J., Taylor, T.M., 2013. Naturally occurring antimicrobials for minimally processed foods. Annual Review of Food Science and Technology 4, 163–190.

Davidson, P.M., Harrison, M.A., 2002. Resistance and adaptation to food antimicrobials, sanitizers, and other process controls. Food Technology 56 (11), 69–78.

Davies, E.A., Bevis, H.E., Delves-Broughton, J., 1997. The use of the bacteriocin, nisin, as a preservative in ricotta-type cheeses to control the food-borne pathogen *Listeria monocytogenes*. Letters in Applied Microbiology 24, 343–346.

Deegan, L.H., Cotter, P.D., Hill, C., Ross, P., 2006. Bacteriocins: biological tools for bio-preservation and shelf-life extension. International Dairy Journal 16, 1058–1071.

Delves-Broughton, J., 2005. Nisin as a food preservative. Food Australia 57, 525–527.

Delves-Broughton, J., Weber, G., 2011. Nisin, natamycin and other commercial fermentates used in food biopreservation. In: Lacroix, C. (Ed.), Protective Cultures, Antimicrobial Metabolites and Bacteriophages for Food and Beverage Biopreservation. Woodhead Publishing, Sawston, Cambridge, UK.

De Vuyst, L., Leroy, F., 2007. Bacteriocins from lactic acid bacteria: production, purification, and food applications. Journal of Molecular Microbiology and Biotechnology 13, 194–199.

De Vuyst, L., Vandamme, E.J., 1994. Lactic acid bacteria and bacteriocins: their practical importance. In: De Vuyst, L., Vandamme, E.J. (Eds.), Bacteriocins of Lactic Acid Bacteria, Microbiology, Genetics and Application. Chapman and Hall Ltd., Glasgow.

Drider, D., Fimland, G., Hechard, Y., McMullen, L.M., Prevost, H., 2006. The continuing story of class IIa bacteriocins. Microbiology and Molecular Biology Reviews 70, 564–582.

Federal Register, 1988. Nisin preparation: affirmation of GRAS status as a direct human food ingredient. Federal Register 54, 11247–11251.

Fields, F.O., 1996. Use of bacteriocins in food: regulatory considerations. Journal of Food Protection (Suppl.), 72–77.

Fimland, G., Jack, R., Jung, G., Nes, I.F., Nissen-Meyer, J., 1998. The bactericidal activity of pediocin PA-1 is specifically inhibited by a 15-mer fragment that spans the bacteriocin from the center toward the C terminus. Applied and Environmental Microbiology 64, 5057–5060.

Franz, C.M.A.P., van Belkum, M.J., Holzapfel, W.H., Abriouel, H., Gálvez, A., 2007. Diversity of enterococcal bacteriocins and their grouping in a new classification scheme. FEMS Microbiological Reviews 31, 293–310.

Galvez, A., Lopez, R.L., Abriouel, H., Valdivia, E., Omar, N.B., 2008. Application of bacteriocins in the control of foodborne pathogenic and spoilage bacteria. Critical Reviews in Biotechnology 28, 125–152.

Garsa, A.K., Kumariya, R., Sood, S.K., Kumar, A., Kapila, S., 2014. Bacteriocin production and different strategies for their recovery and purification. Probiotics and Antimicrobial Proteins 6, 47. http://dx.doi.org/10.1007/s12602-013-9153-z.

Gillor, O., Nigro, L.M., Riley, M.A., 2007. Potential application of bacteriocins as antimicrobials. In: Riley, M.A., Gillor, O. (Eds.), Research and Applications in Bacteriocins. Caister Academic Press, Norfolk, UK.

Goff, J.H., Bhunia, A.K., Johnson, M.G., 1996. Complete inhibition of low levels of *Listeria monocytogenes* on refrigerated chicken meat with pediocin AcH bound to heat-killed *Pediococcus acidilactici* cells. Journal of Food Protection 59, 1187–1192.

Gould, G.W., 2000. Preservation: past, present and future. British Medical Bulletin 56, 84–96.

Gross, E., Morell, J.H., 1971. Structure of nisin. Journal of the American Chemical Society 93, 4634–4635.

Héchard, Y., Sahl, H.G., 2002. Mode of action of modified and unmodified bacteriocins from Gram-positive bacteria. Biochimie 84, 545–557.

Henning, S., Metz, R., Hammes, W.P., 1986. Studies on the mode of action of nisin. International Journal of Food Microbiology 3, 121–134.

Hirsch, A., Grinsted, E., Chapman, H.R., Mattick, A.T.R., 1951. A note on the inhibition of an anaerobic spore-former in Swiss-type cheese by a nisin producing *Streptococcus*. Journal of Dairy Research 18, 205–206.

Hugas, M., Pages, F., Garriga, M., Monfort, J.M., 1998. Application of the bacteriocinogenic *Lactobacilus sakei* CTC494 to prevent growth of *Listeria* in fresh and cooked meat products packed with different atmospheres. Food Microbiology 15, 639–650.

Hurst, A., 1981. Nisin. Advances in Applied Microbiology 27, 85–123.

Jabs, J., Sobal, J., Devine, C.M., 2000. Managing vegetarianism: Identities, norms and interactions. Ecology of Food and Nutrition 39, 375–394.

Janes, M.E., Kooshesh, S., Johnson, M.G., 2002. Control of *Listeria monocytogenes* on the surface of refrigerated, ready-to-eat chicken coated with edible zein film coatings containing nisin and/or calcium propionate. Journal of Food Science 67, 2754–2757.

Jiménez-Díaz, R., Ruiz-Barba, J.L., Cathcart, D.P., Holo, H., Nes, I.F., Sletten, K.H., Warner, P.J., 1995. Purification and partial amino acid sequence of Plantaricin S, a bacteriocin produced by *Lactobacillus plantarum* LPCO10, the activity of which depends on the complementary action of two peptides. Applied and Environmental Microbiology 61, 4459–4463.

Joerger, M.C., Klaenhammer, T.T., 1986. Characterization and purification of helveticin-J and evidence for a chromosomally determined bacteriocin produced by *Lactobacillus helveticus*-841. Journal of Bacteriology 167, 439–446.

Jofré, A., Garriga, M., Aymerich, T., 2008. Inhibition of *Salmonella* sp., *Listeria monocytogenes* and *Staphylococcus aureus* in cooked ham by combining antimicrobials, high hydrostatic pressure and refrigeration. Meat Science 78, 53–59.

Johnson, M.C., Hanlin, M.B., Ray, B., 1992. Low pH and lactate are necessary for conversion of prepediocin to pediocin AcH in *Pediococcus acidilactici* H. In: Annual Meeting 1992, New Orleans, LA, May 26–30. American Society for Microbiology, Washington DC, USA. Abstr. 081.

Jones, L.W., 1974. Effect of butterfat on inhibition of *Staphylococcus aureus* by nisin. Canadian Journal of Microbiology 20, 1257–1260.

Joshi, V.K., Sharma, S., Rana, N.S., 2006. Production, purification, stability and efficacy of bacteriocin from isolates of natural lactic acid fermentation of vegetables. Food Technology and Biotechnology 44, 435–439.

Jung, D.S., Bodyfelt, F.W., Daeshel, M.A., 1992. Influence of fat and emulsifiers on the efficacy of nisin in inhibiting *Listeria monocytogenes* in fluid milk. Journal of Dairy Science 75, 387–393.

Kaiser, A.L., Montville, T.J., 1993. The influence of pH and growth rate on production of the bacteriocin, bavaricin MN, in batch and continuous fermentations. Journal of Applied Bacteriology 75, 536–540.

Kawai, Y., Kemperman, R., Kok, J., Saito, T., 2004. The circular bacteriocins gassericin A and circularin A. Current Protein & Peptide Science 5, 393–398.

Klaenhammer, T.T., 1993. Genetics of bacteriocins produced by lactic acid bacteria. FEMS Microbiological Reviews 12, 39–86.

Knorr, D., 1998. Technological aspects related to microorganisms in functional foods. Trends in Food Science and Technology 9, 295–306.

Kroes, R., Kleiner, J., Renwick, A., 2005. The threshold of toxicological concern concept in risk assessment. Toxicological Science 86, 226–230.

Lindsay, R.C., 2008. Food additives. In: Damodaran, S., Parkin, K.L., Fennema, O.R. (Eds.), Fennema's Food Chemistry. CRC Press, Boca Raton, FL.

Lockie, S., Lyons, K., Lawrence, G., Mummery, K., 2002. Eating "green", motivations behind organic food consumption in Australia. Sociologia Ruralis 42, 23–40.

Lungu, B., Johnson, M.G., 2005a. Potassium sorbate does not increase control of *Listeria monocytogenes* when added to zein coatings with nisin on the surface of full fat turkey frankfurter pieces in a model system at 4°C. Journal of Food Science 70, M95–M99.

Lungu, B., Johnson, M.G., 2005b. Fate of *Listeria monocytogenes* inoculated onto the surface of model turkey frankfurter pieces treated with zein coatings containing nisin, sodium diacetate, and sodium lactate at 4°C. Journal of Food Protection 68, 855–859.

Martinez Viedma, P., Abriouel, H., Ben Omar, N., Lucas López, R., Valdivia, E., Gálvez, A., 2009. Assay of enterocin AS-48 for inhibition of foodborne pathogens in desserts. Journal of Food Protection 72, 1654–1659.

McClintock, M., Serres, L., Marzolf, J.J., Hirsch, A., Mocquot, G., 1952. Ation inhibitrice des streptocoques producteurs de nisine sur le developpement des sporules anaerobies dans le fromage de Gruyere fondu. Journal of Dairy Research 19, 187.

Ming, X., Weber, G.H., Ayres, J.W., Sandine, W.E., 1997. Bacteriocins applied to food packaging materials to inhibit *Listeria monocytogenes* on meats. Journal of Food Science 62, 413–415.

Motlagh, A.M., Bhunia, A.K., Szostek, F., Hansen, T., Johnson, M.C., Ray, B., 1992. Nucleotide and amino acid sequence of *pap*-gene (pediocin AcH production) in *Pediococcus acidilactici* H. Letters in Applied Microbiology 15, 45–48.

Naghmouchi, K., Kheadr, E., Lacroix, C., Fliss, I., 2007. Class I/Class IIa bacteriocin cross-resistance phenomenon in *Listeria monocytogenes*. Food Microbiology 24, 718–727.

Natrajan, N., Sheldon, B.W., 2000. Efficacy of nisin-coated polymer films to inactivate *Salmonella typhimurium* on fresh broiler skin. Journal of Food Protection 63, 1189–1196.

Nes, I.F., Diep, D.B., Havarstein, L.S., Brurberg, M.B., Eijsink, V., Holo, H., 1996. Biosynthesis of bacteriocins in lactic acid bacteria. Antonie Van Leeuwenhoek International Journal of General and Molecular Microbiology 70, 113–128.

Nilsen, T., Nes, I.F., Holo, H., 2003. Enterolysin A, a cell wall-degrading bacteriocin from *Enterococcus faecalis* LMG 2333. Applied and Environmental Microbiology 69, 2975–2984.

O'Bryan, C.A., Sostrin, M.L., Nannapaneni, R., Ricke, S.C., Crandall, P.G., Johnson, M.G., 2009. Sensitivity of *Listeria monocytogenes* Scott A to nisin and diacetyl after starvation in sodium phosphate buffered saline. Journal of Food Science 74, M493–M498.

Ogden, K., 1986. Nisin-a bacteriocin with a potential use in brewing. Journal of the Institute of Brewing 92, 379–383.

O'Mahony, T., Rekhif, N., Cavadini, C., Fitzgerald, G.F., 2001. The application of a fermented food ingredient containing 'variacin', a novel antimicrobial produced by *Kocuria varians*, to control the growth of *Bacillus cereus* in chilled dairy products. Journal of Applied Microbiology 90, 106–114.

O'Sullivan, L., Ross, R.P., Hill, C., 2002. Potential of bacteriocin-producing lactic acid bacteria for improvements in food safety and quality. Biochimie 84, 593–604.

Papagianni, M., 2003. Ribosomally synthesized peptides with antimicrobial properties: biosynthesis, structure, function, and applications. Biotechnology Advances 21, 465–499.

Perez, R.H., Zendo, T., Sonomoto, K., 2014. Novel bacteriocins from lactic acid bacteria (LAB): Various structures and applications. Microbial Cell Factories 13 (Suppl. 1), 1–13.

Post, R.C., 1996. Regulatory perspective of the USDA on the use of antimicrobials and inhibitors in foods. Journal of Food Protection (Suppl.), 78–81.

Ray, B., 1995. *Pediococcus* in fermented foods. In: Hui, Y.H., Khachatourians, G. (Eds.), Food Biotechnology: Microorganisms. Wiley-VCH, Hoboken, NJ.

Rogers, L.A., Whittier, E.O., 1928. Limiting factors in the lactic fermentation. Journal of Bacteriology 16, 211–229.

Rozum, J.J., Maurer, A.J., 1997. Microbiological quality of cooked chicken breasts containing commercially available shelf-life extender. Poultry Science 76, 908–913.

Sahl, H.-G., Jack, R.W., Bierbaum, G., 1995. Biosynthesis and biological activities of lantibiotics with unique post-translational modifications. European Journal of Biochemistry 230, 827–853.

Schillinger, U., Geisen, R., Holzapfel, W.H., 1996. Potential of antagonistic microorganisms and bacteriocins for the biological preservation of foods. Trends in Food Science and Technology 71, 58–64.

Siragusa, G.R., Cutter, C.N., Willett, J.L., 1999. Incorporation of bacteriocin in plastic retains activity and inhibits surface growth of bacteria on meat. Food Microbiology 61, 229–235.

Snyder, A.B., Worobo, R.W., 2014. Chemical and genetic characterization of bacteriocins: antimicrobial peptides for food safety. Journal of the Science of Food and Agriculture 94, 28–44.

Stevens, K.A., Sheldon, B.W., Klapes, N.A., Klaenhammer, T.R., 1991. Nisin treatment for inactivation of *Salmonella* species and other Gram negative bacteria. Applied and Environmental Microbiology 57, 3613–3615.

Stiles, M.E., 1994. Bacteriocins produced by *Leuconostoc* species. Journal of Dairy Science 77, 2718–2724.

Stoyanova, L.G., Ustyugova, E.A., Netrusov, A.I., 2012. Antibacterial metabolites of lactic acid bacteria: their diversity and properties. Applied Biochemistry and Microbiology 48, 229–243.

Suda, S., Cotter, P.D., Hill, C., Ross, R.P., 2012. Lacticin 3147-biosynthesis, molecular analysis, immunity, bioengineering and applications. Current Protein & Peptide Science 13, 193–204.

Sun, L., Song, H., Zheng, W., 2015. Improvement of antimicrobial activity of pediocin PA-1 by site-directed mutagenesis in C-terminal domain. Protein and Peptide Letters 22, 1007–1012.

Tagg, J.R., Dajani, A.S., Wannamaker, L.W., 1976. Bacteriocins of Gram-positive bacteria. Bacteriological Reviews 40, 722–756.

Tessema, G.T., Møretrø, T., Kohler, A., Axelsson, L., Naterstad, K., 2009. Complex phenotypic and genotypic responses *of Listeria monocytogenes* strains exposed to the class IIa bacteriocin sakacin P. Applied and Environmental Microbiology 75, 6973–6980.

Tiwari, B.K., Valdramidis, V.P., O'Donnell, C.P., Muthukumarappan, K., Bourke, P., Cullen, P.J., 2009. Application of natural antimicrobials for food preservation. Journal of Agricultural and Food Chemistry 57, 5987–6000.

von Staszewski, M., Jagus, R.J., 2008. Natural antimicrobials: Effect of Microgard (TM) and nisin against *Listeria innocua* in liquid cheese whey. International Dairy Journal 18, 255–259.

Wood, B.J.B., 1997. Microbiology of Fermented Foods. Blackie A&P, London, UK.

Yildirim, Z., Johnson, M.G., 1998. Characterization and antimicrobial spectrum of bifidocin B, a bacteriocin produced by *Bifidobacterium bifidum* NCFB 1454. Journal of Food Protection 61, 47–51.

Yildirim, Z., Winters, D.K., Johnson, M.G., 1999. Purification, amino acid sequence and mode of action of bifidocin B produced by *Bifidobacterium bifidum* NCFB 1454. Journal of Applied Microbiology 86, 45–54.

Yildirim, Z., Yildirim, M., Johnson, M.G., 2007. Effects of bifidocin b and lactococcin r on the growth of *Listeria monocytogenes* and *Bacillus cereus* on sterile chicken breast. Journal of Food Safety 27, 373–385.

SECTION 4

# *Emerging Food Safety Systems*

CHAPTER 16

# Detection Strategies for Foodborne *Salmonella* and Prospects for Utilization of Whole Genome Sequencing Approaches

**Si Hong Park[1,4], Mushin Aydin[2], Peixin Fan[3], Shinyoung Lee[3], Lin Teng[3], Sun Ae Kim[1], Soohyoun Ahn[3], Steven C. Ricke[1], Zhaohao Shi[1], Kwangcheol C. Jeong[3]**
[1]*University of Arkansas, Fayetteville, AR, United States;* [2]*Arkansas State University, Jonesboro, AR, United States;* [3]*University of Florida, Gainesville, FL, United States;* [4]*Oregon State University, Corvallis, OR, United States*

## *Salmonella* as a Foodborne Pathogen—Food Safety Perspectives

There have been over 200 pathogens identified as being associated with foodborne illnesses, including bacteria, viruses, and parasites (Tauxe, 2002; Scallan et al., 2011). The Centers for Disease Control and Prevention (CDC) reported that approximately 48 million cases of foodborne illnesses, 128,000 hospitalizations, and 3000 deaths occur annually in the United States (US). The top four foodborne pathogens are norovirus (58%), nontyphoid *Salmonella* (11%), *Clostridium perfringens* (10%), and *Campylobacter* (9%), which account for approximately 88% of all foodborne illnesses caused by known agents in the US (Scallan et al., 2011). The CDC report estimates that bacteria are the causative agents of 3.6 million cases of foodborne illness (39%). Among the pathogenic bacteria, *Salmonella* species were the leading cause of foodborne illness-related hospitalizations and deaths, accounting for 19,336 hospitalizations (35% of the total hospitalization cases) and 378 deaths (28% of the total death cases) (Scallan et al., 2011). There is also considerable economic loss associated with the *Salmonella*-related outbreaks in the US, which is estimated to be between $600 million and $3.5 billion each year (Oliver et al., 2005).

*Salmonella* is a facultative anaerobic, Gram-negative, non–spore-forming, rod-shaped bacterium that belongs to the family of Enterobacteriaceae. *Salmonella* is motile except for *Salmonella* Pullorum and *Salmonella* Gallinarum (Bhunia, 2008; Ricke et al., 2013b). This genus is classified as a chemoorganotrophic organism that can utilize a wide range of organic substrates and metabolize nutrients by both respiratory and fermentative pathways (Monadi et al., 2010; Santos et al., 2011). *Salmonella* not only grows optimally at 37°C, which is equal to human body temperature, but also can adapt to extreme environmental conditions including low pH, thermal shock, and minimal nutrient availability (Durant

Food and Feed Safety Systems and Analysis. http://dx.doi.org/10.1016/B978-0-12-811835-1.00016-6

et al., 1999; Kwon and Ricke, 1998; Kwon et al., 2000; Ricke, 2003a,b; Dunkley et al., 2009; Jarvis et al., 2016; Dawoud et al., 2017). Several *Salmonella* strains have the ability to grow in foods stored at 2–4°C, and some others can grow at 54°C (Balamurugan, 2010). The genus *Salmonella* consists of two species, *Salmonella enterica* and *Salmonella bongori*, and *S. enterica* is further divided into six subspecies: enterica (I), salamae (II), arizonae (IIIa), diarizonae (IIIb), houtenae (IV), and indica (VI). Currently, more than 2600 serovars have been identified (Porwollik et al., 2004; Grimont and Weill, 2007; Foley et al., 2011, 2013).

Infections with *Salmonella* have a significant impact on both animal and human health globally. The delineation between human clinical syndromes derived from nontyphoid *Salmonella* and those which are considered systemic in nature is becoming less clear. Numerous *Salmonella* strains are resistant to antibiotics, and this resistance to antibiotics creates treatment challenges in both advanced and developing countries (Ricke and Rivera Calo, 2015). Salmonellosis, the typical foodborne illness caused by *Salmonella*, usually lasts 4–7 days, with most individuals recovering without treatment. However, in certain people, diarrhea and vomiting may be so severe that the patient needs to be hospitalized to prevent dehydration. Most people infected with *Salmonella* develop diarrhea, fever, nausea, and abdominal cramps ranging from 12 to 72 h after infection (Ricke et al., 2013b). Once an individual has ingested *Salmonella*, the bacteria can multiply in the intestine and infect the intestinal tissues leading to the production of enterotoxin, which results in an inflammatory reaction. This is usually followed by the symptoms of stomach pains and diarrhea (Santos et al., 2001). In some cases, an individual will recover from the *Salmonella* infection, but the bacterium will not be completely eliminated from the body. In these cases, *Salmonella* can still be present in the victim's excretions, and the individual evolves into a carrier. *Salmonella* can become associated with food by contamination during improper processing, handling, or by poor hygiene of food handlers with cross-contamination and recontamination occurring (NDSC, 2004; Lynch et al., 2009; Carrasco et al., 2012). Individuals can subsequently develop salmonellosis by eating the *Salmonella*-contaminated food. Beef, pork, poultry, milk, and eggs are the most commonly recognized food vehicles for *Salmonella* (Finstad et al., 2012; Foley et al., 2008, 2011, 2013; Howard et al., 2012; Ricke, 2014). However, fresh produce has also been associated with outbreaks (Hanning et al., 2009; Ricke et al., 2013b; Ricke, 2014) but, in general, it is difficult to preemptively identify contaminated foods because they will more than likely appear normal from a sensory standpoint (Tauxe, 1997). This is particularly problematic for ready-to-eat foods and foods such as fresh produce when consumed raw (Ricke et al., 2013b).

*Salmonella* is one of the representative foodborne pathogens causing clinically severe diseases in humans. Outbreaks associated with *Salmonella* contamination highlights the need for development of rapid and reliable detection methods for preventing *Salmonella* infection. The objective of this chapter is to provide an overview of various approaches to detect and characterize *Salmonella* including culture-, immunological-, and molecular-based methods as well as advances in *Salmonella* identification and detection technologies that are employing whole genome sequencing (WGS).

## *Poultry as a Major Vehicle for Transmission of* Salmonella

There is a wide range of animals, particularly birds both domesticated and wild avian species, that harbor *Salmonella* within their respective gastrointestinal tracts. Poultry are by far one of the primary vehicles for the transmission of *Salmonella* to humans. Some serovars, such as *Salmonella* Typhimurium, *Salmonella* Enteritidis, and *Salmonella* Heidelberg, are capable of infecting a wide range of birds, whereas other *Salmonella* serovars are specific to certain hosts (Foley et al., 2011, 2013; Kaldhone et al., 2016). For example, *S.* Gallinarum and *S.* Pullorum are carried by poultry but usually not by other animals (Foley et al., 2011). *S.* Enteritidis is one of the most frequent causes of foodborne illnesses in the US. It is most commonly associated with eggs and poultry (Olsen et al., 2000; Ricke, 2003a; Shah et al., 2011; Ricke and Gast, 2016).

According to previous studies, 20%–35% of processed poultry are contaminated with *Salmonella*, whereas only 3%–4% of those entering the plant are *Salmonella*-positive (McKee, 2011, 2012). Additionally, the CDC reported that approximately 1 in 10,000 eggs is contaminated by *Salmonella* (CDC, 1990). According to the CDC, during the period from May 1 to November 30, 2010, 1939 *S.* Enteritidis infections were reported that were associated with eggshells (CDC, 2010a). Moreover, 44 persons were infected with *Salmonella* Chester that was associated with a cheesy chicken rice entrée (CDC, 2010b), and 12 persons were infected with *Salmonella* Hadar associated with turkey burgers (CDC, 2011). More recently, as of July 2, 2014, more than 570 people have been infected with different strains of *S.* Heidelberg due to the contamination of chicken products from Foster Farms brand since March 1, 2013 (CDC, 2013).

With growing consumption of poultry and poultry products, the number of salmonellosis cases associated with poultry continues to persist in the US. Because *Salmonella* is a major causative agent for poultry-associated foodborne illnesses, safety of poultry products could be greatly improved by effective monitoring of *Salmonella* contamination. Monitoring of poultry and other related products for *Salmonella* contamination would potentially be improved by employing more rapid and effective detection methods.

## *Detection Technology*

### *Detection Methods for* Salmonella *in Poultry: General Concepts*

As described in the previous section, *Salmonella* species are very important targets for food safety and public health monitoring and subsequent control measures. *Salmonella* outbreaks associated with poultry generate negative economic impacts due to medical costs, recall of contaminated products, expenditures associated with surveillance investigations, and lost productivity due to illness (Gómez-Aldapa et al., 2012). A key to prevent these *Salmonella*-related illnesses is the development of more effective detection methods for *Salmonella* (Park et al., 2014). Although a number of different methods have been developed for detecting *Salmonella* in poultry, there are three major types of detection methods currently available, namely, culture-based (traditional) recovery, identification, and/or enumeration of viable organisms; immunological-based assays for specific cellular antigens; and molecular detection methods (Maciorowski et al., 2005, 2006; Ricke et al., 2013a; Park et al., 2014).

### Culture-Based Methods

The traditional methods are culture-based, so that they require recovering pure cultures of the suspected organism from food matrices followed by tests for the identification of the organism (Ricke et al., 2013b). The details of traditional methods are provided by the US Food and Drug Administration–bacteriological analytical manual (US FDA–BAM) and the US Department of Agriculture–Food Safety and Inspection Service (USDA–FSIS) microbiology laboratory guidebook (FDA, 1998; USDA, 2013). Owing to possible low pathogen concentrations in food samples, selective and nonselective enrichment steps are usually applied to increase the cell numbers to detectable levels, followed by isolation methods with selective/differential media (e.g., brilliant green agar, bismuth sulfite agar, and xylose lysine deoxycholate agar used for isolation and differentiation of *Salmonella*). These methods usually require 3–7 days to provide results. Typically, selective or nonselective enrichment, plating on selective or differential media, and serological as well as biochemical confirmation tests are required (Ricke et al., 2013b). The main advantages of these methods are that *Salmonella* strains can be isolated and only viable cells from poultry samples are identified. Additionally, traditional methods do not require any expensive instrumentation and are generally cost-effective outside the expenses associated with labor and media expenses. However, the culture-based methods can be time-consuming and labor-intensive.

### Immunological-Based Methods

Immunological methods are based on specific binding of antibodies to antigens (March et al., 2005; Kérouanton et al., 2007; Hoorfar, 2011; Maciorowski et al., 2006; Baker et al., 2016). These methods include enzyme-linked immunosorbent assays (ELISA), radioimmunoassays, and immunofluorescent assays (Maciorowski et al., 2006; Baker et al., 2016). Immunological methods are considered to be generally more rapid than the culture-based methods (10–30 h vs. 3–7 days).

However, immunological methods have two main disadvantages: (1) false-negative results that may occur because of denaturation or degradation of the capture antibody and (2) inability to differentiate viable cells from nonviable cells because dead cells also have surface antigens that may bind to the antibodies. Because of these disadvantages, these methods are not desirable in *Salmonella* detection from poultry samples and have been only somewhat adopted as primary detection and identification methods for *Salmonella* in poultry and poultry products. However, immunological methods can be used for *Salmonella* isolation from poultry samples (i.e., immunomagnetic separation) in conjunction with other methods used for identification (Mercanoglu-Taban et al., 2009; Taha et al., 2010). An example of an immunological method that has been employed for bacterial identification and detection for organisms such as foodborne *Salmonella* is the ELISA. The ELISA, as used for *Salmonella* detection in foods, has some limitations including reduced sensitivity due to antigen variability, limited multiplexing ability, false-negative results especially when testing food of complex matrices (containing large amounts of organic substances), limited ability to detect damaged or stressed cells, requirement of preenrichment in a standard medium for production of cell-surface antigens, and cross-reactivity with antigens of closely related bacteria (Maciorowski et al., 2006). These disadvantages have limited the adoption of ELISA-based detection systems as a routine approach for rapid detection and identification of *Salmonella* in poultry and poultry products.

### *Molecular-Based Methods*

Molecular methods are based on the specific composition of the target pathogens' genetic materials (DNA or RNA) and usually depend on binding of oligonucleotide probes to specific sequences of DNA of the targeted bacteria. The sequences of nucleotide bases in DNA vary among species. Research focused on development and optimization for routine application of molecular detection methods in the food industry has improved the identification and detection efficiency for foodborne pathogens. The primary molecular methods for *Salmonella* include polymerase chain reaction (PCR), real-time quantitative PCR (qPCR), microarray, and next-generation sequencing (NGS), all of which have utility for specified applications in food systems (Maciorowski et al., 2005; Ricke et al., 2013b, 2015; Park et al., 2014).

PCR has emerged as one of the most commonly used techniques for detection and characterization of foodborne pathogens and has continued to evolve in both methodology improvements as well as expanded applications (Boyd et al., 2003; Bhagwat, 2003; Kumar et al., 2005; Maciorowski et al., 2005; Park et al., 2014). PCR is essentially an amplification method that uses DNA polymerase and a pair of primers (e.g., commonly called forward and reverse primers) that are complementary to either end of a specific region of target DNA to copy that single targeted region of DNA (Soumet et al., 1999; Szabo and Mackey, 1999). Numerous studies have been conducted to detect and characterize *Salmonella* in poultry and poultry products using PCR by targeting antibiotic resistance and virulence genes along with genus-, species-, and/or serotype-specific genes (Seo et al., 2004; Eyigör et al., 2007; Wise et al., 2009; Levin, 2010). An enrichment step is commonly combined with PCR-based detection methods to enhance the sensitivity of the detection of the respective viable pathogen. For example, although *Salmonella* could not be detected directly from poultry tissue homogenates by PCR (mostly due to the presence of various PCR inhibitors as well as DNases), PCR combined with a 10-h enrichment step was able to detect the presence of *Salmonella* (Myint et al., 2006). Culture enrichment is also advisable to distinguish live cells from dead cells when detecting pathogens with PCR (Cocolin et al., 2011). The quality of the target DNA serving as a PCR template is also a critical factor in the design of a PCR assay for pathogen detection. Although optimized PCR primer design can ensure specificity of the target pathogen detection, this is generally not sufficient in overcoming the effect of PCR inhibitors present in samples. Food products inherently contain PCR-inhibiting components that may bind the DNA template or interfere with DNA polymerase activity (Maciorowski et al., 2005). In addition, the presence of DNA and cells other than those from the targeted organism can affect the efficiency of the PCR method (Wegener et al., 2003).

A multiplex PCR method, amplifying multiple DNA sequences using different primer sets, has also been developed and widely used to simultaneous detection and discrimination of various *Salmonella* species (Hong et al., 2009; McCarthy et al., 2009; Park and Ricke, 2015; Park et al., 2009, 2011; Saeki et al., 2013). More recently, Park and Ricke (2015) developed a novel multiplex PCR assay to detect *Salmonella* genus, *Salmonella* subspecies I, *S*. Enteritidis, *S*. Heidelberg, and *S*. Typhimurium from chicken breast meats using five primer pairs and distinguishable five bands were clearly amplified via a multiplex PCR assay.

In practice, cross-reactivity of primer sets, sensitivity limitations associated with the procedure, and a high number (at least 12) (USDA-FSIS, 2010) of *Salmonella* serovars associated with poultry and poultry products make it quite challenging to use multiplex PCR for reliable and simultaneous detection of *Salmonella* serovars in poultry and poultry products. With the emergence of fluorescence technology that allows detection of targets with increased sensitivity (e.g., intercalating dyes, for example SYBR Green, labeled probes such as TaqMan probes or molecular beacons), limitations of conventional PCR including the errors associated with end-point analyses due to variability in the quantity and quality of the DNA templates and lack of quantification have been overcome (Bohaychuk et al., 2007; Karlen et al., 2007). The "real-time" aspect of real-time PCR (also called qPCR) technology is linked to its ability to label and cumulatively quantify the generated PCR products at each cycle throughout the ongoing amplification process (Corless et al., 2001). However, the enhanced sensitivity of the real-time PCR makes it more susceptible to inhibitors found in complex food matrices, thus it requires particularly effective DNA extraction methods (Pontiroli et al., 2011). Although real-time PCR is an effective tool to detect and identify *Salmonella* with high sensitivity and specificity, it does have several limitations: (1) Multiplexing ability is quite limited and multiplex qPCR requires much more in-depth optimization than standard multiplex PCR. (2) Assay design and optimization requires high technical skills and the testing platform requires proper training and expertise for people who conduct PCR. (3) Equipment cost is higher than conventional PCR. (4) The labile nature of RNA (in viable cell detection) limits adoption of reverse transcriptase (RT)-qPCR application to environmental or food samples. (5) Increased sensitivity can lead to false-positive results due to contaminants present in samples, and DNA contamination of template mRNA for qRT-PCR needs to be carefully monitored.

A novel detection method using most probable number (MPN) combined with qPCR and a shortened incubation time (SIT) (MPN–qPCR–SIT) has also been developed to enumerate *S*. Typhimurium (Kim et al., 2017). The concept of MPN–qPCR–SIT is relatively straightforward and the basic tenets of this procedure are comparable to those associated with conventional MPN. This method retains both the strengths of conventional MPN and qPCR while reducing limitations associated with both methods including the long incubation time of MPN and the poor detection limit with qPCR. The MPN–qPCR–SIT produced high accuracy (The $R^2$ value between the MPN–qPCR–SIT and conventional MPN: 0.9725) and selectivity with a shorter detection time (requiring approximately 7 h) and lower detection limits than previous methods such as conventional plating, conventional MPN, and qPCR only. A key advantage is the ability to use relatively rich nonselective media to support more rapid growth of *Salmonella*. This method has been proposed as a means for rapidly quantifying *S*. Typhimurium by significantly reducing both time and labor (Kim et al., 2017). It is anticipated that these MPN combinations with qPCR will be applicable to other *Salmonella* serovars and perhaps combinations of serovars within the same sample. In time, the ability to quantitate non-*Salmonella* in these same MPN samples may be possible, thus allowing for simultaneous assessment of potential indicator organisms and even overall bacterial population loads.

## Salmonella *Identification Using Whole Genome Sequencing*

### *General Introduction of Whole Genome Sequencing*

Historically, identification and classification of bacteria was mainly conducted based on growth characteristics, colony morphology, Gram staining, sugar assimilation or fermentation, and biochemical tests among other conventional physiological-based approaches (Bertelli and Greub, 2013). Obviously with the advances first with molecular techniques, such as DNA hybridization, denatured gradient gel electrophoresis, conventional PCR, and later more advanced PCR approaches, molecular methods for identification and detection rapidly became the predominant means to identify foodborne pathogens (Ricke et al., 2015). As DNA sequencing technologies improved, WGS emerged as a more practical tool to identify microorganisms due to the ability to generate sufficient high-throughput data to cover the entire genome of a target organism. Not surprisingly, WGS refers to a process that determines the complete DNA sequence of an organism, thus after employment of bioinformatic computational and alignment tools provides a genome profile of the target organism and specific genes of interest (Köser et al., 2012; Ricke et al., 2015). The sequencing technologies have evolved considerably over two decades and faster high-throughput systems have now become routine with capacity increasing as cost per sample declines (Ricke et al., 2015).

In particular, NGS methods have been historically considered to possess considerable potential for microbial identification and characterization (Kwon and Ricke, 2011; Ricke et al., 2015). As NGS technologies have improved, their deployment has enabled the rapid assembly of microbial sequences encoding complete genomes within a day rather than weeks to months with conventional Sanger sequencing methods (Medini et al., 2005; Ricke et al., 2015). NGS has proven to be a versatile and powerful tool in a variety of applications ranging from characterizing the genomic diversity and evolution of *Mycobacterium ulcerans* (Qi et al., 2009) and detecting pathogens such as *Bacillus anthracis* and *Yersinia pestis* (Cummings et al., 2010) to scanning the *S.* Typhimurium genome for conditionally essential genes following transposon mutagenesis (Tn-Seq) (Khatiwara et al., 2012; Kwon et al., 2016). The historical trajectory for the development of sequencing technologies that led to the methodologies currently being employed in research and commercial applications are described in the next section along with a description of the most recent sequencing technologies being used.

### *First-Generation DNA Sequencing Technology*

In 1953, the DNA structure was demonstrated by Watson and Crick as a three-dimensional structure. After that discovery, DNA replication and encoding proteins in nucleic acids continued to be investigated, but mostly for understanding fundamental concepts of genetic regulation and expression. Sequencing of DNA remained relatively incomplete and inefficient. In addition, even in basic sequencing techniques, elucidation of sequencing data remained slow essentially being limited to short lengths of DNA covering a few genes at most (Judson, 1979; Ricke et al., 2015; Heather and Chain, 2016). The breakthrough of first-generation sequencing approaches was led by the Sanger sequencing methodology using enzymatic polymerization known as the chain-termination method or the dideoxynucleotide triphosphate (ddNTP) method (Judson, 1979; Ricke et al., 2015). This method became an important advancement in genomic research for over 40 years because DNA fragments for sequencing could be generated fairly easily and were relatively accurate compared with other methods (França et al., 2002). Likewise, the Maxam and Gilbert (1977) sequencing method offered another major first-generation sequencing approach that fundamentally differed with Sanger's chain-termination sequencing in terms of the incorporation of radioactive $P^{32}$ into the structure of DNA. The method is also referred to as the "chemical-sequencing" method because it used specific chemicals to break the chain of DNA. Based on these initial sequencing approaches, improvements on the previously developed Sanger sequencing method were introduced by using fluorescence-based detection systems and capillary-based electrophoresis (Smith et al., 1985; Prober et al., 1987; Swerdlow and Gesteland, 1990; Luckey et al., 1990).

Despite these advances from earlier versions, limitations of the first-generation sequencing method remained in terms of the length of DNA that could be sequenced, time constraints, and efficacy. This changed dramatically when the human genome project was launched and second-generation DNA sequencing technologies emerged as a result of these efforts (Ricke et al., 2015).

### *Second-Generation Sequencing Technology*

Genetic sequencing technologies have developed dramatically in the past three decades (Metzker, 2010). The human genome project led to the revolution of sequencing technology, with emergence of more economical and higher-throughput NGS technologies, including 454 pyrosequencing, Illumina/Solexa sequencing, Ion Torrent sequencing, and the ABI SOLiD System (Metzker, 2010). These methods enable the library preparation to be performed in a cell-free system, and considerable numbers of generated sequence reads speed up the sequencing process (van Dijk et al., 2014). The principle of second-generation sequencing technology essentially relies on sequence by synthesis (SBS). During sequencing process, clusters of library DNA replicate after amplification and subsequently attach to a solid surface that will be imaged or detected afterward (Mardis, 2013). The outcome of second-generation sequencing dramatically decreases the cost and simplifies the sequencing process. However, compared with Sanger sequencing, read lengths of second-generation sequencing are relatively shorter, which can make the downstream genome assembly more difficult.

#### 454 Pyrosequencing

The 454 pyrosequencing platform was launched in 2005 by 454 Life Sciences and was acquired by Roche in 2007. In 2013, Roche announced closure due to the noncompetitive nature of the technology. However, it cannot be denied that 454 pyrosequencing played a vital role in sequencing history, as it was the first second-generation sequencing platform that successfully reached the commercial market and made key breakthroughs in this field. The 454 pyrosequencing platform completely removed the bacterial cloning step; instead it was based on limiting dilutions and physically compartmentalizing clonal amplification, thus dramatically increasing the sequencing speed and efficiency (Rothberg and Leamon, 2008). The pyrosequencing technology applied by 454 system refers to the detection of light produced whenever pyrophosphate is released during incorporation of nucleotide, which can replace the time-consuming electrophoresis step and enable abundant parallel sequencing reactions carried out on a large scale.

In brief, the library DNA with adapters is initially denatured into single strands that are captured by amplification beads to start the emulsion PCR process. Subsequently, a combination of ATP sulfurylase, luciferase, luciferin, DNA polymerase, and adenosine 5-phosphosulfate (APS along with one of the dNTPs are added to the bases of the template strand and pyrophosphate (PPi) released). The visible light is generated by the transformation of PPi to ATP, which drives the luciferin into oxyluciferin. The pyrosequencing reaction can then be repeated when another dNTP that is complementary to the system is added (Liu et al., 2012). The key advantages of the 454 platform compared with other NGS is its read length and its speed. The sequencing process required approximately 10 h, with 14 G output per run, and each read cycle could reach 700 bp (more than 200,000 reads). However, the high cost and relatively high error rate in terms of polybases longer than 6 bp remain major limitations.

#### Illumina/Solexa Sequencing

In 2006, Solexa released the Genome Analyzer sequencing platform, and Illumina purchased Solexa the following year. The Illumina/Solexa platform used bridge amplification for cluster generation and sequences based on SBS chemistry. All enzymatic processes and imaging steps of the Illumina sequencing occur in a flow cell. First, during library preparation, adapter sequences are introduced to template DNA. Library DNA is subsequently denatured to single strands and transferred to a flow cell, where surface attached forward and reverse oligos that are complementary to the adapter sequences are located. The DNA fragment hybridizes to the surface oligo nucleotides to generate an initial copy of the individual sequencing template, while the initial library DNA is removed afterward, leaving the copied fragment to generate a cluster of identical DNA. This step is called bridge amplification, because the 3′ end of the copied DNA fragments hybridizes to the complementary oligos on the flow cell, which forms a bridge structure. During the amplification, four dNTPs (A, T, C, G) with different fluorescent labels are also incorporated into the growing DNA chain catalyzed by a specific DNA polymerase enzyme, resulting in a fluorescent signal that the detection of each base will ultimately be based on Ansorge (2009).

There are two major Illumina sequencing platforms, HiSeq and MiSeq. The general principle behind the two platforms is similar. The output of the HiSeq platform can reach up to 1 T (4 billion reads) per run after sequencing for 6 days, and the maximal read length is 250 bp. The MiSeq platform was launched in 2011 and was considered a benchtop sequencer, thus being more convenient particularly for bacterial sample sequencing. It can generate 15 GB data (25 million reads) per run, and the maximal read length is 300 bp. Compared with 454 pyrosequencing, the cost of Illumina/Solexa sequencing is considered to be lower, and with combinations of P5/P7 primers and adapters, thousands of samples can be sequenced simultaneously.

#### Ion Torrent Sequencing

Ion Torrent sequencing, also referred as Ion semiconductor sequencing was released in 2010. The technology is also based on of the concept of SBS but differs from other sequencing technologies because no optics is applied during sequencing, and light or fluorescence detection is not required. Instead, the hydrogen ion released when dNTP is incorporated into the template nucleotide is used as the determinant (Loman et al., 2012). In brief, the sequencing process takes place on microwells on the semiconductor chip, which is attached by copies of one single-stranded template DNA. During sequencing, with the aid of a DNA polymerase and specific dNTP, the complementary strand is built, releasing a hydrogen ion that changes the pH of the solution, which can subsequently be detected by an ion-sensitive field-effect transistor.

In general, there are three Ion Torrent sequencing systems, including the ion personal genome machine system, the ion proton system, and the ion S5 system. The output of Ion Torrent sequencing can reach 10 GB, and the maximal read length is 400 bp. The Ion Torrent sequencing system is considered as a rapid and economical benchtop sequencer, due to the avoidance of modified nucleotides and optical measurements. However, its throughput is relatively lower than pyrosequencing and the Illumina HiSeq systems.

#### Applied Biosystems SOLiD

Sequencing by oligonucleotide ligation and detection (SOLiD) was developed by Life Technologies in 2006 and was purchased by Applied Biosystems (ABI) in the same year. The technology is based on ligation sequencing. It differs compared with other second-generation sequencing approaches in that the sequencing by synthesis is driven by a DNA ligase instead of a polymerase. During DNA library preparation, DNA fragments are initially ligated to adapters (P1 and P2) and subsequently bound to beads. At that point, the emulsion PCR starts and generates clonal bead populations, followed by DNA denaturation and deposition of beads onto a glass slide in a random array. During the sequencing, the primers hybridize to the adapter within the DNA template initially, and one of four fluorescently labeled dibase probes can ligate to the primer and the base in turn is detected (Shendure and Ji, 2008). SOLiD's 5500W Series Genetic Analyzer is the major platform for ABI SOLiD. The output and speed of ABI SOLiD is relatively high. Up to 320 GB data can be generated within 8 h per run, and the cost is lower. However, the maximal read length is only 75 bp and is not competitive compared with other second-generation sequencing platforms.

### *Third-Generation Sequencing Technology*

The third-generation DNA sequencing technologies are characterized by their long-range DNA sequencing output, with the commercialization of Pacific Biosciences (PacBio) single-molecule real-time (SMRT) sequencing, the Illumina Tru-seq Synthetic Long-Read technology, and the Oxford Nanopore Technologies sequencing platform (Lee et al., 2016; Schadt et al., 2010). Compared with the second-generation sequencing technologies that generate short reads of a few hundred base pairs, the third-generation sequencing technologies produce long reads of more than 10,000 base pairs (Lee et al., 2016; Schadt et al., 2010). Furthermore, unlike the second-generation sequencing technologies, which can miss some important genes, the single-molecule sequencing (or third-generation sequencing) technologies can cover the entire bacterial genome because of their longer read length. Thus, highly accurate de novo assemblies can be performed to generate gapless chromosomal DNA, which make it possible to identify the structural changes, such as insertions, deletions, and translocations, in the chromosomal architecture.

### *Identification of* Salmonella *Using Whole Genome Sequencing*

The first-generation sequencing methods were applied to *Salmonella* species in numerous studies to understand the specific protein or DNA sequences of *Salmonella* species, especially *S.* Typhimurium (Galakatos et al., 1986; Nichols et al., 1980; Faust et al., 1990). These studies analyzed the protein sequences and revealed the function of the proteins or compared the sequences with other strains to examine the differences between the strains. For example, Christensen et al. (1998) used the chain-termination method to analyze phylogenetic relationships of *Salmonella*. Their study compared 16S and 23S rRNA sequences of *S. enterica, S. bongori*, *Citrobacter freundii*, and *Yersinia enterocolitica* to prove their evolutionary hypothesis.

First-generation sequencing methods played an important role in identifying virulence factors in *Salmonella*. For example, one study demonstrated that one of the genes that occurred in *Salmonella* could actually regulate virulence by controlling RNA polymerase (Fang et al., 1992). Sequencing enabled detailed mutation analyses of *Salmonella* and based on the previously identified sequences, the authors could compare the sequence of the corresponding mutant strains and examined their impact on *Salmonella* virulence expression (Bell et al., 1991; Vogler et al., 1991). Despite limitations associated with first-generation sequencing, such approaches still expanded the range of applications for more inclusive genomic analysis, resulting in a more complete picture of cause and effect between gene sequence and the resulting functionality of the genes in question.

WGS based on second-generation sequencing technology has been successfully applied to identify and characterize serovars *S.* Typhimurium, *S.* Enteritidis, and *Salmonella* Montevideo, which are considered three of the more critical serovars associated with contaminated food resulting in human infections (Holt et al., 2008; Quick et al., 2015; Allard et al., 2012). Among the four major second-generation sequencing platforms, Illumina appeared to be the most popular choice. In general, identification of *Salmonella* using WGS based on second-generation sequencing includes four steps: (1) isolation of *Salmonella* strains from samples, followed by recovery in pure culture; (2) extraction of bacterial DNA from the *Salmonella* culture; (3) preparation of a DNA library following the protocol depending on the corresponding sequencing platform; and (4) loading of the library DNA to the corresponding sequencer of choice and initiation of the sequencing process. With the WGS data, species information can be initially confirmed using NCBI or CGE database. SeqSero (www.denglab.info/SeqSero), a Web-based tool, has been demonstrated to possess speed and accuracy for determining the respective serotype of interest by using either the raw sequencing data or genome assemblies. With the complete genome sequence, virulence factors, antibiotic resistance genes, and any other genes of interest are easily identified using PATRIC tools (https://www.patricbrc.org) and comprehensive antibiotic resistance database (CARD, http://arpcard.mcmaster.ca).

As WGS technologies continue to improve and evolve, they will offer considerable advantages for determining the transmission route and the origin of a foodborne *Salmonella* outbreak. Previous methods, such as serotyping, phage typing, pulsed-field gel electrophoresis, and multilocus variable number of tandem repeat analysis, were useful before WGS technologies became available, but they lacked resolution for differentiating highly clonal serovars that only vary by thousands of single nucleotide polymorphisms (SNPs), whereas WGS combined with SNP-based phylogenetic analysis may provide far greater resolution power (Leekitcharoenphon et al., 2014). This SNP-based approach has been shown to reliably cluster outbreak samples and in turn quantitate the genetic distance between outbreak-associated and non–outbreak-associated isolates. Holt et al. (2008) conducted WGS for 19 *Salmonella* Typhi strains using a combination of 454 (Roche) and Solexa (Illumina) sequencing technologies and identified the evolutionary dynamics of Typhi using SNP-based phylogenetic analysis. In another study, *S.* Montevideo strains were sequenced using Roche 454 GS Titanium NGS technology, and phylogenetic relatedness with reference strains that were associated with a spiced-meat outbreak event in the US were determined based on SNP analysis (Allard et al., 2012). Similarly, after WGS of 20 *S.* Montevideo strains isolated from human sources during the period of the *S.* Montevideo outbreak along with another 27 control isolates from human, food, animal, and environmental sources before the outbreak, phenotype characteristics of outbreak strains were predicted, and 4 putative case isolates were found clustered with the outbreak strains (den Bakker et al., 2011).

An increasing number of complete genome sequences of *Salmonella* have been acquired using third-generation sequencing technologies. Achievements made using PacBio SMRT technology include the first fully closed genome sequence of *Salmonella* Cubana and the first complete genome sequence of *S.* Typhimurium ATCC 13311 (NCTC 74) (Hoffmann et al., 2014; Terabayashi et al., 2014).

## Conclusions and Future Prospects

Certainly, these more recent studies incorporating NGS for *Salmonella* characterization support the use of this overall approach for high-throughput, multiplexing, and accurate diagnostic assessment of pathogens including *Salmonella*. Although NGS provides the most comprehensive data for rapidly attaining microbial sequences (DNA or gene expression), this burgeoning technology has notable limitations that restrict its adoption for *Salmonella* detection in foods: (1) costly instrumentation; (2) requirement for bioinformatic technical support and data analysis infrastructure to process, store, and analyze large data sets; (3) requirement for a preenrichment process to ensure viability of pathogens; and (4) purification and amplification of samples prior to NGS analysis. Advances in technology have improved methods to detect and differentiate foodborne *Salmonella*; however, as described in the previous sections, most of these sequencing approaches still have disadvantages, which make them suboptimal for the food industry.

- Speed: rapid generation of information on pathogens that may be associated with raw and ready-to-eat foods.
- Specificity: ability to detect particular pathogens that are present along with nonpathogenic organisms in food matrices or other complex biological environments.
- Sensitivity: ability to detect a small number of pathogens in food or other complex samples that could be sufficient as an infectious dose.
- Cost-effectiveness: multiple serotypes of *Salmonella* can be simultaneously detected and labor costs can be significantly reduced due to high-throughput ability and potential automation of the assay.

Therefore, developments and attempts to improve on NGS technology and potential applications for *Salmonella* are still needed. As these technologies evolve and become more routine, the cost will decrease accordingly and such equipment will become more readily available to smaller budget laboratories. This will most certainly not only include laboratories in academic settings but commercial food company microbiology and quality control laboratories as well. However, although tremendous amounts of data can be generated from the use of these sequencers, the true challenge will not only be analyzing and managing the resulting large data sets but more importantly what questions the data will be used to address.

## References

Allard, M.W., Luo, Y., Strain, E., Li, C., Keys, C.E., Son, I., Stones, R., Musser, S.M., Brown, E.W., 2012. High resolution clustering of *Salmonella enterica* serovar Montevideo strains using a next-generation sequencing approach. BMC Genomics 13 (1), 32.

Ansorge, W.J., 2009. Next-generation DNA sequencing techniques. New Biotechnology 25 (4), 195–203.

Baker, C.A., Rubinelli, P.M., Park, S.H., Ricke, S.C., 2016. Immuno-based detection of Shiga toxin-producing pathogenic *Escherichia coli* in food–A review on current approaches and potential strategies for optimization. Critical Reviews in Microbiology 42 (4), 656–675.

Balamurugan, S., 2010. Growth temperature associated protein expression and membrane fatty acid composition profiles of *Salmonella enterica* serovar Typhimurium. Journal of Basic Microbiology 50 (6), 507–518.

Bell, D.A., Levine, J.G., DeMarini, D.M., 1991. DNA sequence analysis of revertants of the hisD3052 allele of *Salmonella typhimurium* TA98 using the polymerase chain reaction and direct sequencing: application to 1-nitropyrene-induced revertants. Mutation Research/Environmental Mutagenesis and Related Subjects 252 (1), 35–44.

Bertelli, C., Greub, G., 2013. Rapid bacterial genome sequencing: methods and applications in clinical microbiology. Clinical Microbiology and Infection 19 (9), 803–813.

Bhagwat, A.A., 2003. Simultaneous detection of *Escherichia coli* O157: H7, *Listeria monocytogenes* and *Salmonella* strains by real-time PCR. International Journal of Food Microbiology 84 (2), 217–224.

Bhunia, A.K., 2008. Foodborne Microbial Pathogens. Springer Science+Business Media, LLC, New York, NY.

Bohaychuk, V.M., Gensler, G.E., McFall, M.E., King, R.K., Renter, D.G., 2007. A real-time PCR assay for the detection of *Salmonella* in a wide variety of food and food-animal matrices. Journal of Food Protection 70 (5), 1080–1087.

Boyd, E.F., Porwollik, S., Blackmer, F., McClelland, M., 2003. Differences in gene content among *Salmonella enterica* serovar Typhi isolates. Journal of Clinical Microbiology 41 (8), 3823–3828.

Carrasco, E., Morales-Rueda, A., García-Gimeno, R.M., 2012. Cross-contamination and recontamination by *Salmonella* in foods: a review. Food Research International 45 (2), 545–556.

Centers for Disease Control, Prevention (CDC), 2013. Multistate outbreak of multidrug-resistant *Salmonella* Heidelberg infections linked to foster farms brand chicken. In: *Salmonella* Outbreaks. United States Centers for Disease Control and Prevention. Available from: http://www.cdc.gov/salmonella/heidelberg-10-13/index.html.

Centers for Disease Control, Prevention (CDC), 2011. Investigation announcement: multistate outbreak of *Salmonella* Hadar infections associated with Turkey Burgers. In: *Salmonella* Outbreaks. United States Centers for Disease Control and Prevention. Available from: http://www.cdc.gov/salmonella/hadar0411/040411/index.html.

Centers for Disease Control, Prevention (CDC), 2010a. Investigation update: multistate outbreak of human *Salmonella* enteritidis infections associated with shell eggs. In: *Salmonella* Outbreaks. United States Centers for Disease Control and Prevention. Available from: http://www.cdc.gov/salmonella/enteritidis/index.html.

Centers for Disease Control, Prevention (CDC), 2010b. Investigation update: multistate outbreak of human *Salmonella* chester infections. In: *Salmonella* Outbreaks. United States Centers for Disease Control and Prevention. Available from: http://www.cdc.gov/salmonella/enteritidis/index.html.

Centers for Disease Control, Prevention (CDC), 1990. 'Epidemiologic notes and reports update: *Salmonella* enteritidis infections and shell eggs – United States. MMWR 39 (50), 909–912.

Christensen, H., Nordentoft, S., Elmerdahl Olsen, J., 1998. Phylogenetic relationships of *Salmonella* based on rRNA sequences. International Journal of Systematic Bacteriology 48, 605–610.

Cocolin, L., Rajkovic, A., Rantsiou, K., Uyttendaele, M., 2011. The challenge of merging food safety diagnostic needs with quantitative PCR platforms. Trends in Food Science & Technology 22, S30–S38.

Corless, C.E., Guiver, M., Borrow, R., Edwards-Jones, V., Fox, A.J., Kaczmarski, E.B., 2001. Simultaneous detection of *Neisseria meningitidis, Haemophilus influenzae*, and *Streptococcus pneumoniae* in suspected cases of meningitis and septicemia using real-time PCR. Journal of Clinical Microbiology 39 (4), 1553–1558.

Cummings, C.A., Chung, C.A.B., Fang, R., Barker, M., Brzoska, P., Williamson, P.C., Beaudry, J., Matthews, M., Schupp, J., Wagner, D.M., Birdsell, D., Vogler, A.J., Furtado, M.R., Keim, P., Budowle, B., 2010. Accurate, rapid and high-throughput detection of strain-specific polymorphisms in *Bacillus anthracis* and *Yersinia pestis* by next-generation sequencing. Investigative Genetics 1, 5.

Dawoud, T.M., Davis, M.L., Park, S.H., Kim, S.A., Kwon, Y.M., Jarvis, N., O'Bryan, C.A., Shi, Z., Crandall, P.G., Ricke, S.C., 2017. The potential link between thermal resistance and virulence in *Salmonella*: a review. Frontiers in Veterinary Science 4 (93). http://dx.doi.org/10.3389/fvets.2017.00093.

den Bakker, H.C., Switt, A.I.M., Cummings, C.A., Hoelzer, K., Degoricija, L., Rodriguez-Rivera, L.D., Wright, E.M., Fang, R., Davis, M., Root, T., Schoonmaker-Bopp, D., 2011. A whole genome SNP based approach to trace and identify outbreaks linked to a common *Salmonella enterica* subsp. *enterica* serovar Montevideo Pulsed Field Gel Electrophoresis type. Applied and Environmental Microbiology 77 (24), 8648–8655.

Dunkley, K.D., Callaway, T.R., Chalova, V.I., McReynolds, J.L., Hume, M.E., Dunkley, C.S., Kubena, L.F., Nisbet, D.J., Ricke, S.C., 2009. Foodborne *Salmonella* ecology in the avian gastrointestinal tract. Anaerobe 15 (1), 26–35.

Durant, J.A., Corrier, D.E., Byrd, J.A., Stanker, L.H., Ricke, S.C., 1999. Feed deprivation affects crop environment and modulates *Salmonella* enteritidis colonization and invasion of leghorn hens. Applied and Environmental Microbiology 65 (5), 1919–1923.

Eyigör, A., Goncagül, G., Carli, K.T., 2007. A PCR-ELISA for the detection of *Salmonella* from chicken intestine. Journal of Biological and Environmental Sciences 1 (1), 45–49.

Fang, F.C., Libby, S.J., Buchmeier, N.A., Loewen, P.C., Switala, J., Harwood, J., Guiney, D.G., 1992. The alternative sigma factor katF (rpoS) regulates *Salmonella* virulence. Proceedings of the National Academy of Sciences 89 (24), 11978–11982.

Faust, L.R., Connor, J.A., Roof, D.M., Hoch, J.A., Babior, B.M., 1990. Cloning, sequencing, and expression of the genes encoding the adenosylcobalamin-dependent ethanolamine ammonia-lyase of *Salmonella* typhimurium. Journal of Biological Chemistry 265 (21), 12462–12466.

FDA, 1998. Bacteriological Analytical Manual (BAM). United States Food and Drug Administration. Available from: http://www.fda.gov/food/foodscienceresearch/laboratorymethods/ucm2006949.htm.

Finstad, S., O'Bryan, C.A., Marcy, J.A., Crandall, P.G., Ricke, S.C., 2012. *Salmonella* and broiler processing in the United States: relationship to foodborne salmonellosis. Food Research International 45 (2), 789–794.

Foley, S.L., Johnson, T.J., Ricke, S.C., Nayak, R., Danzeisen, J., 2013. *Salmonella* pathogenicity and host adaptation in chicken-associated serovars. Microbiology and Molecular Biology Reviews 77 (4), 582–607.

Foley, S.L., Nayak, R., Hanning, I.B., Johnson, T.J., Han, J., Ricke, S.C., 2011. Population dynamics of *Salmonella enterica* serotypes in commercial egg and poultry production. Applied and Environmental Microbiology 77 (13), 4273–4279.

Foley, S.L., Lynne, A.M., Nayak, R., 2008. *Salmonella* challenges: prevalence in swine and poultry and potential pathogenicity of such isolates. Journal of Animal Science 86, E149–E162.

França, L.T., Carrilho, E., Kist, T.B., 2002. A review of DNA sequencing techniques. Quarterly Reviews of Biophysics 35 (02), 169–200.

Galakatos, N.G., Daub, E., Botstein, D., Walsh, C.T., 1986. Biosynthetic alr alanine racemase from *Salmonella* typhimurium: DNA and protein sequence determination. Biochemistry 25 (11), 3255–3260.

Gómez-Aldapa, C.A., Torres-Vitela, M.D.R., Villarruel-López, A., Castro-Rosas, J., 2012. The role of foods in *Salmonella* infections. In: Mahmoud Barakat, B.S.M. (Ed.), *Salmonella* – A Dangerous Foodborne Pathogen. InTech, pp. 21–46. http://dx.doi.org/10.5772/28316. (Chapter 2). Available from: http://www.intechopen.com/books/salmonella-a-dangerous-foodborne-pathogen/the-role-of-foods-in-salmonella-infections.

Grimont, P.A.D., Weill, F.X., 2007. Antigenic Formulae of the *Salmonella* Serovars, ninth ed. Institut Pasteur, Paris, France.

Hanning, I.B., Nutt, J.D., Ricke, S.C., 2009. Salmonellosis outbreaks in the United States due to fresh produce: sources and potential intervention measures. Foodborne Pathogens and Disease 6 (6), 635–648.

Heather, J.M., Chain, B., 2016. The sequence of sequencers: the history of sequencing DNA. Genomics 107 (1), 1–8.

Hoffmann, M., Muruvanda, T., Pirone, C., Korlach, J., Timme, R., Payne, J., Evans, P., Meng, J., Brown, E.W., Allard, M.W., 2014. First fully closed genome sequence of *Salmonella enterica* subsp. *enterica* serovar Cubana associated with a food-borne outbreak. Genome Announcements 2 (5), e01112–e01114.

Holt, K.E., Parkhill, J., Mazzoni, C.J., Roumagnac, P., Weill, F.X., Goodhead, I., Rance, R., Baker, S., Maskell, D.J., Wain, J., Dolecek, C., 2008. High-throughput sequencing provides insights into genome variation and evolution in *Salmonella* Typhi. Nature Genetics 40 (8), 987–993.

Hong, Y., Liu, T., Lee, M.D., Hofacre, C.L., Maier, M., White, D.G., Ayers, S., Wang, L., Berghaus, R., Maurer, J., 2009. A rapid screen of broth enrichments for *Salmonella* enterica serovars enteritidis, Hadar, Heidelberg, and Typhimurium by Using an allelotyping multiplex PCR that targets O- and H-antigen alleles. Journal of Food Protection 72 (10), 2198–2201.

Hoorfar, J., 2011. Rapid detection, characterization, and enumeration of foodborne pathogens. Acta Pathologica Microbiologica et Immunologica Scandinavia 119 (s133), 1–24.

Howard, Z.R., O'Bryan, C.A., Crandall, P.G., Ricke, S.C., 2012. *Salmonella Enteritidis* in shell eggs: current issues and prospects for control. Food Research International 45 (2), 755–764.

Jarvis, N.A., O'Bryan, C.A., Dawoud, T.M., Park, S.H., Kwon, Y.M., Crandall, P.G., Ricke, S.C., 2016. An overview of *Salmonella* thermal destruction during food processing and preparation. Food Control 68, 280–290.

Judson, H.F., 1979. The Eighth Day of Creation – The Makers of the Evolution in Biology. Simon and Schuster, Simon & Schuster, New York, NY.

Kaldhone, P.R., Foley, S.L., Ricke, S.C., 2016. *Salmonella* Heidelberg in layer hens and egg production: incidence and potential issues. In: Ricke, S.C., Gast, R.K. (Eds.), Producing Safe Eggs: Microbial Ecology of *Salmonella*. Academic Press, Cambridge, MA (Chapter 12).

Karlen, Y., McNair, A., Perseguers, S., Mazza, C., Mermod, N., 2007. Statistical significance of quantitative PCR. BMC Bioinformatics 8 (1), 131.

Kérouanton, A., Hennekinne, J.A., Letertre, C., Petit, L., Chesneau, O., Brisabois, A., De Buyser, M.L., 2007. Characterization of *Staphylococcus aureus* strains associated with food poisoning outbreaks in France. International Journal of Food Microbiology 115 (3), 369–375.

Kim, S.A., Park, S.H., Lee, S.I., Ricke, S.C., 2017. Development of a rapid method to quantify *Salmonella* Typhimurium using a combination of MPN with qPCR and a shortened time incubation. Food Microbiology 65, 7–18.

Khatiwara, A., Jiang, T., Sung, S.S., Dawoud, T., Kim, J.N., Bhattacharya, D., Kim, H.B., Ricke, S.C., Kwon, Y.M., 2012. Genome scanning for conditionally essential genes in *Salmonella enterica* serotype Typhimurium. Applied and Environmental Microbiology 78 (9), 3098–3107.

Köser, C.U., Ellington, M.J., Cartwright, E.J., Gillespie, S.H., Brown, N.M., Farrington, M., Holden, M.T., Dougan, G., Bentley, S.D., Parkhill, J., Peacock, S.J., 2012. Routine use of microbial whole genome sequencing in diagnostic and public health microbiology. PLoS Pathogens 8 (8), e1002824.

Kumar, S., Balakrishna, K., Singh, G.P., Batra, H.V., 2005. Rapid detection of *Salmonella typhi* in foods by combination of immunomagnetic separation and polymerase chain reaction. World Journal of Microbiology & Biotechnology 21 (5), 625–628.

Kwon, Y.M., Ricke, S.C. (Eds.), 2011. High-throughput Next Generation Sequencing: Methods and Applications, Methods in Molecular Microbiology. Springer Protocols, Humana Press, New York, NY.

Kwon, Y.M., Ricke, S.C., 1998. Induction of acid resistance of *Salmonella* Typhimurium by exposure to short-chain fatty acids. Applied and Environmental Microbiology 64 (9), 3458–3463.

Kwon, Y.M., Ricke, S.C., Mandal, R.K., 2016. Transposon sequencing: methods and expanding applications. Applied Microbiology and Biotechnology 100 (1), 31–43.

Kwon, Y.M., Park, S.Y., Birkhold, S.G., Ricke, S.C., 2000. Induction of resistance of *Salmonella typhimurium* to environmental stresses by exposure to short-chain fatty acids. Journal of Food Science 65 (6), 1037–1040.

Lee, H., Gurtowski, J., Yoo, S., Nattestad, M., Marcus, S., Goodwin, S., McCombie, W.R., Schatz, M., 2016. Third-generation sequencing and the future of genomics. bioRxiv 048603.

Leekitcharoenphon, P., Nielsen, E.M., Kaas, R.S., Lund, O., Aarestrup, F.M., 2014. Evaluation of whole genome sequencing for outbreak detection of *Salmonella enterica*. PLoS One 9 (2), e87991.

Levin, R.E., 2010. Rapid Detection and Characterization of Foodborne Pathogens by Molecular Techniques. CRC Press, Taylor & Francis Group, Boca Raton, Florida.

Liu, L., Li, Y., Li, S., Hu, N., He, Y., Pong, R., Lin, D., Lu, L., Law, M., 2012. Comparison of next-generation sequencing systems. Biomed Research International 2012, 11.

Loman, N.J., Misra, R.V., Dallman, T.J., Constantinidou, C., Gharbia, S.E., Wain, J., Pallen, M.J., 2012. Performance comparison of benchtop high-throughput sequencing platforms. Nature Biotechnology 30 (5), 434–439.

Luckey, J.A., Drossman, H., Kostichka, A.J., Mead, D.A., D'Cunha, J., Norris, T.B., Smith, L.M., 1990. High speed DNA sequencing by capillary electrophoresis. Nucleic Acids Research 18 (15), 4417–4421.

Lynch, M.F., Tauxe, R.V., Hedberg, C.W., 2009. The growing burden of foodborne outbreaks due to contaminated fresh produce: risks and opportunities. Epidemiology and Infection 137 (03), 307–315.

Maciorowski, K.G., Herrera, P., Jones, F.T., Pillai, S.D., Ricke, S.C., 2006. Cultural and immunological detection methods for *Salmonella* spp. in animal feeds–a review. Veterinary Research Communications 30 (2), 127–137.

Maciorowski, K.G., Pillai, S.D., Jones, F.T., Ricke, S.C., 2005. Polymerase chain reaction detection of foodborne *Salmonella* spp. in animal feeds. Critical Reviews in Microbiology 31 (1), 45–53.

March, C., Manclús, J.J., Abad, A., Navarro, A., Montoya, A., 2005. Rapid detection and counting of viable beer-spoilage lactic acid bacteria using a monoclonal chemiluminescence enzyme immunoassay and a CCD camera. Journal of Immunological Methods 303 (1), 92–104.

Mardis, E.R., 2013. Next-generation sequencing platforms. Annual Review of Analytical Chemistry 6, 287–303.

Maxam, A.M., Gilbert, W., 1977. A new method for sequencing DNA. Proceedings of the National Academy of Sciences 74 (2), 560–564.

McCarthy, N., Reen, F.J., Buckley, J.F., Frye, J.G., Boyd, E.F., Gilroy, D., 2009. Sensitive and rapid molecular detection assays for *Salmonella enterica* serovars Typhimurium and Heidelberg. Journal of Food Protection 72 (11), 2350–2357.

McKee, S.R., 2012. Pre-harvest Survey for Best Practices in Pathogen Control. Presented at the International Poultry Scientific Forum, Atlanta, GA).

McKee, S.R., 2011. *Salmonella* and *Campylobacter* Control during Poultry Processing. Presented at the International Poultry Scientific Forum, Atlanta, GA).

Medini, D., Donati, C., Tettelin, H., Masignani, V., Rappuoli, R., 2005. The microbial pan-genome. Current Opinion in Genetics & Development 15 (6), 589–594.

Mercanoglu-Taban, B., Ben, U., Aytac, S.A., 2009. Rapid detection of *Salmonella* in milk by combined immuno-magnetic separation-polymerase chain reaction assay. Journal of Dairy Science 92 (6), 2382–2388.

Metzker, M.L., 2010. Sequencing technologies—the next generation. Nature Reviews Genetics 11 (1), 31–46.

Monadi, A.R., Mirzaei, H., Javadi, A., Hosseinzade, N., Amjadi, Y., 2010. Effect of some probiotics on *Salmonella typhi* during associated growth in milk. African Journal of Microbiology Research 4 (24), 2708–2711.

Myint, M.S., Johnson, Y.J., Tablante, N.L., Heckert, R.A., 2006. The effect of pre-enrichment protocol on the sensitivity and specificity of PCR for detection of naturally contaminated *Salmonella* in raw poultry compared to conventional culture. Food Microbiology 23 (6), 599–604.

NDSC – National Disease Surveillance Centre, 2004. Preventing Foodborne Disease: A Focus on the Infected Food Handler. Health Protection Surveillance Centre, Government of Ireland. Available from: https://www.hpsc.ie/A-Z/Gastroenteric/FoodborneIllness/Publications/File,871,en.pdf.

Nichols, B.P., Miozzari, G.F., van Cleemput, M., Bennett, G.N., Yanofsky, C., 1980. Nucleotide sequences of the trpG regions of *Escherichia coli*, *Shigella dysenteriae*, *Salmonella typhimurium* and *Serratia marcescens*. Journal of Molecular Biology 142 (4), 503–517.

Oliver, S.P., Jayaro, B.M., Almeida, R.A., 2005. Foodborne pathogens, mastitis, milk quality, and dairy food safety. NMC Annual Meeting Proceedings 1, 3–27.

Olsen, S.J., MacKinnon, L.C., Goulding, J.S., Bean, N.H., Slutsker, L., 2000. Surveillance for foodborne-disease outbreaks—United States, 1993–1997. MMWR 49, 1–62.

Park, S.H., Ricke, S.C., 2015. Development of multiplex PCR assay for simultaneous detection of *Salmonella* genus, *Salmonella* subspecies I, *Salm.* Enteritidis, *Salm.* Heidelberg and *Salm.* Typhimurium. Journal of Applied Microbiology 118 (1), 152–160.

Park, S.H., Aydin, M., Khatiwara, A., Dolan, M.C., Gilmore, D.F., Bouldin, J.L., Ahn, S., Ricke, S.C., 2014. Current and emerging technologies for rapid detection and characterization of *Salmonella* in poultry and poultry products. Food Microbiology 38, 250–262.

Park, S.H., Hanning, I., Jarquin, R., Moore, P., Donoghue, D.J., Donoghue, A.M., Ricke, S.C., 2011. Multiplex PCR assay for the detection and quantification of *Campylobacter* spp, *Escherichia coli* O157:H7, and *Salmonella* serotypes in water samples. FEMS Microbiology Letters 316 (1), 7–15.

Park, S.H., Kim, H.J., Cho, W.H., Kim, J.H., Oh, M.H., Kim, S.H., Lee, B.K., Ricke, S.C., Kim, H.Y., 2009. Identification of *Salmonella enterica* subspecies I, *Salmonella enterica* serovars Typhimurium, Enteritidis and Typhi using multiplex PCR. FEMS Microbiology Letters 301 (1), 137–146.

Pontiroli, A., Travis, E.R., Sweeney, F.P., Porter, D., Gaze, W.H., Mason, S., Hibberd, V., Holden, J., Courtenay, O., Wellington, E.M.H., 2011. Pathogen quantitation in complex matrices: a multi-operator comparison of DNA extraction methods with a novel assessment of PCR inhibition. PLoS One 6 (3), e17916.

Porwollik, S., Boyd, E.F., Choy, C., Cheng, P., Florea, L., Proctor, E., McClelland, M., 2004. Characterization of *Salmonella enterica* subspecies I genovars by use of microarrays. Journal of Bacteriology 186 (17), 5883–5898.

Prober, J.M., Trainor, G.L., Dam, R.J., Hobbs, F.W., Robertson, R.W., Zagursky, R.J., Cocuzza, A.J., Jensen, M.A., Baueister, K., 1987. A system for rapid DNA sequencing with fluorescent chain-terminating dideoxynucleotides. Science 238 (4825), 336–341.

Qi, W., Käser, M., Röltgen, K., Yeboah-Manu, D., Pluschke, G., 2009. Genomic diversity and evolution of *Mycobacterium ulcerans* revealed by next-generation sequencing. PLoS Pathogens 5 (9), e1000580.

Quick, J., Ashton, P., Calus, S., Chatt, C., Gossain, S., Hawker, J., Nair, S., Neal, K., Nye, K., Peters, T., De Pinna, E., 2015. Rapid draft sequencing and real-time nanopore sequencing in a hospital outbreak of *Salmonella*. Genome Biology 16 (1), 114.

Ricke, S.C., 2014. Application of molecular approaches for understanding foodborne *Salmonella* establishment in poultry production. Advances in Biology 813275. http://dx.doi.org/10.1155/2014/813275. 25 p.

Ricke, S.C., 2003a. The gastrointestinal tract ecology of *Salmonella* Enteritidis colonization in molting hens. Poultry Science 82 (6), 1003–1007.

Ricke, S.C., 2003b. Perspectives on the use of organic acids and short chain fatty acids as antimicrobials. Poultry Science 82 (4), 632–639.

Ricke, S.C., Rivera Calo, J., 2015. Antibiotic resistance in pathogenic *Salmonella*. In: Chen, C.Y., Yan, X., Jackson, C. (Eds.), Antimicrobial Resistance and Food Safety: Methods and Techniques. Academic Press/ Elsevier Science, Amsterdam, The Netherlands, pp. 37–53 (Chapter 3).

Ricke, S.C., Gast, R.K. (Eds.), 2016. Producing Safe Eggs: Microbial Ecology of *Salmonella*. Academic Press, Cambridge, MA.

Ricke, S.C., Dawoud, T., Kwon, Y.M., 2015. Application of molecular methods for traceability of foodborne pathogens in food safety systems. In: Ricke, S.C., Donaldson, J.R., Phillips, C.A. (Eds.), Food Safety: Emerging Issues, Technologies and Systems. Elsevier, Oxford, UK, pp. 37–63 (Chapter 4).

Ricke, S.C., Khatiwara, A., Kwon, Y.M., 2013a. Application of microarray analysis of foodborne *Salmonella* in poultry production: a review. Poultry Science 92 (9), 2243–2250.

Ricke, S.C., Koo, O.K., Foley, S., Nayak, R., 2013b. *Salmonella*. In: Labbé, R., García, S. (Eds.), Guide to Foodborne Pathogens, second ed. Wiley-Blackwell, Oxford, UK, pp. 112–137 (Chapter 7).

Rothberg, J.M., Leamon, J.H., 2008. The development and impact of 454 sequencing. Nature Biotechnology 26 (10), 1117–1124.

Saeki, E.K., Alves, J., Bonfante, R.C., Hirooka, E.Y., Oliveira, T.C.R.M., 2013. Multiplex PCR (mPCR) for the detection of *Salmonella* spp and the differentiation of the Typhimurium and Enteritidis serovars in chicken meat. Journal of Food Safety 33 (1), 25–29.

Santos, R.L., Zhang, S., Tsolis, R.M., Kingsley, R.A., Adams, L.G., Bäumler, A.J., 2001. Animal models of *Salmonella* infections: enteritis versus typhoid fever. Microbes and Infection 3 (14), 1335–1344.

Santos, S.A.D., Anrade Júnior, D.R.D., Andrade, D.R.D., 2011. TNF-a production and apoptosis in hepatocytes after *Listeria monocytogenes* and *Salmonella* Typhimurium invasion. Revista do Instituto de Medicina Tropical de São Paulo 53 (2), 107–112.

Scallan, E., Hoekstra, R.M., Angulo, F.J., Tauxe, R.V., Widdowson, M.A., Roy, S.L., Jones, J.L., Griffin, P.M., 2011. Foodborne illness acquired in the United States—major pathogens. Emerging Infectious Diseases 17 (1), 7–15.

Schadt, E.E., Turner, S., Kasarskis, A., 2010. A window into third-generation sequencing. Human Molecular Genetics 19 (R2), R227–R240.

Seo, K.H., Valentin-Bon, I.E., Brackett, R.E., Holt, P.S., 2004. Rapid, specific detection of *Salmonella* Enteritidis in pooled eggs by real-time PCR. Journal of Food Protection 67 (5), 864–869.

Shah, D.H., Zhou, X., Addwebi, T., Davis, M.A., Orfe, L., Call, D.R., Guard, J., Besser, T.E., 2011. Cell invasion of poultry-associated *Salmonella enterica* serovar Enteritidis isolates is associated with pathogenicity, motility and proteins secreted by the type III secretion system. Microbiology 157 (5), 1428–1445.

Shendure, J., Ji, H., 2008. Next-generation DNA sequencing. Nature Biotechnology 26 (10), 1135–1145.

Smith, L.M., Fung, S., Hunkapiller, M.W., Hunkapiller, T.J., Hood, L.E., 1985. The synthesis of oligonucleotides containing an aliphatic amino group at the 5′ terminus: synthesis of fluorescent DNA primers for use in DNA sequence analysis. Nucleic Acids Research 13 (7), 2399–2412.

Soumet, C., Ermel, G., Rose, V., Rose, N., Drouin, P., Salvat, G., Colin, P., 1999. Identification by a multiplex PCR-based assay of *Salmonella* Typhimurium and *Salmonella* Enteritidis strains from environmental swabs of poultry houses. Letters in Applied Microbiology 29 (1), 1–6.

Swerdlow, H., Gesteland, R., 1990. Capillary gel electrophoresis for rapid, high resolution DNA sequencing. Nucleic Acids Research 18 (6), 1415–1419.

Szabo, E.A., Mackey, B.M., 1999. Detection of *Salmonella* enteritidis by reverse transcription-polymerase chain reaction (PCR). International Journal of Food Microbiology 51 (2), 113–122.

Taha, E.G., Mohamed, A., Srivastava, K.K., Reddy, P.G., 2010. Rapid detection of *Salmonella* in chicken meat using immunomagnetic separation, CHROMagar, ELISA and real-time polymerase chain reaction (RT-PCR). International Journal of Poultry Science 9 (831), e835.

Tauxe, R.V., 2002. Emerging foodborne pathogens. International Journal of Food Microbiology 78 (1), 31–41.

Tauxe, R.V., 1997. Emerging foodborne diseases: an evolving public health challenge. Emerging Infectious Diseases 3 (4), 425.

Terabayashi, Y., Juan, A., Tamotsu, H., Ashimine, N., Nakano, K., Shimoji, M., Shiroma, A., Teruya, K., Satou, K., Hirano, T., 2014. First complete genome sequence of *Salmonella enterica* subsp. *enterica* serovar Typhimurium strain ATCC 13311 (NCTC 74), a reference strain of multidrug resistance, as achieved by use of PacBio single-molecule real-time technology. Genome Announcements 2 (5) e00986–14.

USDA-FSIS, 2013. Microbiology Laboratory Guidebook. United States Department of Agriculture. Available from: http://www.fsis.usda.gov/wps/portal/fsis/topics/science/laboratories-and-procedures/guidebooks-and-methods/microbiology-laboratory-guidebook/microbiology-laboratory-guidebook.

USDA-FSIS, 2010. Serotypes Profile of *Salmonella* Isolates from Meat and Poultry Products, January 1998 through December 2009. United States Department of Agriculture. Available from: http://www.fsis.usda.gov/Science/Serotypes_Profile_Salmonella_Isolates/index.asp.

van Dijk, E.L., Auger, H., Jaszczyszyn, Y., Thermes, C., 2014. Ten years of next-generation sequencing technology. Trends in Genetics 30 (9), 418–426.

Vogler, A.P., Homma, M., Irikura, V.M., Macnab, R.M., 1991. *Salmonella* typhimurium mutants defective in flagellar filament regrowth and sequence similarity of FliI to F0F1, vacuolar, and archaebacterial ATPase subunits. Journal of Bacteriology 173 (11), 3564–3572.

Wegener, H.C., Hald, T., Wong, D.L.F., Madsen, M., Korsgaard, H., Bager, F., Gerner-Smidt, P., Mølbak, K., 2003. *Salmonella* control programs in Denmark. Emerging Infectious Diseases 9 (7), 774–780.

Wise, M.G., Siragusa, G.R., Plumblee, J., Healy, M., Cray, P.J., Seal, B.S., 2009. Predicting *Salmonella enterica* serotypes by repetitive sequence-based PCR. Journal of Microbiological Methods 76 (1), 18–24.

CHAPTER 17

# Use and Impacts of Whole Genome Sequencing in Retail Food Safety

**Clara Assisi, Andrea J. Etter, Haley F. Oliver**
*Purdue University, West Lafayette, IN, United States*

## History of Whole Genome Sequencing

Frederick Sanger developed the first sequencing technology in 1977. "Sanger Sequencing" is an enzymatic chain termination process and is the foundational technology for next-generation sequencing (NGS) technologies. Currently, Sanger sequencers (e.g., ABI3730xl by Applied Biosystems, Foster City, CA) have the capacity to read 1000–1200 base pairs (bp) of DNA template (Zhang et al., 2011). Sanger technology requires a specific primer to initiate sequencing (Heather and Chain, 2016). Initially, four radiolabeled nucleotides were used in independent reactions and purified by 2D gel chromatography (gel electrophoresis) (Heather and Chain, 2016). Later, fluorescently labeled dideoxynucleotides were adopted (Swerdlow and Gesteland, 1990), and in 1987, the first automatic capillary electrophoresis-based sequencer was introduced by Applied Biosystems (Foster City, CA), making the sequencing process more rapid and accurate (Liu et al., 2012). Sanger technology was the dominant sequencing method until approximately 2005.

To read longer sequences of DNA template, shotgun sequencing was developed. In this approach, the DNA template is broken down into smaller fragments and cloned into sequencing vectors (Zhang et al., 2011). The clones are then amplified in *Escherichia coli* and individually sequenced (Zhang et al., 2011). The complete sequence is generated by aligning and assembling the fragments based on the overlap of the partial sequences (Zhang et al., 2011; Heather and Chain, 2016).

The genome of *Haemophilus influenza* was the first complete bacterial genome published (Fleischmann et al., 1995). In 1996, the first *unicellular* eukaryote, *Saccharomyces cerevisiae*, was sequenced (Goffeau et al., 1996); *Caenorhabditis elegans* was sequenced in 1998 (Anonymous, 1998). In 1999, the entire DNA sequence of human chromosome 22, the shortest human autosome, was completed and published (Dunham et al., 1999). Genome sequences for *Drosophila melanogaster* (Adams et al., 2000) and the first plant genome (*Arabidopsis thaliana*) were fully sequenced in 2000 (Anonymous, 2000). In 2002, the

Food and Feed Safety Systems and Analysis. http://dx.doi.org/10.1016/B978-0-12-811835-1.00017-8

genome of the laboratory mouse *Mus musculus* was completed (Waterston et al., 2002), and in 2001, the Human Genome Project published a draft composite (compiled from multiple donors) human genome (Lander et al., 2001; Venter et al., 2001; Levy et al., 2007). Three years later, the Human Genome Project published a finished version (99% coverage) of the human genome published in 2001 (Anonymous, 2004), and in 2007, the first individual human diploid sequence was published (Levy et al., 2007). The techniques and practices developed for the Human Genome Project were instrumental in developing the massive parallel sequencing used in NGS (Zhang et al., 2011; Venter et al., 2003; Shendure et al., 2005).

## Current and Future Platforms

NGS, also known as high-throughput sequencing or massively parallel sequencing, has facilitated large-scale whole genome sequencing (WGS) (Zhang et al., 2011). NGS platforms sequence millions of shorter reads (50–500 bp) that are aligned to resolve the WGS. The average number of times each nucleotide is sequenced determines the depth of coverage, which is a measurement of the confidence in the accuracy of the nucleotide resolution.

Commercialized NGS technologies that were or are widely used include Roche GS FLX 454 (Branford, CT), Applied Biosystems SOLiD (Foster City, CA), Ion Torrent Personal Genome Machine (San Francisco, CA), Pacific Bioscience single-molecule real time (SMRT) (Menlo Park, CA), Oxford Nanopore Technologies MinION (Oxford, UK), and Illumina Solexa Genome Analyzer (San Diego, CA). These technologies simplified sample preparation and produced rapid, massively parallel amplification and sequencing of the DNA template compared with Sanger Sequencing (Zhang et al., 2011). The performance of the next-generation DNA sequencers is summarized in Table 17.1 and select sequencing technologies are detailed below.

## Single-Molecule, Real-Time Sequencing

SMRT sequencing was commercialized by Pacific Bioscience (PacBio; Menlo Park, CA) in 2011. The two key innovations in this sequencing technique are the utilization of zero-mode waveguide (ZMW) and the use of phospholinked instead of base-linked nucleotides with distinctive fluorophores (Eid et al., 2009). The ZMW is a nanowell facilitating capture of the light signal produced when a single nucleotide is incorporated into a read (PacBio, 2016a). The dimensions of the ZMW allow light to be emitted from the bottom of the well, excluding background noise; DNA sequencing is real time and continuous (PacBio, 2016a). Advantages of this sequencing technology include that (1) it does not require amplification of the DNA library before sequencing, (2) it has a long read length (up to 15 kb) compared with most other commercial sequencers, and (3) it is the fastest sequencer to date (PacBio, 2016a,b). It is able to produce high-quality reads (99.999% accuracy) with 30S coverage (PacBio, 2016b).

The DNA library is prepared by fragmenting DNA template and addition of adapters. The library is added to the ZMW where each molecule will bind to a single DNA polymerase and create an immobilized complex (PacBio, 2016a; Eid et al., 2009). Signal is captured in the ZMW as each phospholinked nucleotide incorporates to the DNA molecule (PacBio, 2016a). The mean read length of PacBio RS II sequencer is 10–15 kb with output data of 0.5–1 Gb and a run time of 4 h (PacBio, 2016b).

## Oxford Nanopore (MinION)

In 2014, Oxford Nanopore Technologies launched the first commercial sequencer that uses nanopores in lieu of the sequencing by synthesis (SBS) approach used by most NGS technologies (Quick et al., 2014). MinION uses "strand sequencing" technology; this technology is based on real-time detection of the DNA strand as it passed though protein nanopores embedded in a fabricated polymer membrane, such as silicon nitride or graphene (Loman et al., 2012; Technologies, 2016). The advantages of MinION are (1) its ability to sequence DNA in real time, (2) no amplification step of the DNA library before sequencing leading to reduced bias and preparation time, (3) portable, and (4) low capital cost (Lavera et al., 2015).

The DNA library strand is prepared by attaching a hairpin adapter to the end of the DNA strands before mixing with helicase, resulting in one long single-strand DNA (consisting of the forward strand followed by a hairpin strand and reverse strand) (Lavera et al., 2015; Technologies, 2016). The processive enzyme inserts it through the nanopore, where the sequencing takes place (Technologies, 2016; Lavera et al., 2015; Quick et al., 2014). The base calling of the nucleotides in the DNA strand is a function of the disruption to the electrical current of the membrane; these disruptions are distinctive for each nucleotide (A, T, C, G) (Technologies, 2016). The MinION has the ability to read 230–300 kb from 512 nanopore sequencing channels and is able to generate up to 2.5 M reads with a final output of up to 25 Mb in 48 h (Technologies, 2016). The accuracy of MinION is 96%, relatively low compared with the other commercial sequencers (see Table 17.1). Oxford Nanopore is currently preparing to release a new platform, PromethION, which has a higher capacity and bigger scale (48 flow cells and 144,000 nanopore channels) and will be able to generate more reads (up to 700 Mb) in 48 h (Technologies, 2016).

## Illumina Platforms

The Solexa Genome Analyzer (Solexa GA; Illumina) was introduced in 2006 and uses SBS technology similar to the Roche 454 (Mardis, 2008; Liu et al., 2012; Zhang et al., 2011). Illumina yields equivalent data quality to Roche and SOLiD platforms, with the advantage of a faster processing time and simplified library preparation (Zhang et al., 2011; Liu et al., 2012).

Initially, the Solexa GA had an output of 1 Gb/run and a sequencing time of 4 days (Mardis, 2008). In 2010, Illumina launched the HiSeq 2000, which was initially able to produce a

Table 17.1: Comparison of select whole genome sequencing platforms.

| Platform | 454 GS FLX titanium XL+ | SOLiD 5500xl | Ion PGM | Single-molecule real time | MinION | Illumina HiSeq 2500 | Illumina MiSeq |
|---|---|---|---|---|---|---|---|
| Sequencing chemistry | Pyrosequencing | Sequencing by ligation | Short read length | Sequencing by synthesis (SBS) | Strand sequencing | SBS | SBS |
| Instrument cost | $500K | $495K | $80K | $695K | $1000 | $654K | $128K |
| Sequence yield per run | 700 Mb | 10–15 Gb/day | 30–100 Mb on 314 chip, 300 Mb-1 Gb on 316 chip, 600 Mb-2 Gb on 318 chip | 0.5–1 Gb | 25–45 Gb | 1500 Gb | 15 Gb |
| Sequencing cost per Gb | $10 (per Mb) | $150 | $1000 (318 chip) | $2000 | $40 (per Mb) | $41 | $502 |
| Run time | 23 h | 7 days | 2.3–3.7 h on 314 chip, 3.0–4.9 h on 316 chip, 4.4–7.3 h on 318 chip | 4 h | 48 h | 7 h–6 days | 4–55 h |
| Reported accuracy | 99.997% | 99.99% | 99.80% | 100.00% | 96% | 99.95% | 99.90% |
| Read length/reads | 700 bp–1 kb | 75 bp | 200 bp | 10 bp–15 kb | 230–300 kb | 2 × 150 bp | 2 × 300 bp |
| Paired reads | No | Yes | Yes | Yes | Yes | Yes | Yes |
| Typical DNA requirements | – | – | 100–1000 ng | ~1 μg | 10 pg–1 μg | 50–1000 ng | 50–1000 ng |
| Advantage | Long read length, minimize sample preparation bias, rapid | Rapid throughput, best data quality, highest accuracy, cost-effective | High throughput, rapid detection, lower cost, and more compact instrument | Real-time sequencing, no DNA library amplification, long read length, fastest sequencer | Real-time sequencing, no DNA library amplification, extreme long read length, portable, and low cost | Large-scale genomics, higher read number compared with MiSeq (5 billion) | Targeted and small-scale sequencing, benchtop sequencer, longer reads compared with HiSeq 2500 |
| Disadvantage | Error rate with same base more than 6, high cost, low yield | Short read length, tedious library preparation | Short read length | Low output | Relatively low accuracy | Shorter read length compared with MiSeq | Lower depth compared with HiSeq 2500 (25 million) |
| References | Sequencing, Liu et al. (2012), Zhang et al. (2011) and Mardis (2008) | Liu et al. (2012), Mardis (2008) and Zhang et al. (2011) | Quail et al. (2012), Liu et al. (2012) and Flusberg et al. (2010) | Quail et al. (2012), PacBio, (2016a,b) and Zhang et al. (2011) | Technologies (2016), Quick et al. (2014), Loman et al. (2012), and Lavera et al. (2015) | Quail et al. (2012), Liu et al. (2012), Zhang et al. (2011) and Mardis (2008) | |

read length of 2 × 100 bp (paired-end) and generated 200 Gb/run with raw base accuracy above 99.95% (Zhang et al., 2011). Later, Illumina launched HiSeq 2500 and the output was improved to 1500 Gb/run with longer read lengths (2 × 150 bp) and a run time of 7 h to 6 days (Quail et al., 2012). In 2011, Illumina launched a benchtop sequencer known as MiSeq (Zhang et al., 2011). It uses SBS technology similar to HiSeq but has a longer read length (2 × 300 bp), a more limited output of 15 Gb, and a run time of 4–55 h (Anonymous, 2016). The improvement of longer read length by MiSeq results in improved read assembly of the sequences compared to HiSeq; however, the HiSeq provides a greater depth of coverage (Liu et al., 2012).

Library DNA is prepared by breaking the DNA template into fragments, then attaching adapter sequences to both ends (Mardis, 2008). The DNA library is then denatured to single strands and hybridized onto the surface of the flow cell (Mardis, 2008). This is followed by bridge amplification to generate clusters of sequence clones (Mardis, 2008). During amplification, all four nucleotides, plus DNA polymerase, are added simultaneously into the flow cell channels (Zhang et al., 2011). When the dideoxynucleotide triphosphates, each containing a different cleavable fluorescent dye, complement the template strands of the clusters, the fluorescent signal is captured by a camera.

## *Whole Genome Sequencing Has Changed the Epidemic Curve for Food Safety*

Molecular epidemiology's ability to reveal foodborne pathogens' source and transmission depends on the resolution of the technology used (Grad et al., 2011). Starting in 1996, PulseNet, a national laboratory network coordinated by CDC, has collected DNA fingerprints to detect both local and multistate outbreaks (CDC, 2015b; 2016). Until recently, DNA fingerprinting was achieved by pulsed-field gel electrophoresis (PFGE), a method that uses restriction digests of genomic DNA followed by electrophoresis to generate a unique banding pattern or "fingerprint" (Graves and Swaminathan, 2001). However, since 2013, PulseNet has begun using WGS to assist with epidemiology (Barkley et al., 2016).

WGS has become the preferred tool for outbreak investigations primarily because of its higher discriminatory ability compared with PFGE (CDC, 2013; Barkley et al., 2016; Wang et al., 2015); however, it also provides much more information than PFGE. WGS can be used to detect single nucleotide polymorphisms (SNPs), inversions, and insertions or deletions (InDels), as well as antimicrobial resistance genes, pathogen serotype, and mutations in virulence factors (Stasiewicz et al., 2015; Wang et al., 2015; Schmitz-Esser et al., 2015; Hoffmann et al., 2014). With its higher discriminatory ability, WGS significantly increases the confidence level for determining the source of an outbreak, inclusion of isolates within an outbreak, and relatedness of outbreaks (CDC, 2011a,b; Rohde et al., 2011; Mellmann et al., 2011; Grad et al., 2011).

WGS made its first significant contribution to outbreak investigations during the outbreak of O104:H4 *E. coli* from contaminated sprouts, which occurred in Europe in May to July 2011 (CDC, 2011b; Mellmann et al., 2011). In Germany, the outbreak caused approximately 3000 cases of bloody diarrhea, 855 cases of hemolytic uremic syndrome (HUS), and 32 HUS-associated deaths from May to July 2011 (WHO, 2011; RKI, 2011; Frank et al., 2011). In France, the outbreak strain caused 10 cases of bloody diarrhea and 7 cases of HUS (Gault et al., 2011; Grad et al., 2011). Overall, the outbreak involved more than 4000 illnesses and caused 50 deaths across 16 countries (WHO, 2011); it was later attributed to the contaminated sprouts (RKI, 2011).

Conventional molecular epidemiology approaches at the time were not able to detect substantial diversity among the isolates, hindering attempts to use molecular epidemiology to find the origin of the outbreak; they also struggled to make sense of the isolates' hybrid enteroaggregative *E. coli* (EAEC)/enterohemorrhagic *E. coli* (EHEC) nature (Grad et al., 2011; Rohde et al., 2011; Mellmann et al., 2011). Consequently, several research groups used WGS approaches to study the outbreak.

Grad et al. (2011) utilized Illumina, 454, and Pacific Bioscience–generated sequences to compare the outbreak isolates. They found that the outbreak isolates from Germany were less diverse than the French outbreak isolates and represented a subclade within the cluster of French outbreak isolates (Grad et al., 2011). The authors hypothesized that this pattern of genetic diversity was potentially due to a genetic bottleneck, the distribution pattern of the seeds, or varying handling and storage procedures for the contaminated seeds (Grad et al., 2011). Meanwhile, Rohde et al. (2011) used Illumina platforms together with Ion Torrent PGM to investigate the genetic origins of the *E. coli* O104:H4 outbreak strain in an open-source genomic research program. This research program integrated the use of NGS platforms, open-source data release, and crowd-sourced bioinformatics analysis (Rohde et al., 2011). A de novo assembly revealed that the outbreak isolate (from a German patient) belonged to an EAEC lineage that had acquired genes for Shiga-toxin production and antibiotic resistance (Rohde et al., 2011). Phylogenetic analysis showed that it resembled strains previously found in two other continents (Rohde et al., 2011). Mellmann et al. also used the Ion Torrent PGM to characterize the genetic origins of several German *E. coli* O104:H4 outbreak isolates and found that they were (1) highly related to a both a 2001 O104:H4 EHEC isolate and a previously sequenced EAEC isolate and that (2) both the EAEC isolate and the outbreak isolates likely evolved from an EHEC common ancestor, with differing gains and losses of genes and plasmids (Mellmann et al., 2011).

Den Bakker et al. (2011) used a whole genome SNP-based evolutionary approach to identify and trace back a 2010 multistate outbreak of *Salmonella* Montevideo. This outbreak was linked to the consumption of salami made with contaminated spices (black pepper and red crushed pepper); however, the outbreak strain's PFGE pattern was the most common pattern identified for *Salmonella* Montevideo isolates in the PulseNet database, which complicated the outbreak investigation (CDC, 2010). The WGS–SNP approach with SOLiD provided better resolution, demonstrating (1) a clear connection to the actual pepper-related outbreak for just 20% of the putative outbreak cases, (2) several human isolates were connected with a pistachio-associated outbreak, and (3) the presence of several additional potential outbreaks among the control isolates (Den Bakker et al., 2011).

A more recent example is the use of WGS to solve a small outbreak of listeriosis in Rhode Island (Barkley et al., 2016). Three patients had become ill after eating at the same restaurant; however, the outbreak isolates all had a common PFGE pattern, which resulted in a lower confidence trace-back for the PFGE-based investigation (Barkley et al., 2016). The CDC used a whole-genome multilocus sequence typing (wgMLST) approach to analyze the 2014 outbreak isolates, a clinical isolate from 2013, and a food sample from the restaurant (Barkley et al., 2016). The analysis conclusively connected the three 2014 outbreak cases with the 2013 clinical isolate and the restaurant food isolate, and the restaurant made improvements to prevent further illnesses (Barkley et al., 2016).

## Whole Genome Sequencing Projects

The food safety community is rapidly adopting WGS to better identify, understand, and control microorganisms in food systems. To assist that goal, several initiatives to sequence large numbers of food isolates and develop databases are underway. Three of the initial WGS projects related to food safety are described below.

### The 100,000 Pathogen Genome Project

Originally proposed by Professor Bart C. Weimer of the University of California Davis (UC Davis), the 100K Pathogen Genome Project was founded by academic (UC Davis), public (FDA), and private (Agilent Technology) partners (Weimer). Launched in July 2012, this 5-year project aims to collect whole genome sequences for 100,000 isolates from foodborne bacteria (including *Salmonella enterica*, *Listeria monocytogenes*, *E. coli*) and viruses to build a library of foodborne isolates for epidemiological trace-backs (Chen et al., 2017; Weimer). The isolates for the 100K Pathogen Genome Project come from various foods, multiple environments, and varied sources around the globe (Weimer). Project partners use NGS technology instruments (primarily ~10 units of HiSeq 2000) to sequence genomes of isolates from their collections (Weimer). Ultimately, the collaborators hope this database will be used to accelerate the identification of foodborne illnesses, shorten outbreak duration, and increase food security as well as identify genes correlated with traits such as persistence and pathogenesis (Weimer). As of early 2017, 306 *L. monocytogenes* genome sequences have been published (Chen et al., 2017).

### GenomeTrakr

GenomeTrakr is a WGS-based network built by the FDA (FDA, 2017b). Originally, the FDA collected the genome sequences of foodborne pathogens along with their geographical data and uploaded them to the GenomeTrakr database at the National Center for Biotechnology Information (NCBI) (FDA, 2017b). In 2013, the Centers for Disease Control and Prevention (CDC) joined the GenomeTrakr network to add sequences from human clinical isolates in the United States (Sauders, 2013). Currently, the FDA labs, the CDC's Enteric Diseases Laboratory in Atlanta, 14 federal labs, 14 state health and university laboratories, the U.S. Department of Agriculture Food Safety and Inspection Service's lab in Athens, GA, 9 international labs, Brigham and Women's Hospital in Boston, MA, and several academic researchers collaborate to integrate the analysis of clinical, food, and environmental samples of foodborne pathogens (FDA, 2017b; Sauders, 2013). More than 9700 isolates have been sequenced and 175 genomes have been closed by the GenomeTrakr network to date, and the network's current capacity is 3500 isolates per month (FDA, 2017a). Ultimately, GenomeTrakr will aid in speeding up the identification of foodborne illnesses, tracing the origins of the infectious disease to shorten the outbreaks, and increasing food safety (Whitworth, 2015).

## Current Applications of Whole Genome Sequencing in the Retail Food System

The advanced technology of NGS with its rapid performance and higher resolution allows for more accurate analysis of InDels and SNPs, assisting in the differentiation of foodborne pathogens. Consequently, WGS has the potential to be used not only as a rapid and efficient tool to identify and stop outbreaks but also as a routine tool to track environmental microbial contamination originating from retail food systems.

Currently, WGS-based multilocus sequence typing (wgMLST) is used by food safety regulators and investigators as a tool to investigate the origins of foodborne outbreaks (Barkley et al., 2016). wgMLST compares genome sequences at thousands of gene-coding loci in a genome to a set of reference genomes, generating unique genetic fingerprints that can be used to determine relationships between isolates (Barkley et al., 2016). The CDC and FDA have used WGS-based analyses to investigate several multistate *Salmonella* outbreaks, which were not able to be well elucidated by PFGE, including a *Salmonella* Enteritidis outbreak linked to bean sprouts (CDC, 2015b; Anonymous, 2014), and an outbreak linked to multidrug-resistant *Salmonella* I 4,[5],12:i:- and *Salmonella* Infantis in pork (CDC, 2015a), and a *Salmonella* Poona outbreak in cucumbers (CDC, 2015c). In 2014, the CDC also used wgMLST to determine the source of a small cluster of listeriosis cases in Rhode Island, linking both them and a sporadic case of listeriosis 2013 back to a single restaurant (Barkley et al., 2016). The CDC and partnering state public health laboratories now sequence all the clinical isolates they collect, due to wgMLST's superior discriminatory power compared with PFGE (Barkley et al., 2016).

Initial work by Stasiewicz et al. (2015) used WGS to improve the definition of persistent *L. monocytogenes* strains in retail food systems. The aims were to (1) subtype the longitudinally collected *L. monocytogenes* by utilizing a WGS–SNPs-based analysis, (2) use the SNP count to differentiate persistent and sporadic strains, and (3) to identify genetic determinants that contribute to *L. monocytogenes* persistence (Stasiewicz et al., 2015). Their research revealed prophage regions differentiating similar populations in isolates considered identical using PFGE (≤3 bands difference) (Stasiewicz et al., 2015). Persistent isolates (i.e., isolates from the same subtype from the same deli) also had fewer SNP differences, with a median of 2–22 SNP differences in their genome (Stasiewicz et al., 2015). However, for some pulsotypes (unique PFGE AscI–ApaI combinations), isolates differing by very few SNPs could be found across multiple delis in multiple states, suggesting repeated reintroduction or more importantly that some *L. monocytogenes* strains are common among distinct geographic regions (Stasiewicz et al., 2015). WGS–SNPs analysis was not able to identify genes that were more common among persistent isolates compared with sporadic isolates (Stasiewicz et al., 2015).

## Future Applications of Whole Genome Sequencing in Retail

WGS has tremendous potential as a tool for food monitoring (surveillance) in food production facilities, retails, grocery stores, and restaurants as well as for epidemiology. Early detection of persistent of foodborne pathogens in the retail food system can help prevent contamination that will lead to food outbreaks. The retail industry can also use WGS to identify stores with persistent *L. monocytogenes* contamination, which is a significant public health risk. These facilities need to be identified for enhanced cleaning, facilities redesign, equipment replacement, employee training, or other necessary interventions (Hammons et al., 2017). WGS also aids in early detection of foodborne illness clusters, ultimately reducing the number of foodborne illnesses. As the number of defined listeriosis outbreaks has increased and the average size of outbreaks decreased since the onboarding of WGS, it is increasingly important for food safety experts in retail to interface with WGS systems. Interpreting WGS data, understanding the analytical approach, and even adopting this technology are critical for the retail food industry that is able to identify and solve challenges and manage liability.

## References

Adams, M.D., Celniker, S.E., Holt, R.A., Evans, C.A., Gocayne, J.D., Amanatides, P.G., Scherer, S.E., Li, P.W., Hoskins, R.A., Galle, R.F., George, R.A., Lewis, S.E., Richards, S., Ashburner, M., Henderson, S.N., Sutton, G.G., Wortman, J.R., Yandell, M.D., Zhang, Q., Chen, L.X., Brandon, R.C., Rogers, Y.H., Blazej, R.G., Champe, M., Pfeiffer, B.D., Wan, K.H., Doyle, C., Baxter, E.G., Helt, G., Nelson, C.R., Gabor, G.L., Abril, J.F., Agbayani, A., An, H.J., Andrews-Pfannkoch, C., Baldwin, D., Ballew, R.M., Basu, A., Baxendale, J., Bayraktaroglu, L., Beasley, E.M., Beeson, K.Y., Benos, P.V., Berman, B.P., Bhandari, D., Bolshakov, S., Borkova, D., Botchan, M.R., Bouck, J., Brokstein, P., Brottier, P., Burtis, K.C., Busam, D.A., Butler, H., Cadieu, E., Center, A., Chandra, I., Cherry, J.M., Cawley, S., Dahlke, C., Davenport, L.B., Davies, P., de Pablos, B., Delcher, A., Deng, Z., Mays, A.D., Dew, I., Dietz, S.M., Dodson, K., Doup, L.E., Downes, M., Dugan-Rocha, S., Dunkov, B.C., Dunn, P., Durbin, K.J., Evangelista, C.C., Ferraz, C., Ferriera, S., Fleischmann, W., Fosler, C., Gabrielian, A.E., Garg, N.S., Gelbart, W.M., Glasser, K., Glodek, A., Gong, F., Gorrell, J.H., Gu, Z., Guan, P., Harris, M., Harris, N.L., Harvey, D., Heiman, T.J., Hernandez, J.R., Houck, J., Hostin, D., Houston, K.A., Howland, T.J., Wei, M.H., Ibegwam, C., et al., 2000. The genome sequence of *Drosophila melanogaster*. Science 287, 2185–2195.

Anonymous, 1998. Genome sequence of the nematode *C. elegans*: a platform for investigating biology. Science 282, 2012–2018.

Anonymous, 2000. Analysis of the genome sequence of the flowering plant *Arabidopsis thaliana*. Nature 408, 796–815.

Anonymous, 2004. Finishing the euchromatic sequence of the human genome. Nature 431, 931–945.

Anonymous, 2014. Recalled bean sprouts linked to 2 *Listeria* Deaths, 3 hospitalizations. Food Safety News. [Online], 2016. Available: http://www.foodsafetynews.com/2014/11/recalled-bean-sprouts-linked-to-2-listeria-deaths-3-hospitalizations/#.VxggKtWrSM8.

Anonymous, 2016. Sequencing Power for Every Scale. [Online], Illumina. Available: http://www.illumina.com/systems/sequencing.html.

Barkley, J.S., Gosciminski, M., Miller, A., 2016. Whole-genome sequencing detection of ongoing *Listeria* contamination at a restaurant, Rhode Island, USA, 2014. Emerging Infectious Disease 22, 1474–1476.

CDC, 2010. Multistate Outbreak of Human Salmonella Montevideo Infections (Final Update). [Online]. Available: http://www.cdc.gov/salmonella/2010/montevideo-5-4-2010.html.

CDC, 2011a. Multistate Outbreak of Human Salmonella Enteritidis Infections Linked to Alfalfa Sprouts and Spicy Sprouts (Final Update). [Online]. Available: http://www.cdc.gov/salmonella/2011/alfalfa-spicy-sprouts-7-6-2011.html.

CDC, 2011b. Outbreak of Shiga Toxin-producing *E. coli* O104 (STEC O104:H4) Infections Associated with Travel to Germany (Final Update). [Online]. Available: http://www.cdc.gov/ecoli/2011/travel-germany-7-8-11.html.

CDC, 2013. Multistate Outbreak of Listeriosis Linked to Crave Brothers Farmstead Cheeses (Final Update). [Online]. Available: http://www.cdc.gov/listeria/outbreaks/cheese-07-13/index.html.

CDC, 2015a. Multistate Outbreak of Multidrug-resistant *Salmonella* I 4,[5],12:i:- and *Salmonella* Infantis Infections Linked to Pork (Final Update). [Online]. Available: http://www.cdc.gov/salmonella/pork-08-15/index.html.

CDC, 2015b. Multistate Outbreak of *Salmonella* Enteritidis Infections Linked to Bean Sprouts (Final Update). [Online]. Available: http://www.cdc.gov/salmonella/enteritidis-11-14/index.html.

CDC, 2015c. Multistate Outbreak of *Salmonella* Poona Infections Linked to Imported Cucumbers (Final Update). [Online]. Available: http://www.cdc.gov/salmonella/poona-09-15/index.html.

CDC, 2016. About PulseNet. [Online]. Available: http://www.cdc.gov/pulsenet/about/index.html.

Chen, P., Kong, N., Huang, B., Thao, K., Ng, W., Storey, D.B., Arabyan, N., Foutouhi, A., Foutouhi, S., Weimer, B.C., 2017. 100K Pathogen Genome Project: 306 *Listeria* draft genome sequences for food safety and public health. American Society for Microbiology 5, 1–7.

Den Bakker, H.C., Switt, A.I.M., Cummings, C.A., Hoelzer, K., Degoricija, L., Rodriguez-Rivera, L.D., Wright, E.M., Fang, R., Davis, M., Root, T., Schoonmaker-Bopp, D., Musser, K.A., Villamil, E., Waechter, H.N., Kornstein, L., Furtado, M.R., Wiedmann, M., 2011. A whole genome SNP based approach to trace and identify outbreaks linked to a common *Salmonella enterica* subsp. *enterica* serovar Montevideo pulsed field gel electrophoresis type. Applied and Environmental Microbiology 77, 8648–8655.

Dunham, I., Shimizu, N., Roe, B.A., Chissoe, S., Hunt, A.R., Collins, J.E., Bruskiewich, R., Beare, D.M., Clamp, M., Smink, L.J., Ainscough, R., Almeida, J.P., Babbage, A., Bagguley, C., Bailey, J., Barlow, K., Bates, K.N., Beasley, O., Bird, C.P., Blakey, S., Bridgeman, A.M., Buck, D., Burgess, J., Burrill, W.D., O'brien, K.P., 1999. The DNA sequence of human chromosome 22. Nature 402, 489–495.

Eid, J., Fehr, A., Gray, J., Luong, K., Lyle, J., Otto, G., Peluso, P., Rank, D., Baybayan, P., Bettman, B., Bibillo, A., Bjornson, K., Chaudhuri, B., Christians, F., Cicero, R., Clark, S., Dalal, R., Dewinter, A., Dixon, J., Foquet, M., Gaertner, A., Hardenbol, P., Heiner, C., Hester, K., Holden, D., Kearns, G., Kong, X., Kuse, R., Lacroix, Y.,

Lin, S., Lundquist, P., Ma, C., Marks, P., Maxham, M., Murphy, D., Park, I., Pham, T., Phillips, M., Roy, J., Sebra, R., Shen, G., Sorenson, J., Tomaney, A., Travers, K., Trulson, M., Vieceli, J., Wegener, J., Wu, D., Yang, A., Zaccarin, D., Zhao, P., Zhong, F., Korlach, J., Turner, S., 2009. Real-time DNA sequencing from single polymerase molecules. Science 323, 133–138.

FDA, 2017a. GenomeTrakr Fast Facts. [Online]. Available: http://www.fda.gov/Food/FoodScienceResearch/WholeGenomeSequencingProgramWGS/ucm403550.htm.

FDA, 2017b. GenomeTrakr Network. [Online]. Available: http://www.fda.gov/Food/FoodScienceResearch/WholeGenomeSequencingProgramWGS/ucm363134.htm.

Fleischmann, R.D., Adams, M.D., White, O., Clayton, R.A., Kirkness, E.F., Kerlavage, A.R., Bult, C.J., Tomb, J.F., Dougherty, B.A., Merrick, J.M., 1995. Whole-genome random sequencing and assembly of *Haemophilus influenzae* Rd. Science 269, 496–512.

Flusberg, B.A., Webster, D.R., Lee, J.H., Travers, K.J., Olivares, E.C., Clark, T.A., Korlach, J., Turner, S.W., 2010. Direct detection of DNA methylation during single-molecule, real-time sequencing. Nature Methods 7, 461–465.

Frank, C., Faber, M.S., Askar, M., Bernard, H., Fruth, A., Gilsdorf, A., Höhle, M., Karch, H., Krause, G., Prager, R., Spode, A., Stark, K., Werber, D., Team, O. B. O. T. H. I., 2011. Large and ongoing outbreak of haemolytic uraemic syndrome, Germany, May 2011. Eurosurveillance 16.

Gault, G., Weill, F.X., Mariani-Kurkdjian, P., Jourdan-da Silva, N., King, L., Aldabe, B., Charron, M., Ong, N., Castor, C., Mace, M., Bingen, E., Noel, H., Vaillant, V., Bone, A., Vendrely, B., Delmas, Y., Combe, C., Bercion, R., D'andigne, E., Desjardin, M., de Valk, H., Rolland, P., 2011. Outbreak of haemolytic uraemic syndrome and bloody diarrhoea due to *Escherichia coli* O104:H4, south-west France, June 2011. Eurosurveillance 16.

Goffeau, A., Barrell, B.G., Bussey, H., Davis, R.W., Dujon, B., Feldmann, H., Galibert, F., Hoheisel, J.D., Jacq, C., Johnston, M., Louis, E.J., Mewes, H.W., Murakami, Y., Philippsen, P., Tettelin, H., Oliver, S.G., 1996. Life with 6000 genes. Science 274 (546), 563–567.

Grad, Y.H., Lipsitch, M., Feldgarden, M., Arachchi, H.M., Cerqueira, G.C., Fitzgerald, M., Godfrey, P., Haas, B.J., Murphy, C.I., Russ, C., Sykes, S., Walker, B.J., Wortman, J.R., Young, S., Zeng, Q., Abouelleil, A., Bochicchio, J., Chauvin, S., Desmet, T., Gujja, S., Mccowan, C., Montmayeur, A., Steelman, S., FRIMODT-Møller, J., Petersen, A.M., Struve, C., Krogfelt, K.A., Bingen, E., Weill, F.X., Lander, E.S., Nusbaum, C., Birren, B.W., Hunga, D.T., Hanage, W.P., 2011. Genomic epidemiology of the *Escherichia coli* O104:H4 outbreaks in Europe, 2011. Proceedings of the National Academy of Sciences 109, 3065–3070.

Graves, L.M., Swaminathan, B., 2001. PulseNet standardized protocol for subtyping *Listeria monocytogenes* by macrorestriction and pulsed-field gel electrophoresis. International Journal of Food Microbiology 65, 55–62.

Hammons, S.R., Etter, A.J., Wang, J., Wu, T., Ford, T., Howard, M.T., Oliver, H.F., 2017. Evaluation of third-party deep cleaning as a *Listeria monocytogenes* control strategy in retail delis. Journal of Food Protection.

Heather, J.M., Chain, B., 2016. The sequence of sequencers: the history of sequencing DNA. Genomics 107, 1–8.

Hoffmann, M., Zhao, S., Pettengill, J., Luo, Y., Monday, S.R., Abbott, J., Ayers, S.L., Cinar, H.N., Muruvanda, T., Li, C., Allard, M.W., Whichard, J., Meng, J., Brown, E.W., Mcdermott, P.F., 2014. Comparative genomic analysis and virulence differences in closely related *Salmonella enterica* serotype Heidelberg isolates from humans, retail meats, and animals. Genome Biology and Evolution 6, 1046–1068.

Lander, E.S., Linton, L.M., Birren, B., Nusbaum, C., Zody, M.C., Baldwin, J., Devon, K., Dewar, K., Doyle, M., Fitzhugh, W., Funke, R., Gage, D., Harris, K., Heaford, A., Howland, J., Kann, L., Lehoczky, J., Levine, R., Mcewan, P., Mckernan, K., Meldrim, J., Mesirov, J.P., Miranda, C., Morris, W., Naylor, J., Raymond, C., Rosetti, M., Santos, R., Sheridan, A., Sougnez, C., Stange-Thomann, Y., Stojanovic, N., Subramanian, A., Wyman, D., Rogers, J., Sulston, J., Ainscough, R., Beck, S., Bentley, D., Burton, J., Clee, C., Carter, N., Coulson, A., Deadman, R., Deloukas, P., Dunham, A., Dunham, I., Durbin, R., French, L., Grafham, D., Gregory, S., Hubbard, T., Humphray, S., Hunt, A., Jones, M., Lloyd, C., Mcmurray, A., Matthews, L., Mercer, S., Milne, S., Mullikin, J.C., Mungall, A., Plumb, R., Ross, M., Shownkeen, R., Sims, S., Waterston, R.H., Wilson, R.K., Hillier, L.W., Mcpherson, J.D., Marra, M.A., Mardis, E.R., Fulton, L.A., Chinwalla, A.T., Pepin, K.H., Gish, W.R., Chissoe, S.L., Wendl, M.C., Delehaunty, K.D., Miner, T.L., Delehaunty, A., Kramer, J.B., Cook, L.L., Fulton, R.S., Johnson, D.L., Minx, P.J., Clifton, S.W., Hawkins, T., Branscomb, E., Predki, P., Richardson, P., Wenning, S., Slezak, T., Doggett, N., Cheng, J.F., Olsen, A., Lucas, S., Elkin, C., Uberbacher, E., Frazier, M., et al., 2001. Initial sequencing and analysis of the human genome. Nature 409, 860–921.

Lavera, T., Harrisona, J., O'neilla, P.A., Moorea, K., Farbosa, A., Paszkiewicza, K., Studholmea, D.J., 2015. Assessing the performance of the Oxford nanopore technologies MinION. Biomolecular Detection and Quantification 3, 1–8.

Levy, S., Sutton, G., Ng, P.C., Feuk, L., Halpern, A.L., Walenz, B.P., Axelrod, N., Huang, J., Kirkness, E.F., Denisov, G., Lin, Y., Macdonald, J.R., Pang, A.W., Shago, M., Stockwell, T.B., Tsiamouri, A., Bafna, V., Bansal, V., Kravitz, S.A., Busam, D.A., Beeson, K.Y., Mcintosh, T.C., Remington, K.A., Abril, J.F., Gill, J., Borman, J., Rogers, Y.H., Frazier, M.E., Scherer, S.W., Strausberg, R.L., Venter, J.C., 2007. The diploid genome sequence of an individual human. PLoS Biology 5.

Liu, L., Li, Y., Li, S., Hu, N., He, Y., Pong, R., Lin, D., Lu, L., Law, M., 2012. Comparison of next-generation sequencing systems. Journal of Biomedicine and Biotechnology 2012, 1–11.

Loman, N.J., Constantinidou, C., Chan, J.Z., Halachev, M., Sergeant, M., Penn, C.W., Robinson, E.R., Pallen, M.J., 2012. High-throughput bacterial genome sequencing: an embarrassment of choice, a world of opportunity. Nature Reviews Microbiology 10, 599–606.

Mardis, E.R., 2008. The impact of next-generation sequencing technology on genetics. Trends in Genetics 24, 133–141.

Mellmann, A., Harmsen, D., Cummings, C.A., Zentz, E.B., Leopold, S.R., Rico, A., Prior, K., Szczepanowski, R., Ji, Y., Zhang, W., Mclaughlin, S.F., Henkhaus, J.K., Leopold, B., Bielaszewska, M., Prager, R., Brzoska, P.M., Moore, R.L., Guenther, S., Rothberg, J.M., Karch, H., 2011. Prospective genomic characterization of the German enterohemorrhagic *Escherichia coli* O104:H4 outbreak by rapid next generation sequencing technology. PLoS One 6, 1–9.

PacBio, 2016a. Advance Genomics with Single Molecule, Real-Time (SMRT) Sequencing. [Online]. Available: http://www.pacb.com/smrt-science/smrt-sequencing/.

PacBio, 2016b. Pacific Biosciences. [Online]. Available: http://allseq.com/knowledge-bank/pacific-biosciences/.

Quail, M.A., Smith, M., Coupland, P., Otto, T.D., Harris, S.R., Connor, T.R., Bertoni, A., Swerdlow, H.P., Gu, Y., 2012. A tale of three next generation sequencing platforms: comparison of Ion Torrent, Pacific Biosciences and Illumina MiSeq sequencers. BMC Genomics 13.

Quick, J., Quinlan, A.R., Loman, N.J., 2014. A reference bacterial genome dataset generated on the MinION™ portable single-molecule nanopore sequencer. Gigascience 3.

RKI, 2011. Final Report – EHEC/HUS O104:H4 Outbreak, September 2011. Supplement update June 2014.

Rohde, H., Qin, J., Cui, Y., Li, D., Loman, N.J., Hentschke, M., Chen, W., Pu, F., Peng, Y., Li, J., Xi, F., Li, S., Li, Y., Zhang, Z., Yang, X., Zhao, M., Wang, P., Guan, Y., Cen, Z., Zhao, X., Christner, M., Kobbe, R., Loos, S., Oh, J., Yang, L., Danchin, A., Gao, G.F., Song, Y., Li, Y., Yang, H., Wang, J., Xu, J., Pallen, M.J., Wang, J., Aepfelbacher, M., Yang, R., Consortium, E. C. O. H. G. A. C.-S., 2011. Open-source genomic analysis of shiga-toxin-producing *E. coli* O104:H4. The New England Journal of Medicine 365, 718–724.

Sauders, B., 2013. CDC/FDA Partnership Targets Whole Genome Sequencing of *Listeria Monocytogenes*. [Online]. Food Safety News. Available: http://www.foodsafetynews.com/2013/11/cdc-begins-real-time-whole-genome-sequencing-of-listeria-monocytogenes/#.Vv16nUCDHm5.

Schmitz-Esser, S., Muller, A., Stessl, B., Wagner, M., 2015. Genomes of sequence type 121 *Listeria monocytogenes* strains harbor highly conserved plasmids and prophages. Frontiers in Microbiology 6, 380.

Shendure, J., Porreca, G.J., Reppas, N.B., Lin, X., Mccutcheon, J.P., Rosenbaum, A.M., Wang, M.D., Zhang, K., Mitra, R.D., Church, G.M., 2005. Accurate multiplex polony sequencing of an evolved bacterial genome. Science 309, 1728–1732.

Stasiewicz, M.J., Oliver, H.F., Wiedmann, M., Den Bakker, H.C., 2015. Whole-genome sequencing allows for improved identification of persistent *Listeria monocytogenes* in food-associated environments. Applied and Environmental Microbiology 81, 6024–6037.

Swerdlow, H., Gesteland, R., 1990. Capillary gel electrophoresis for rapid, high resolution DNA sequencing. Nucleic Acids Research 18, 1415–1419.

Technologies, O.N., 2016. DNA: Nanopore Sequencing. [Online]. Available: https://nanoporetech.com/applications/dna-nanopore-sequencing.

Venter, J.C., Adams, M.D., Myers, E.W., Li, P.W., Mural, R.J., Sutton, G.G., Smith, H.O., Yandell, M., Evans, C.A., Holt, R.A., Gocayne, J.D., Amanatides, P., Ballew, R.M., Huson, D.H., Wortman, J.R., Zhang, Q.,

Kodira, C.D., Zheng, X.H., Chen, L., Skupski, M., Subramanian, G., Thomas, P.D., Zhang, J., Gabor Miklos, G.L., Nelson, C., Broder, S., Clark, A.G., Nadeau, J., Mckusick, V.A., Zinder, N., Levine, A.J., Roberts, R.J., Simon, M., Slayman, C., Hunkapiller, M., Bolanos, R., Delcher, A., Dew, I., Fasulo, D., Flanigan, M., Florea, L., Halpern, A., Hannenhalli, S., Kravitz, S., Levy, S., Mobarry, C., Reinert, K., Remington, K., ABU-Threideh, J., Beasley, E., Biddick, K., Bonazzi, V., Brandon, R., Cargill, M., Chandramouliswaran, I., Charlab, R., Chaturvedi, K., Deng, Z., di Francesco, V., Dunn, P., Eilbeck, K., Evangelista, C., Gabrielian, A.E., Gan, W., Ge, W., Gong, F., Gu, Z., Guan, P., Heiman, T.J., Higgins, M.E., Ji, R.R., Ke, Z., Ketchum, K.A., Lai, Z., Lei, Y., Li, Z., Li, J., Liang, Y., Lin, X., Lu, F., Merkulov, G.V., Milshina, N., Moore, H.M., Naik, A.K., Narayan, V.A., Neelam, B., Nusskern, D., Rusch, D.B., Salzberg, S., Shao, W., Shue, B., Sun, J., Wang, Z., Wang, A., Wang, X., Wang, J., Wei, M., Wides, R., Xiao, C., Yan, C., et al., 2001. The sequence of the human genome. Science 291, 1304–1351.

Venter, J.C., Levy, S., Stockwell, T., Remington, K.A., Halpern, A., 2003. Massive parallelism, randomness and genomic advances. Nature Genetics 33, 219–227.

Wang, Q., Holmes, N., Martinez, E., Howard, P., HILL-Cawthorne, G., Sintchenko, V., 2015. It is not all about single nucleotide polymorphisms: comparison of mobile genetic elements and deletions in *Listeria monocytogenes* genomes links cases of hospital-acquired listeriosis to the environmental source. Journal of Clinical Microbiology 53, 3492–3500.

Waterston, R.H., Lindblad-Toh, K., Birney, E., Rogers, J., Abril, J.F., Agarwal, P., Agarwala, R., Ainscough, R., Alexandersson, M., An, P., Antonarakis, S.E., Attwood, J., Baertsch, R., Bailey, J., Barlow, K., Beck, S., Berry, E., Birren, B., Bloom, T., Bork, P., Botcherby, M., Bray, N., Brent, M.R., Brown, D.G., Brown, S.D., Bult, C., Burton, J., Butler, J., Campbell, R.D., Carninci, P., Cawley, S., Chiaromonte, F., Chinwalla, A.T., Church, D.M., Clamp, M., Clee, C., Collins, F.S., Cook, L.L., Copley, R.R., Coulson, A., Couronne, O., Cuff, J., Curwen, V., Cutts, T., Daly, M., David, R., Davies, J., Delehaunty, K.D., Deri, J., Dermitzakis, E.T., Dewey, C., Dickens, N.J., Diekhans, M., Dodge, S., Dubchak, I., Dunn, D.M., Eddy, S.R., Elnitski, L., Emes, R.D., Eswara, P., Eyras, E., Felsenfeld, A., Fewell, G.A., Flicek, P., Foley, K., Frankel, W.N., Fulton, L.A., Fulton, R.S., Furey, T.S., Gage, D., Gibbs, R.A., Glusman, G., Gnerre, S., Goldman, N., Goodstadt, L., Grafham, D., Graves, T.A., Green, E.D., Gregory, S., Guigo, R., Guyer, M., Hardison, R.C., Haussler, D., Hayashizaki, Y., Hillier, L.W., Hinrichs, A., Hlavina, W., Holzer, T., Hsu, F., Hua, A., Hubbard, T., Hunt, A., Jackson, I., Jaffe, D.B., Johnson, L.S., Jones, M., Jones, T.A., Joy, A., Kamal, M., Karlsson, E.K., et al., 2002. Initial sequencing and comparative analysis of the mouse genome. Nature 420, 520–562.

Weimer, B.C. 100K pathogen genome project.

Whitworth, J., 2015. GenomeTrakr Sequencing Average of 800 Isolates per month. [Online]. Available: http://www.foodqualitynews.com/Lab-Technology/FDA-GenomeTrakr-sequencing-stats-updated.

WHO, 2011. Outbreaks of *E. coli* O104:H4 Infection: Update 30.

Zhang, J., Chiodini, R., Badr, A., Zhang, G., 2011. The impact of next-generation sequencing on genomics. Journal of Genetics and Genomics 38, 95–109.

CHAPTER 18

# Historical, Current, and Future Prospects for Food Safety in Poultry Product Processing Systems

**Rachael E. Blevins[1], Sun Ae Kim[1], Si Hong Park[1,3], Rafael Rivera[2], Steven C. Ricke[1]**
[1]*University of Arkansas, Fayetteville, AR, United States;* [2]*U.S. Poultry & Egg Association, Tucker, GA, United States;* [3]*Oregon State University, Corvallis, OR, United States*

## Introduction

Microbial quality and contamination is of great concern to poultry processors. Although much of this might be attributed to the increasing regulatory standards regarding *Campylobacter* and *Salmonella*, shelf-life and the increase in consumer interest are also major contributing factors. With the recent increase in *Salmonella* and *Campylobacter* regulations on postchill bird carcasses since 2011 and the ever-changing economy affecting profit margins, poultry processors are looking for more efficient and more cost-effective means of reducing the microbial numbers on poultry products (USDA-FSIS, 2011, 2014a, 2015a). This means that there is a need for additional methods for reducing bacterial loads without adding more chemicals, extra equipment, or cost to the process. These concerns and these issues have led processors and researchers to take a more in-depth look at the poultry slaughter process to better understand the effects that each processing step has on bacterial contamination. Having a better understanding of how each component of the system affects the final microbial population levels before entering the second processing will allow processors to fine-tune what a facility is already doing to reduce microbial numbers in a cost-effective and efficient manner. The objective of this review is to discuss the microbial ecology and steps in poultry processing (from live hang to postchill including scalder, picker, evisceration, inside/outside bird wash (IOBW), and on-line reprocessing (OLR) and chilling), past and ongoing governmental regulations regarding safety control of poultry, and approaches for monitoring microbial contamination currently used in poultry slaughter facility. Finally both current and future approaches for assessing the composition of bacterial populations will be discussed.

Food and Feed Safety Systems and Analysis. http://dx.doi.org/10.1016/B978-0-12-811835-1.00018-X

## *Salmonella and Campylobacter as Major Pathogens in Poultry*

*Salmonella* and *Campylobacter* are two pathogens that should be given consideration when processing raw poultry. Both organisms are responsible for serious gastrointestinal illnesses that have been linked to the consumption or handling of raw or undercooked poultry products; both organisms are found naturally on raw poultry (Fung, 1987; Barbut, 2002; Horrocks et al., 2009; Owens et al., 2010; Cox et al., 2011; Finstad et al., 2012).

Salmonellosis, the disease caused by *Salmonella* infection, is characterized by diarrhea, fever, and stomach cramps. These symptoms can occur anywhere from 12 to 72 h after ingestion of the bacterial cells. It is considered self-limiting and symptoms usually diminish after 4–7 days. In some cases, usually in very young children, in the elderly, or in immunocompromised individuals, the disease can be fatal (Cunningham and Cox, 1987; Fung, 1987; Jay, 1996; Foley et al., 2011; Finstad et al., 2012). Campylobacteriosis, the disease caused by *Campylobacter*, has symptoms very similar to that of *Salmonella*. Diarrhea, stomach cramping, and fever are common symptoms. However, nausea and vomiting can also accompany *Campylobacter* infection. This normally does not occur with *Salmonella*. *Campylobacter* infection is also considered self-limiting and symptoms usually subside within 2–5 days. Again, in very young children, in the elderly, and in individuals who are considered to be immunocompromised, the disease can be deadly if not diagnosed and treated with antibiotics in a timely manner (Fung, 1987; Horrocks et al., 2009; Forsythe, 2010; Rowe and Madden, 2014).

According to the Centers for Disease Control and Prevention (CDC) in 1996, there were 12,450 confirmed cases of *Salmonella*. This accounted for 55.1% of all foodborne disease for the year. There were also two deaths that were attributed to *Salmonella* in 1996. There were 101 cases of *Campylobacter*, which accounted for 0.4% of all foodborne disease in 1996 (CDC, 2000). According to the 2009–10 report, *Salmonella* accounted for 243 outbreaks or 24% of all foodborne outbreaks. Additionally, *Salmonella* caused the most hospitalizations that could be contributed to a foodborne outbreak (49%). This is a substantial increase since 1996. *Campylobacter* was the cause of 40 outbreaks or 4% of all foodborne outbreaks for 2009 and 2010 (CDC, 2013a). According to this report, the number of foodborne illnesses caused by *Salmonella* did not meet the goals of the Healthy People 2010 program set out by the US Department of Health and Human Services (CDC, 2013b). "Healthy People" was a program developed by the US Department of Health and Human Services to improve the health of Americans. The program is based on goals or objectives that are monitored and reassessed every 10 years, on which new goals are established (CDC, 2015).

In 2014, the CDC released its *Surveillance for Foodborne Disease Outbreaks, United States*: 2012 Annual Report. This was the first annual report that was taken after most of the industry had implemented the 2010 standards and regulations. According to this report, *Salmonella* was responsible for 113 of 831 foodborne outbreaks (20%) and 3394 of 14,972 foodborne illnesses (28%) (CDC, 2014). As mentioned previously, the *Campylobacter* standards were introduced in 2010 and implemented by most processors by 2011. According to the CDC, in 2012 *Campylobacter* was responsible for 30 of 425 confirmed foodborne outbreaks (7.1%) and 434 of 10,396 confirmed illnesses (4.2%) (CDC, 2014). This report shows that there has been a decline in both *Salmonella* outbreaks and illnesses since the implementation of the most recent standards. However, the incidences of *Campylobacter* outbreaks and infections have remained unchanged since 2010 (CDC, 2012, 2015).

## Poultry Product Shelf-Life and Spoilage Microbiota

Spoilage organisms can diminish the shelf-life of raw poultry products and thus adversely impact profits (Blackburn, 2006). As raw poultry is stored in a chilled environment, the growth of normal mesophilic bacteria is slowed. However, the growth of psychrophiles and psychrotrophs becomes a concern (Fung, 1987; Russell et al., 1995). Psychrophiles prefer refrigerated temperatures, anywhere from 12 to 15°C (53.6–59°F), whereas the optimum temperature for mesophiles is from 30 to 45°C (86–113°F). Psychrotrophs are mesophilic bacteria that are able to grow in refrigerated environments (Rao et al., 1998; Forsythe, 2010). Psychrophiles and psychrotrophs are responsible for the off-odors and the breakdown of the protein within the product, which is essentially spoilage. *Acinetobacter* and *Pseudomonas* spp. are common spoilage organisms that are found in poultry (Barnes, 1972; Fung, 1987; Russell et al., 1995; Cox et al., 1998; Hinton et al., 2004). Once these organisms reach a certain threshold, approximately $10^7$ colony-forming units (cfu)/$cm^2$, the bacteria deplete the carbohydrates available in the meat. Once the carbohydrates are depleted, the bacteria begin to consume the amino acids that are present. As the bacteria metabolize the amino acids the off-odors that are associated with spoilage are produced (Russell et al., 1995).

## Bacterial Contamination/Reduction During Poultry Processing Steps

### Before Scalder

Broilers arrive at a poultry processing facility colonized by any number of microorganisms. The amount and the species of microorganisms present on the broilers upon arrival can vary depending on many variables, i.e., hatchery, geography, grower, transport conditions, as well as many other factors (Mead, 2004). The processing facility has little control over what conditions the birds are exposed to before entering the slaughter facility. The birds not only rub against one another during both transport and processing, but carcasses come in contact with shared equipment. Shackles, framework, guides, blades, water, and belts are considered some of the most common points of contact (Mead et al., 1994). This can permit the spread of bacteria not only from bird to bird but also from flock to flock as the day progresses within the facility (Mead et al., 1994; Zweifel and Stephan, 2012). Moreover, without an effective sanitation program in place, the contamination can continue to persist further from day to day (Schuler and Badenhop, 1972). Because of these conditions, processors must try to accommodate for the worst-case scenario in the most efficient, effective, and cost-effective way.

Broilers are placed on a communal conveyor on entering the processing facility. Subsequently, the birds are hung upon the overhead shackles. At this point, the birds are still essentially in the same state as at the farm. Nothing has been washed or rinsed in any matter and any fecal material or dirt/soil that was present during transportation is still present on the bird (Mead, 1974, 1995; Owens et al., 2010). The birds are killed, bled, and subsequently scalded. Just before the scalder, some facilities have installed what is known as a prescald brush (Alter, 2017). Studies have shown that adding a brush before the scald tanks may not necessarily provide a practical reduction in microbial populations before scalding, as the reduction in populations is often not significant through processing (Berrang and Bailey, 2009). However, these brushes have been proven to be able to reduce the amount of physical debris from the birds before entering the scald process. This reduction in physical debris, which can include bedding and feces, can potentially aid in reducing the microbial populations further in the process (Berrang and Bailey, 2009; Pacholewicz et al., 2016). Reducing the amount of organic material on the bird before entering the scalder is important, as it is the first of many communal dips, baths, and washes.

### Scalder

The purpose of the scalder is to prepare the carcasses for the removal of the feathers. It also serves to remove the cuticle in some processes depending on the temperature of the scald (Bowker et al., 2014). Scalding is generally referred to as soft scald or hard scald. The difference between the two types of scalding is determined by time and temperature (Parry, 1995; Barbut, 2002). According to Parry (1995), soft scald consists of 50.0–51.5°C for up to 3.5 min. Soft scalding allows for the retention of the cuticle. Hard scald consists of water temperatures of 56.0–60.0°C for 2.0–2.5 min (Barbut, 2002). Scalding is a necessary part of the process; however, it is often perceived as a potential point of cross-contamination because the scalder water is visibly dirty as soon as the first birds of the day pass through the tanks (Mead, 1974, 1995; Owens et al., 2010). Despite these perceptions, studies have shown that aerobic plate count (APC) bacteria, coliform, and *Salmonella* and *Campylobacter* populations on broiler carcasses are generally reduced through the scalder (Buhr et al., 2005; Cason et al., 1999).

Numerous practices have been implemented within the scalder to reduce the possibility of cross-contamination. Some of these practices include the following: a counterflow water system (i.e., the birds will go from an area of dirty water to cleaner water), using a multistage counterflow scalder (i.e., each stage is a separate tank and the water inside is set up as counterflow, so none of the dirty water from the previous stage is able to intermingle with the water in the tank proceeding it). According to Cason et al. (1999), the amount of total solids and APC populations in a three-stage scalder were significantly lower in the third stage of a multistage scalder compared with the total solids and the APC populations found within the end of a single-stage scalder. This could lead to the conclusion that the carcasses would then be less contaminated, because the microbial populations of the water that the carcasses are passing through have been significantly reduced. Additionally, the use of scald additives or antimicrobials has also shown promising reductions in *Salmonella* levels (McKee et al., 2008; Russell, 2008) and *Campylobacter* populations (Okrend et al., 1986).

### *Picker*

Feather removal or plucking is performed in the pickers. Pickers consist of large bays of rubberized "fingers." These fingers aggressively push on the carcasses as the flock is moved through each bay. These "fingers" remove the feathers that the scalder has loosened (Parry, 1995; Sams and McKee, 2010). Feather removal has been proven to be a prime location for cross-contamination (Berrang and Dickens, 2000; Allen et al., 2003). As the birds pass through the bays of pickers, they rub against each other as well as the equipment. The pickers are equipped with a rinse to help push the feathers away from the carcasses as they are removed. This addition of water in conjunction with the sheer force of the picker fingers can aerosolize the bacteria present on the birds (Allen et al., 2003). Studies have shown that microbial numbers associated with poultry carcasses can increase significantly as the carcasses are moved through the pickers (Berrang and Dickens, 2000; Cason et al., 2004). Furthermore, Allen et al. (2003) were able to demonstrate how widespread the contamination from the pickers can be. "Seeder" birds were inoculated on the outside of the carcass with a generic marker organism just before entering the pickers. On exiting the pickers, birds that had preceded the "seeder" carcass through the pickers were found to be positive for the marker organism as well as the birds that followed. Positive birds were found as far ahead of the "seeder" as 30 carcasses and as far behind the "seeder" as 200 carcasses.

According to Berrang and Dickens (2000), *Campylobacter* numbers can increase by as much as 3 logs after passing through the picking bays. The rubber fingers on the picking machines, as stated earlier, rub aggressively over the feathers and skin of all the birds. Arnold (2009) established that the rubber in new picker fingers resist bacterial attachment. However, over time, the picker fingers wear down due to the nature of their function as well as the exposure to either sanitation chemicals or chemicals used for microbial intervention. This causes the rubber to acquire cracks and crevices that can become harborage areas for spoilage organisms as well as pathogens (Arnold and Silvers, 2000). Bacteria can remain undisturbed once they are allowed to settle into the compromised areas of the picker fingers. This, in turn, allows the formation of biofilms (Arnold and Silvers, 2000; Arnold, 2009). Biofilms are a polysaccharide coating that acts as a protective covering to the bacteria. Biofilms make it much more difficult to completely remove the bacteria from the equipment (Arnold, 2009; Sofos et al., 2013). Because biofilms often accumulate in areas that cannot be physically scrubbed, the bacteria can persist for long periods as most chemicals and sanitizers are ineffective (Steenackers et al., 2012). Heat is also largely ineffective in the destruction of bacteria that is embedded in a biofilm. In this case, biofilms are usually removed or loosened during production, thus contaminating the carcasses passing over it (Arnold and Silvers, 2000; Arnold, 2009; Forsythe, 2010). This can lead to the spread of psychrophiles and psychrotrophs as they are commonly found on the feathers and skin of the birds (Barnes, 1972; Sofos et al., 2013). This can also lead to the spread of both *Salmonella* (Steenackers et al., 2012) and *Campylobacter* (Arnold, 2009).

Another possibility for contamination through the pickers is due to the amount of bacteria that are released from the intestinal contents of the birds as the picker fingers forcefully push on the abdomen of the carcasses. The ceca, the crop, and the cloaca all carry significant quantities of bacteria. This includes the pathogens *Campylobacter* and *Salmonella* (Smith and Berrang, 2006). According to Musgrove et al. (2001), 95%–99% of broiler crops and 63%–100% of ceca were contaminated with *Campylobacter*, with average population levels of 3.6 $\log_{10}$ cfu/gram and 6.8 $\log_{10}$ cfu/gram, respectively. Hargis et al. (1995) reported *Salmonella* prevalence in crops and ceca to be 52% and 14.6%, although the samples were not enumerated to determine the loading of *Salmonella* in the crops and ceca. Processors can extend the amount of time the birds are in the pickers to ensure the extent of feather removal. However, extending the amount of time that the birds are in the picker bays can prolong the extent to which carcasses can be contaminated by fecal material that is released by the bird or by birds that have passed through the pickers previously (Allen et al., 2003; Cason et al., 2004; Burfoot et al., 2007).

Considerable research has been conducted in this area of processing (Musgrove et al., 1997; Cason et al., 1999, 2004; Allen et al., 2003; Buhr et al., 2003; Berrang et al., 2011a). However, there has been very little progress as far as how to mitigate contamination. Some studies have gone so far as to suggest plugging the cloaca before the birds enter the pickers. This method did help reduce the amount of cross-contamination (Musgrove et al., 1997). Yet, this is not a practical solution as there is not an effective or an efficient way to implement this practice within commercial processing. Other studies have stated the importance of reducing or minimizing the amount of organic loading before entering the pickers, thus reducing the contamination potential in the pickers (Allen et al., 2003). The application of antimicrobials postpick has been recommended (Berrang et al., 2011b) and also keeping the entire process of feather removal separate from other processing areas (Mead, 1995; Barbut, 2002). The heads and feet of the birds are also usually removed in the same area as defeathering. In addition, the carcasses are subsequently transferred to the evisceration shackles. This "rehang" process can either happen manually or automatically (Parry, 1995).

## *Evisceration*

Evisceration is the process by which the gastrointestinal tracts are removed from the bird carcass. This is a multistep process. This area also includes the subsequent washing/rinsing cabinets leading to the chiller area. Evisceration, similar to the picking area, is an area where extensive research has been conducted in regard to cross-contamination because the potential for such contamination to occur is obvious (Russell and Walker, 1997; Berrang and Dickens, 2000; Li et al., 2002; Northcutt et al., 2003; Hinton et al., 2004; Gill et al., 2006; Smith et al., 2007; Reiter et al., 2007; Berrang and Bailey, 2009; Guerin et al., 2010; Berrang et al., 2011a; Liang et al., 2013; Seliwiorstow et al., 2016). Numerous factors in the evisceration process can increase the potential for cross-contamination, but with proper management of equipment operations, facility hygiene, and employee hygiene, this can be minimized (Barbut, 2002).

In facilities with an automatic process, there are several equipment components at this step of processing including the oil gland remover, the venter or cutter, the eviscerator, the pack puller, the cropper, and the IOBW that all touch the inside of each and every bird within the process (Sams and McKee, 2010). The oil gland remover consists of a rotating blade that excises the gland from the tail (Barbut, 2002). The venter/opener opens the carcass up in preparation for removal of the intestines. The venter probes open the carcass and pull the lower intestine out of the carcass. If the machine is set up properly and managed to accommodate bird size, the intestine stays intact. If not, the intestines can rip, causing contamination of the machine (Parry, 1995). Chlorinated sprays are often used on the equipment to reduce contamination (Sams and McKee, 2010). Most automatic eviscerators are comprised of a mechanism that opens up the carcass as well as a second mechanism that pulls the viscera from the inside of the carcass. If at any time the intestinal packs are torn, which can occur in both manual and automatic evisceration, the packs can release the contents of the intestinal tract onto the birds and onto the equipment (Barbut, 2002; Smith et al., 2007; Sams and McKee, 2010). Thus each subsequent carcass that passes through the equipment can become contaminated. This is a major concern as the intestines are known to be a key reservoir for pathogens such as *Salmonella* and *Campylobacter*, as well as numerous other organisms (Hargis et al., 1995; Berrang and Dickens, 2000; Musgrove et al., 2001; Hinton et al., 2004; Smith and Berrang, 2006). The giblets may or may not be collected at this point in the process. This will depend on the customer market of the processing facility. After the intestines are removed, the crop is removed by the cropper. This consists of a serrated probe that passes through the carcass as it rotates. Once the probe is extended through the neck, the crop is dropped out of the bird (Sams and McKee, 2010).

The crop can also be a concern for contamination. The crop is a compartment of the digestive tract in birds. It is used to store food before entering the gizzard for digestion (Svihus, 2014). Feed can bypass the crop if birds have food readily available as it would be when fed *ad libitum*. However, when birds were fed at specified times, the crop contained significantly higher crop content or feed (Svihus, 2014). As previously stated, Musgrove et al. (2001) found *Campylobacter* in 95%–99% of crops sampled. While Hargis et al. (1995) found 52% of crops to be contaminated with *Salmonella*, additionally, they noted that crops were more likely to tear than ceca. These data support the conclusion that proper management of the cropper is important to the hygiene of the carcasses. Croppers are often equipped with an antimicrobial rinse to help mitigate contamination (Sams and McKee, 2010).

### Inside/Outside Bird Wash and On-Line Reprocessing

After all of the inner components of the carcass have been removed, the birds are further transported through the IOBW. The IOBW is designed to wash visual contamination from the inside of the carcass as well as the outside of the carcass. Typically, IOBWs are set up as cabinets that consist of probes, which may or may not have small rubber fingers that are lowered into the carcasses. As these probes spray the inside of the carcass, spray bars rinse the outside of the carcass with potable water. Because the IOBW is more focused on the removal of organic material and does not necessarily focus specifically on microbial population reductions, a potable water rinse is very common. However, IOBWs can be equipped with antimicrobial sprays (Parry, 1995; Barbut, 2002; Sams and McKee, 2010). The probes may or may not touch the insides of the carcass. However, if the probes do touch the inside of the carcass and there is intestinal content remaining within the bird, the sprayers found inside the probes can then propel associated microbial contaminants onto each subsequent bird. The IOBW often does not include any type of antimicrobial; thus the rinsing action of the water is the only bacterial deterrent. The water is also frequently set at ambient temperatures. Therefore, it is not cold enough to prevent the proliferation of *Salmonella*, being not less than 5°C (Graziani et al., 2017). It is also not warm enough to inhibit the growth of or reduce *Campylobacter* populations. Li et al. (2002) reported that carcasses that were inoculated with *Campylobacter* showed no reduction in *Campylobacter* populations when exposed to 20°C water in an IOBW. When the temperature of the water was increased to 55°C, *Campylobacter* populations were reduced by 0.6 $\log_{10}$cfu/cm$^2$. Yet, in that same study, when 50 parts per million chlorine were added to the IOBW, *Campylobacter* populations on the carcasses were reduced by 0.5 $\log_{10}$cfu/cm$^2$ (Li et al., 2002). This illustrates the importance of considering the use of antimicrobials. Northcutt et al. (2003) reported that despite the removal of visual contamination through the IOBW, reductions in bacterial populations were not consistent. This was due to the inconsistencies of application of the IOBW in the three facilities that were sampled.

OLR has become an industry standard in recent years. The use of an OLR relieves a plant from the constraints of off-line reprocessing, which consists of having to pull visibly contaminated birds from the shackle lines and manually washing the carcass (USDA-FSIS, 2014d). This process is labor-intensive, and there is no guarantee that each bird that has visible fecal contamination is pulled from the line and washed. An OLR system is generally a cabinet or dip tank that contains an antimicrobial. Every bird on the processing line passes through this system. This ensures to a higher degree of certainty that if a bird is visibly contaminated, it will be properly washed. For an establishment to install an OLR system, it must go through a validation to prove to US Department of Agriculture (USDA) that their system is effective in reducing the microbial counts found on carcasses that are visually contaminated (Bilgili et al., 2002).

In addition to the IOBW and OLRs, many facilities also utilize other wash (spray or dip) and brush cabinets. The effectiveness of these applications at reducing bacterial loads is often dependent on many variables. One of the most important factors is where in the process the application is located, the antimicrobial used, the concentration of the antimicrobial, the amount of time the antimicrobial is able to stay in contact with the carcass, as well as the upkeep of the brushes/spray bars that are utilized (Cords et al., 2005).

## *Chiller*

Broilers proceed to the chiller system, after passing through the OLR. The most common chill system used in the United States is the immersion chill system (Carroll and Alvarado, 2008; Sams and McKee, 2010). Immersion chill systems are set up to immerse the broilers in chilled water to lower the temperatures of the carcasses to less than 40°F (4.4°C). Most modern immersion chillers are set up as multiple tanks with water that flows in a counterflow manner. Clean water is added at the exit end of the chiller and flows back toward the entrance of the chiller. This is set up to ensure that as the birds are exposed to the cleanest water possible on exiting the chiller (Petrak et al., 1999). The process is often set up as follows: the prechiller tank, the main chiller (which usually consists of multiple tanks), and a postchill tank (Sams and McKee, 2010). The prechiller has two main functions. It is used to aid in the chilling process by gradually lowering the carcass temperature. Secondly, it is used to wash any remaining organic material from the carcass of the bird before entering the main chiller (Barbut, 2002). The carcasses are generally in the prechiller for 10–15 min and subsequently transferred to the main chiller (Sams and McKee, 2010).

Chilling of carcasses is regulated by the USDA and the process has to be monitored closely by the processing facilities (USDA-FSIS, 2014b). However, immersion chillers are used for much more than reducing the temperatures of the birds. The broiler carcasses are in the chiller for a prolonged period—often anywhere from 45 to 110 min (Sams and McKee, 2010). This ample dwell time gives processors the opportunity to introduce antimicrobials to the birds and allow them to have adequate contact time with these antimicrobials to effectively lower the levels of pathogens and spoilage organisms to a point that is almost undetectable (Mead and Thomas, 1973; Allen et al., 2000). Numerous studies and reviews have been conducted in regard to which parameters and antimicrobials are the most effective for the chiller application (Blood and Jarvis, 1974; Thomas and McMeekin, 1980; Allen et al., 2000; Buhr et al., 2005;

James et al., 2006; Northcutt et al., 2006, 2008a,b; Demirok et al., 2013). Immersion chillers have been criticized because of the potential for cross-contamination (Cunningham and Cox, 1987; Bilgili et al., 2002). If contaminated carcasses enter the chill system, there is a possibility of those contaminants spreading to other carcasses (Petrak et al., 1999; Carroll and Alvarado, 2008; Zheng et al., 2011). Chlorine is commonly used in the main chiller as an effective antimicrobial in the United States. With the addition of chlorine and the proper parameters for counterflow chilling and makeup water, the immersion chill system can reduce APC by as much as 1.0 log cfu/mL and can reduce *Salmonella* incidence and *Campylobacter* incidence as well (Bilgili et al., 2002). In recent years, processors have had to look for other antimicrobial options due to fact that the Food Safety Inspection Service (FSIS) microbiological standards for postchill broiler carcasses have become stricter in recent years (USDA-FSIS, 2011). This has caused other antimicrobials such as peracetic acid (PAA) to gain popularity (Sukted et al., 2017). PAA has been proven to be an effective antimicrobial in the chiller system. Nagel et al. (2013) reported that poultry chillers treated with PAA exhibited promising reductions in both *Salmonella* and *Campylobacter*, at approximately 2 $\log_{10}$ cfu/mL for both pathogens. Other studies have demonstrated increased shelf-life in poultry products that were treated with PAA during the chiller process (Bauermeister et al., 2008a,b).

## Food Safety Regulations Regarding Pathogen Reduction

In 1996 the USDA FSIS introduced requirements that meat and poultry facilities must follow to reduce the number and incidence of pathogenic bacteria within poultry facilities. These requirements were established to reduce the number of foodborne illnesses that were associated with the consumption of these products. Thus, the Hazard Analysis and Critical Control Point (HACCP) system was introduced and implemented in the meat industry. Each facility must establish a HACCP system or plan to fit the unique process for each product that is produced. Each critical control point (CCP) that is established in the HACCP plan will have a corresponding check. Checks are simply the documented monitoring of the established CCPs. Checks are used to ensure that the system is working (USDA-FSIS, 1996).

Possible sources of contamination on broilers during production would be from bacteria found on the feathers and skin of the birds (Sofos et al., 2013), as well as the bacteria that is found in the intestinal tract. Ingesta and fecal material are commonly thought to be a source of contamination on broilers within a slaughter facility (Bilgili et al., 2002; Smith and Berrang, 2006; Smith et al., 2007). FSIS Directive 6420.2 states, "In slaughter establishments, fecal contamination of carcasses is the primary avenue for contamination by pathogens. Pathogens may reside in fecal material, both in the gastrointestinal tract and on the exterior surfaces of the animal or bird going to slaughter" (USDA-FSIS, 2004). It is because of this that FSIS enforces a "zero tolerance" policy for any visible fecal material or ingesta on raw poultry. According to the Code of Federal Regulations, each facility must establish a check within the HACCP system that will prevent contaminated carcasses from entering the chill system. The plant will subsequently perform this check at an established frequency. The USDA inspector at the plant will also inspect for fecal contamination just before the chiller by inspecting 10 prechill carcasses from each evisceration line. This occurs at least two times per shift. Any findings will be evaluated to determine whether the finding was an isolated incident or the root cause can be linked to other findings. This helps the USDA to determine whether there is a possible flaw within the process or the HACCP system (USDA-FSIS, 2004).

Temperature can also be considered a CCP and is monitored immediately after the carcasses exit the chiller. The 9 CFR 381.66 (USDA-FSIS, 2014b) regulation states that carcasses should be adequately chilled after slaughter or upon exiting the chiller to prevent the proliferation of pathogens. Until October 2014, this regulation stated that the internal temperature of the carcasses must be no greater than 40°F (4.4°C). However, the regulation no longer specifies the temperature and allows each slaughter facility to establish a chilling policy based on scientific validations of the process being followed. This policy must be documented in the establishment's HACCP plan, a sanitation standard operating procedure (SSOP), or another prerequisite program.

In 1996, along with the introduction of HACCP, the USDA also established the *Salmonella* performance standard. This standard was based on the baseline data that were collected by the USDA and it mandated that all poultry facilities be able to meet the baseline performance for *Salmonella*. At that time, the standard established that the number of positive postchill whole bird carcass rinses in a broiler slaughter facility could not exceed 20% or 12 positives of 51 broiler carcasses tested. According to the 1996 Final Rule, the purpose of this standard along with the implementation of HACCP was to reduce the incidence and number of harmful bacteria, thus drastically reducing the incidence of foodborne illness that had been attributed to these products in the past (USDA-FSIS, 1996). In 2006, in an effort to reduce the incidence of *Salmonella* in young chickens, the USDA initiated a grading system based on the 1996 performance standard. This grading system divided establishments into three categories based on how well each establishment was meeting the standard. Category 1 facilities were any plants that had a positive rate of less than or equal to 10%, which is 50% of the 20% standard implemented in 1996. Category 2 facilities were any facilities that were above 10%, but less than the standard 20%. Category 3 facilities were any facilities that were above the 20% standard established in 1996. In 2008, the FSIS began publishing the names of any facilities that were in either category 2 or category 3 (USDA-FSIS, 2008, 2010a). A new *Salmonella* baseline was conducted in 2007 and 2008. This baseline concluded that the prevalence of *Salmonella* in young chickens was 7.5%. In 2010 the USDA announced that it would be implementing a new performance standard based on the findings of the 2007 baseline. In 2011, the new standard was implemented. Category 1 now included plants that were at or below 4% positive (2 of 51). Category 2 was now considered to be above 4%, but at or below 9.8% positive. Category 3 now included any establishments that exceeded 9.8% (5 of 51). Establishments that exceeded the new standard (9.8%) were published on the FSIS website in a quarterly report. The new standard of 9.8% was established because it was determined that each facility should be able to meet the 7.5% standard 80% of the time if the facility can stay at or below the 9.8% limit (USDA-FSIS, 2011).

In addition to the tighter *Salmonella* standard, FSIS also introduced a standard for *Campylobacter* (USDA-FSIS, 2011). Until 2010, *Campylobacter* testing had not been included. The 2011 performance standard for *Campylobacter* is set up differently than the performance standard for *Salmonella*. It is comprised of two parts. First, 1 mL of a 400 mL rinse is directly plated and then enumerated. Second, if the 1 mL sample is negative, a 30 mL sample is enriched and subsequently direct plated. The 30 mL sample would subsequently be determined to be either positive or negative. The 2006 and 2007 baseline established that the prevalence of *Campylobacter* within the industry was at 10.4%. This estimate was based on the results of the 1 mL samples. Therefore, to guide the industry to lower the rate of *Campylobacter* within each facility, FSIS set the standard at 15.7% (8 of 51) for 1 mL samples. However, each facility is allowed no more than 52.9% (27 of 51) for a combination of 1 mL and 30 mL samples. FSIS had intended to implement a category system much like the system associated with the *Salmonella* performance standard, but that has not come to fruition at this time (USDA-FSIS, 2011). In 2014, FSIS discontinued the testing of the 30 mL samples for qualitative *Campylobacter* testing stating that the 1 mL sampling was sufficient to determine process control within an establishment (USDA-FSIS, 2014c).

The effects of these regulations and performance standards have been monitored by the US Department of Health and Human Services since implementation in 1996 (CDC, 2012, 2015). All the regulations and the performance standards discussed previously are in place to mitigate the risk of contamination by pathogens and other bacteria or to slow the rate of proliferation of these organisms. The *Salmonella* and *Campylobacter* performance standards, specifically, were meant to encourage processors to reduce the prevalence of these pathogens in poultry facilities to help reduce the rate of foodborne illness due to the consumption of poultry products (USDA-FSIS, 1996, 2010b, 2011). The CDC monitors and reports the incidence and the rate of foodborne illness in the reports of Surveillance for Foodborne Disease Outbreaks, United States. The CDC states that its foodborne disease surveillance program serves three purposes—disease prevention and control, knowledge of disease causation, and administrative guidance. Analysis of these reports can provide evidence as to whether these regulations and performance standards are promoting progress in food safety efforts (CDC, 2000, 2014).

In January 2015, the USDA introduced new measures to further reduce the risk of both *Salmonella* and *Campylobacter*. During 2012, the USDA conducted a baseline study to establish the rate of percent positive for both *Salmonella* and *Campylobacter* in raw chicken parts. This would include chicken breasts, thighs, wings, legs, necks, backs, half or quarter carcasses, and internal organs. The rate of *Salmonella* in chicken parts was found to be 26.3% and the rate for *Campylobacter* was 21.4%. The results of this baseline were compared with the results from the 2007–08 young chicken baseline (chickens that are approximately 6–10 weeks old). These results were compared as a whole and as individual parts. *Salmonella* and *Campylobacter* percent positive rates in young chickens were 5.9% and 10.6%, respectively (USDA-FSIS, 2015b). Obviously, there is a significant difference between the whole bird baseline and the parts baseline. These findings have resulted in the implementation of both chemical and physical interventions into second processing.

Effective SSOPs and good manufacturing processes are critical to the effectiveness of the microbial intervention strategy and the HACCP plan (Barbut, 2002; Forsythe, 2010; Bilgili, 2010; Davis et al., 2010). SSOPs are the written details regarding how a plant will start each production run under sanitary conditions, how the plant will produce product under sanitary conditions, and how the plant will be restored to sanitary conditions after the production run has ended. The Code of Federal Regulations Title 9 part 416 states, "Each official establishment must be operated and maintained in a manner sufficient to prevent the creation of insanitary conditions and to ensure that product is not adulterated." Part 416 addresses the creation of written SSOPs, the implementation, maintenance, corrective action, proper documentation, and FSIS verification of those SSOPs. It is imperative that a facility not only develops written SSOPs that address sanitation (operational and preoperational), but that these SSOPs have been established as effective (Davis et al., 2010).

## Methods of Monitoring Sanitation Effectiveness in the Poultry Slaughter Facility

Preoperational swabs are used to measure and verify the effectiveness of the sanitation procedures (Mead, 1995; Downes and Ito, 2001; Cramer, 2013). Preoperational swabbing programs often consist of either ATP (adenosine triphosphate swabs) or bacterial plate swabs (Downes and Ito, 2001). Both of these types of swabs are used to focus on non–product contact surfaces as well as product contact surfaces within the processing area.

To measure the presence or the amount of viable bacteria present, a bacterial plate swab is used (Forsythe, 2010). These cotton-tipped swabs can either be taken before sanitizer application or after the application of a sanitizer, but just before the setup of production (Downes and Ito, 2001). If taken after the application of sanitizer, the swab must be moistened in a buffer that is capable of neutralizing the sanitizer applied (Downes and Ito, 2001). This will prevent any residual sanitizer from reducing the bacteria that is present on the swab. These swabs are generally streaked onto APC agar. Results are generally reported as cfu per square inch or cfu per square centimeter. Because bacterial colonies can be enumerated and further identified if necessary, these swabs allow food safety personnel within the plants to determine potential hygiene risks of the processing environment. The drawback of bacterial swabs is that results are in a time lag. Results are generally not available for at least 24–48 h after collection (Forsythe, 2010; Cramer, 2013). These numbers are critical in establishing the effectiveness of SSOPs, essentially ensuring that a facility is starting the production day under sanitary conditions.

Adenosine triphosphate swabs, also referred to as bioluminescence swabs, are effective at exposing any residual films or buildup from prior shifts of production before production commences. Adenosine triphosphate swabs measure the amount of relative light units emitted by ATP that is found on the surface of the equipment (Aycicek et al., 2006; Cramer, 2013). While this is helpful in determining if there is residue present, ATP swabs do not show whether these films are made up of protein residues or of viable bacteria (Forsythe, 2010; Shama and Malik, 2013; Cramer, 2013). Although ATP swabs cannot provide data regarding bacterial counts, they can provide the knowledge that there is a potential for bacterial growth, either due to bacterial populations that are present or due to the presence of a reside that can potentially promote the growth of bacteria (Aycicek et al., 2006; Forsythe, 2010; Shama and Malik, 2013; Cramer, 2013). Adenosine triphosphate swabs are convenient as they provide results within minutes as opposed to bacterial plate swabs that usually provide results the following day at the earliest as previously stated. This allows the facility to take immediate action to correct the situation. The area or piece of equipment is able to be recleaned before the product can be affected (Cramer, 2013). However, care must be taken when performing ATP swabs, as the presence of residual detergents and sanitizers can adversely affect the results (Forsythe, 2010; Shama and Malik, 2013). This can potentially cause processors to make decisions based on erroneous data.

Operational swabs are taken to ensure that sanitary conditions are being maintained throughout the production day. Obviously, the results of these swabs are not able to be directly compared with the results of the preoperational swabs as the product contact surfaces have been exposed to raw poultry, which is not sterile. Because these swabs will have higher counts, it is necessary to make serial dilutions to ensure a countable result. Also, if a facility establishes a baseline of the microbiological counts on the equipment that is swabbed during operations, future results can be compared with the baseline to establish whether there is an upward trend that is indicating that a microbiological issue such as spoilage exists.

## *Current and Future Prospects for Microbial Profiling in the Poultry Slaughter Facility*

Identification of overall microbial populations in poultry production processing steps have always been considered an important monitoring tool for assessing sanitizer effectiveness and the corresponding responses of bacteria load levels on poultry carcasses. The ability to generate nonpathogen bacterial data during all aspects of processing has utility for detecting microbial shifts that may be occurring during the various stages of processing (Sofos et al., 2013; Giombelli and Gloria, 2014; Handley et al., 2015). Conceptually, estimating microbial levels represents a quantitative means to assess impact of applied antimicrobial interventions, temperature shifts occurring during and after processing, and potential cross-contamination that alters bacterial load (Handley et al., 2010, 2015). In addition, the use of specific indicator microorganisms that behave in a manner similar to specific foodborne pathogens provides a more continuous means to evaluate and predict the effectiveness of an antimicrobial against the corresponding pathogen if it were present on a chicken carcass (Cason et al., 1997; Handley et al., 2015). Hygienic indicator bacteria such as total APC, coliform, and *Enterobacteriaceae* have historically been used to evaluate both quality and safety of meat products (Hutchison et al., 2006; Sofos et al., 2013; Handley et al., 2015). However, some of these culture methods such as total aerobic bacteria estimated by most probable number have been found to not be suitable for predicting *Salmonella* or *Campylobacter* levels on broiler carcasses (Cason et al., 1997). Likewise, Hutchison et al. (2006) after enumerating aerobic bacteria, *Enterobacteriaceae*, and *Pseudomonas* in poultry slaughter facilities concluded that there was only a weak relationship between presumptive hygienic processing bacterial indicator numbers and process hygiene. External factors can be a factor as well. For example, it has been reported that *Pseudomonas* populations differ in fresh compared with frozen poultry and considerable phenotypic and genotypic variability even occurs within *Pseudomonas* characterized from spoiled poultry fillets (Arnaud-Rollier et al., 1999; Morales et al., 2016).

Finding the most appropriate indicator bacteria in poultry products may play an important role to improve their quality and safety. However, as Rouger et al. (2017) have pointed out poultry product matrices can be fairly heterogeneous and meat products in general can harbor rather diverse microbial populations. Therefore, focusing on single microorganisms or groups of microorganisms as general indicators may not be all that informative. Newer methods to characterize microbial contamination without selective plating or enrichment cultures that favor individual microorganisms such as nucleic acid–based methods have been proposed and examined to some extent (Sofos et al., 2013; Handley et al., 2015). Certainly, molecular methods such as polymerase chain reaction (PCR) assays for detecting the presence and/or estimating levels of pathogens such as *Salmonella* and pathogenic *E. coli* have been used by the meat industry for some time (Park et al., 2014; Baker et al., 2016). However, overall microbial community assessments in poultry processing have only recently been explored using molecular approaches. Handley et al. (2015) suggested that next-generation sequencing technology based on 16S rRNA gene sequences could be utilized for characterizing and profiling of microbiota in a wide range of samples. Ribosomal gene-based sequencing technology can provide comprehensive taxonomic information of the resident bacterial community from phylum to genus levels, resulting in a bacterial census of the bacterial population. Based on this, Kim et al. (2017a) investigated bacterial population compositional shifts in chicken carcass rinsate samples collected from various poultry processing steps as well as before and after specific antimicrobial treatments. They sequenced the microbiome from these samples using an Illumina MiSeq platform. From the analysis of these data, they observed that the abundance of the phylum Firmicutes increased during poultry processing while that of Proteobacteria that are considered to be phylogenetically related to major pathogens decreased during processing after exposure to antimicrobial treatments (Kim et al., 2017a). In addition, they compared microbiome populations of the rinsate solutions with the APC colonies that had originated from these same rinsate solutions. After comparing microbiome analyses, they noted that phyla were generally similar between the two populations but different when genera groupings were compared.

Handley et al. (2015) also suggested that sequencing technology could not only provide census of the general microbial populations present on poultry carcasses but could be used to identify indicator microorganisms that were most representative of foodborne pathogens potentially present on these carcasses. This is important because the occurrence of foodborne pathogens may be relatively infrequent, thus necessitating the substitution of an indicator organism that is much more likely to be consistently present and also behaves in a similar fashion as the corresponding pathogen when exposed to different conditions such as antimicrobial interventions. To achieve identification of optimal indicator candidates requires a better understanding of the microbial distribution of bacterial groups that align with the presence or absence of specific foodborne pathogens. There are a couple of approaches for accomplishing this alignment. One approach is illustrated in Fig. 18.1. Here the concept essentially involves recovery and pooling of colonies from general nonselective plates such as APC media and comparing their sequenced microbial populations with colonies from pathogen selective plates. When the populations are compared, the goal is to identify nonpathogenic bacteria that overlap and thus are present on all three plates, thus assuming that these bacteria would behave in a similar fashion as the corresponding pathogen because they are present on both the nonselective and selective plates. Although this outcome is entirely hypothetical and depends on the assumption that not all colonies originating from the selective plate are actually the targeted pathogenic organism, there is some evidence to support this

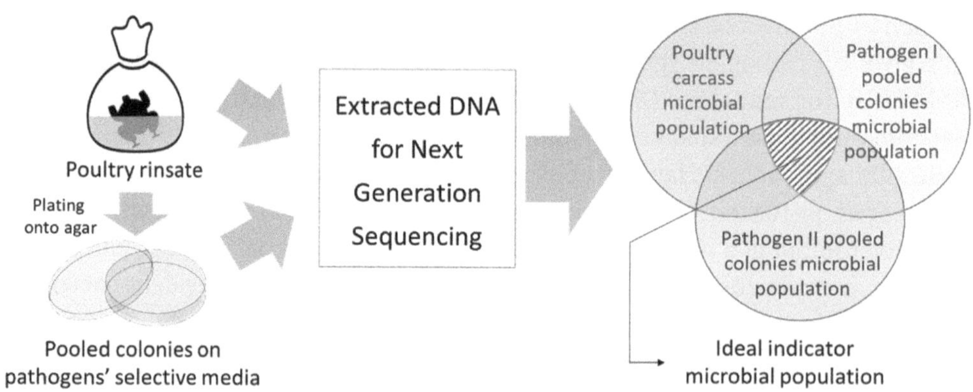


**Figure 18.1**
Microbiome sequencing strategy to identify ideal indicator bacteria both from poultry rinsate and from pooled colonies in selective media of pathogens with next-generation sequencing technology.

concept. For example, Kim et al. (2017a) noted that *Campylobacter* selective plate colonies were not all *Campylobacter* when the colonies were pooled and several non-*Campylobacter* genera were identified. This is consistent with earlier pyrosequencing results by Oakley et al. (2012) that compared different *Campylobacter* selective media and demonstrated that non-*Campylobacter* taxa could be identified in all of them. Whether this holds true for other foodborne pathogens such as *Salmonella* remains to be determined.

Another approach was developed by Kim et al. (2017b) and involved nonselective media combined with a most probable number (MPN) dilution series. Here their idea was to generate MPN results from bacterial growth in a nonselective media that should support growth of both pathogenic and nonpathogenic bacteria simultaneously from chicken breast samples inoculated with a *Salmonella* Typhimurium marker strain. Once bacterial growth becomes sufficiently detectable to estimate the quantity of microorganisms based on MPN calculations, DNA can be extracted from all of wells to conduct quantitative PCR to estimate the quantity of the corresponding pathogen (in this case *S.* Typhimurium) based on the MPN dilutions. They also used the DNA for 16S rRNA–based sequencing to generate a microbial community analyses of the bacterial taxa present in these MPN wells. Based on these results, they noted that there was consistent relationship between the level of *S.* Typhimurium and the abundance of *Enterobacteriaceae* proportion of the general bacterial population taxa, suggesting that this group may be a candidate quantitative indicator organism for the presence of *Salmonella*. While considerable follow-up research remains once candidate indicator organisms have been validated, detection systems constructed for their routine use as indicator organisms could be developed.

## Conclusions

Microbial contamination of poultry products generally consists of foodborne pathogens and nonpathogenic bacteria. There are many opportunities for contamination of the poultry product by pathogenic and nonpathogenic bacteria during processing from live birds to final products. While the foodborne pathogens represent a few groups of microorganisms such as certain *Salmonella* serovars and *Campylobacter jejuni*, the nonpathogenic bacterial population can be quite diverse. Food safety regulations continue to evolve as the goals for foodborne pathogen reduction become more refined. Reducing nonpathogenic bacterial loads in carcasses is also important to limit spoilage and extend shelf-life of the poultry product. Consequently, the development of effective antimicrobials and sanitizers that are effective against both foodborne pathogens and nonpathogenic bacteria continues to be of interest.

As a part of this strategy, numerous efforts have been attempted and assays developed to detect specific foodborne pathogens on poultry products. In addition to developing accurate detection methods for foodborne pathogens, identifying optimal indicator organisms that are associated with poultry quality and best emulate pathogen responses remains a critical area of research. Such efforts have remained somewhat imprecise because of the rather diverse nature of the indigenous microbiota associated with the poultry carcass. The development of 16S rRNA–based microbiome sequencing offers a comprehensive approach to generate microbial profiles of these diverse poultry carcass microbial populations throughout processing. As more sequencing data are generated, it is anticipated that standard microbial profiles can be established and representative microorganisms identified that can serve as optimal signature indicator bacteria for evaluating poultry processing performance.

## References

Allen, V.M., Tinker, D.B., Hinton, M.H., Wathes, C.M., 2003. Dispersal of micro-organisms in commercial defeathering systems. British Poultry Science 44, 53–59.

Allen, V.M., Corry, J.E.L., Burton, C.H., Whyte, R.T., Mead, G.C., 2000. Hygiene aspects of modern poultry chilling. International Journal of Food Microbiology 58, 39–48.

Alter, T., 2017. Prevention and mitigation strategies for *Campylobacter*. In: Klein, G. (Eds.), Academic Press, New York, NY, pp. 111–129 (Chapter 6).

Arnaud-Rollier, L., De Zutter, L., Van Hoof, J., 1999. Identities of the *Pseudomonas* spp. in flora from chilled chicken. International Journal of Food Micorbiology 48 (2), 87–96.

Arnold, J.W., Silvers, S., 2000. Comparison of poultry processing equipment surfaces for susceptibility to bacterial attachment and biofilm formation. Poultry Science 79, 1215–1221.

Arnold, J.W., 2009. Biofilms in poultry processing. In: Fratamico, P.M., Annous, B.A., Gunther, N.W. (Eds.), Biofilms in the Food and Beverage Industries. Woodhead Publishing, Cambridge, UK, pp. 455–473.

Aycicek, H., Oguz, U., Karci, K., 2006. Comparison of results of ATP bioluminescence and traditional hygiene swabbing methods for the determination of surface cleanliness at a hospital kitchen. International Journal of Hygiene and Environmental Health 209, 203–206.

Baker, C.A., Rubinelli, P.M., Park, S.H., Carbonero, F., Ricke, S.C., 2016. Shiga toxin-producing *Escherichia coli* in food: incidence, ecology, and detection strategies in the United States. Food Control 59, 407–419.

Barbut, S., 2002. Poultry Products Processing: An Industry Guide. CRC Press, Boca Raton, FL.

Barnes, E.M., 1972. Food poisoning and spoilage bacteria in poultry processing. The Veterinary Record 90, 720–723.

Bauermeister, L.J., Bowers, J.W., Townsend, J.C., McKee, S.R., 2008a. Validating the efficacy of peracetic acid mixture as an antimicrobial in poultry chillers. Journal of Food Protection 71, 1119–1122.

Bauermeister, L.J., Bowers, J.W., Townsend, J.C., McKee, S.R., 2008b. The microbial quality properties of poultry carcasses treated wtih peracetic acid as an antimicrobial treatment. Poultry Science 87, 2390–2398.

Berrang, M.E., Smith, D.P., Meinersmann, R.J., 2011a. Variations on standard broiler processing in an effort to reduce *Campylobacter* numbers on postpick carcasses. Journal of Applied Poultry Research 20, 197–202.

Berrang, M.E., Windham, W.R., Meinersmann, R.J., 2011b. *Campylobacter, Salmonella*, and *Escherichia* coli on broiler carcasses subjected to a high pH scald and low pH postpick chlorine dip. Poultry Science 90, 896–900.

Berrang, M.E., Bailey, J.S., 2009. On-line brush and spray washers to lower numbers of *Campylobacter* and *Escherichia coli* and presence of *Salmonella* on broiler carcasses during processing. Journal of Applied Poultry Research 18, 74–78.

Berrang, M.E., Dickens, J.A., 2000. Presence and level of *Campylobacter* spp. on broiler carcasses throughout the processing plant. Journal of Applied Poultry Research 9, 43–47.

Bilgili, S.F., 2010. Poultry meat inspection and grading. In: Owens, C.M., Alvarado, C.Z., Sams, A.R. (Eds.), Poultry Meat Processing, second ed. CRC Press Taylor and Francis Group, Boca Raton, FL USA, pp. 67–99.

Bilgili, S.F., Waldroup, A.L., Zelenka, D., Marion, J.E., 2002. Visible ingesta on prechill carcasses does not affect the microbiological quality of broiler carcasses after immersion chilling. Journal of Applied Poultry Research 11, 233–238.

Blackburn, C.D.W., 2006. In: Blackburn, C.D.W. (Ed.), Introduction. Pages Xvii–Xxiii in Food Spoilage Microorganisms. Woodhead Publishing Ltd, Cambridge, UK.

Blood, R.M., Jarvis, B., 1974. Chilling of poultry: the effects of process parameters on the level of bacteria in spin-chiller waters. International Journal of Food Science and Technology 9, 157–169.

Bowker, B.C., Zhuang, H., Buhr, R.J., 2014. Impact of carcass scalding and chilling on muscle proteins and meat quality of broiler breast fillets. LWT–Food Science and Technology 59, 156–162.

Buhr, R.J., Bourassa, D.V., Northcutt, J.K., Hinton Jr., A., Ingram, K.D., Cason, J.A., 2005. Bacteria recovery from genetically feathered and featherless broiler carcasses after immersion chilling. Poultry Science 84, 1499–1504.

Buhr, R.J., Berrang, M.E., Cason, J.A., 2003. Bacterial recovery from breast skin of genetically feathered and featherless broiler carcasses immediately following scalding and picking. Poultry Science 82, 1641–1647.

Burfoot, D., Whyte, R.T., Tinker, D.B., Hall, K., Allen, V.M., 2007. A novel method for assessing the role of air in the microbiological contamination of poultry carcasses. International Journal of Food Microbiology 115, 48–52.

Carroll, C.D., Alvarado, C.Z., 2008. Comparison of air and immersion chilling on meat quality and shelf life of marinated broiler breast filets. Poultry Science 87, 368–372.

Cason, J.A., Hinton Jr., A., Buhr, R.J., 2004. Impact of feathers and feather follicles on broiler carcass bacteria. Poultry Science 83, 1452–1455.

Cason, J.A., Buhr, R.J., Dickens, J.A., Musgrove, M.T., Stern, N.J., 1999. Carcass microbiological quality following intermittent scalding and defeathering. Poultry Science 8, 368–373.

Cason, J.A., Bailey, J.S., Stern, N.J., Whittemore, A.D., Cox, N.A., 1997. Relationship between aerobic bacteria, salmonellae, and *Campylobacter* on broiler carcasses. Poultry Science 76, 1037–1041.

CDC, 2015. About Healthy People | Healthy People 2020. National Center for Health Statistics.

CDC, 2014. Surveillance for Foodborne Disease Outbreaks, United States, 2012, Annual Report. .

CDC, 2013a. Surveillance for foodborne disease outbreaks, United States, 2009–10. MMWR. Morbidity and Mortality Weekly Report 62.

CDC, 2013b. Incidence and trends of infections with pathogens transmitted commonly through food-foodborne diseases active surveillance network, 10 U.S. sites, 1996–2012. MMWR. Morbidity and Mortality Weekly Report 62 (15), 283–287.

CDC, 2012. Healthy People 2010 Final Review. Available at: https://www.cdc.gov/nchs/data/hpdata2010/hp2010_final_review.pdf.

CDC, 2000. Surveillance for Foodborne-Disease Outsbreaks-United States, 1993–97.

Cords, B.R., Burnett, S.L., Hilgren, J., Finley, M., Magnuson, J., 2005. Sanitizers: halogens, surface-active agents and peroxides. In: Davidson, P.M., Sofos, J.N., Branen, A.L. (Eds.), Antimicrobials in Food, third ed. Taylor and Francis, 600 Broken Sound Parkway NW, Suite 300, Boca Raton, FL 33487–2742, pp. 507–565.

Cox, N., Cason, J., Richardson, L., 2011. Minimization of *Salmonella* contamination on raw poultry. Annual Review of Food Science and Technology 2, 75–95.

Cox, N.A., Russell, S.M., Bailey, J.S., 1998. The microbiology of stored poultry. In: Davies, A., Board, R. (Eds.), The Microbiology of Meat and Poultry. Blackie Academic & Professional, London, UK, pp. 266–287.

Cramer, M.M., 2013. Food Plant Sanitation: Design, Maintenance, and Good Manufacturing Practices, second ed. CRC Press, Boca Raton, FL.

Cunningham, F.E., Cox, N.A., 1987. The Microbiology of Poultry Meat Products. Academic Press, Inc., Orlando, FL.

Davis, M.A., Singh, M., Conner, D.E., 2010. Poultry-borne pathogens: plant considerations. In: Owens, C.M., Alvarado, C.Z., Sams, A.R. (Eds.), Poultry Meat Processing, second ed. CRC Press Taylor and Francis Group, Boca Raton, FL, pp. 175–203.

Demirok, E., Veluz, G., Stuyvenberg, W.V., Castañeda, M.P., Byrd, A., Alvarado, C.Z., 2013. Quality and safety of broiler meat in various chilling systems. Poultry Science 92, 1117–1126.

Downes, F.P., Ito, K., 2001. Microbiological monitoring of the food processing environment. In: Downes, F.P., Ito, K. (Eds.), The Compendium of Methods for the Microbiological Examination of Foods, fourth ed. American Public Health Association, Washington, DC, pp. 27–28.

Finstad, S., O'Bryan, C.A., Marcy, J.A., Crandall, P.G., Ricke, S.C., 2012. *Salmonella* and broiler processing in the United States: relationship to foodborne salmonellosis. Food Research International 45, 789–794.

Foley, S.L., Nayak, R., Hanning, I.B., Johnson, T.J., Han, J., Ricke, S.C., 2011. Population dynamics of *Salmonella enterica* serotypes in commercial egg and poultry production. Applied and Environmental Microbiology 77, 4273–4279. http://dx.doi.org/10.1128/AEM.00598-11.

Forsythe, S.J., 2010. The Microbiology of Safe Food, second ed. Wiley-Blackwell, Chichester, West Sussex, PO19 8SQ, United Kingdom.

Fung, D.Y.C., 1987. Types of microorganisms. In: Cunningham, F.E., Cox, N.A. (Eds.), The Microbiology of Poultry Meat Products. Academic Press, Inc., Orlando, FL, pp. 5–27.

Gill, C.O., Moza, L.F., Badoni, M., Barbut, S., 2006. The effects on the microbiological condition of product of carcass dressing, cooling, and portioning processes at a poultry packing plant. International Journal of Food Microbiology 110, 187–193.

Giombelli, A., Gloria, M.B.A., 2014. Prevalence of *Salmonella* and *Campylobacter* on broiler chickens from farm to slaughtr and effciency of methods to remove visible fecal contamination. Journal of Food Protection 77, 1851–1859.

Graziani, C., Losasso, C., Luzzi, I., Ricci, A., Scavia, G., Pasquali, P., 2017. *Salmonella*. In: Dodd, C.E.R., Aldsworth, T., Stein, R.A., Cliver, D.O., Riemann, H.P. (Eds.), Foodborne Diseases, third ed. Academic Press, London, UK, pp. 133–169.

Guerin, M.T., Sir, C., Sargeant, J.M., Waddell, L., O'Connor, A.M., Wills, R.W., Bailey, R.H., Byrd, J.A., 2010. The change in prevalence of *Campylobacter* on chicken carcasses during processing: a systematic review. Poultry Science 89, 1070–1084.

Handley, J.A., Park, S.H., Shi, Z., Dawoud, T.M., Kwon, Y.M., Ricke, S.C., 2015. *Salmonella* and the potential role for microbial process indicators on chicken carcasses (Chapter 6). In: Ricke, S.C., Donaldson, J.R., Phillips, C.A. (Eds.), Food Safety: Emerging Issues, Technologies and Systems. Elsevier, Oxford, UK, pp. 81–104.

Handley, J.A., Hanning, I., Ricke, S.C., Johnson, M.G., Jones, F.T., Apple, R.O., 2010. Temperature and bacterial profile of post chill poultry carcasses stored in processing combo held at room temperature. Journal of Food Science 75, M515–M520.

Hargis, B.M., Caldwell, D.J., Brewer, R.L., Corrier, D.E., DeLoach, J.R., 1995. Evaluation of the chicken crop as a source of *Salmonella* contamination for broiler carcasses. Poultry Science 74, 1548–1552.

Hinton Jr., A., Cason, J.A., Ingram, K.D., 2004. Tracking spoilage bacteria in commercial poultry processing and refrigerated storage of poultry carcasses. International Journal of Food Microbiology 91, 155–165.

Horrocks, S.M., Anderson, R.C., Nisbet, D.J., Ricke, S.C., 2009. Incidence and ecology of *Campylobacter jejuni* and *coli* in animals. Anaerobe 15, 18–25.

Hutchison, M.L., Walters, L.D., Mead, G.C., Howell, M., Allen, V.M., 2006. An assessment of sampling methods and micrbiological hygiene indicators for process verification in poultry slaughterhouses. Journal of Food Protection 69, 145–153.

James, C., Vincent, C., Lima, T.I., James, S.J., 2006. The primary chilling of poultry carcasses-a review. International Journal of Refrigeration 29, 847–862.

Jay, J.M., 1996. Modern Food Microbiology, fifth ed. Chapman & Hall, New York, NY.

Kim, S.A., Park, S.H., Lee, S.I., Owens, C.M., Ricke, S.C., 2017a. Assessment of individual chicken carcass microbiome responses during processing in the presence of commercial antimicrobials using a next generation sequencing approach. Scientific Reports 7, 43354.

Kim, S.A., Park, S.H., Lee, S.I., Rick, S.C., 2017b. Development of a rapid method to quantify *Salmonella* Typhimurium using a combination of MPN with qPCR and a shortened time incubation. Food Microbiology 65, 7–18.

Li, Y., Yang, H., Swem, B.L., 2002. Effect of high-temperature inside-outside spray on survival of *Campylobacter jejuni* attached to prechill chicken carcasses. Poultry Science 81, 1371–1377.

Liang, R., Tian, J., She, R., Meng, H., Xiao, P., Chang, L., 2013. Airborne microbial composition in a high-throughput poultry slaughtering facility. Journal of Food Protection 76, 413–419.

McKee, S.R., Townsend, J.C., Bilgili, S.F., 2008. Use of a scald additive to reduce levels of *Salmonella* Typhimurium during poultry processing. Poultry Science 87, 1672–1677.

Mead, G.C., 2004. Microbial hazards in production and processing. In: Mead, G.C. (Ed.), Poultry Meat Processing and Quality. Woodhead Publishing Limited and CRC Press LLC, Boca Raton, FL, pp. 232–257.

Mead, G.C., 1995. Hygiene problems and control of process contamination. In: Mead, G.C. (Ed.), Processing of Poultry. Chapman & Hall, Bristol, UK, pp. 183–220.

Mead, G.C., 1974. Bacteriological control in the processing of poultry. The Veterinary Record 95, 569–572.

Mead, G.C., Hudson, W.R., Hinton, M.H., 1994. Use of a marker organism in poultry processing to identify sites of cross contamination and evaluate possible control measures. British Poultry Science 35, 345–354.

Mead, G.C., Thomas, N.L., 1973. Factors affecting the use of chlorine in the spin-chilling of eviscerated poultry. British Poultry Science 14, 99–117.

Morales, P.A., Aguirre, J.S., Troncoso, M.R., Figueroa, G.O., 2016. Phenotypic and genotypic characterization of *Pseudomonas* spp. present in spoiled poultry fillets sold in retail settings. LWT–Food Science Ant Technology 73, 609–614.

Musgrove, M.T., Berrang, M.E., Byrd, J.A., Stern, N.J., Cox, N.A., 2001. Detection of *Campylobacter* spp. in ceca and crops with and without enrichment. Poultry Science 80, 825–828.

Musgrove, M.T., Cason, J.A., Fletcher, D.L., Stern, N.J., Cox, N.A., Bailey, J.S., 1997. Effect of cloacal plugging on microbial recovery from partially processed broilers. Poultry Science 76, 530–533.

Nagel, G.M., Bauermeister, L.J., Bratcher, C.L., Singh, M., McKee, S.R., 2013. *Salmonella* and *Campylobacter* reduction and quality characteristics of poultry carcasses treated with various antimicrobials in a post-chill immersion tank. International Journal of Food Microbiology 165, 281–286.

Northcutt, J.K., Cason, J.A., Ingram, K.D., Smith, D.P., Buhr, R.J., Fletcher, D.L., 2008a. Recovery of bacteria from broiler carcasses after immersion chilling in different volumes of water, Part 2. Poultry Science 87, 573–576.

Northcutt, J.K., Smith, D., Huezo, R.I., Ingram, K.D., 2008b. Microbiology of broiler carcasses and chemistry of chiller water as affected by water reuse. Poultry Science 87, 1458–1463.

Northcutt, J.K., Cason, J.A., Smith, D.P., Buhr, R.J., Fletcher, D.L., 2006. Broiler carcass bacterial counts after immersion chilling using either a low or high volume of water. Poultry Science 85, 1802–1806.

Northcutt, J.K., Berrang, M.E., Smith, D.P., Jones, D.R., 2003. Effect of commercial bird washers on broiler carcass microbiological characteristics. Journal of Applied Poultry Research 12, 435–438.

Oakley, B.B., Morales, C.A., Line, J.E., Seal, B.S., Hiett, K.L., 2012. Application of high-throughput sequencing to measure the performance of commonly used selective cultivation methods for the foodborne pathogen *Campylobacter*. FEMS Microbiology Ecology 79, 327–336.

Okrend, A.J., Johnston, R.W., Moran, A.B., 1986. Effect of acetic acid on the death rates at 52°C of *Salmonella newport, Salmonella typhimurium* and *Campylobacter jejuni* in poultry scald water. Journal of Food Protection 49, 500–503.

Owens, C.M., Alvarado, C.Z., Sams, A.R., 2010. Poultry Meat Processing. CRC Press, 6000 Broken Sound Parkway, NW, Suite 300, Boca Raton, FL 33487, USA.

Pacholewicz, E., Lipman, L.J.A., Swart, A., Havelaar, A.H., Heemskerk, W.J.C., 2016. Pre-scald brushing for removal of solids and associated broiler carcass bacterial contamination. Poultry Science 95, 2979–2985.

Park, S.H., Aydin, M., Khatiwara, A., Dolan, M.C., Gilmore, D.F., Bouldin, J.L., Ahn, S., Ricke, S.C., 2014. Current and emerging technologies for rapid detection and characterization of *Salmonella* in poultry and poultry products. Food Microbiology 38, 250–262.

Parry, R.T., 1995. Technological developments in pre-slaughter handling and processing. In: Mead, G.C. (Ed.), Processing of Poultry. Chapman & Hall, Bristol, UK, pp. 65–101.

Petrak, T., Kalodera, Z., Novakovic, P., Karolyi, L.G., 1999. Bacteriological comparison of parallel and counter flow water chilling of poultry meat. Meat Science 53, 269–271.

Rao, D.N., Nair, K.K.S., Sakhare, P.Z., 1998. Meat microbiology and spoilage in tropical countries. In: Davies, A., Board, R. (Eds.), The Microbiology of Meat and Poultry. Blackie Academic & Professional, London, UK, pp. 220–265.

Reiter, M.G.R., Fiorese, M.L., Moretto, G., López, M.C., Jordano, R., 2007. Prevalence of *Salmonella* in a poultry slaughterhouse. Journal of Food Protection 70, 1723–1725.
Rouger, A., Remenant, B., Prévost, H., Zagorec, M., 2017. A method to isolate bacterial communities and characterize ecosytems from food products: validation and utilization as a reproducible chicken meat model. International Journal of Food Microbiology 247, 38–47.
Rowe, M.T., Madden, R.H., 2014. *Campylobacter* | introduction. In: Batt, C.A., Tortorello, M.L. (Eds.), Encyclopedia of Food Microbiology, second ed. Academic Press, Oxford, UK, pp. 351–356.
Russell, S.M., 2008. The effect of an acidic, copper sulfate-based commercial sanitizer on indicator, pathogenic, and spoilage bacteria associated with broiler chicken carcasses when applied at various intervention points during poultry processing. Poultry Science 87, 1435–1440.
Russell, S.M., Walker, J.M., 1997. The effect of evisceration on visible contamination and the microbiological profile of fresh broiler chicken carcasses using the nu-tech evisceration system or the conventional streamlined inspection system. Poultry Science 76, 780–784.
Russell, S.M., Cox, N.A., Fletcher, D.L., 1995. Spoilage bacteria of fresh broiler chicken carcasses. Poultry Science 74, 2041–2047.
Sams, A.R., McKee, S.R., 2010. First processing: slaughter through chilling. In: Owens, C.M., Alvarado, C.Z., Sams, A.R. (Eds.), Poultry Meat Processing, second ed. CRC Press, Boca Raton, FL, pp. 25–50.
Schuler, G.A., Badenhop, A.F., 1972. Microbiology survey of equipment in selected poultry processing plants. Poultry Science 51, 830–835.
Seliwiorstow, T., Baré, J., Berkvens, D., Van Damme, I., Uyttendaele, M., De Zutter, L., 2016. Identification of risk factors for *Campylobacter* contamination levels on broiler carcasses during the slaughter process. International Journal of Food Microbiology 226, 26–32.
Shama, G., Malik, D.J., 2013. The uses and abuses of rapid bioluminescence-based ATP assays. International Journal of Hygiene and Environmental Health 216, 115–125.
Smith, D.P., Northcutt, J.K., Cason, J.A., Hinton, A., Buhr, R.J., Ingram, K.D., 2007. Effect of external or internal fecal contamination on numbers of bacteria on prechilled broiler carcasses. Poultry Science 86, 1241–1244.
Smith, D.P., Berrang, M.E., 2006. Prevalence and numbers of bacteria in broiler crop and gizzard contents. Poultry Science 85, 144–147.
Sofos, J.N., Flick, G., Nychas, G., O'Bryan, C.A., Ricke, S.C., Crandall, P.G., 2013. Meat, poultry, and seafood. In: Doyle, M.P., Buchanan, R.L. (Eds.), Food Microbiology: Fundamentals and Frontiers, fourth ed. ASM Press, Washington, DC, pp. 111–167.
Steenackers, H., Hermans, K., Vanderleyden, J., De Keersmaeker, S.C.J., 2012. *Salmonella* biofilms: an overview on occurrence, structure, regulation and eradication. Food Research International 45, 502–531.
Sukted, N., Tuitemwong, P., Tuitemwong, K., Poonlapdecha, W., Erickson, L.E., 2017. Inactivation of *Campylobacter* during immersion chilling of chicken carcasses. Journal of Food Engineering 202, 25–33.
Svihus, B., 2014. Function of the digestive system. Journal of Applied Poultry Research 23, 306–314.
Thomas, C.J., McMeekin, T.A., 1980. Contamination of broiler carcass skin during commercial processing procedures: an electron microscopic study. Applied and Environmental Microbiology 40, 133–144.
USDA-FSIS, 2015a. Consumer Protection Standards: Raw Products. Title 9. Code of Federal Regulations 441.1.
USDA-FSIS, 2015b. The Nationwide Microbiological Baseline Data Collection Program: Raw Chicken Parts Survey.
USDA-FSIS, 2014a. Isolation and Identification of Salmonella from Meat, Poultry, Pasteurized Eggs, and Catfish Products and Carcass and Environmental Sponges. Microbiological Laboratory Guidebook 4.08.
USDA-FSIS, 2014b. Poultry Products Inspection Regulations. Title 9. Code of Federal Regulations 381.66.
USDA-FSIS, 2014c. Discontinuation of the qualitative (30 mL) *Campylobacter* analysis for young chickens. Federal Register 79, 35.
USDA-FSIS, 2014d. Poultry Products Inspection Regulations. Title 9. Code of Federal Regulations 381.91.
USDA-FSIS, 2011. New Performance Standards for Salmonella and Campylobacter in Young Chicken and Turkey Slaughter Establishments: A Response to Comments and Announcement of Implementation Schedule. 76.
USDA-FSIS, 2010a. New Performance Standards for Salmonella and Campylobacter in Young Chicken and Turkey Slaughter Establishments; New Compliance Guide Federal Register, vol. 75 93, pp. 27288–27294.

USDA-FSIS, 2010b. New Performance Standards for Salmonella and Campylobacter in Young Chicken and Turkey Slaughter Establishments; New Compliance. 75.
USDA-FSIS, 2008. Salmonella Verification Sampling Program: A Response to Comments and New Agency Policies. FR 73. No. 18, pp. 4767–4774.
USDA-FSIS, 2004. Verification of Procedures for Controlling Fecal Material, Ingesta, and Milk in Slaughter Operations.
USDA-FSIS. 1996. Pathogen reduction: a hazard analysis and a critical control point (HACCP) systems; final rule. 61, 38806–38944.
Zheng, L., Jeong, J.Y., Janardhanan, K.K., Ryser, E.T., Kang, I., 2011. Microbiological quality of water immersion–chilled and air-chilled broilers. Journal of Food Protection 74, 1531–1535.
Zweifel, C., Stephan, R., 2012. 3-Microbial decontamination of poultry carcasses. In: Demirci, A., Ngadi, M.O. (Eds.), Microbial Decontamination in the Food Industry. Woodhead Publishing, Cambridge, UK, pp. 60–95.

# CHAPTER 19

# Unraveling Food Production Microbiomes: Concepts and Future Directions

**Steven C. Ricke[1], Jessica C. Hacker[1], Kristin L. Yearkey[1], Zhaohao Shi[1], Si Hong Park[1,2], Chase E. Rainwater[1]**
*[1]University of Arkansas, Fayetteville, AR, United States; [2]Oregon State University, Corvallis, OR, United States*

## Introduction

Each year, a segment of the human population in the United States is impacted in some fashion by foodborne diseases resulting not only in hospitalizations and in some cases deaths but being an economic burden that can exceed several billion dollars (Scallan et al., 2011; Scharff, 2012; Painter et al., 2013; Batz et al., 2014). Such outbreaks place considerable concern on the safety of the food supply by consumers and leads policy makers and regulatory agencies to increase efforts to not only understand and track foodborne pathogen movement but encourage development of better control measures. However, the complexity of food production continues to offer challenges in the development of control measures that limit the impact of the presence of foodborne pathogens in all phases of food production from growth of vegetables in the fields and housing of food animals to processing and retail of the resulting food products. In live animal production, considerable efforts have been directed toward controlling foodborne pathogenic bacteria and simultaneously improving gastrointestinal health by promoting an adult microbiota that supports optimal gut function and enhances host animal performance. With the removal of antibiotics in animal feeds, efforts have intensified to develop alternative dietary amendments that potentially elicit similar responses in the animal. Likewise, in postharvest food–processing efforts to limit foodborne pathogens has continued to be developed via the application of novel antimicrobial compounds. In addition, assessment of nonpathogenic microbiota and the subsequent impact on spoilage and potential shelf life are becoming important economic considerations for in-depth microbial analyses.

For nearly a century, food microbiology and microbiology in general were largely reliant on culturing bacterial growth–based methods for characterizing microbial populations on foods and using bacterial enrichment or selective isolation as a means for detection of foodborne pathogens (Rodriguez-R and Konstantinidis, 2014; Escobar-Zapeda et al., 2015). Such approaches represented inherent limitations by being restricted to what organisms could be successfully recovered

Food and Feed Safety Systems and Analysis. http://dx.doi.org/10.1016/B978-0-12-811835-1.00019-1

after employment of various enrichment and selective bacterial growth media. Characterization of indigenous bacteria from the gastrointestinal tracts (GITs) of food animals offered even more challenges due to the strict anaerobic physiological nature of many of the inhabitants that required highly specialized media and culturing techniques resulting in the recovery of only a fraction of the total microbial population (Ricke and Pillai, 1999; Ricke, 2015a).

However, with the emergence of genomics and the molecular tools that resulted from these advancements in the past few decades, the food industry has become much less dependent on culture-based methods for characterizing microbial populations (Ricke et al., 2015). The availability of techniques such as polymerase chain reaction (PCR) amplification of specific microbial sequences has revolutionized the analytical capabilities both in terms of specificity and sensitivity for routine foodborne pathogen detection and quantitation (Maciorowski et al., 2005; Kuchta et al., 2014; Park et al., 2014a; Baker et al., 2016). More recently, conducting whole genome sequencing (WGS) in a more routine manner has resulted in the ability to precisely identify specific strains of foodborne pathogens and in turn track and trace the origin of specific outbreak isolates (Ricke et al., 2015). Likewise, the application of 16S ribosomal RNA (rRNA) gene–based sequencing has led to molecular profiling of microbial communities and generation of taxonomic microbial census data for comparing complex microbial communities in different environments and ecosystems. However, as these microbiome data sets are generated, the true challenge has been not just the data analyses but data management and depending on the circumstances, data cyber security. Therefore the field of bioinformatics which involves the conversion of raw sequence data into interpretable biological information has become critical to the effective utilization of the molecular laboratory tools for routine employment. The focus of this chapter will be on microbiome applications in food production systems along with an overview of the bioinformatic programs currently available and the information that can be generated from such programs as well as potential new developments in additional programmatic tools that are becoming available.

## *Whole Genome Sequencing and the Food Industry*

By the beginning of the 21st century equipment technologies for processing and generating sequences from DNA began to evolve tremendously in part due to the rapid progress made with the development of high-throughput sequencing technologies associated with the Human Genome project (Quail et al., 2012; Ricke et al., 2015; Hamdoun and Ehsan, 2017). Consequently, several commercial next-generation sequencer platforms such as Illumina MiSeq, Ion Torrent, and Pacific Biosciences became commonly available and were utilized for genomic and molecular research (Quail et al., 2012; Fichot and Norman, 2013; Escobar-Zapeda et al., 2015; Ricke et al., 2015; Comeau et al., 2017; Roumpeka et al., 2017). Each of these platforms utilize somewhat different approaches for accomplishing sequencing and each possess strengths and weaknesses in terms of cost, sequencing yield, run time (hours vs. days), accuracy, read length, and observed raw error rate (Quail et al., 2012; Garrido-Cardenas et al., 2017). Of these systems, Illumina systems that use a sequencing by synthesis approach have gained considerable popularity (Quail et al., 2012). A feature of Illumina systems is that in vitro clonal amplification is via bridge PCR for multiplying the number of bases to be sequenced and sequencing is based on fluorescent nucleotide analogs serving as reversible terminators during amplification (Morozova et al., 2009; Garrido-Cardenas et al., 2017). Third-generation sequencing approaches involving nanosensor-based technologies have also begun to emerge for single-molecule real-time long-read sequencing that does not require library preparation or sequencing reagents (Coupland et al., 2012; Quail et al., 2012; Garrido-Cardenas et al., 2017).

As these sequencing technologies improve, it is anticipated that not only will the capacity increase but the cost per sample will decline making sequencing more available for routine uses over a much wider range of applications. This certainly includes a multitude of applications for food microbiology and foodborne disease. Obviously, WGS for classifying and identification of foodborne pathogens has received tremendous interest for several reasons. Not only does sequencing the entire genome offer the opportunity to differentiate individual strains for taxonomic assignment, but this genomic precision provides the means for tracking individual isolates during foodborne disease outbreaks and identifying potential food production sources. As expected, this has diagnostic utility for epidemiology and limiting the dissemination and spread of the outbreak organism and offers a viable replacement for older methods such as pulsed field gel electrophoresis (PFGE). In several recent outbreak events, WGS has successfully been used to identify the causative strain of foodborne pathogen (Vongkamjn and Weidemann, 2015; Ricke et al., 2015; Kaldhone et al., 2016).

The application of WGS also allows for in-depth analysis and delineation among otherwise closely related isolates. For example, Hoffmann et al. (2014) implemented single nucleotide polymorphism analysis to distinguish *Salmonella*, Heidelberg isolates from an outbreak that had been considered nearly identical according to PFGE analysis. More recently, Emond-Rheault et al. (2017) have formed the *Salmonella* Syst-OMICS consortium to conduct an extensive compilation of the *Salmonella* genome information from 4500 *Salmonella* genomes to ultimately generate a comprehensive metadata base for the dissemination and prevalence of phenotypes associated with virulence genes and antibiotic resistance genes. They have proposed that the outcome of this consortium effort will be to not only better understand *Salmonella* evolution but also enhance diagnostic and control efforts as well as improve epidemiology and surveillance via identification of additional markers. Certainly, a WGS undertaking at this scale offers tremendous opportunities to match the ability to dissect the pattern of a specific outbreak with the myriad of external seemingly unrelated factors associated with the multitude of interactive components that are present during the complex nature of food production, processing, transportation, storage, and retail that now occurs on a global scale. It will be interesting to assess how these comprehensive efforts evolve as WGS commercial platforms advance and their corresponding resolution improves.

In addition, WGS can be used to develop an understanding of gene functionality either by direct interpretation of the DNA genome sequence or sequencing the RNA (RNA-seq) to conduct transcriptomic analysis and assess gene expression levels in response to some experimental variable of interest. Quantifying gene expression levels before WGS became readily available had relied largely on RNA isolation and PCR amplification of individual specific genes of interest or the use of DNA microarrays to gain more genome coverage (Sirsat et al., 2010; Ricke et al., 2013). Other approaches to identify gene functionality have involved the use of transposon mutagenesis methods that cover the entire genome by creating libraries of mutants for subsequent negative selection to identify essential genes after exposure to an environmental condition of interest (Kwon et al., 2010, 2016). Sequencing of the recovered transposon mutant pools (Tn-seq) reveals genes that would be essential for the foodborne pathogen to survive under a particular condition or treatment in question (Kwon et al., 2016). For example, Khatiwara et al. (2012) used a combination of a modified EZ:Tn5 transposon and massively parallel Illumina sequencing to identify and map conditionally essential genes associated with responding to the presence of bile or elevated body temperature that occurs during *S.* Typhimurium infection of an avian host. Based on their analysis, they were able to identify over 100 protein encoding genes that could be considered conditionally essential for at least one of these stressors, and they used competition with constructed deletion mutants to confirm the predicted phenotype for several of these genes. As more of these types of approaches are used, the opportunity to add detailed phenotype-genotype information into the comprehensive data sets being proposed should enhance the understanding of *Salmonella* evolution and these techniques should be helpful for non-*Salmonella* foodborne pathogens as well.

Application of WGS has also proved useful for modeling the dissemination patterns of foodborne pathogens in experimentally controlled environments. For example, Yang et al. (2017) generated "barcoded" *S.* Enteritidis strains which were essentially the same strain except each genetically modified (isogenic) strain contained a unique DNA sequence segment inserted in the genome at a chromosomal site that had been identified as phenotypically nonessential and noninfluential on *S.* Enteritidis physiology and pathogenesis. Using WGS to differentiate the individual barcoded strains, the constructed isogenic *S.* Enteritidis marker strains were used to track the dissemination patterns and potential infectious dosage impact on their respective population levels recovered from the avian GIT and feet of chickens exposed to these strains (Yang et al., 2017). As more becomes known on these dissemination patterns, there is a possibility that *Salmonella* pure cultures present as a colony or in liquid culture may be more heterogeneous than previously thought and this in turn may explain some of the variability associated with *Salmonella*-host relationships and infectious doses.

This could be critical information for a wide range of applications ranging anywhere from assigning risk evaluations to certain *Salmonella* serovars to developing optimal vaccines that ensure the most effective protection to the host.

## Characterizing Food System Microbiota

Much of the microbiological focus in food processing has understandably been on the detection of foodborne pathogens and development of interventions to limit their contamination in food production. Obviously, their occurrence and frequency on food products during and after processing remain critical concerns. However, the presence of nonpathogenic bacteria can have a major impact on food production as well. The highly diverse nature of indigenous bacteria presents diagnostic challenges due to their wide taxonomic range, varying metabolism, and physiology. Traditionally bacterial growth–based culture methods were used to isolate and enumerate selective groups that were assumed to represent certain aspects of microbial quality including spoilage organisms, overall bacterial load, and indicator organisms for fecal contamination just to name a few (Sofos et al., 2013; Handley et al., 2015). Although generation of bacterial plate count data certainly served to provide some idea of bacterial population patterns during processing, limitations inherent in the selectivity also biased the data toward those organisms favored to grow and form colonies under these conditions. Consequently, the assumption has to be made that the trends associated with the bacteria recovered on plate media such as aerobic plate count media paralleled the remainder of microbial population on the food product not recovered on plates.

In food animal production, the GIT microbial populations are also important in part because the GIT serves as a niche for foodborne pathogens such as *Salmonella*, *Campylobacter*, and pathogenic *Escherichia coli* (Dunkley et al., 2009; Horrocks et al., 2009; Foley et al., 2011, 2013; Baker et al., 2016). As expected numerous chemical feed amendments such as organic acids, biological control agents including bacteriophage, and botanicals have been employed in attempts to decrease the foodborne pathogen population levels in the GIT (Ricke, 2003; Ricke et al., 2005, 2012; Sirsat et al., 2009; O'Bryan et al., 2015; Calo et al., 2015). Although these approaches are designed to remove pathogens already established in the GIT, interest has grown in the administration of feed additives that prevent initial colonization by foodborne pathogens. These include bacterial cultures commonly referred to as either probiotics, competitive exclusion cultures, or direct fed microbials which consist of single or combinations of organisms that can potentially become established in the GIT and thus serve as a barrier to foodborne pathogen colonization (Callaway and Ricke, 2012; Siragusa and Ricke, 2012; Hanning et al., 2015; Ricke and Saengkerdsub, 2015; Callaway et al., 2016). Prebiotics are feed additives that generally occur in some form of complex carbohydrate that favor or select certain groups of GIT bacteria already present to proliferate, and like probiotic cultures, serve as potential barriers to the establishment of foodborne pathogens (Park et al., 2013; Ricke, 2015b, 2016; Roto et al., 2015; Ricke et al., 2017).

As more of these GIT-acting products came into the commercial market, it was realized that the overall indigenous GIT microbial populations could potentially be influenced by the presence of these additives. Consequently, methods for characterizing the GIT population were needed that could provide more coverage of all the different groups that are present in the GIT. This required the use of specific culture methods that allowed for the successful isolation and growth of strict anaerobic microorganisms (Bryant, 1972; Hungate, 1950, 1966, 1969; Holdeman and Moore, 1972; Holdeman et al., 1977; Macy et al., 1972; Miller and Wolin, 1974; Balch and Wolfe, 1976; Levett, 1991). Traditional culture–based approaches originating from rumen microbiology had been designed to recover the strict anaerobes known to dominate or were presumed functionally important in the GIT ecosystems in the rumens of ruminants as well as the lower GIT and ceca of nonruminants (Hungate, 1950, 1966; Bryant, 1959; Mackie and White, 1997a,b; Ricke and Pillai, 1999; Saengkerdsub and Ricke, 2014; Ricke, 2015a). However, only a small percentage were recoverable with these techniques, and it became clear that many groups were minimally recovered and/or not accounted for when such data were reported (Ricke and Pillai, 1999). Therefore although extensive characterization of individual GIT isolates was accomplished leading to some appreciation of the metabolic complexities of GIT ecosystems, no methods existed for analyzing an entire GIT microbial population that possessed the capability for accounting of all groups of GIT microorganisms.

## Sequencing Food System Microbiomes

Advancements in molecular biology methodology in parallel with the concept that conserved gene sites such as ribosomal genes were optimal candidates for separating out microbial evolutionary differences among groups of organisms opened the door for molecular characterization of microbial ecosystems and the sequencing approaches that followed (Woese and Fox, 1977; Pace et al., 1985; Schloss et al., 2009; Escobar-Zapeda et al., 2015). Initial efforts focused on the use of gel electrophoresis approaches such as denaturing gradient gel electrophoresis (DGGE) that involved separation of DNA based on visually distinct banding patterns followed by isolation of specific bands and subsequent sequencing of those bands (Yu and Morrison, 2004; Hanning and Ricke, 2011; Escobar-Zapeda et al., 2015; Ricke et al., 2015). Extensive use of DGGE in species such as chickens led to the establishment of diet and bird maturity impacting GIT microbial populations resulting in shifts in GIT populations that could be detected from both in in vitro cecal incubations and in vivo samples (Hume et al., 2003; Dunkley et al., 2007a,b). Handley et al. (2010) also used DGGE to delineate changes in microbial populations in postchill processed chicken carcasses during storage at room temperature. In all these cases, shifts in banding patterns were used to indicate changes in microbial populations. However, it still remained virtually unknown which bacterial taxonomic groups contributed to these changes. Despite relying strictly on unknown DNA-banding patterns some bacterial identification could still be achieved with DGGE. For example,

Park et al. (2014b) and Lee et al. (2016) used DGGE profiling along with an ABI 3100 capillary analyzing system for DNA sequencing of particular bands extracted from the gels to examine the impact of a yeast-based prebiotic compound on the cecal populations in pasture and conventionally raised broilers. Others (Koo et al., 2012; Mertz et al., 2014) used a similar DGGE-sequencing approach to identify groups of bacteria occurring at different sites in meat deli slicers and establish the overall prominence of *Pseudomonas*. Based on this overall microbial profiling approach, Mertz et al. (2014) suggested that *Pseudomonas* might be an optimal indicator organism for assessing effectiveness of cleaning and sanitizer regimens for retail deli slicers.

With the advent of next-generation sequencing (NGS), the idea that systematic DNA sequencing could be used to combine microbial taxonomy 16S rRNA gene analyses with sequencing microbial populations became the method of choice for assessing microbial community diversity (Escobar-Zapeda et al., 2015; Ricke et al., 2015). As the methodology developed, definitions and terminology of what constituted a microbial population became established with "microbiota" being defined as the representative microbial diversity within a given microbial community and "microbiome" as the descriptive term for genetic diversity of these same microbial populations (AAM, 2013; Roto et al., 2015). As expected, these technologies were rapidly introduced to food systems to characterize microbial communities associated with all phases of food animal production and food product processing. Early efforts focused on the use of 454 pyrosequencing as the method of choice since it was one of the first sequencing methodologies available (Petrosino et al., 2009). In live animal production for example, pyrosequencing was successfully used to differentiate and identify GIT (crop, ilea, and ceca) microbial populations in broilers (Park et al., 2015), laying chicks and hens (Callaway et al., 2009; Escarcha et al., 2012), and rumens and feces of cattle (Dowd et al., 2008; Callaway et al., 2010). Pyrosequencing has also been used to analyze microbial populations in retail meat products as well (Koo et al., 2016a,b) along with attempts to directly compare sequences recovered from DGGE bands with pyrosequencing of the same microbial communities. For example, diversity of microbial populations in ground beef retail samples taken from different geographical locations in South Korea were compared by Koo et al. (2016a). Their 454 pyrosequencing analysis on a subset of samples revealed that the sequenced bands from PCR-DGGE were also present in the 454 pyrosequencing results and consistently yielded taxa that included *Pseudomonas* among others (Koo et al., 2016a).

The more recent introduction of benchtop models of sequencers, such as the Illumina MiSeq has paved the way for more extensive characterization of microbial communities in food animals and food processing. Proliferation of these sequencing platforms increased the range and variety of microbial ecosystems being profiled offering new insights into microbiota patterns as a function of a wide range of factors such as changing environmental conditions, alterations in diet, age of animal, introduction of external stressors such as antimicrobials and multiple factors applied simultaneously. Although some microbiome work has been done with postharvest applications such as poultry processing (Oakley et al., 2013; Rothrock et al., 2016; Kim et al., 2017), most of the research emphasis has been on preharvest GIT microbiome work. Given the interest in animal GIT microbiota response to dietary alterations and feed amendments, it is not surprising that much of the focus of Illumina-based sequencing has focused on characterizing the microbiome during exposure to such treatments. Much of this has been directed toward incorporation of prebiotics, and identifying microbial taxa shifts in GIT microbiota in response to their introduction in the diets of the birds. In vitro studies utilizing cecal inocula grown anaerobically in the presence of the feed additive of interest have demonstrated utility for screening a wide range of compounds and determining effective candidates. Park et al. (2017a) used such an approach to demonstrate that age of bird that the cecal contents was retrieved from and used as an inoculum in the in vitro incubation system exhibited a major impact on microbiome taxonomy and diversity. In vivo studies on different broiler breeds and management environments have also been explored. For example, Park et al. (2016) examined microbiome patterns in naked neck pasture flock broilers fed a yeast-based prebiotic and noted changes in taxa groups among the control group, antibiotic-fed group, and the yeast prebiotic–fed birds. In a second study, Park et al. (2017b) fed either fructooligosaccharide, galactooligosaccharide, or plum fiber to Cornish Cross White Plymouth Rock commercial broiler chicks also raised under pasture flock conditions. Here, they used an Illumina HiSeq for sequencing cecal populations and applied synthetic learning in microbial ecology analysis to interpret the 16S rRNA microbiome data and identify specific bacterial taxa that could serve as predictors for microbial responses to specific prebiotics. Based on these in-depth data analyses, they demonstrated unique impacts for each dietary additive on the cecal microbiome and noted that certain taxa groups appeared to be predictive of *Campylobacter*. Although this was perceived as a preliminary result, the implications from a GIT microbial ecology standpoint could certainly support the concept of a symbiotic relationship between *Campylobacter* and GIT microbiota as suggested by Indikova et al. (2015). Obviously, more detailed future studies are needed to confirm this and determine what types of ecological and/or metabolic relationships this might indicate. Microbiome sequencing may also indicate a relationship between *Salmonella* and the cecal microbiota as well. Park et al. (2017c) compared microbial communities vaccinated with genetically quite different *S.* Typhimurium constructed strains but derived from the same parent strain and observed that microbial populations were distinct for each of the vaccinated groups of birds. This begs the question as to what types of interactions may exist between the bird's immune system, the respective *Salmonella* vaccine strain introduced to the young bird, and the indigenous GIT microbiota as it develops with the progressive maturity of the bird. To answer, these questions will require sophisticated analytics using bioinformatic programs that can achieve sufficient data mining to sort out which factors have the most impact. The following sections will describe some of the current and potential sequencing approaches, data processing, and data interpretation programs that are available.

## Molecular-Based Microbial Diversity Profiling

Microbial diversity has always been the hallmark of microbial ecology research. However, the term "microbial diversity" is very dependent on the method(s) used to identify and characterize the individual members of a microbial community. Consequently, the concept of microbial diversity can not only evolve as techniques to measure it change but also remain somewhat nebulous as it becomes redefined. Traditionally, microbial diversity was perceived as differences in bacterial cell morphology, metabolism, cell structure, cell wall composition, and other trademarks of the classically defined phenotype (Holdeman et al., 1977). With the arrival of genetic analyses, this obviously has altered the meaning of microbial diversity to the point of being primarily about distinctive differences at the genomic level. Although initial molecular techniques indirectly assessed differences as a function of chromosomal DNA hybridization between microorganisms, or nucleic acid content (guanine and cytosine), the emergence of sequencing radically altered the philosophy of what constituted a unique organism (Rodriguez-R and Konstantinidis, 2014).

Escobar-Zapeda et al. (2015) has suggested that sequencing can be used to determine microbial diversity by two fundamentally different approaches and has provided a comprehensive review of both methods which will not be discussed in detail here. Briefly, one approach which has already been discussed to some extent in the current review involves amplicon sequencing analysis or metaprofiling where the analysis across a microbial community is a function of the comparison of the sequence of a single gene that is considered a robust bacterial taxonomical marker such as the 16S ribosomal gene (Escobar-Zepeda et al., 2015). An alternative approach involves shotgun cloning and is referred to as shotgun metagenomics where large fragments and/or whole genomes are recovered that include phylogenetic markers for classification along with reconstruction of functional gene pathways (Escobar-Zapeda et al., 2015). Obviously, knowing the gene sequence is only part of the answer as further data analyses are required to achieve gene prediction and annotation (Escobar-Zapeda et al., 2015). Functional metagenomic analyses and screening can lead to the identification of numerous genes and gene families considered critical for understanding relationships of groups of organisms in various ecosystems such as the animal GIT (Yoon et al., 2015). Therefore the ability to not only generate the data but also convert that data into meaningful biological information that can be interpreted to address questions of microbial diversity and genomic functionality is critical. The analytics used to accomplish this is often referred to as bioinformatics and essentially involves the development and implementation of computer programs designed to convert raw sequencing data to its corresponding biologically interpretable outcome. Although bioinformatics can encompass a variety of definitions and interpretations (Smith, 2015; Vincent and Charette, 2015), the ultimate goal of such efforts still remains the same.

As expected once sequencing became more routine, the generation of large quantities of raw data permeated the scientific literature and presented challenges for integration of these data sets into meaningful applications. This is certainly true for the food industry where the ongoing discourse has continued regarding the potential and future use of such data for food systems and food safety. However, the volume of data generated continues to increase leading to the perception of the "Big Data" concept which Whitford (2017) describes as data sets being so large that traditional computational systems are not equipped to process such data. And what quantitatively constitutes "Big" is a moving target as well. To put this in perspective, Whitford (2017) noted that in 2017 internet activity is now greater than 1 zettabyte (1 sextillion bytes). In food production, Strawn et al. (2015) have also defined "Big Data" as a function of the sheer volume of the data involved and have suggested that computational strategies for integrating independent data sets represents a tremendous boon for the food industry in areas such as improved food safety management. Certainly, WGS and what such data offer in terms of precise identification of foodborne pathogen strains accompanied by more effective tracking in food production systems would seem to fit well with a larger data set that integrates several independent data compilations generated by a food company.

Likewise, the microbiome sequencing data currently being generated in food animal production and processing would seem to also qualify as a contributor to these larger integrated data sets. However, unlike WGS and other data sets, the interpretation and potential practical utility for microbiome profiling are still somewhat elusive. Part of the problem lies with interpretation and subsequent application of microbiome data. Fundamentally, there is uncertainty associated with the biological relevancy of microbiome-defined microbial diversity in the ecosystems of food production and processing. In practical terms, it is a question of not just which microorganisms are present at various stages of food animal production and processing, but what constitutes their ecological role in the microbial community. For example, how do they contribute to the presence/absence of foodborne pathogens in the animal GIT or participate in the initiation of spoilage and subsequent decrease in the shelf life of a particular food product? Furthermore, how do the microbiome data align with nonmolecular indicators?

Some of the uncertainty associated with food systems generated microbiome data is also due to the different computational methods that are available for bioinformatic analyses and the level of achievable detail. Metaprofiling and metagenomics as discussed previously provide information on which organisms are present and in the case of metagenomics the genes and gene families present in the microbial communities. They also provide information depending on what is being sampled and how microbial communities shift over time or respond to an external factor being introduced. Ideally, such information would be even more useful if it could be used to predict outcomes of future events based on the current microbiome profile. For example, could the microbial community microbiome profile currently present in a sample be used to predict when foodborne pathogen GIT colonization would most likely take place or when spoilage in a fresh food product would occur? To address the possibilities for answering these questions, the following sections describe the data bioinformatic analytical programs commonly being used for microbiome analyses followed by a description of computational programs that offer an opportunity for potential prediction of particular outcomes.

## Current Sequencing Strategies and Bioinformatics Software Programs

### Microbiome Sequencing Strategies

Microbiome sequencing and delineating taxa within that microbiota community requires several steps including choice of target gene(s) for sequencing, extraction of DNA followed by sequencing, and finally data analyses, taxa identification, and statistical assessment of diversity for comparisons within and among microbiomes of microbial ecosystems of interest. Obviously the first step is selecting a representative set of genes that meet the requirements of not only always being present but possessing the right combination of conserved sequence with sufficient variability to allow taxa distinction without introducing bias due to genomic plasticity. The 16S rRNA gene is a component of the 30S small subunit of the ribosome and is the usual target for sequencing since is universally found in all bacteria and archaea (Amato, 2017). The advantage of this choice for microbiome sequencing and taxa classification is two-fold. The 16S rRNA gene possesses conserved genomic sequence regions shared by all taxa which offer constant and predictable priming targets for amplification but, just as importantly, it also contains nine variable regions (V1–V9) that are unique to specific taxa groups and therefore can be used for classification (Robinson et al., 2016). Hypothetically, one would desire to sequence the entire 16S gene (1500 bps), but this is not possible in practice since most commercial high-throughput sequencing platforms are short read based (eg, 100 to 350 bps on the Illumina systems) and cannot provide sufficient coverage (Robinson et al., 2016; Amato, 2017). Consequently, specific V1–V9 region(s) are chosen for targeted sequencing. However, Comeau et al. (2017) has noted that the choice of which V1–V9 region(s) to sequence is critical because not all variable regions yield similar PCR amplification efficiencies nor are they equally representative to provide similar resolution among all taxa. This is supported by previous work specifically comparing different V regions for either DGGE microbiome profiles or constructing sequence similarity dendrograms among different pathogenic bacteria (Yu and Morrison, 2004; Chakravorty et al., 2007). In both studies, it was demonstrated that different V regions do not equally differentiate all groups of bacteria and that certain V regions were more optimal choices (Yu and Morrison, 2004; Chakravorty et al., 2007). With this in mind, one or more variable regions are typically used for most microbiome sequencing with V1–V3 and/or V4 being the most likely choices because they can generally detect both bacteria and archaea with less bias from most microbiota communities with V4 being preferred for some commercial sequencing platforms such as Illumina (Robinson et al., 2016; Amato, 2017). Robinson et al. (2016) cautions that because technologies and platforms can change rapidly, making comparisons across independent studies is difficult; thus it is important to use data sequenced on the same platform with the same variable region and subsequently reprocess the data with currently available bioinformatic tools.

Once the sequencing is completed on the microbial community, extracted DNA and raw DNA sequence data are generated, and analyses for taxonomic assignments is required. Taxonomic identification or picking operational taxonomic units (OTUs) from the sequence reads allows for further classification of the members of the microbial community. Once sequences are assembled into single OTU clusters that possess a certain level of similarity usually at least 97% they can be further identified (Schloss, 2010; Kozich et al., 2013; Robinson et al., 2016; Amato, 2017). At this stage, clustered sequences can be assigned to their respective classification, namely, phyla, class, order, family, genus, and species by comparing with databases of already compiled 16S sequenced organisms such as the Greengenes database (DeSantis et al., 2006; Jovel et al., 2016). Although most of the upper taxonomic levels are relatively well defined and therefore assignable, species identification is much more difficult given the paucity of sequenced organisms in the databases currently available. Therefore much of the classification at the species level remains as "OTUs" with no organism specifically named (Amato, 2017). When processing raw microbial community data derived from sequencing of 16S rRNA genes, several assessments of microbiome profiles are needed to conduct comparative analyses of the corresponding microbial communities. This involves the use of bioinformatics, a term and field of study which Luscombe et al. (2001) defined as the use of computational-based analyses to understand and subsequently organize information generated from biological data or more specifically for purposes of the current discussion, converting sequence data into digital data. As they point out the goals for application of bioinformatic tools are to create accessible data information storage for ongoing entry of new data, provide tools for analyzing the deposited data, and finally the ability to interpret the data for drawing biological conclusions.

Once the microbiome taxa have been identified and phylogenetic assignments made, two types of diversity measurements are sought for comparing microbiome data. Gensollen and Blumberg (2017) defined microbiome diversity by two terms: (1) richness or total number of different species in a microbial community and (2) evenness or equality in species abundance. This more in-depth comparison of microbiome data requires statistical tools to differentiate phylogenetic patterns and systematically screen for levels of relatedness versus distinguishable taxonomic differences within a microbial community. It also requires a comparison of composite profiles among different microbial communities to assess how taxonomically related they may be. Alpha diversity comparisons enable assessments of the microbial taxonomic diversity within a sample or single microbial community. These comparisons reveal the level of microbial complexity in a given sample by presenting the total number of OTUs that are recognizable as being genetically distinct. This can be used to assign some sort of diversity index associated with whichever pipeline analysis program is being used. This allows for comparisons between such variables as how dietary treatments or animal age differences might impact GIT microbiome development as a function of diversity. For example, questions such as does animal development and eventually reaching maturity increase microbial community diversity in the GIT?

For comparisons among samples, beta diversity analyses can be done by accounting for the shared taxa among the sample microbial populations versus missing taxa among these same samples. These comparisons allow for pairwise distance appraisal of how similar microbiomes are by assessing the presence or absence of OTUs and/or abundance of OTUs (Robinson et al., 2016). From these comparisons, statistical analyses and interpretations can be conducted to determine significant differences. Further assessment can be done by generating three-dimensional plots referred to as principle coordinates analysis. To accomplish these diversity comparisons, there are several customized software programs or analysis pipelines based on various algorithms that are available for assembling and organizing data into the respective diversity comparisons and visualization graphic presentations. As Nilakanta et al. (2014) have pointed out, the general purpose of these analytical pipelines is to streamline sequencing curation, filtering, analysis, and output. Although more programs continue to be developed, two of the more commonly used pipeline analysis programs used for microbiome data analyses, mothur and QIIME (Schloss et al., 2009; Schloss, 2010; Caporaso et al., 2010; Gonzalez and Knight, 2012; Huse et al., 2014; Nilakanta et al., 2014), will be discussed in the following sections.

### *Mothur*

The mothur bioinformatics pipeline is an open source project that was originally intended to address bioinformatics requirements for the microbial ecology community. It was developed by a research group led by the Department of Microbiology and Immunology at the University of Michigan in response to the dramatic increase in sequence data occurring at that time and the need for software with more efficient algorithms that could accommodate these larger data sets (Schloss et al., 2009). The mothur pipeline basically reimplemented numerous preexisting algorithms from previous computing tools into one high performance executable program (Kuczynski et al., 2012; Amato, 2017). Consequently, mothur was capable of functioning regardless of a particular user's operating system. The original goal of mothur was to have access to a single resource that would be able to analyze molecular data and make the software free to the public and open to modification by researchers (Schloss et al., 2009). The mothur program was developed with the ability to go from a raw sequence to visualization tools rapidly, which made the mothur pipeline attractive for routine use due to its capacity and speed as a pipeline analysis program (Albert, 2014). The mothur pipeline program was written in C and C++ because they were considered compiled languages (Schloss et al., 2009). This was meant to ensure that once the code was compiled, another program would not be required to run it. The program itself was essentially driven by the command line. There were three main modes that mothur could operate in. The default mode was utilized as the interactive mode. There was also a command line mode and a batch mode. The command line mode was considered helpful for developing a pipeline that included mothur. Batch mode required text files that contained all the commands and thus allowed the program to execute all of them simultaneously.

## QIIME

The Quantitative Insights into Microbial Ecology (QIIME) (pronounced "chime") software pipeline was also generated as an open source software project intended to analyze the microbiome from raw high-throughput DNA sequencing data (Caporaso et al., 2010; QIIME development team, 2015). Like some of the other pipelines, QIIME was developed to support graphical displays of the data allowing interaction by the researcher and the ability to assess diversity within and between microbial communities (Caporaso et al., 2010). Although much of the functionality within QIIME was written by QIIME developers, much more of the code was written by other research groups as well. However, in order to be able to use the QIIME pipeline, there remained a need to download large quantities of dependent programs which were all independently created by the outside sources. Although QIIME could be used with laptop computers as well as high performance computing clusters (Caporaso et al., 2010), in practice it was shown to work best with Mac OS X as it required a Virtual Box to run on Windows or Linux for installation (Nilakanta et al., 2014). QIIME consisted of a Python interface that, in turn, possessed the capability of connecting all the separate programs. The interface itself converted inputs and outputs of the individual programs and helped the user to be able to transition data between the packages. This allowed the user to feed one program into another and therefore create a pipeline. QIIME also possessed the ability to create visualization from raw sequencing data, and it was also driven by the command line.

## QIIME Versus Mothur

Nilakanta et al. (2014) conducted a systematic review of several software packages available at that time for 16S rRNA gene sequencing of microbial communities. They evaluated the respective programs by a number of criteria including among others, being cost free, the pipeline had to be self-contained, and the programs had to be publicly accessible but allow for sufficient data security. They concluded that mothur and QIIME represented the best available pipelines due to their comprehensive array of functions/procedures and the accompanying documentation. In short, the primary differences between mothur and QIIME are considered fairly cosmetic and are pretty much associated with how the tools were initially developed. The fact that each of the dependent programs within QIIME were independently developed may imply that QIIME offers a higher performance than what is implemented in the mothur software pipeline, but mothur has one unified interface which makes it more consistent and better documented (Albert, 2014). Nilakanta et al. (2014) also pointed out that the mothur software package offered straightforward instructional information and accompanying tutorials. The important point is that QIIME and mothur essentially accomplish the same end point result, regardless of the type of pipeline software initially used. Speed at which certain applications within each program are completed may vary, but each program is characterized by several strengths and weaknesses (Schloss, 2016). For example, because mothur was written in C/C++, it runs much faster than QIIME since QIIME is written in Python. However, because QIIME is written in Python which is currently considered a highly popular, well-developed language, there has been extensive input from numerous contributors.

It is highly debated among research groups in the microbial ecology scientific community which program is better, but any attempt to make direct comparisons would be difficult. However, some recent attempts using different approaches have been made. Plummer et al. (2015) compared a subset of fecal samples from preterm human infants born at less than 32 weeks gestation and receiving either a placebo or a probiotic cocktail (*Bifidobacterium* and *Streptococcus* combination). The DNA from the collected samples was extracted and subjected to pyrosequencing on Roche 454 Genome Sequencer. They did not detect statistical differences in QIIME versus mothur assessment of microbial communities at the phyla taxonomic level. However, they did observe detectable differences at the genus level especially with regard to the organisms belonging to the *Enterobacteriacae* family although they attributed some of this to the employment of the 16S rRNA V3–V5 region as their sequencing target which they suggested may not be the optimal variable region for this specific bacterial family. In terms of the use of the respective pipelines, they noted that QIIME was more user friendly than mothur, presented commands that were easier to understand, and appeared to be preferable for very large data sets due to its rapid analysis time. However, they suggested that mothur appeared to be more flexible and may have advantages for conducting complex amplicon analysis. Plummer et al. (2015) also noted that their comparisons were limited by originating from only one sample type and that multiple sources of samples should probably be compared to factor in potential differences in sample types.

To overcome potential errors generated during actual laboratory sequencing that cannot be overcome by software alone, Golob et al. (2017) generated in-silico 100 synthetic human stool microbiomes and after simulated sequencing compared accuracies of mothur and QIIME by evaluating OTU generation accuracy, taxonomic classification, alpha and beta diversity measures with different settings in the pipelines. Based on introduction of even minor sequencing errors to the simulated sequencing, the resulting correct classification of OTUs was dramatically reduced by the respective pipeline. The authors concluded that when microbiome data are reported, it is critical for reproducibility and interpretability of the pipeline analysis to provide details on the pipeline including OTU generation strategy, reference sets, clustering algorithm, and specific software implementation.

## Machine Learning and Potential Applications for Bioinformatics

### General Concepts

As more sequencing projects are taken on and completed, the amount of microbiome and metagenome information will continue to climb. Certainly, current pipelines used to analyze such data will continue to have considerable utility. In addition, newer bioinformatic pipelines are emerging that will provide the means for more targeted metagenomic analyses along with WGS tools for more rapid web interfacing to detect specific gene groupings such as antimicrobial resistance genes (Rowe et al., 2015; Fosso et al., 2017; Siegold et al., 2017). Incorporating databases such as metaproteomics to understand complex GIT microbial communities and their interaction with the host will require further development of even more encompassing databases (Petriz and Franco, 2017). As sequencing technology develops further, the amount of data generated is expected to increase immensely and potentially become less manageable. Pallen (2016) has pointed out that this will further strain current sequence depositories as more and more identical or nearly identical sequences from commonly sequenced bacterial species such as *E. coli* are sequenced and the need for development of newer approaches to discern true nonredundancies in these databases. As the challenges of curation and metadata annotation grow, Pallen (2016) has suggested that application of new tools for bioinformatics such as machine learning may need to be applied.

As these bioinformatic tools develop, the ability to glean or "mine" more detailed and in-depth information not only from ongoing studies but from existing raw data sets that have already been deposited from published research will become more attractive to identify trends or patterns. Data mining is essentially the process of extracting useful information from a dataset. Machine learning is the process of learning from data using algorithms with no reliance on rule-based programming. It emerged as a subfield of Artificial Intelligence and has been applied in a wide range of fields as diverse as indoor air quality monitoring and early warning analysis for food safety (Yip et al., 2013; Vanderplas et al., 2016; Adeleke et al., 2017; Geng et al., 2017). Yip et al. (2013) in reviewing its application for genome annotation described machine learning as the construction of a mathematical model directed to a specific concept based on certain characteristics of observed data. However, machine learning is only useful when large quantities of complex data come into play. Under these conditions, machine learning can serve as a useful tool for translating otherwise unmanageable volume and the corresponding convoluted nature of large data sets into something more feasible and applicable. Yip et al. (2013) summarized the attractiveness for machine learning in bioinformatics by its ability to automatically ferret out patterns in data sets especially where knowledge is incomplete or inaccurate, data sets that are too large to be dealt with by conventional manual approaches, or where exceptions to general cases occur.

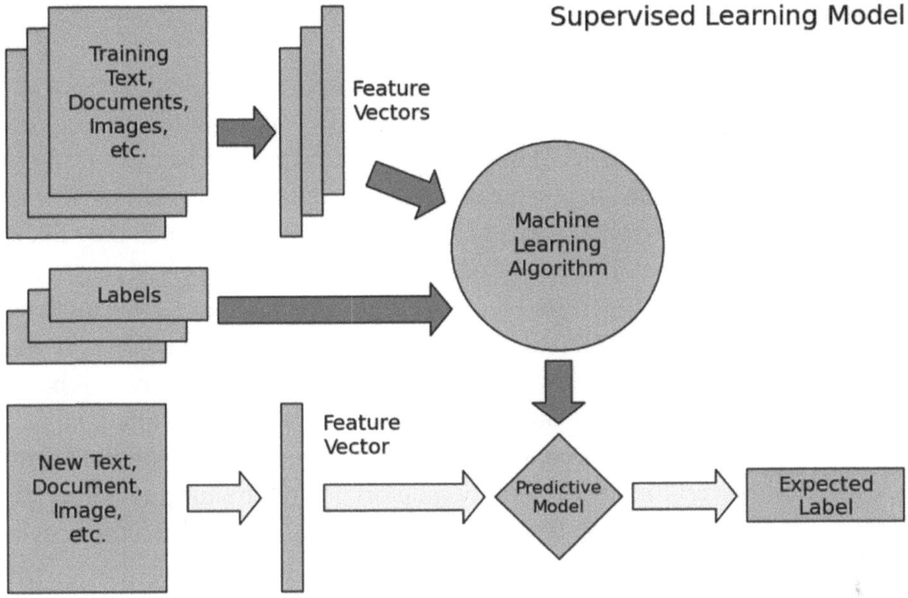


**Figure 19.1**
Supervised learning model. http://www.astroml.org/sklearn_tutorial/general_concepts.html.

Machine learning models create algorithms that help the computer make decisions based on generalizing rather than storing and retrieving data as a traditional computer system would do. This discipline is built with several different building blocks, but it begins with classical statistics. As seen in Fig. 19.1, machine learning algorithms are created by two different inputs: a set of training material and labels. From the training set, feature vectors are found to help the algorithm learn from the training set. Feature vectors are a set of aspects that describe one instance of the data. A label is the category that an instance falls in. These components together allow the algorithm to learn what past data instances were labeled as respectively and in turn create a predictive model. The user can subsequently apply new data entries to the model where the algorithm will attempt to predict the new data entry's label based on what it learned from the training set. The predictive model is only as good as its training set.

Machine learning is split into three basic categories: supervised, unsupervised, and reinforcement learning. Fig. 19.1 describes the supervised learning approach to machine learning. Unsupervised learning has the same basic flow of information except it does not take in labels and the output can be anything from a probability to a cluster identification. Unsupervised learning attempts to look at the data and find trends within the training data in order to predict potential outcomes. These trends may not always make logical sense when compared to real world scenarios (Malisiewicz, 2015). The following sections provide a summary of each of these categories.

## Supervised Learning

Supervised machine learning is based on a set of data that possesses both features and labels. The goal of supervised learning is to learn a classification system that the user has created and predict the label of an object when given a set of features. Supervised learning trains itself to come up with realistic predictions in response to any new data introduced to it. For example, given a list of movies that a person has watched and a rating on how much they liked those movies, a supervised learning algorithm would predict a movie that person may want to watch. As seen from Fig. 19.2, supervised learning is broken down into two smaller categories: regression and classification. The difference between these two categories is whether the label is discrete or continuous. For example, determining whether a patient will have a heart attack is discrete because the possible outputs are true or false, but determining a patient's age would be continuous because time is continuous. A common problem found in supervised learning is overfitting. Overfitting is essentially memorizing the training set rather than learning a more general classification technique (AI Horizon, 2015). If the training set has an extensive set of examples of one way of thinking about a problem and another way of

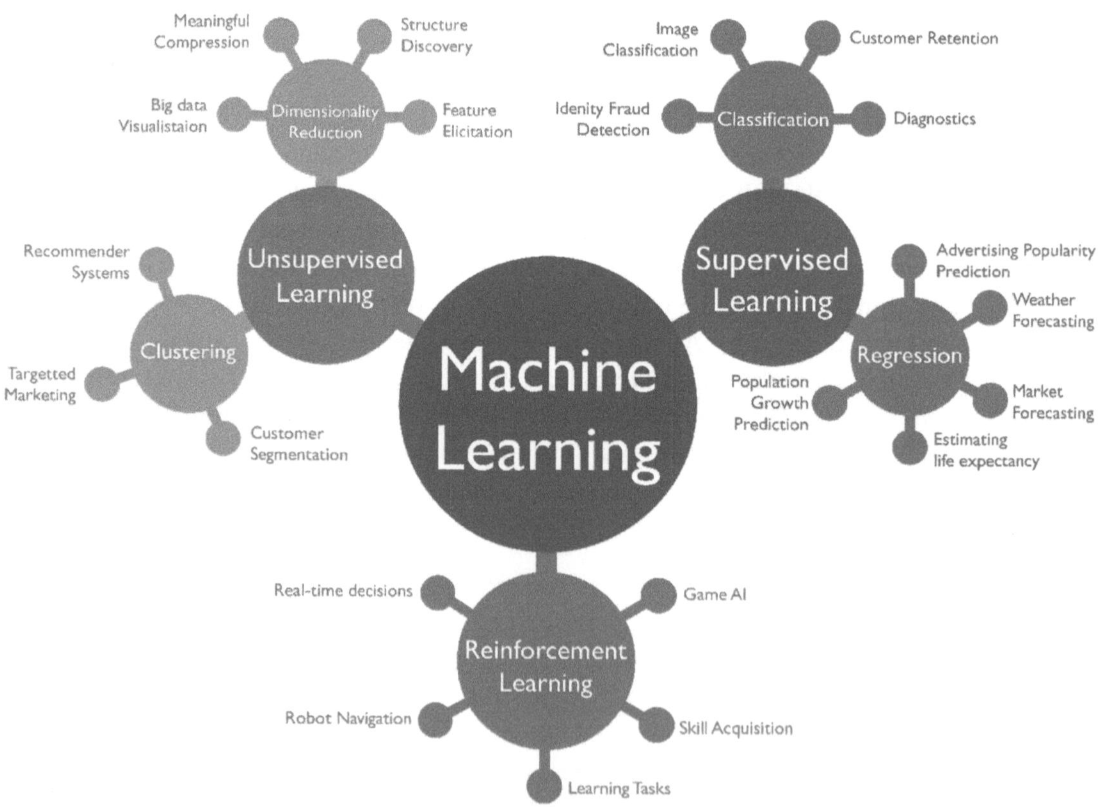


**Figure 19.2**
Machine learning. http://isaziconsulting.co.za/machinelearning.php.

thinking about the problem is never introduced, the algorithm will only learn one way of thinking and all other ways would be wrong in its eyes. That is what it was trained to do.

These techniques could include other applications of supervised learning where the classifier tried to label new unclassified data based on previous data that it had learned from. This would add another dimension of analysis to the array of experiments being conducted. It would help to make all experiments comparable so that it could be seen if the control group was able to maintain the attributes that distinguished it as the control group throughout any experiment conducted. For bioinformatics, Yip et al. (2013) suggested that semisupervised machine learning might be useful where training examples are limited or only available for certain classes and thus could be applied to provide a rough estimate of the feature space that belongs to each class. Once the rough approximation is established Yip et al. (2013) concluded that the distribution of feature values of unlabeled genomic areas could be used to estimate the boundaries of the areas.

### Unsupervised Learning

Unsupervised learning is based on a set of data that does not have any labels. The goal of unsupervised learning is to find similarities within the data in order to detect labels from the data itself. It is supposed to learn from the data without having any specific instructions on how to do it. As presented in Fig. 19.2, unsupervised learning is also broken down into two categories: dimensionality reduction and clustering. Clustering is defined as finding similarities within the data and grouping similar data sets. However, it will not be able to reveal what these groupings are called. These algorithms can use the clusters to assign new data points to a cluster (AI Horizon, 2015). Amazon.com uses this type of algorithm to recommend books by using social information and information about what their customers have already read.

Unsupervised learning algorithms could be applied to look for other trends within the data that may not be initially apparent to the researchers. A machine learning algorithm could also begin to cluster and learn unidentified OTUs. This could help to recognize corresponding OTUs within the Greengenes database which would help to expand the knowledge of the microbiota. Yip et al. (2013) have suggested that unsupervised learning might be used in bioinformatics when classes are not already predefined but instead want to identify natural clusters of genomic regions based on some unique feature distribution. Dimensionality reduction is a method of scaling down the number of random variables that are being considered. The algorithm obtains an uncorrelated set of principle variables.

### Reinforcement Learning

A third type of machine learning is referred to as reinforcement learning. Some researchers have merged this with unsupervised learning because the algorithm is not told what to do. The goal of this algorithm is to maximize the amount of reward. Reinforcement learning algorithms must attempt different actions to test which actions return the most reward. This type of learning is usually broken down into two different characteristics: trial and error search and delayed reward (Sutton and Barto, 2005). An example of this type of machine learning is an algorithm that plays chess. It must make a move while anticipating the costs and benefits of any possible replies and counterreplies.

## Conclusions and Future Directions

It is anticipated that DNA sequencing technologies will continue to make advances which in turn will make generation of WGS and microbiome data less expensive and more commonplace. Availability of these expanding data bases will potentially become overwhelming making biological interpretation and comparisons across independent studies ever more difficult. A further potential complication is the application of various omics methods such as transcriptomics and proteomics that have the potential to generate considerable multidimensional data on microbial communities along with their corresponding metabolic activities and the concomitant host responses such as immune and nutritional physiology systems. Consequently, the definition of what quantitatively constitutes "Big Data" will probably continue to be a moving target.

However, the continued refinement of current bioinformatic tools and development of new computational approaches offers the opportunity to better integrate these separate data sets and derive global microbiota-host interactive networks. Likewise, gaining a better understanding of the microbiome responses at the food-processing stage of food production has important implications for not only food safety applications but shelf life and spoilage issues. Traditionally, much of the sequencing efforts have been focused on simply identifying who the members of the microbial community are and if they shift or change as external factors such as antimicrobial interventions are introduced. However, with the improvements in metagenomics the ability to not only still identify the members of the microbial community, but target specific gene groups such as antimicrobial resistance gene(s) will provide more understanding of the microbial ecology and functionality. This is important because the composition of the microbial community may not reveal detectable shifts of individual members when the 16S rRNA gene–based microbiome is sequenced before and after exposure to some form of external stress, but their physiological responses could adapt and profile differently when assessed by transcriptomic analyses, for example. Identifying these responses to antimicrobial interventions or other external alterations such as rapid temperature fluctuations may be helpful for deciding which antimicrobials to use, when to use them, and what combinations might be most optimal for achieving a maximum multiple hurdle impact.

Although the generation of routine large data sets of sequencing information or metagenomic data sets may become more commonplace from a research and academic standpoint, whether these technologies would become routine in the food industry remains unclear at this particular time period. Part of the issue is determining which of these extensive data sets would be sufficiently useful for there to be investment in their development for routine use. Certainly, some aspects such as WGS of foodborne pathogens have some obvious implications for tracking foodborne disease outbreaks and identifying sources and transmission routes. Identifying nonpathogens may be less obvious but could be helpful for determining the best indicator organisms for potential assessment of effectiveness against organisms that behave similar to the corresponding pathogens if they were present (Handley et al., 2015). In reality, a more nuanced application of these technologies is likely in the food industry either in the form of limited numbers of samples collected and/or limited sequencing based on prioritization of critical areas of the food production cycle with regards to overall bacterial contamination or risk-based assumptions for foodborne pathogens. If that becomes the case, computer machine learning–based approaches may become attractive for handling incoming data due to their ability to predict responses from limited data entries. For this to occur, computer programs will no doubt need to become more user friendly and as Pallen (2016) points out there will need to be development of a skilled labor force not only in the food industry but in the affiliated government food regulatory agencies.

## Acknowledgments

A portion of this review originated from an undergraduate Honors College thesis written by coauthor Jessica Hacker in the Department of Industrial Engineering, College of Engineering, University of Arkansas, Fayetteville, AR.

## References

AAM, 2013. Human Microbiome FAQ. American Academy of Microbiology of American Society for Microbiology, Washington DC.

Adeleke, J.A., Moodley, D., Rens, G., Adewumi, A.O., 2017. Integrating statistical machine learning in a semantic sensor web for proactive monitoring and control. Sensors 17 (807), 1–23. http://dx.doi.org/10.3390/s17040807.

AI Horizon, 2015. AI Horizon: Introduction to Machine Learning. Available from: http://www.aihorizon.com/essays/generalai/supervised_unsupervised_machine_learning.htm.

Albert, I., 2014. QIIME vs Mothur: Why Use One Over the Other? Available from: https://www.biostars.org/p/94869/.

Amato, K.R., 2017. An introduction to microbiome analysis for human biology applications. American Journal of Human Biology 29 (e22931), 1–17. http://dx.doi.org/10.1002/ajhb.22931.

Baker, C.A., Rubinelli, P.M., Park, S.H., Carbonero, F., Ricke, S.C., 2016. Shiga toxin-producing *Escherichia coli* in food: incidence, ecology, and detection strategies in the United States. Food Control 59, 407–419.

Balch, W.E., Wolfe, R.S., 1976. A new approach to the cultivation of methanogenic bacteria: 2-mercaptoethanesulfonic acid (HS-CoM)-dependent growth of *Methanobacterium ruminantium* in a pressurized atmosphere. Advances in Applied Microbiology 32, 781–791.

Batz, M., Hoffmann, S., Morris Jr., J.G., 2014. Disease-outcome trees, EQ-5D scores, and estimated annual losses of quality-adjusted life years (QALYs) for 14 foodborne pathogens in the United States. Foodborne Pathogens and Disease 11, 395–402.

Bryant, M.P., 1959. Bacterial species of the rumen. Bacteriological Reviews 23, 125–153.

Bryant, M.P., 1972. Commentary on the Hungate technique for culture of anaerobic bacteria. American Journal of Clinical Nutrition 25, 1324–1328.

Callaway, T.R., Ricke, S.C. (Eds.), 2012. Direct Fed Microbials/Prebiotics for Animals: Science and Mechanisms of Action. Springer Science, New York, NY. 206 p.

Callaway, T.R., Dowd, S.E., Wolcott, R.D., Sun, Y., McReynolds, J.L., Edrington, T.S., Byrd, J.A., Anderson, R.C., Krueger, N., Nisbet, D.J., 2009. Evaluation of the bacterial diversity in cecal contents of laying hens fed various molting diets by using bacterial tag-encoded FLX amplicon pyrosequencing. Poultry Science 88, 298–302.

Callaway, T.R., Dowd, S.E., Edrington, T.S., Anderson, R.C., Krueger, N., Bauer, N., Kononoff, P.J., Nisbet, D.J., 2010. Evaluation of bacterial diversity in the rumen and feces of cattle fed different levels of dried distillers grains plus soluble using bacterial tag-encoded FLX amplicon pyrosequencing. Journal of Animal Science 88, 3977–3983.

Callaway, T.R., Edrington, T.S., Byrd, J.A., Nisbet, D.J., Ricke, S.C., 2016. Use of direct fed microbials in layer hen production - performance response and *Salmonella* control. In: Ricke, S.C., Gast, R.K. (Eds.), Producing Safe Eggs-the Microbial Ecology of *Salmonella*. Elsevier, Inc., San Diego, CA, pp. 301–322 (Chapter 15).

Calo, J.R., Crandall, P.G., O'Bryan, C.A., Ricke, S.C., 2015. Essential oils as antimicrobials in food systems – a review. Food Control 54, 111–119.

Caporaso, J.G., Kuczynski, J., Stombaugh, J., Bittinger, K., Bushman, F.D., Costella, E.K., Fier, N., Peña, A.G., Goodrich, J.K., Gordon, J.I., Huttley, G.A., Kelley, S.T., Knights, D., Koenig, J.E., Ley, R.E., Lazupone, C.A., McDonald, D., Muegge, B.D., Pirrung, M., Reeder, J., Sevinsky, J.R., Turnbaugh, P.J., Walters, W.A., Widmann, J., Yatsunenko, T., Zaneveld, J., Knight, R., 2010. QIIME allows analysis of high-throughput community sequencing data. Nature Methods 7, 335–336.

Chakravorty, S., Helb, D., Burday, M., Connell, N., Alland, D., 2007. A detailed analysis of 16S ribosomal RNA gene segments for the diagnosis of pathogenic bacteria. Journal of Microbiological Methods 69 (2), 330–339. http://dx.doi.org/10.1016/j.mimet.2007.02.005.

Comeau, A.M., Douglas, G.M., Langille, M.G.I., 2017. Microbiome helper: a custom and streamlined workflow for microbiome research. mSystems 2. http://dx.doi.org/10.1128/mSystems 00127–16. e00127–16.

Coupland, P., Chadra, T., Quail, M., Reek, W., Swerdlow, H., 2012. Direct sequencing of small genomes on the Pacific Biosciences RS without library preparation. Biotechniques 53, 365–372.

DeSantis, T.Z., Hugenholz, P., Larsen, N., Rojas, M., Brodie, E.L., Keller, K., Huber, T., Dalevi, D., Hu, P., Andersen, G.L., 2006. Greengenes, a chimera-checked 16S rRNA gene database and workbench compatible with ARB. Applied and Environmental Microbiology 72, 5069–5072.

Dowd, S.E., Callaway, T.R., Wolcott, R.D., Sun, Y., McKeehan, T., Hagevoort, G.R.G., Edrington, T.S., 2008. Evaluation of the bacterial diversity in the feces of cattle using 16S rDNA bacterial tag-encoded FLX amplicon pyrosequencing (bTEFAP). BMC Microbiology 8, 125.

Dunkley, K.D., Dunkley, C.S., Njongmeta, N.L., Callaway, T.R., Hume, M.E., Kubena, L.F., Nisbet, D.J., Ricke, S.C., 2007a. Comparison of in vitro fermentation and molecular microbial profiles of high-fiber feed substrates (HFFS) incubated with chicken cecal inocula. Poultry Science 86, 801–810.

Dunkley, K.D., McReynolds, J.L., Hume, M.E., Dunkley, C.S., Callaway, T.R., Kubena, L.F., Nisbet, D.J., Ricke, S.C., 2007b. Molting in *Salmonella* Enteritidis challenged laying hens fed alfalfa crumbles II. Fermentation and microbial ecology response. Poultry Science 86, 2101–2109.

Dunkley, K.D., Callaway, T.R., Chalova, V.I., McReynolds, J.L., Hume, M.E., Dunkley, C.S., Kubena, L.F., Nisbet, D.J., Ricke, S.C., 2009. Foodborne *Salmonella* ecology in the avian gastrointestinal tract. Anaerobe 15, 26–35.

Emond-Rheault, J.-G., Freschi, L., Kukavica-Ibrulj, I., Boyle, B., Dupont, M.-J., Colavecchio, A., Barrere, V., Cadieux, B., Arya, G., Bekal, S., Berry, C., Burnett, E., Cavestri, C., Chapin, T.K., Crouse, A., Daigle, F., Danyluk, M.D., Delaquis, P., Dewar, K., Doualla-Bell, F., Fliss, I., Fong, K., Fournier, E., Franz, E., Garduno, R., Gill, A., Gruenheid, S., Harris, L., Huang, C.B., Huang, H., Johnson, R., Joly, Y., Kerhoas, M., Kong, N., Lapointe, G., Larivière, L., Loignon, S., Malo, D., Moineau, S., Mottawea, W., Mukhopadhyay, K., Nadon, C., Nash, J., Feze, I.N., Ogunremi, D., Perets, A., Pilar, A.V., Reimer, A.R., Robertson, J., Rohde, J., Sanderson, K.E., Song, L., Stephan, R., Tamber, S., Thomassin, P., Tremblay, D., Usongo, V., Vincent, C., Wang, S., Weadge, J.T., Wiedmann, M., Wijnands, L., Wilson, E.D., Wittum, T., Yoshida, C., Youfsi, K., Zhu, L., Weimer, B.C., Goodridge, L., Levesque, R.C., 2017. A Syst-OMICS approach to ensuring food safety and reducing the economic burden of salmonellosis. Frontiers in Microbiology 8, 996. http://dx.doi.org/10.3389/fmicb.2017.00996.

Escarcha, J.F., Callaway, T.R., Byrd, J.A., Miller, D.N., Edrington, T.S., Anderson, R.C., Nisbet, D.J., 2012. Effects of dietary alfalfa inclusion on *Salmonella* Typhimurium populations in growing layer chicks. Foodborne Pathogens and Disease 9, 945–951.

Escobar-Zapeda, A., Vera-Ponce de León, A., Sanchez-Flores, A., 2015. The road to metagenomics: from microbiology to DNA sequencing technologies and bioinformatics. Frontiers in Genetics 6 (Article 348), 1–15.

Fichot, E.B., Norman, R.S., 2013. Microbial phylogenetic profiling with the Pacific Biosciences sequencing platform. Microbiome 1 (10). http://dx.doi.org/10.1186/2049-2618-1-10.

Foley, S., Nayak, R., Hanning, I.B., Johnson, T.L., Han, J., Ricke, S.C., 2011. Population dynamics of *Salmonella enterica* serotypes in commercial egg and poultry production. Applied and Environmental Microbiology 77, 4273–4279.

Foley, S.L., Johnson, T.J., Ricke, S.C., Nayak, R., Danzeisen, J., 2013. *Salmonella* pathogenicity and host adaptation in chicken-associated serovars. Microbiology and Molecular Biology Reviews 77, 582–607.

Fosso, B., Santamaria, M., Antonio, M.D., lovero, D., Corrado, G., Vizza, E., Passaro, N., Garbuglia, A.R., Capobianchi, M.R., Crescenzi, M., Valiente, G., Pesole, G., 2017. MetaShot: an accurate workflow for taxon classification of host-associated microbiome from shotgun metagenomic data. Bioinformatics 33 (11), 1730–1732. http://dx.doi.org/10.1093/bioinformatics/btx036.

Garrido-Cardenas, J.A., Garcia-Maroto, F., Alvarez-Bermejo, J.A., Manzano-Agugliaro, F., 2017. DNA sequencing sensors: an overview. Sensors 17, 588. http://dx.doi.org/10.3390/s17030588.

Geng, Z., Zhao, S., Tan, G., Han, Y., 2017. Early warning modeling and analysis based on analytic hierarchy process integrated extreme learning machine (AHP-ELM): application to food safety. Food Control 78, 33–42.

Gensollen, T., Blumberg, R.S., 2017. Correlation between early-life regulation of the immune system by microbiota and allergy development. Journal of Allergy and Clinical Immunology 139, 1084–1091.

Golob, J.L., Margolis, E., Hoffman, N.G., Fredricks, D.N., 2017. Evaluating the accuracy of amplicon-based microbiome computational pipelines on simulated human gut microbial communities. BMC Bioinformatics 18, 283. http://dx.doi.org/10.1186/s12859-017-1690-0.

Gonzalez, A., Knight, R., 2012. Advancing analytical algorithms and pipelines for billions of microbial sequences. Current Opinion in Biotechnology 23 (1), 64–71.

Hamdoun, Z., Ehsan, H., 2017. Aftermath of the human genome project: an era of struggle and discovery. Turkish Journal of Biology 41, 403–418.

Handley, J.A., Hanning, I., Ricke, S.C., Johnson, M.G., Jones, F.T., Apple, R.O., 2010. Temperature and bacterial profile of post chill poultry carcasses stored in processing combo held at room temperature. Journal of Food Science 75, M515–M520.

Handley, J.A., Park, S.H., Shi, Z., Dawoud, T.M., Kwon, Y.M., Ricke, S.C., 2015. *Salmonella* and the potential role for microbial process indicators on chicken carcasses. In: Ricke, S.C., Donaldson, J.R., Phillips, C.A. (Eds.), Food Safety: Emerging Issues, Technologies and Systems. Elsevier, Oxford, UK, pp. 81–104 (Chapter 6).

Hanning, I., Ricke, S.C., 2011. Prescreening methods of microbial populations for the assessment of sequencing potential. In: Kwon, Y.M., Ricke, S.C. (Eds.), Methods in Molecular Microbiology 733-High-Throughput Next Generation Sequencing: Methods and Applications. Springer Protocols, Humana Press, New York, NY, pp. 159–170 (Chapter 11).

Hanning, I., Lingbeck, J., Ricke, S.C., 2015. Cardiovascular health and disease prevention: association with foodborne pathogens and potential benefits of probiotics. In: Watson, R.R., Preedy, V.R. (Eds.), Bioactive Foods in Promoting Health: Probiotics-Prebiotics, second ed. Elsevier, San Diego, CA, pp. 793–806 (Chapter 60).

Hoffmann, M., Zhao, S., Pettengill, J., Luo, Y., Monday, S.R., Abbott, J., Ayers, S.L., Cinar, H.N., Muruvanda, T., Li, C., Allard, M.W., Whichard, J., Meng, J., Brown, E.W., McDermott, P.F., 2014. Comparative genomic analysis and virulence differences in closely related *Salmonella enterica* serotype Heidelberg isolates from humans, retail meats, and animals. Genome Biology and Evolution 6, 1046–1068.

Holdeman, L.V., Moore, W.E.C., 1972. Roll-tube techniques for anaerobic bacteria. American Journal of Clinical Nutrition 25, 1314–1317.

Holdeman, L.V., Cato, E.P., Moore, W.E.C. (Eds.), 1977. Anaerobe Laboratory Manual, fourth ed. Virginia Polytechnic Institute and State University, Blacksburg, VA.

Horrocks, S.M., Anderson, R.C., Nisbet, D.J., Ricke, S.C., 2009. Incidence and ecology of *Campylobacter* in animals. Anaerobe 15, 18–25.

Hume, M.E., Kubena, L.F., Edrington, T.S., Donskey, C.J., Moore, R.W., Ricke, S.C., Nisbet, D.J., 2003. Poultry digestive microflora diversity as indicated by denaturing gradient gel electrophoresis. Poultry Sciecne 82, 1100–1107.

Hungate, R.E., 1950. The anaerobic mesophilic cellulolytic bacteria. Bacteriological Reviews 14, 1–49.

Hungate, R.E., 1966. The Rumen and its Microbes. Academic Press, New York, NY.

Hungate, R.E., 1969. A roll tube method for cultivation of strict anaerobes. In: Norris, J.R., Ribbons, D.W. (Eds.), Methods in Microbiology. vol. 3B. Academic Press, New York, NY.

Huse, S.M., Mark Welch, D.B., Voorhis, A., Shipunova, A., Morrison, H.G., Eren, A.M., Sogin, M.I., 2014. VAMPS: a website for visualization and analysis of microbial population structures. BMC Bioinformatics 15 (1), 41.

Indikova, I., Humphrey, T.J., Hilbert, F., 2015. Survival with a helping hand: *Campylobacter* and microbiota. Frontiers in Microbiology 6, 1266.

Jovel, J., Patterson, J., Wang, W., Hotte, N., O'Keefe, S., Mitchel, T., Perry, T., Kao, D., Mason, A.L., Madsen, K.L., Wong, G.K.-S., 2016. Characterization of the gut microbiome using 16S or shotgun metagenomics. Frontiers in Microbiology 7 (Article 459), 17. http://dx.doi.org/10.3389/fmicb.00459.2016.

Kaldhone, P.R., Foley, S.L., Ricke, S.C., 2016. *Salmonella* Heidelberg in layer hens and egg production: incidence and potential issues. In: Ricke, S.C., Gast, R.K. (Eds.), Producing Safe Eggs-the Microbial Ecology of *Salmonella*. Elsevier, Inc., San Diego, CA, pp. 235–256 (Chapter 12).

Khatiwara, A., Jiang, T., Sung, S.-S., Dawoud, T., Kim, J.N., Bhattacharya, D., Kim, H.-B., Ricke, S.C., Kwon, Y.M., 2012. Genome scanning for conditionally essential genes in *Salmonella*. Applied and Environmental Microbiology 78, 3098–3107.

Kim, S.A., Park, Si H., In Lee, S., Owens-Hanning, C., Ricke, S.C., 2017. Assessment of chicken carcass microbiome responses during processing in the presence of commercial antimicrobials using a next generation sequencing approach. Scientific Reports 7, 43354. http://dx.doi.org/10.1038/srep43354.

Koo, O.K., Mertz, A.W., Akins, E.L., Sirsat, S.A., Neal, J.A., Morawicki, R., Crandall, P.G., Ricke, S.C., 2012. Analysis of microbial diversity on deli slicers using polymerase chain reaction and denaturing gradient gel electrophoresis technologies. Letters in Applied Microbiology 56, 111–119.

Koo, O.-K., Kim, H.-J., Baker, C.A., Park, S.H., Ricke, S.C., 2016a. Comparison of the microbial diversity in Korean ground beef products with PCR denaturing gradient gel electrophoresis (DGGE) and 454 pyrosequencing analyses. Food Biotechnology 30, 63–77.

Koo, O.-K., Baker, C.A., Kim, H.J., Park, S.H., Ricke, S.C., 2016b. Metagenomic assessment of the microbial diversity in ground pork products from South Korean Markets. Journal of Environmental Science and Health B 51, 622–627.

Kozich, J.J., Westcott, S.L., Baxter, N.T., Highlander, S.K., Schloss, P.D., 2013. Development of a dual-index sequencing strategy and curation pipeline for analyzing amplicon sequence data on the MiSeq Illumina sequencing platform. Applied and Environmental Microbiology 79, 5112–5120.

Kuchta, T., Knutsson, R., Fiore, A., Kudirkiene, E., Höhl, A., Tomic, D.H., Gotcheva, V., Pöpping, B., Scaramagli, S., To Kim, A., Wagner, M., De Medici, D., 2014. A decade with nucleic acid-based microbiological methods in safety control of foods. Letters in Applied Microbiology 59, 263–271.

Kuczynski, J., Lauber, C., Walters, W.A., Parfrey, L.W., Clemente, J.C., Gevers, D., Knight, R., 2012. Experimental and analytical tools for studying the human microbiome. Nature Reviews in Genetics 13, 47–58.

Kwon, Y.M., Bhattacharya, D., Ricke, S.C., 2010. Genomic approaches to bacterial pathogens using transposon mutagenesis: food - safety applications. In: Ricke, S.C., Jones, F.T. (Eds.), Perspectives on Food Safety Issues of Food Animal Derived Foods. University of Arkansas Press, Fayetteville, AR, pp. 165–176.

Kwon, Y.M., Ricke, S.C., Mandal, R.K., 2016. Transposon sequencing: methods and expanding applications. Applied Microbiology and Biotechnology 100, 31–43.

Lee, S.I., Park, S.H., Ricke, S.C., 2016. Assessment of cecal microbiota, integrons, fermentation, and *Salmonella* frequency in conventionally raised broilers in response to a commercial prebiotic. Poultry Science 95, 144–153.

Levett, P.N. (Ed.), 1991. Anaerobic Microbiology: A Practical Approach. Oxford University Press, New York, NY.

Luscombe, N.M., Greenbaum, D., Gerstein, M., 2001. What is bioinformatics? A proposed definition and overview of the field. Methods of Information in Medicine 40, 346–358.

Maciorowski, K.G., Pillai, S.D., Jones, F.T., Ricke, S.C., 2005. Polymerase chain reaction detection of foodborne *Salmonella* spp. in animal feeds. Critical Reviews in Microbiology 31, 45–53.

Mackie, R.I., White, B.A. (Eds.), 1997a. Gastrointestinal Ecosystems and Fermentations, vol. 1. Chapman and Hall, New York, NY.

Mackie, R.I., White, B.A. (Eds.), 1997b. Gastrointestinal Microbes and Host Interactions, vol. 2. Chapman and Hall, New York, NY.

Macy, J.M., Snellen, J.E., Hungate, R.E., 1972. Use of syringe methods for anaerobiosis. American Journal of Clinical Nutrition 25, 1318–1323.

Malisiewicz, T., 2015. Deep Learning vs Machine Learning vs Pattern Recognition. Available from: http://www.computervisionblog.com/2015/03/deep-learning-vs-machine-learning-vs.html.

Mertz, A.W., Koo, O.K., O'Bryan, C.A., Morawicki, R., Sirsat, S.A., Neal, J.A., Crandall, P.G., Ricke, S.C., 2014. Microbial ecology of meat slicers as determined by denaturing gradient gel electrophoresis. Food Control 42, 242–247.

Miller, T.L., Wolin, M.J., 1974. A serum bottle modification of the Hungate technique for cultivating obligate anaerobes. Advances in Applied Microbiology 27, 985–987.

Morozova, O., Hirst, M., Marra, M.A., 2009. Applications of new sequencing technologies for transcriptome analysis. Annual Review of Genomics and Human Genetics 10, 135–151.

Nilakanta, H., Drews, K.L., Firrell, S., Foulkes, M.A., Jablonski, K.A., 2014. A review of software for analyzing molecular sequences. BMC Research Notes 7 (830), 1–9.

Oakley, B.B., Morales, C.A., Line, J., Berrang, M.E., Meinersmann, R.J., Tillman, G.E., Wise, M.G., Siragusa, G.R., Hiett, K.L., Seal, B.S., 2013. The poultry-associated microbiome: network analysis and farm-to-fork characterizations. PLoS One 8 (2), e57190. http://dx.doi.org/10.1371/journal.pone.0057190.

O'Bryan, C.A., Pendleton, S.J., Crandall, P.G., Ricke, S.C., 2015. Potential of plant essential oils and their components in animal agriculture – *in vitro* studies on antibacterial mode of action. Frontiers in Veterinary Science 2 (Article 35), 1–8. http://dx.doi.org/10.3389/fvets.2015.00035.

Pace, N.R., Stahl, D.A., Lane, D.J., Olsen, G.J., 1985. Analyzing natural microbial populations by rRNA sequences. ASM News 51, 4–12.

Painter, J.A., Hoekstra, R.M., Ayers, T., Tauxe, R.V., Braden, C.R., Angulo, F.J., Griffin, P.M., 2013. Attribution of foodborne illnesses, hospitalizations, and deaths to food commodities by using outbreak data, United States, 1998–2008. Emerging Infectious Diseases 19, 407–415.

Pallen, M.J., 2016. Microbial bioinformatics 2020. Microbial Biotechnology 9 (5), 681–686. http://dx.doi.org/10.1111/1751-7915.12389.

Park, S.H., Hanning, I., Perrota, A., Bench, B.J., Alm, E., Ricke, S.C., 2013. Modifying the gastrointestinal ecology in alternatively raised poultry and the potential for molecular and metabolomic assessment. Poultry Science 92, 546–561.

Park, S.H., Aydin, M., Khatiwara, A., Dolan, M.C., Gilmore, D.F., Bouldin, J.L., Ahn, S., Ricke, S.C., 2014a. Current and emerging technologies for rapid detection and characterization of *Salmonella* in poultry and poultry products. Food Microbiology 38, 250–262.

Park, S.H., Gibson, K., Almeida, G., Ricke, S.C., 2014b. Assessment of gastrointestinal microflora in pasture raised chickens fed two commercial prebiotics. Journal of Probiotics and Health 2, 122.

Park, S.H., Dowd, S.E., McReynolds, J.L., Byrd, J.A., Nisbet, D.J., Ricke, S.C., 2015. Evaluation of feed grade sodium bisulfate impact on gastrointestinal tract microbiota ecology in broilers via a pyrosequencing platform. Poultry Science 94 (12), 3040–3047.

Park, S.H., Lee, S.I., Ricke, S.C., 2016. Microbial populations in naked neck chicken ceca raised on pasture flock fed with commercial yeast cell wall prebiotics via an Illumina MiSeq platform. PLoS One 11 (3), e0151944. http://dx.doi.org/10.1371/journal.pone.0151944.

Park, S.H., Kim, S.A., Lee, S.I., Rubinelli, P.M., Roto, S.M., Pavlidis, H.O., McIntyre, D.R., Ricke, S.C., 2017a. Original XPC™ effect on *Salmonella* Typhimurium and cecal microbiota from three different ages of birds when incubated in an anaerobic *in vitro* culture system. Frontiers in Microbiology 8 (Article 1070). http://dx.doi.org/10.3389/fmoicb.201701070.

Park, S.H., Perrotta, A., Hanning, I., Diaz-Sanchez, S., Pendleton, S., Alm, E., Ricke, S.C., 2017b. The chicken gut microbiome changes in response to prebiotics and plum fibers. Poultry Science 96, 1820–1830.

Park, S.H., Kim, S.A., Rubinelli, P.M., Roto, S.M., Ricke, S.C., 2017c. Microbial compositional changes in broiler chicken cecal contents from birds challenged with different *Salmonella* vaccine candidate strains. Vaccine 35, 3204–3208.

Petriz, B.A., Franco, O.L., 2017. Metaproteomics as a complementary approach to gut microbiota in health and disease. Frontiers in Chemistry 5 (Article 4), 1–7. http://dx.doi.org/10.3389/chem.2017.00004.

Petrosino, J.F., Highlander, S., Luna, R.A., Gibbs, R.A., Versalovic, J., 2009. Metagenomic pyrosequencing and microbial identification. Clinical Chemistry 55, 856–866.

Plummer, E., Twin, J., Bulach, D.M., Garland, S.M., Tabrizi, S.N., 2015. A comparison of three bioinformatics pipelines for the analysis of preterm gut microbiota using 16S rRNA gene sequencing data. Journal of Proteomics & Bioinformatics 8, 283–291. http://dx.doi.org/10.4172/jpb.1000381.

QIIME development team, 2015. What Is QIIME? Available from: http://QIIME.org/.

Quail, M.A., Smith, M., Coupland, P., Otto, T.D., Harris, S.R., Connor, T.R., Bertoni, A., Swerdlow, H.P., Gu, Y., 2012. A tale of three next generation sequencing platforms: comparison of Ion torrent, Pacific Biosciences and Illumina MiSeq sequencers. BMC Genomics 12, 341.

Ricke, S.C., 2003. Perspectives on the use of organic acids and short chain fatty acids as antimicrobials. Poultry Science 82, 632–639.

Ricke, S.C., 2015a. Application of anaerobic microbiology laboratory training and writing comprehension to food safety education. In: Ricke, S.C., Donaldson, J.R., Phillips, C.A. (Eds.), Food Safety: Emerging Issues, Technologies and Systems. Elsevier, Oxford, UK, pp. 395–419 (Chapter 19).

Ricke, S.C., 2015b. Potential of fructooligosaccharide prebiotics in alternative and nonconventional poultry production systems. Poultry Science 94, 1411–1418.

Ricke, S.C., 2016. Gastrointestinal ecology of *Salmonella* Enteritidis in laying hens and intervention by prebiotic and non-digestible carbohydrate dietary supplementation. In: Ricke, S.C., Gast, R.K. (Eds.), Producing Safe Eggs-the Microbial Ecology of *Salmonella*. Elsevier, Inc., San Diego, CA, pp. 323–345 (Chapter 16).

Ricke, S.C., Pillai, S.D., 1999. Conventional and molecular methods for understanding probiotic bacteria functionality in gastrointestinal tracts. Critical Reviews in Microbiology 25, 19–38.

Ricke, S.C., Saengkerdsub, S., 2015. *Bacillus* probiotics and biologicals for improving animal and human health: current applications and future prospects. In: Ravishankar Rai, V., Jamuna Bai, A. (Eds.), Beneficial Microbes in Fermented and Functional Foods. CRC Press/Taylor & Francis Group, Boca Raton, FL, pp. 341–360 (Chapter 19).

Ricke, S.C., Kundinger, M.M., Miller, D.R., Keeton, J.T., 2005. Alternatives to antibiotics: chemical and physical antimicrobial interventions and foodborne pathogen response. Poultry Science 84, 667–675.

Ricke, S.C., Hererra, P., Biswas, D., 2012. Bacteriophages for potential food safety applications in organic meat production. In: Ricke, S.C., Van Loo, E.J., Johnson, M.G., O'Bryan, C.A. (Eds.), Organic Meat Production and Processing. Wiley Scientific/IFT, New York, NY, pp. 407–424 (Chapter 23).

Ricke, S.C., Khatiwara, A., Kwon, Y.M., 2013. Application of microarray analysis of foodborne *Salmonella* in poultry production: a review. Poultry Science 92, 2243–2250.

Ricke, S.C., Dawoud, T.M., Kwon, Y.M., 2015. Application of molecular methods for traceability of foodborne pathogens in food safety systems. In: Ricke, S.C., Donaldson, J.R., Phillips, C.A. (Eds.), Food Safety: Emerging Issues, Technologies and Systems. Elsevier, Oxford, UK, pp. 37–63 (Chapter 4).

Ricke, S.C., Perumalla, A.V.S., Hettiarachchy, N.S., 2017. Alternatives to antibiotics in preventing zoonoses and other pathogens in poultry: prebiotics and related compounds. In: Ricke, S.C. (Ed.), Achieving Sustainable Production of Poultry Meat. Safety, Quality and Sustainability, vol. 1. Burleigh Dodd Publishing, Cambridge, UK, pp. 87–108 (Chapter 5).

Robinson, C.K., Brotman, R.M., Ravel, J., 2016. Intricacies of assessing the human microbiome in epidemiological studies. Annals of Epidemiology 26, 311–321.

Rodrigues-R, L.M., Konstaninidis, K.T., 2014. Bypassing cultivation to identify bacterial species. Microbe 9, 111–118.

Rothrock Jr., M.J., Locatelli, A., Glenn, T.C., Thomas, J.C., Caudill, A.C., Kiepper, B.H., Hiett, K.L., 2016. Assessing the microbiomes of scalder and chiller tank waters throughout a typical commercial poultry processing day. Poultry Science 95, 2372–2382.

Roto, S.M., Rubinelli, P.M., Ricke, S.C., 2015. An introduction to the avian gut microbiota and the effects of yeast-based prebiotic compounds as potential feed additives. Frontiers in Veterinary Science 2 (Article 28), 1–18. http://dx.doi.org/10.3389/fvets.2015.00028.

Roumpeka, D.D., Wallace, R.J., Escalettes, F., Fotheringham, I., Watson, M., 2017. A review of bioinformatics tools for bio-prospecting from metagenomic sequence data. Frontiers in Genetics 8 (Article 23), 1–10. http://dx.doi.org/10.3389/fgene.2017.00023.

Rowe, W., Baker, K.S., Verner-Jeffreys, D., Baker-Austin, C., Ryan, J.J., Maskell, D., Pearce, G., 2015. Search engine for antimicrobial resistance: a cloud compatible pipeline and web interface for rapidly detecting antimicrobial resistance genes directly from sequence data. PLoS One 10 (7), e0133492. http://dx.doi.org/10.1371/journal.pone.013492. 1–12.

Saengkerdsub, S., Ricke, S.C., 2014. Ecology and characteristics of methanogenic archaea in animals and humans. Critical Reviews in Microbiology 40, 97–116.

Scallan, E., Hoekstra, R.M., Angulo, F.J., Tauxe, R.V., Widdowson, M.-A., Roy, S.L., Jones, J.L., Griffin, P.M., 2011. Foodborne illness acquired in the United States—major pathogens. Emerging Infectious Diseases 17, 7–15.

Scharff, R.L., 2012. Economic burden from health losses due to foodborne illness in the United States. Journal of Food Protection 75, 123–131.

Schloss, P.D., 2010. The effects of alignment quality, distance calculation method, sequence filtering and region on the analysis of 16S rRNA gene-based studies. PLoS Computational Biology 6, e1000844.

Schloss, P., 2016. Mothur and QIIME. Available from: http://blog.mothur.org/2016/01/12/mothur-and-QIIME/.

Schloss, P.D., Westcott, S.L., Ryabin, T., Hall, J.R., Hartmann, M., Hollister, E.B., Lesniewski, R.A., Oakley, B.D., Parks, D.H., Robinson, C.J., Sahl, J.W., Stres, B., Thallinger, G.G., Van Horn, D.J., Weber, C.F., 2009. Introducing mothur: open-source, platform-independent, community-supported software for describing and comparing microbial communities. Applied and Environmental Microbiology 75, 7537–7541.

Siegold, L., Touzet, H., Lemoine, Y., Hot, D., Audebert, C., Caboche, S., 2017. Assessment of common and emerging bioinformatics pipelines for targeted metagenomics. PLoS One 12 (1), 1–26. http://dx.doi.org/10.1371/journal.pone.0169563.

Siragusa, G.R., Ricke, S.C., 2012. Probiotics as pathogen control agents for organic meat production. In: Ricke, S.C., Van Loo, E.J., Johnson, M.G., O'Bryan, C.A. (Eds.), Organic Meat Production and Processing. Wiley Scientific/IFT, New York, NY, pp. 331–349 (Chapter 20).

Sirsat, S.A., Muthaiyan, A., Ricke, S.C., 2009. Antimicrobials for pathogen reduction in organic and natural poultry production. Journal of Applied Poultry Research 18, 379–388.

Sirsat, S.A., Muthaiyan, A., Dowd, S.E., Kwon, Y.M., Ricke, S.C., 2010. The potential for application of foodborne *Salmonella* gene expression profiling assays in postharvest poultry processing. In: Ricke, S.C., Jones, F.T. (Eds.), Perspectives on Food Safety Issues of Food Animal Derived Foods. University of Arkansas Press, Fayetteville, AR, pp. 195–222.

Smith, D.R., 2015. Broadening the definition of a bioinformatician. Frontiers in Genetics 6, 258 DOI:10.3389fgene2015.00258.

Sofos, J.N., Flick, G., Nychas, G.-J., O'Bryan, C.A., Ricke, S.C., Crandall, P.G., 2013. Meat, poultry, and seafood. In: Doyle, M.P., Buchanan, R.L. (Eds.), Food Microbiology-Fundamentals and Frontiers, fourth ed. American Society for Microbiology, Washington, D.C., pp. 111–167 (Chapter 6).

Strawn, L.K., Brown, E.W., David, J.R.D., Den Bakker, H.C., Vangay, P., Yiannas, F., Wiedemann, M., 2015. Big data in food. Food Technology 69 (2), 42–49.

Sutton, R.S., Barto, A.G., 2005. Reinforcement Learning: An Introduction. Available from: https://webdocs.cs.ualberta.ca/~sutton/book/ebook/node7.html.
Vanderplas, J., Grisel, O., Grobler, J., Varoquaux, G., 2016. Machine Learning 101: General Concepts. Available from: http://www.astroml.org/sklearn_tutorial/general_concepts.html.
Vincent, A.T., Charette, S., 2015. Who qualifies to be a bioinformatician? Frontiers in Genetics 6, 164. http://dx.doi.org/10.3389/fgene.2015.00164.
Vongkamjn, K., Weidmann, M., 2015. Starting from the bench – prevention and control of foodborne and zoonotic diseases. Preventive Veterinary Medicine 118 (203), 189–195.
Whitford, W., 2017. The era of digital biomanufacturing. BioProcess International 15, 12–18.
Woese, C.R., Fox, G.E., 1977. Phylogenetic structure of the prokaryotic domain: the primary kingdoms. Proceedings of the National Academy of Sciences of the United States of America 74, 5088–5090.
Yang, Y., Ricke, S.C., Tellez, G., Kwon, Y.M., 2017. Quantitative tracking of *Salmonella* Enteritidis transmission routes using barcode-tagged isogenic strains in chickens – proof of concept study. Frontiers in Veterinary Science 4, 15. http://dx.doi.org/10.3389/fvets.2017.0001.
Yip, K.Y., Cheng, C., Gerstein, M., 2013. Machine learning and genome annotation: a match meant to be? Genome Biology 14, 205.
Yoon, S.S., Kim, E.-K., Lee, W.-J., 2015. Functional genomic and metagenomic approaches to understanding gut microbiota-animal mutualism. Current Opinion in Microbiology 24, 38–46.
Yu, Z., Morrison, M., 2004. Comparisons of different hypervariable regions of *rrs* genes for use in fingerprinting of microbial communities by PCR-denaturing gradient gel electrophoresis. Applied and Environmental Microbiology 70, 4800–4806.

CHAPTER 20

# Student Cross-Training Opportunities for Combining Food, Transportation, and Critical Infrastructure Cybersecurity Into an Academic Food Systems Education Program

**Dale R. Thompson, Chase E. Rainwater, Jia Di, Steven C. Ricke**
*University of Arkansas, Fayetteville, AR, United States*

## Introduction

### Security Challenges in the United States

The security challenges facing the United States are increasing and evolving in complexity. Digital technology is now impacting almost all sectors of the US economy. For example, Whitford (2017) has pointed out that the field of digital biomanufacturing is now connecting product and process designers in manufacturing engineering and using the resulting information to develop strategies for greater efficiencies, productivities, and growth in this segment of the industry. This type of scenario is being played out in a wide range of commercial enterprises. A connected world means that data and cybersecurity reach almost every industry. Therefore, industries that drive the US economy leave the nation vulnerable to new threats. Among these major industries are the fields of transportation ($1.5 trillion contribution to GDP and 4 million jobs (Wang, 2015)), infrastructure, and food (contribution of over 17 million jobs and over $800 billion to US GDP (Lewrene Glaser, 2016)). From a governmental perspective, each of these industries is supported by its agency counterpart (e.g., Department of Transportation, Food and Drug Administration, and US Department of Agriculture). Each of these agencies faces and must rise to the challenges of the citizens it serves and protects. Among the largest of the challenges faced by government organizations is that data/cyber influences are prevalent and daunting due to the numerous motives for bad actors to profit/gain from the information gathered. Unfortunately, these new challenges are met with an increasingly insufficient security workforce. In traditional cybersecurity alone, 209,000 jobs

Food and Feed Safety Systems and Analysis. http://dx.doi.org/10.1016/B978-0-12-811835-1.00020-8

went unfilled in the United States (Intel, 2016) and 1 million jobs went unfilled worldwide (Cisco, 2015). For analytics positions focused on data and information, it is predicted that by 2018 an additional 150,000 jobs will go unfilled (http://blog.trendmicro.com, 2015). Existing professionals cite lack of education as a key reason that this problem exists. As recently as August 2016, a survey showed that only 23% of existing professionals believe that existing security education programs are effective in preparing future security workers (http://www.eab.com, 2016). However, it is also interesting to note that as many as 75% of current industry professionals would prefer students who received a degree in a nonsecurity field that was simply related (http://blog.trendmicro.com, 2015). It is becoming more apparent that the desire for education beyond traditional security calls for a new, interdisciplinary, approach to next-generation security education. In this review, the needs for food cybersecurity training and education will be identified and the means for constructing academic programs to meet those needs will be discussed.

## Food Security—General Concepts

Although food and water are among the most vital resources required as well as influenced by human activities and changing climate, their security problems have both similarities and differences. Food security historically focused on the threat of accidental contamination of food (also referred to as food safety) and therefore was considered the primary objective for food-related research and training and the corresponding policy of government regulatory agencies (Osborne et al., 2006; Rushing and Walsh, 2006; Ricke, 2010). Since the early 2000s, the specter of intentional food contamination has become a more legitimate concern (Norton, 2003; Crutchley et al., 2007). In some aspects, food security still has most of the same basic research needs as those from conventional sources of contamination, including rapid detection of pathogens, access to extensive databases, and the corresponding interpretation of complex data sets (Hanning et al., 2012; Ricke, 2010; Ricke et al., 2015). However, in other respects such as anticipated human behavior and origin of contamination, the types of analysis tools required are quite different. Not only do the potential sites where intentional contamination of biological contaminant increase, but also the logical critical control points where cross-contamination has been identified as frequently occurring still remain. A critical component is how the results will be shared and/or managed among government and private commercial entities (Parker, 2006; Monaghan, 2006). For example, although foodborne pathogen research has remained fairly accessible, tighter restrictions and more accountability of cataloging and sharing of strains among institutions have now become increasingly difficult to manage particularly as faculty undergo generational hiring shifts and academic program emphasis changes (Enquist, 2015; Ricke, 2015).

Probably the greatest obstacle in dealing with food security and the corresponding data management systems is the fact that food security threats can occur up and down the food supply chain: not only at the primary sites of harvest, processing, and transportation, but also, as sources of food supply become more diverse, at the collection points where rural, industrial, and urban sources interface with each other (Parker, 2006; Premier and Ledger, 2006; Hanning et al., 2012; Ricke, 2010; Ricke et al., 2015). The reason that food security suddenly has become more of a critical issue is that the entire food industry is evolving rapidly to meet economic and political requirements associated with consumer perceptions and market demands (Auler et al., 2017). This change impacts the retail sector as well as the supply and sources of food commodities. Compounding the increasing complexity in the food supply chain are the global issues of sustainable production and consumption of food. The global population is expected to approach 9.7 billion by 2050 and 11.2 billion by 2100, which will require more than doubling of the quantity of food produced (World Population Prospects, 2015). Additionally, the United Nations Food and Agriculture Organization reported that roughly one-third of the food produced is lost or wasted globally (Gustavsson et al., 2011), which represents a significant opportunity for increasing food availability. Therefore, developing technologies for monitoring, documenting, and mitigating these inefficiencies in the food supply will become increasingly important.

Consequently, the complexity/diversity of food supply systems is increasing almost exponentially for a variety of reasons. Part of it is simple economics. As food companies continue to grow via mergers, their line of food products almost invariably becomes more diverse at the retail level, which is often the economic driver in the first place. As personnel, food product lines, and retail markets are merged/assimilated, organizational structure and control systems are severely challenged. This adds new layers of production, processing, and transportation complexities that preclude the conventional linear cause-and-effect scenarios currently used to assess food security issues. Likewise, when security breaches do occur, because of the sheer sizes and volume sales of these merged companies, the negative impact is much greater and longer lasting economically. This scenario becomes even more complicated with increased international trade and importation of certain food products, for example, vegetables (Monaghan, 2006; Premier and Ledger, 2006; Rushing and Walsh, 2006). Thus, traceability and country of origin become an important part of food security and data management (Monaghan, 2006; Crandall et al., 2013; Ricke et al., 2015). Another economics-based reason is caused by political forces, such as the rising demand for organic and "natural" sources of food products (Ricke et al., 2012). This is fueled mostly by perception that these food sources represent a variety of perceived better quality features, which has resulted in the growth of local markets in the food chain. In turn this has impacted large food companies to appeal to the changing marketing demographics by incorporating more "local" sources of food products into their distribution systems, hence adding yet another complication in their supply chain management systems. Likewise, the animal welfare movement has grown extensively in political clout and this has now altered the way food animals are produced, for example, cage-free layer chickens for egg production have now become the preferred means of production (Holt et al., 2011; Gast, 2016; Ricke, 2017). Consequently, documentation, record keeping, and other forms of tangible data gathering will become a part of the landscape for the food and meat industries. Given the human/political activist nature of this particular, and somewhat inflammatory/contentious issue, once again security documentation will be of paramount importance.

## An Example of Data Generation and Food Security—The Integrated Poultry Industry

The poultry industry has evolved over the past century from a small farmer-oriented production local retail system to a highly sophisticated vertically integrated supply chain set of operations (Owens et al., 2010). This includes the commercial hatcheries, feed mills, breeder farms, live bird grower facilities, transportation of finished birds to processing plants, processing and conversion of bird carcasses to poultry meat products, and shipping of frozen, cooked, and fresh meat products and ingredients to grocery store and restaurant retail establishments (Owens et al., 2010; Kabinlapat and Sutthachai, 2017). In addition to the primary functions of poultry processing, affiliated commercial operations such as the generation of ingredients from by-products for pet food represents further integration (Morawicki, 2010; Meeker and Meisinger. 2015). Not surprisingly, as poultry industry operations become more integrated, the demand for more complex data increases as well. This includes not only the classical economic data concerning costs and market revenue but other components as well as the myriad of environmental inputs needed to generate life cycle assessment (LCA) for reducing the environmental footprint of the industry. The application of LCA has mostly been considered for the live production side of the poultry industry, but opportunities for generating data on energy and water use along with other environmental inputs on processing are also a potential focal point (Pelletier, 2008; Boggia et al., 2010; Luján-Rhenals et al., 2017a,b). The development of further and more frequent data gathering capabilities will almost certainly add to the complexities of these data sets and require more detailed analyses and interpretation. Depending on the purpose and source of the data, its corresponding cyberprotection will also be critical and require personnel trained not only to conduct the analyses but also to protect its confidentiality and integrity.

The potential for generation of large data sets in the poultry industry is emerging from other production aspects as well. This is particularly true for food safety and microbiological quality control. Food safety requirements and regulations have continued to evolve in response to the ever-changing epidemiological impact of foodborne disease organisms such as *Salmonella*, *Listeria*, pathogenic *Escherichia coli*, and *Campylobacter* along with the corresponding public health demands (Scallan et al., 2011; Scharff, 2012). The industry continues to improve on control measures and monitoring capabilities. This is in part due to the improvements in bacterial pathogen detection technologies both as a function of increased sensitivity and more high-throughput capacity, enabling more samples to be analyzed faster (Kuchta et al., 2014; Park et al., 2014; Baker et al., 2016a,b). A more recent development that has begun to impact the food and poultry industry and food safety has been the emergence of molecular next-generation sequencing technologies (Ricke et al., 2015). Although traditionally sequencing had been the domain of the academic research community where much of the advancements in basic laboratory approaches were developed, it has now evolved into instrumentation that is more universally available for nonacademic research and commercial applications (Metzker, 2010; Glenn, 2011; Kwon and Ricke, 2011; Quail et al., 2012; Ricke et al., 2015). This has made whole genome sequencing of foodborne pathogens possible and subsequently resulted in complete sequences being established for a multitude of foodborne pathogen species and strains within a species (Betts and Jones, 2017). As the sequencing resolution improves, the ability to pinpoint specific strains related to a particular foodborne disease outbreak has become possible allowing for trace-back to source or origin of the organism and the route it travels to ultimately cause disease (Betts and Jones, 2017). However, as this database grows, issues of ownership and cyberprotection and the corresponding consequences have emerged as ongoing concerns by the various parties involved (Betts and Jones, 2017). Consequently, potential commercial and regulatory implementation will probably require certain levels of safeguards, which in turn will mean more advanced personnel training not just in the raw data laboratory generation and analyses aspect of such data but developing appropriate security mechanisms and protocols to protect the resulting data and interpretation.

Other types of sequencing also are beginning to impact the poultry and food industry as well (Handley et al., 2015; Haynes and Hudson, 2017). For example, the poultry industry routinely monitors bacterial load on poultry carcasses during various stages of poultry processing to assess effectiveness of various antimicrobial treatments being applied, which is referred to as biomapping and essentially involves microbial plating and enumeration of a variety of microorganisms that have been classified as indicator organisms (Handley et al., 2015). This is important because pathogen loads on carcasses are generally low and infrequent and therefore do not make for good process controls for routine data collection. With the emergence of next-generation sequencing technologies, the opportunity has now risen to move away from selective plating of a few microorganisms to a more detailed analysis of the entire microbial community of a poultry carcass (Handley et al., 2015; Ricke et al., 2015). This involves sequencing based on 16S rRNA genes and analysis of the raw sequence data to identify the microbial populations taxonomically using bioinformatics programs. This approach not only represents a dynamic shift in the poultry industry's approach to monitoring process controls but also presents a data collection and analysis challenge. The few studies published thus far indicate that sequencing data sets will offer a "microbiome" map of the bacterial communities occurring on carcasses as they proceed through the processing steps to become poultry meat (Oakley et al., 2013; Rothrock et al., 2016; Kim et al., 2017). Once standardized, such data will allow the industry to greatly enhance their capabilities to track and pinpoint sources of microbial contamination throughout the production chain. Based on the work done thus far, the future direction of the poultry industry will likely begin to rely more on the generation of sequence data from the microbial communities in poultry processing and develop tools to implement it into a microbiome mapping platform for practical use. However, there will be a need to compare traditional microbial data with sequence data, develop data models for process controls from the sequence data, and establish routine approaches for biomapping based on microbiome data. To accomplish these objectives will require a proof-of-concept plan for establishing microbiome mapping as means for assessing process controls in poultry processing. From a personnel standpoint, this will include collection of carcass rinsate samples from commercial poultry processing sites and plating for microbial counts as well as sequencing the rinsates and working in conjunction with microbiological laboratory personnel to collect sequence data and perform analysis using bioinformatics programs to identify individual taxonomic groups and assess diversity of identified populations and establish the best indicators for process control. To pull these complex data sets and the accompanying interpretative translation into a routine microbiome-based mapping exercise will require an educational level at least at the more senior levels with advanced understanding of not only classical microbiology because this is still a part of the routine monitoring that currently occurs in poultry plants but also the ability to statistically analyze and evaluate molecular sequence–based data and bioinformatics interpretations. Similar to the whole genome sequence data, it is quite possible that cybersecurity measures would need to be inserted into these data sets, particularly if any length of storage time is required. Given the complexities associated with these types of approaches, it is clear that the demand for well-trained graduate-level candidates, who are effectively "cross-trained" to manage not only the microbial "wet lab" component of a commercial microbial quality laboratory but also the tremendous wealth of data that will be generated, are needed. In addition, policies related to food production and retail service are predicted to begin to be driven by the availability of "big data" and more linkages between urbanization and food systems are identified (Seto and Ramankutty, 2016; Athey, 2017). This will certainly further increase the urgency to generate a workforce in the food sector that is comfortable with large complex data sets derived from a multitude of sources in the production chain that may require identification of influential factors that impact the entire food system chain. The next sections address how such graduate cross-training can be accomplished in academic institutions that would serve as potential feeder sources for recruitment of students who can multitask across basic tenets of food microbiology, food science, computer science, transportation, and critical infrastructure, to integrate into a cumulative food systems and cybersecurity concentration background to meet the demands of the food industry and the federal agencies involved in regulating these entities.

## The Need for Graduate-Level Students

It is becoming apparent in the food industry that food security no longer consists of just simple isolated cause-and-effect events but a much more complex series of highly integrated actions that result in tremendous challenges for food-related data generation, management, and security. Existing trained labor pools from traditional BS-level and even some MS-level programs are rapidly losing relevancy for careers in the corporate food realm because of the inherent narrow specialization and focus on standard classroom performance. Consequently, two divergent career paths are emerging. One avenue is entry-level positions where much of the training is done on-site at the employer's expense after graduation. While there remains a demand for these types of hires, the demand for more extensively trained individuals is becoming more the norm. In particular, MS and PhD graduate students are becoming more attractive for industry employers. For one reason, the pool of PhD students available is increasing as opportunities for faculty positions continue to dwindle and, in turn, become more competitive while overall employment opportunities for PhD students have been relatively volatile over the past two decades (Mervis, 2016). Therefore, industry careers are becoming more appealing to such candidates (Fuhrmann et al., 2012; National Science Foundation, 2012). In addition, this has also been spurred on by tremendous growth in academically linked start-up companies that have emerged from university–industry hosted research commercialization incubators (Sharp, 2014). Another reason is that the food industry has an increasing need of analytical integrative skills obtained at the graduate level to retain competitiveness and at the same time advance into new markets through innovation. Increasing these integrative approaches in the personnel being hired will involve development of skill sets beyond traditional wet laboratory research and data analysis. The key to this is the concept of integrative "cross-training" for graduate students in academia. The increasing complexity of the food supply chain and the global reach of agricultural production make the training of students in systems thinking an increasingly important consideration. Graduate students are the best targets for this type of training because their lengths of education and competency expectations offer sufficient flexibility to alter the training philosophy. Equally attractive is the fact that graduate students are more likely to have a greater long-term impact to become team leaders and decision makers in industry.

A contemporary education in cybersecurity often calls for an understanding of computer systems, software vulnerabilities, and data storage protection. Although rare, some programs outside of computing offer studies in homeland security or military defense. These programs are valuable but result in students with a broad understanding of existing cybersecurity threats and little domain expertise or experience to be useful to government agencies and companies serving the areas. Therefore, the key theme of this review is that a domain-driven education in cybersecurity is vital for producing effective professionals in the field. Although no program can provide training in every discipline, academic institutions are poised to offer a comprehensive experience that will provide expertise to government agencies that serve citizens and industries, which contribute to a significant percentage of the US annual GDP: food, transportation, and critical infrastructure. Each of these areas contributes differently to the educational problem confronting the current food industry and associated regulatory agencies. Classes in transportation offer insights into asset data, logistics systems, and operations that make use of countless amounts of customer data and financial systems that support the nation's transportation activities. Moreover, fundamental skills in transportation classes, such as optimization and decision models, are consistent with required security skills. While typically not thought of in the cybersecurity realm, the areas of food production, transportation, and consumption pose significant security threats. As the global population is expected to approach 9.7 billion by 2050 and 11.2 billion by 2100, more than doubling of the quantity of food currently produced will be required, which will induce a huge economic impact not only within food production itself, but also up and down the food supply chain. The safety and security of the products people consume has always been of concern. However, advancements in sensors, advanced analytics, and large-scale data storage systems have made improved safety and cybersecurity platforms a reality. As recently as 2015, leaders in Asia envisioned the "Internet of Things," as applied to food production and supply chain, to be the next frontier in the agriculture and food delivery domain (Internet of Things, 2016). Enabling advancements in food security requires the collection, storage, and analysis of information for traceability, contamination detection, database access, and complex data set interpretation. The security of such systems is a primary concern because the information that will help ensure food safety is also of great interest to their adversaries. The Department of Homeland Security has labeled this threat as agroterrorism (NCI at Excelsior, 2016) and has warned against those that seek to exploit an increase in system access points (e.g., sensors and databases) to modify the process of food production and delivery. Eliminating these topics from academic education programs only damages their ability to produce cybersecurity professionals for federal, state, and local governments.

## Education Strategies

### Interdisciplinary Education

While the educational program and curricula components of academic institutions have been slow to respond to cyber and physical security, the operational side of such institutions certainly has had to face these more and more on a regular basis. The awareness of the need for improved security in a wide range of activities that occur in academia particularly in the biological sciences has become an ongoing concern with the increase in potential threats both externally and internally (Berger, 2016). The impact of data, analyses, and eventually the need to protect that data has also already entered into educational institutions. For example, the generation and analyses of large data sets to develop predictors of future academic scientific discovery and the corresponding potential impact is an emerging interdisciplinary field referred to as the "science of science" (Clauset et al., 2017). It has been suggested by Clauset et al. (2017) that this type of interdisciplinary data gathering and analytical approach will be used to drive funding priorities, faculty hiring decisions, and advancements, along with publishers of scientific research. As this becomes a reality across educational institutions, the need for cyberprotection of sensitive data, particularly personnel evaluations, will be critical. Likewise, it is not difficult to imagine that, as such analytical approaches become better developed and more mature, they will be readily adapted by private industries such as food corporations for similar applications including personnel decisions, as well as product development and safety concerns. In addition, given the trend toward increases in private industry expenditures on agricultural research and development (Fuglie et al., 2012), the need to protect the resulting information and intellectual investment will certainly become critical.

Consequently, university education curricula, practices, and focus will need to adjust dramatically to equip future graduates with the tools and background to cope with this changing landscape. Specifically, universities will need to change their education strategies to provide interdisciplinary data-enabled security research and education to scholars through on-campus resources and surrounding support. This requires establishing an interdisciplinary group or institute with cross-cutting research and education thrusts. One such institute that is lacking in many universities is one that combines food security, cybersecurity, transportation security, and critical infrastructure security although at least one does exist at the University of Arkansas (ASCENT, 2017). This single institute can be strengthened by coordinating and partnering with on-campus research centers that often already exist in the areas of food, cybersecurity, transportation, and critical infrastructure. Leveraging existing centers that are well supported by industry and government in conducting cutting-edge research and providing interdisciplinary education to students will provide a strong foundation for interdisciplinary data-enabled security research and education.

## Proposed Curriculum

Generating curricula concentrated in a particular focus area can be difficult and slow if classes have to be newly created. However, academia should be able to build on existing graduate classes in cybersecurity to create an interdisciplinary graduate program to cross-train students in food, cyber, transportation, and critical infrastructure security. This graduate program could be a graduate concentration or a graduate certificate. An advantage of a graduate certificate is that each of the participating departments, such as food science, computer science, computer engineering, industrial engineering, and electrical engineering, would have the flexibility to accept one or more courses in the graduate certificate program to also count toward the corresponding MS or PhD program. In addition, this program would be attractive for industry in cross-training their employees in cybersecurity while retaining their domain-specific area of interest.

A cybersecurity graduate certificate commonly consists of 12–24 h of courses, with 12 h being common (cyberdegrees.org, 2017). Often, all students in the program require an introductory computer security or similarly named course and it is designed to provide a good theoretical knowledge of security in the digital world and how security issues relate to various other fields. A pool of courses satisfies the remaining credit hours, most from computer science/computer engineering, but other departments may offer some of the courses covering privacy or legal topics.

Computer security is an introductory course that covers a wide range of security topics, including cryptography, key management, authentication, access control, network security, email security, malware, Web security, mobile security, and privacy. It assumes prerequisite topics that a computer science or computer engineering major would cover in their undergraduate program including a programming course and an operating systems course. To create an interdisciplinary cybersecurity graduate certificate, this course has to be designed such that it is accessible to majors outside a computer science/computer engineering department to majors that include food science, industrial engineering, and electrical engineering. Existing cybersecurity graduate certificate programs already make this course available to majors outside computer science and engineering to include students from philosophy, business, economics, and law programs (cyberdegrees.org, 2017). To strengthen the interdisciplinary nature of the proposed curriculum, this introductory course should include domain-specific food science, cybersecurity-related activities for transportation, and critical infrastructure.

For a traditional cybersecurity graduate certificate, the remaining three or more courses are typically taken from a pool of courses inside a computer science/computer engineering department and include technical classes such as network security, wireless security, database security, computer forensics, applied cryptography, penetration testing, and secure systems administration (SA) (cyberdegrees.org, 2017). However, the pool of classes may also include courses related to managing security such as auditing, security risk assessment, security policies, disaster recovery, ethics, and cyber law, which may be taught outside the computer science/computer engineering department (cyberdegrees.org, 2017).

It is proposed to expand this pool of courses in the cybersecurity graduate certificate to include domain-specific courses to cross-train students. For example, a transportation cybersecurity student may include an introduction to optimization course that discusses system-level optimization, which includes variables such as risk and security. Or a critical infrastructure cybersecurity student may include an advanced power distribution systems course that discusses security and reliability in industrial control systems (NSA, 2017). Finally, a food cybersecurity student may include a course or courses in foodborne disease epidemiology or more specifically on foodborne pathogens in poultry production processing to understand how to defend against intentional contamination. These courses would include the requirement of tracing the foodborne pathogens in a food system by sampling at various points in the system, uploading the data into a database, analyzing the results, and taking corrective measures as well as gaining wet lab training in handling microorganisms and integrative writing exercises adapted from undergraduate courses in food microbiology (Maciorowski and Ricke, 2000; Dittmar et al., 2006; Howard et al., 2006; O'Bryan et al., 2009, 2010; Sirsat et al., 2014; Ricke et al., 2015; Ricke, 2015; Howton et al., 2016). This track and tracing of the foodborne pathogens uses all aspects of a cyber system and requires that the data be secured and, more importantly, a distributed database that logs the samples with timestamps into a ledger that can be audited by third parties and is difficult to modify without being detected (IBM Blockchain, 2017). Common cybersecurity themes in food, cyber, transportation, and critical infrastructure are to identify the risks, protect assets using appropriate safeguards, detect the occurrence of a cybersecurity event, respond to an event, and recover from an event (NICE, 2016).

Combining food systems, transportation, and critical infrastructure into an academic food systems education program will require a phased-in approach. First, a cybersecurity graduate certificate program is established with the introductory cybersecurity course having examples from food, transportation, and critical infrastructure. The remaining courses are subsequently taken from a pool of primarily cyber courses that also contains one or two courses from each of the food, transportation, and critical infrastructure, which have a related cybersecurity component or compelling motivation to add cybersecurity. An example cybersecurity graduate certificate combining food, transportation, and critical infrastructure into an academic food systems education program is shown in Table 20.1 below.

**Table 20.1: Cybersecurity graduate certificate combining transportation, critical infrastructure, and food.**

| Cybersecurity graduate certificate student scholar (12-h or 4 security-related courses) | | | | |
|---|---|---|---|---|
| **Area** | **Cyber** | **Transportation** | **Critical infrastructure** | **Food** |
| Common course | Computer security | | | |
| At least two cybersecurity courses from this list | • Computer forensics<br>• Database security<br>• Network security<br>• Privacy enhancing technologies<br>• Security and privacy in mobile and distributed systems<br>• Wireless systems security<br>• Advanced cryptography<br>• Computer architecture security<br>• Secure digital system design | | | |
| At least one course from this list | • One course from the list of cybersecurity courses above | • Design of industrial experiments<br>• Introduction to optimization theory<br>• Network optimization in transportation logistics and decision models | • Design of advanced power distribution systems<br>• Power systems analysis<br>• Control of electrical power systems<br>• Protecting aging substation from cyber attacks | • Foodborne pathogen epidemiology<br>• Molecular biology of foodborne pathogens<br>• Food product development<br>• Food law<br>• Food processing<br>• Immunology |

### Mapping to a Standard Cybersecurity Framework

The need for expanding cybersecurity into the workforce has led to national initiatives in cybersecurity. The National Institute of Standards and Technology created the National Initiative for Cybersecurity Education (NICE) to create partnerships between government, academia, and the private sector for cybersecurity education, training, and workforce development (NICE, 2016). The goal of NICE is to develop cybersecurity programs to train the workforce and secure the US nation. NICE developed the NICE Cybersecurity Workforce Framework (known as the NICE Framework) to provide a common and standard set of specialty areas and work roles (NICE, 2016). The NICE Framework also identifies common knowledge, skills, and abilities (KSAs) for each work role. Therefore, a cybersecurity graduate certificate should follow a set of guidelines such as the NICE Framework.

For example, under the NICE Framework, students completing the computer security course would align with the Operate and Maintain category that has specialty areas of data administration, knowledge management, customer service and technical support, network services, SA, and systems analysis. These areas include the following five specific KSAs for the federal government workforce:

- KSA0001: Knowledge of computer networking concepts and protocols, and network security methodologies.
- KSA0002: Knowledge of risk management processes (e.g., methods for assessing and mitigating risk).
- KSA0003: Knowledge of national and international laws, regulations, policies, and ethics as they relate to cybersecurity.
- KSA0004: Knowledge of cybersecurity principles.
- KSA0005: Knowledge of cyber threats and vulnerabilities.

## Recruitment

The students for the cybersecurity graduate certificate combining transportation, critical infrastructure, and food program can be recruited from both academia and industry. The students from academia could be from undergraduate food, cyber, transportation, and critical infrastructure such as food science, computer science/computer engineering, industrial engineering, and electrical engineering. Some of these students will use the graduate certificate to obtain valuable skills so that they can enter the workplace. Other students from academia will complete the graduate certificate in addition to an MS or PhD in their particular area by counting some courses toward both programs. Each MS or PhD program can decide whether to permit counting one or more courses in the graduate certificate toward their program. Students from industry can be recruited to complete the cybersecurity graduate certificate to expand their skills into cybersecurity within their domain-specific area. A university can leverage their existing relationships with food science, cyber, transportation, and critical infrastructure industries to recruit professionals into the cybersecurity graduate certificate program.

## Placement

This program develops a pipeline of well-trained food, cyber, transportation, critical, and infrastructure security professionals in the nation, ensuring that government agencies and industry have the capability to secure the transportation, critical infrastructure, and food systems. This exposure to security should enhance students' interest in security careers, helping to close the security labor gap experienced across the United States. Some students could be placed in federal, state, and local government agencies. A recent review of job postings for federal employees listed cybersecurity positions in not only the expected agencies such as the Department of Defense, Army, Air Force, and Navy but also included the Department of Agriculture and the Department of Transportation (usajobs.gov, 2017). A review of job postings in Texas revealed that there were over 500 job postings for cybersecurity, 101 postings for cybersecurity and critical infrastructure, 61 postings for cybersecurity and transportation, and 61 postings for food defense (WorkInTexas.com, 2017). Finally, a review of nationwide job postings for industry listed over 35,000 postings for cybersecurity, 1800 postings for cybersecurity and critical infrastructure, 600 postings for cybersecurity and transportation, and 80 postings for cybersecurity and food (indeed.com, 2017).

## Conclusions

The lack of security training for food, transportation, and critical infrastructure areas leaves the United States vulnerable to the corresponding attacks. In response, this effort of cross-training students for combining food, transportation, and critical infrastructure cybersecurity into an academic food systems education program rapidly expands the pool of well-trained individuals entering government agencies and industry who also understand security threats and defense techniques so that transportation, critical infrastructure, and food systems can be much better protected. As other government agencies and companies consider security as one of the top concerns, a concentrated study approach as part of university academic curricula could serve as a rich source of programmatic information. As these programs evolve, the programmatic processes to further refine these programs so that the methods and processes are more efficient and effective to meet workforce requirements.

## References

ASCENT, 2017. http://ascent.uark.edu.

Athey, S., 2017. Beyond prediction: using big data for policy problems. Science 355, 483–485.

Auler, D.P., Teixeira, R., Nardi, V., 2017. Food safety as a field in supply chain management studies: a systematic literature review. International Food and Agribusiness Management Review 20, 99–112.

Baker, C.A., Rubinelli, P.M., Park, S.H., Ricke, S.C., 2016a. Immuno-based detection of shiga toxin-producing pathogenic *Escherichia coli* – a review on current approaches and potential strategies for optimization. Critical Reviews in Microbiology 42, 656–675.

Baker, C.A., Rubinelli, P.M., Park, S.H., F.Carbonero, Ricke, S.C., 2016b. Shiga toxin-producing *Escherichia coli* in food: incidence, ecology, and detection strategies in the United States. Food Control 59, 407–419.

Berger, K.M., 2016. What life scientists should know about security threats. Science 354, 1237–1239.

Betts, R., Jones, G., 2017. The impact of DNA sequencing methods on food safety and quality. International Food & Meat Topics 28, 11–12.

Boggia, A., Paolotti, L., Castellini, C., 2010. Environmental impact evaluation of conventional, organic and organic-plus poultry production systems using life cycle assessment. World's Poultry Science Journal 66 (1), 95–114.

Cisco, 2015. http://www.cisco.com/c/dam/en/us/products/collateral/security/cybersecurity-talent.pdf.

Clauset, A., Larremore, D.B., Sinatra, R., 2017. Data-driven predictions in the science of science. Science 355, 477–480.

Crandall, P.G., O'Bryan, C.A., Babu, D., Jarvis, N., Davis, M.L., Buser, M., Adam, B., Marcy, J., Ricke, S.C., 2013. Whole-chain traceability, is it possible to know which steer your hamburger came from? Meat Science 95, 137–144.

Crutchley, T.M., Rodgers, J.B., Whiteside Jr., H.P., Vanier, M., Terndrup, T.M., 2007. Agroterrorism: where are we in the ongoing war on terrorism? Journal of Food Protection 70, 791–804.

Cyberdegrees.org, 2017. http://cyberdegrees.org.

Dittmar, R.S., Kundinger, M.M., Woodward, C.L., Donalson, L.M., Golbach, J.L., Kim, W.K., Chalova, V., Ricke, S.C., 2006. Quality control laboratory design project for poultry science undergraduate students enrolled in an advanced food microbiology course. Journal of Consumer Protection and Food Safety 1, 77–82.

Enquist, L.W., 2015. Passing the torch and the samples. Microbe 10, 46–47.

Fuglie, K., Heisey, P., King, J., Pray, C.E., Schimmelpfennig, D., 2012. The contribution of private industry to agricultural innovation research funding. Science 338, 1031–1032.

Gast, R.K., 2016. Microbiology of shell egg production in the United States. In: Ricke, S.C., Gast, R.K. (Eds.), Producing Safe Eggs-The Microbial Ecology of *Salmonella*. Elsevier, Inc., San Diego, CA, pp. 25–44 (Chapter 2).

Glenn, T.C., 2011. Field guide to next-generation DNA sequencers. Molecular Ecology Resources 11, 759–769.

Global Study Reveals Businesses, Countries Vulnerable Due to Shortage of Cybersecurity Talent, 2016. Retrieved from: https://newsroom.intel.com/news-releases/global-study-reveals-businesses-countries-vulnerable-due-shortage-cybersecurity-talent/.

Gustavsson, J., Cederberg, C., Sonesson, U., van Otterdijk, R., Meybeck, A., 2011. Global Food Losses and Food Waste. http://www.fao.org/docrep/014/mb060e/mb060e00.pdf.

Handley, J.A., Park, S.H., Shi, Z., Dawoud, T.M., Kwon, Y.M., Ricke, S.C., 2015. *Salmonella* and the potential role for microbial process indicators on chicken carcasses. In: Ricke, S.C., Donaldson, J.R., Phillips, C.A. (Eds.), Food Safety: Emerging Issues, Technologies and Systems. Elsevier, Oxford, UK, pp. 81–104 (Chapter 6).

Hanning, I.B., O'Bryan, C.A., Crandall, P.G., Ricke, S.C., 2012. Food safety and security. Nature Education Knowledge 3 (7), 9.

Haynes, E., Hudson, A., 2017. Next generation sequencing – the future of food microbiology. International Food & Meat Topics 28, 23–25.

Hobin, J.A., Fuhrmann, C., Lindstaedt, B., Clifford, P., Sept. 7, 2012. You need a game plan. Science.

Holt, P.S., Davies, R.H., Dewulf, J., Gast, R.K., Huwe, J.K., Jones, D.R., Waltman, D., Willian, K.R., 2011. The impact of different housing systems on egg safety and quality. Poultry Science 90, 251–262.

http://blog.trendmicro.com/the-challenges-of-cyber-security-education-and-training-in-2015/.

https://www.eab.com/daily-briefing/2016/08/05/theres-a-cybersecurity-expert-shortage-colleges-can-help.

https:www.un.org/development/desa/publications/world-population-prospects-2015-revision.html.

Howard, Z.R., Donalson, L.M., Kim, W.K., Li, X., Zabala Diaz, I.B., Landers, K.L., Maciorowski, K.G., Ricke, S.C., 2006. Development of research paper writing skills of poultry science undergraduate students taking food microbiology. Poultry Science 85, 352–358.

Howton, J., Keifer, E., Murphy, C.A., Sirsat, S., O'Bryan, C.A., Ricke, S.C., Crandall, P.G., Neal, J., 2016. A comparison of online food safety training programs using the customizable tool for online training evaluation. Food Control 59, 82–87.

IBM Blockchain, 2017. https://www.ibm.com/blockchain/.

Indeed.com, 2017. Retrieved from: www.indeed.com.

Internet of Things: Food Safety Apps Set to Emerge, 2016. IoT Agenda. Retrieved from: http://internetofthingsagenda.techtarget.com/feature/Internet-of-Things-Food-safety-apps-set-to-emerge.

Kabinlapat, P., Sutthachai, S., 2017. An application of activity-based costing in the chicken processing industry: a case of joint products. International Food and Agribusiness Management Review 20, 85–97.

Kim, S.A., Park, S.H., Lee, S.In., Owens-Hanning, C., Ricke, S.C., 2017. Assessment of chicken carcass microbiome responses during processing in the presence of commercial antimicrobials using a next generation sequencing approach. Scientific Reports 7, 43354. http://dx.doi.org/10.1038/srep43354.

Kuchta, T., Knutsson, R., Fiore, A., Kudirkiene, E., Höhl, A., Tomic, D.H., Gotcheva, V., Pöpping, B., Scaramagli, S., To Kim, A., Wagner, M., De Medici, D., 2014. A decade with nucleic acid-based microbiological methods in safety control of foods. Letters in Microbiology 59, 263–271.

Kwon, Y.M., Ricke, S.C. (Eds.), 2011. High-throughput Next Generation Sequencing: Methods and Applications. Methods in Molecular Microbiology 733-Springer Protocols. Humana Press, New York, NY. 308 p.

Lewrene Glaser, R., 2016. USDA ERS – Ag and Food Sectors and the Economy. Retrieved from: Ers.usda.govhttp://www.ers.usda.gov/data-products/ag-and-food-statistics-charting-the-essentials/ag-and-food-sectors-and-the-economy.aspx.

Luján-Rhenals, D., Morawicki, R., Van Loo, E.J., Ricke, S.C., 2017a. Energy and water use in poultry processing. In: Ricke, S.C. (Ed.), Achieving Sustainable Production of Poultry Meat. Safety, Quality and Sustainability, vol. 1. Burleigh Dodd Publishing, Cambridge, UK, pp. 389–409 (Chapter 19).

Luján-Rhenals, D., Morawicki, R., Van Loo, E.J., Ricke, S.C., 2017b. Waste management and emissions in poultry processing. In: Ricke, S.C. (Ed.), Achieving Sustainable Production of Poultry Meat. Safety, Quality and Sustainability, vol. 1. Burleigh Dodd Publishing, Cambridge, UK, pp. 411–431 (Chapter 20).

Maciorowski, K.G., Ricke, S.C., 2000. Improving content writing instruction in an undergraduate food bacteriology class. Dairy, Food and Environmental Sanitation 20, 196–204.

Meeker, D.L., Meisinger, J.L., 2015. Rendered ingredients significantly influence sustainability, quality, and safety of pet food. Journal of Animal Science 93, 835–847.

Mervis, J., 2016. 'Employment crisis' for new Ph.D.s is an illusion. Science 352, 880.

Metzker, M.L., 2010. Sequencing technologies—the next generation. Nature Reviews Genetics 11, 31–46.

Monaghan, J.M., 2006. United Kingdom and European approach to fresh produce food safety and security. HortTechnology 16, 559–562.

Morawicki, R., 2010. Coproducts and by-products from poultry processing. In: Owens, C.M., Alvarado, C.Z., Sams, A.R. (Eds.), Poultry Meat Processing, second ed. CRC Press, Boca Raton, FL, pp. 381–396 (Chapter 19).

National Science Foundation, 2012. Science and Engineering Indicators. (Chapter 5).

Draft NIST Special Publication 800-181. NICE Cybersecurity Workforce Framework (NCWF), November 2016. National Institute of Standards and Technology (NIST).

NCI at Excelsior, 2016. What Does Cybersecurity Have to Do with Our Food Supply? Cyper Experts Blog at National Cybersecurity Institute.

Norton, R.A., 2003. Symposium: agro-terrorism: biological threats and security measures. Food security issues—a potential comprehensive plan. Poultry Science 82, 958–963.

NSA/DHS National Centers of Academic Excellence for Cyber Defense Focus Areas, 2017. https://www.iad.gov/NIETP/documents/Requirements/CAE-CD_Focus_Areas.pdf.

Oakley, B.B., Morales, C.A., Line, J., Berrang, M.E., Meinersmann, R.J., Tillman, G.E., Wise, M.G., Siragusa, G.R., Hiett, K.L., Seal, B.S., 2013. The poultry-associated microbiome: network analysis and farm-to-fork characterizations. PLoS One 8 (2), e57190. http://dx.doi.org/10.1371/journal.pone.0057190.

O'Bryan, C.A., Dittmar, R.S., Chalova, V.I., Kundinger, M.M., Crandall, P.G., Ricke, S.C., 2010. Assessment of a food microbiology senior undergraduate course as a potential food safety distance education course for poultry science majors. Poultry Science 89, 2542–2545.

O'Bryan, C.A., Crandall, P.G., Shores-Ellis, K., Johnson, D.M., Ricke, S.C., Marcy, J., 2009. Optimizing web based instruction – a case study using poultry processing unit operations. Journal of Food Science Education 8, 93–100.

Osborne, D.J., Sanders, D.C., Ward, D.R., Rushing, J.W., 2006. Gaining producer buy-in for food safety and security in the United States. HortTechnology 16, 563–565.

Owens, C.M., Alvarado, C.Z., Sams, A.R., 2010. Poultry Meat Processing. CRC Press, Boca Raton, FL. 441 p.

Park, S.H., Aydin, M., Khatiwara, A., Dolan, M.C., Gilmore, D.F., Bouldin, J.L., Ahn, S., Ricke, S.C., 2014. Current and emerging technologies for rapid detection and characterization of *Salmonella* in poultry and poultry products. Food Microbiology 38, 250–262.

Parker, R., 2006. A chainstore approach to food safety and emerging pathogens. HortTechnology 16, 555–559.

Pelletier, N., 2008. Environmental performance in the US broiler poultry sector: life cycle energy use and greenhouse gas, ozone depleting, acidifying and eutrophying emissions. Agricultural Systems 98 (2), 67–73.

Premier, R., Ledger, S., 2006. Good agricultural practices in Australia and Southeast Asia. HortTechnology 16, 552–555.

Quail, M.A., Smith, M., Coupland, P., Otto, T.D., Harris, S.B., Connor, T.R., Bertoni, A., Swerdlow, H.P., Gu, Y., 2012. A tale of three next generation sequencing platforms: comparison of Ion Torrent, Pacific Biosciences and Illumina MiSeq sequencers. BMC Genomics 13, 341 http://www.biomedcentral.com/1471-2164/13/341.

Ricke, S.C., 2017. Microbial ecology of eggs: a focus on *Salmonella* and microbial contamination in post-harvest table shell egg production. In: Sant'Ana, A.de.S. (Ed.), Quantitative Microbiology in Food Processing – Modeling the Microbial Ecology. John Wiley & Sons, Ltd., Oxford, UK, pp. 416–441 (Chapter 22).

Ricke, S.C., 2015. Anaerobic microbiology training. In: Ricke, S.C., Donaldson, J.R., Phillips, C.A. (Eds.), Food Safety: Emerging Issues, Technologies and Systems. Elsevier, Oxford, UK, pp. 395–419 (Chapter 19).

Ricke, S.C., 2010. Future prospects for advancing food – safety research in food animals. In: Ricke, S.C., Jones, F.T. (Eds.), Perspectives on Food Safety Issues of Food Animal Derived Foods. University of Arkansas Press, Fayetteville, AR, pp. 335–350.

Ricke, S.C., Dawoud, T.M., Kwon, Y.M., 2015. Application of molecular methods for traceability of foodborne pathogens in food safety systems. In: Ricke, S.C., Donaldson, J.R., Phillips, C.A. (Eds.), Food Safety: Emerging Issues, Technologies and Systems. Elsevier, Oxford, UK, pp. 37–63 (Chapter 4).

Ricke, S.C., Van Loo, E.J., Johnson, M.G., O'Bryan, C.A. (Eds.), 2012. Organic Meat Production and Processing. Wiley Scientific/IFT, New York, NY. 444 p.

Rothrock Jr., M.J., Locatelli, A., Glenn, T.C., Thomas, J.C., Caudill, A.C., Kiepper, B.H., Hiett, K.L., 2016. Assessing the microbiomes of scalder and chiller tank waters a typical commercial poultry processing day. Poultry Science 95, 2372–2382.

Rushing, J.W., Walsh, C.S., 2006. The United States Food and Drug Administration's approach to training international suppliers in food safety and security. HortTechnology 16, 566–569.

Scallan, E., Hoekstra, R.M., Angulo, F.J., Tauxe, R.V., Widdowson, M.-A., Roy, S.L., Jones, J.L., Griffin, P.M., 2011. Foodborne illness acquired in the United States—major pathogens. Emerging Infectious Diseases 17, 7–15.

Scharff, R.L., 2012. Economic burden from health losses due to foodborne illness in the United States. Journal of Food Protection 75, 123–131.

Seto, K.C., Ramankutty, N., 2016. Hidden linkages between urbanization and food systems. Science 352, 943–945.
Sharp, P.A., 2014. Meeting global challenges: discovery and innovation through convergence – integrate biology, physics, engineering, and social science to innovate. Science 346, 1468–1471.
Sirsat, S.A., Kim, K., Gibson, K.E., Crandall, P.G., Ricke, S.C., Neal, J.A., 2014. Tracking microbial contamination in retail environments using Glo Germ™ powder - a retail delicatessen environment example. Journal of Visualized Experiments 85, e51402.
USAjobs.gov, 2017. Retrieved from: http://usajobs.gov.
Wang, X., 2015. Transportation Industry Job Demand and Skills Data and Implications for Workforce Development. Presentation, Transportation Learning Center.
Whitford, W., 2017. The era of digital biomanufacturing. BioProcess International 15, 12–18.
WorkInTexas, 2017. Retrieved: https://wit.twc.state.tx.us.

# Index

'*Note*: Page numbers followed by "f" indicate figures, "t" indicate tables.'

## T

MIX
Papier | Fördert
gute Waldnutzung
FSC® C083411

Kolibri 360 GmbH
Ferdinand-Jühlke-Straße 7
99095 Erfurt, Deutschland
produktsicherheit@kolibri360.de